PLATE I.—Macculloch's Tree submerged in Lava (p. 111).

BRITISH GEOLOGICAL SURVEY

# Tertiary and post-Tertiary geology of Mull, Loch Aline and Oban

Memoir for parts of geological sheets 43, 44, 51 and 52
(Scotland)

E. B. Bailey, MC, BA, C. T. Clough, LlD, MA,
W. B. Wright, BA, J. E. Richey, MC, BA,
and G. V. Wilson, BSc

*with contributions by*
E. M. Anderson, MA, BSc, H. B. Mauffe, BA, G. W. Lee, DSc,
B. Lightfoot, MC, BA, T. O. Bosworth, MA, DSc, and G. A. Burnett, BSc

*with petrology by*
H. H. Thomas, MA, ScD, and E. B. Bailey, MC, BA

*chemical analyses by*
E. G. Radley, FCS, and F. R. Ennos, BA, BSc, AIC

*and palaeobotany by*
A. C. Seward, MA, ScD, FRS, and R. E. Holttum, MA

EDINBURGH   HIS MAJESTY'S STATIONERY OFFICE   1924
*1987 reprint*

**HER MAJESTY'S STATIONERY OFFICE**

HMSO publications are available from:

**HMSO Publications Centre**
(Mail and telephone orders)
PO Box 276, London SW8 5DT
Telephone orders (01) 622 3316
General enquiries (01) 211 5656
*Queueing system in operation for both numbers*

**HMSO Bookshops**
49 High Holborn, London WC1V 6HB
 (01) 211 5656 (Counter service only)
258 Broad Street, Birmingham B1 2HE
 (021) 643 3740
Southey House, 33 Wine Street, Bristol
 BS1 2BQ (0272) 264306
9 Princess Street, Manchester M60 8AS
 (061) 834 7201
80 Chichester Street, Belfast BT1 4JY
 (0232) 238451
71–73 Lothian Road, Edinburgh
 EH3 9AZ (031) 228 4181

**HMSO's Accredited Agents**
(see Yellow Pages)

*And through good booksellers*

**BRITISH GEOLOGICAL SURVEY**
Keyworth, Nottinghamshire NG12 5GG
 Plumtree (060 77) 6111
Murchison House, West Mains Road,
Edinburgh EH9 3LA (031) 667 1000

The full range of Survey publications is available through the Sales Desks at Keyworth and Murchison House. Selected items are stocked by the Geological Museum Bookshop, Exhibition Road, London SW7 2DE; all other items may be obtained through the BGS London Information Office in the Geological Museum ((01) 589 4090). All the books are listed in HMSO's Sectional List 45. Maps are listed in the BGS Map Catalogue and Ordnance Survey's Trade Catalogue. They can be bought from Ordnance Survey Agents as well as from BGS.

*The British Geological Survey carries out the geological survey of Great Britain and Northern Ireland (the latter as an agency service for the government of Northern Ireland), and of the surrounding continental shelf, as well as its basic research projects. It also undertakes programmes of British technical aid in geology in developing countries as arranged by the Overseas Development Administration.*

*The British Geological Survey is a component body of the Natural Environment Research Council.*

*First published 1924 by HMSO*
*Second impression 1987 issued by BGS*

Printed in the UK for BGS by Derry and Sons Ltd, Nottingham

C5 6/87

ISBN 0 85272 096 3

# PREFACE.

The work of the Geological Survey in Mull began in the year 1902, when Dr. Harker examined a small area on the south coast near Malcolm's Point in connection with the preparation of Sheet 36 (Kilmartin). It was not, however, till 1907 that an actual start was made. In that year Mr. Cunningham Craig was engaged in mapping the south part of the Ross of Mull in Sheet 35 (Colonsay); while at the same time Mr. Maufe, in Lismore, and Mr. Bailey, near Loch Don in Mull, began the Survey of Sheet 44, which contains a large part of east and central Mull. In 1905 the West Highland district was placed under the charge of Dr. Clough, who acted as District Geologist there till his death in 1916. From 1907 onwards till the outbreak of war in 1914, three or four geologists and one District Geologist were occupied in Mull during part of each summer. The geology of Central Mull proved excessively complex, and progress was unavoidably slow, but in 1914 Sheet 44 was practically completed. The portions of Mull lying in other Sheets, viz.: 43, 51 and 52, were finished in 1920. Mr. Bailey succeeded Dr. Clough as District Geologist in this area in 1919. With personal knowledge of the whole district he has supervised the preparation of the maps, and has compiled this memoir, taking advantage of maps, notes, and descriptions furnished by the other Officers who took part in the work. The task has been rendered additionally heavy by the death of Dr. Clough, who surveyed a considerable portion of Central Mull, and by the fact that several of those who shared in the survey left the Service before the work was completed. The area surveyed by each Officer may be inferred from the initials that appear at the end of the paragraphs descriptive of his ground. The palæobotanical chapter by Professor Seward and Mr. Holttum and the petrological sections by Dr. Thomas are similarly indicated.

The successive stages of the work can be traced in the reports printed in the "Summary of Progress of the Geological Survey" for each year since 1902.

A memoir dealing with the pre-Tertiary rocks of Mull, Loch Aline and Oban is at present in the press; and this, together with Explanations of Sheets 43, 51 and 52 will complete the record of the Geological Survey's work in the island.

J. S. FLETT,
*Director.*

GEOLOGICAL SURVEY OFFICE,
   28 JERMYN STREET,
      LONDON, S.W.1.

12*th June* 1924.

# CONTENTS.

| CHAPTER. | PAGE. |
|---|---|
| I. Introduction, | 1 |
| II. History of Research, | 40 |
| III.–IV. Sediments, | |
|     III. Field-Relations, | 53 |
|     IV. Palæobotany, | 67 |
| V.–X. Basalt and Mugearite Lavas, | |
|     V. Introduction, | 92 |
|     VI. Outside the Limit of Pneumatolysis (Sheet 44), | 100 |
|     VII. Outside the Limit of Pneumatolysis (Sheets 43, 51, 52), | 107 |
|     VIII. Between the Limit of Pneumatolysis and the Central Calderas, | 120 |
|     IX. Central Calderas, | 131 |
|     X. Petrology, | 136 |
| XI. Various Dolerites and Gabbros, | 156 |
| XII. Early Granophyres, | 165 |
| XIII. Tectonics, | 172 |
| XIV. Syenite, Trachytes, and Bostonites, | 185 |
| XV.–XVI. Explosion-Phenomena, | |
|     XV. Surface-Agglomerates, | 195 |
|     XVI. Vent-Agglomerates, | 199 |
| XVII. Three Early Felsites, | 211 |
| XVIII. Augite-Diorites, | 216 |
| XIX. Intermediate and Acid Cone-Sheets, | 221 |
| XX. Loch Uisg Granophyre and Gabbro, | 230 |
| XXI. Early Basic Cone-Sheets, | 234 |
| XXII. Three Main Olivine-Gabbros, | 242 |
| XXIII.–XXVII. Sheets exclusive of Cone-Sheets, | |
|     XXIII.–XXVI. South-West Mull, | |
|         XXIII. Introduction and Loch Scridain Sheets, | 257 |
|         XXIV. Loch Scridain Xenoliths, | 268 |
|         XXV. Loch Scridain Petrology, | 280 |
|         XXVI. Gribun and Ben More, | 288 |
|     XXVII. Elsewhere in Mull and Neighbourhood, | 292 |
| XXVIII. Late Basic Cone-Sheets, | 296 |
| XXIX.–XXXII. Ring-Dykes and Central Intrusions, | |
|     XXIX. Ring-Dykes of Glen More and Beinn Chàisgidle, | 306 |
|     XXX. Gravitational Differentiation, | 320 |
|     XXXI. Glen Cannel Granophyre, | 331 |
|     XXXII. Ring-Dykes of Loch Bà, Beinn a' Ghràig, and Knock, | 337 |
| XXXIII. Hybrids, | 351 |
| XXXIV.–XXXV. Dykes, | |
|     XXXIV. Dykes of known Tertiary Age, | 356 |
|     XXXV. Camptonite-Dykes of uncertain Age, | 377 |
| XXXVI.–XXXVII. Post-Igneous History, | |
|     XXXVI. Pre-Glacial, | 383 |
|     XXXVII. Glacial and Post-Glacial, | 392 |
| XXXVIII. Economics, | 415 |
| Appendix I. Photographs and Lantern-Slides, | 420 |
| Appendix II. Bibliography, | 421 |
| Index, | 427 |

# LIST OF ILLUSTRATIONS.

## LIST OF FIGURES IN THE TEXT.

| | | | PAGE. |
|---|---|---|---|
| Fig. | 1. | Stress-Diagram, illustrating mode of formation of Cone-Sheets and Ring-Dykes, | 12 |
| ,, | 2. | Variation-Diagram : Normal Mull Magma-Series, | 14 |
| ,, | 3. | Variation-Diagrams : Alivalite-Eucrite Magma-Series ; and Porphyritic Central Magma-Type, | 22 |
| ,, | 4. | Variation-Diagram : Mull Alkaline Magma-Series, | 26 |
| ,, | 5. | *Equisetum Campbelli* Forbes, | 72 |
| ,, | 6. | *Equisetum* sp., | 73 |
| ,, | 7. | *Onoclea hebridica* (Forbes), spores of, | 73 |
| ,, | 8. | *Pinites* sp., | 75 |
| ,, | 9. | *Cupressites MacHenryi* Bail., | 75 |
| ,, | 10. | *Sequoiites* (?) *Langsdorfi* (Brong.), | 75 |
| ,, | 11. | *Pagiophyllum Sternbergi* (Goepp.), | 76 |
| ,, | 12. | *Corylites hebridica* sp. nov., | 79 |
| ,, | 13. | Nut (?), | 80 |
| ,, | 14. | *Platanus hebridica* (Forbes), | 83 |
| ,, | 15. | *Plantæ Incertæ sedis*, | 87 |
| ,, | 16. | Columnar Jointing in Basalt-Lava, Staffa, | 110 |
| ,, | 17. | Tertiary Lavas overlying a Knob of Sandstone, Bloody Bay, | 116 |
| ,, | 18. | Map showing Distribution of Pillow-Lavas, Mull, | 133 |
| ,, | 19. | Microsections : Plateau Basalt-Lava, and Segregation-Vein, | 137 |
| ,, | 20. | Map showing Zeolite-Localities, Ben More, | 142 |
| ,, | 21. | Microsections : Pillow-Lavas, Mull, | 151 |
| ,, | 22. | Map showing distribution of Big-Felspar, and Small-Felspar Dolerites, Sgùrr Dearg, | 159 |
| ,, | 23. | Microsections : Small-Felspar Dolerite, | 163 |
| ,, | 24. | Map of Glas Bheinn Granophyre, Cnoc na Faoilinn, | 166 |
| ,, | 25. | Serial Sections across the Folds of Eastern Mull, | 174 |
| ,, | 26. | View of Inninmore Fault, Sound of Mull, | 182 |
| ,, | 27. | Microsection Alkali-Syenite, Gamhnach Mhòr, | 189 |
| ,, | 28. | Microsections : Trachytes and Bostonite, | 191 |
| ,, | 29. | Map showing Distribution of Gneiss-Fragments in Mull Agglomerates | 201 |
| ,, | 30. | Map of Vents in Neighbourhood of Sgùrr Dearg, | 204 |
| ,, | 31. | Map and Section of Beinn Mheadhon Felsite, | 213 |
| ,, | 32. | Microsections : Augite-Diorite, | 218 |
| ,, | 33. | Microsections : Craignurite and Allied Granophyre | 225 |
| ,, | 34. | Section showing Loch Uisg Granophyre cutting folded Lavas | 231 |
| ,, | 35. | Sections showing Early Basic Cone-Sheets cutting the Loch Spelve Anticline, near Sgùrr Dearg, | 237 |
| ,, | 36. | Map showing probable Displacement of Early Basic Cone-Sheets at Loch Bà Felsite, | 238 |
| ,, | 37. | Map of Beinn Bheag Gabbro, | 244 |
| ,, | 38. | Section showing Dual Relationship of Ben Buie Gabbro to Vent-Agglomerates, | 247 |
| ,, | 39. | Microsections : Allivalites of Ben Buie Complex, | 249 |
| ,, | 40. | Microsections : Ben Buie Gabbro (Eucrite), | 251 |
| ,, | 41. | Microsections : Craig Porphyrite, | 254 |
| ,, | 42. | Map of South-West Mull showing Distribution of Sills, | 258 |
| ,, | 43. | Map of Pitchstone-Sills near Loch Scridain, | 261 |
| ,, | 44. | Sketch showing Sheath-and-Core Structure in Sill, | 262 |
| ,, | 45. | Section across Composite and Xenolithic Sill, Rudh' a' Chromain, | 267 |

|  |  | PAGE. |
|---|---|---|
| Fig. 46. | Microsections : Xenoliths of Loch Scridain District, | 275 |
| ,, 47. | Microsections : Leidleites, | 282 |
| ,, 48. | Microsections : Innimorites, | 283 |
| ,, 49. | Section along Gaodhail River, showing Late Basic Cone-Sheets, | 297 |
| ,, 50. | Microsections : Talaidh Type of Quartz-Dolerite, | 302 |
| ,, 51. | Microsections : Variolites, Cruachan Dearg, | 305 |
| ,, 52. | Map showing Ring-Dykes of Allt Molach, | 308 |
| ,, 53. | Map showing Ring-Dykes, Maol nam Fiadh, | 312 |
| ,, 54. | Map of Differentiated Ring-Dyke, Glen More, | 322 |
| ,, 55. | Diagram to illustrate Relation of Density to Altitude in Differentiated Ring-Dyke, Glen More, | 323 |
| ,, 56. | Microsections : Differentiated Ring-Dyke, Glen More, | 326 |
| ,, 57. | Microsections : Glen Cannel and Beinn a' Ghràig Granophyres, | 334 |
| ,, 58. | Map of Loch Bà Ring-Dyke, | 338 |
| ,, 59. | Microsections : Knock Granophyre and Loch Bà Felsite, | 347 |
| ,, 60. | Map showing Distribution of Dykes in the South-West Highlands, | 357 |
| ,, 61. | Map showing Agglomerate-Vents along Multiple Dyke, Loch Feochan, | 364 |
| ,, 62. | Microsections : Tholeiites of Salen and Brunton Types, | 371 |
| ,, 63. | Microsections : Camptonites, | 381 |
| ,, 64. | Map showing Superficial Deposits and Glaciation of Central Mull, | 393 |
| ,, 65. | Map illustrating the General Glaciation of Mull and District, | 395 |
| ,, 66. | Map showing Gravel-Fan and Eskers, Glen Forsa, | 404 |

# LIST OF PLATES.

|  |  |  | TO FACE PAGE. |
|---|---|---|---|
| Plate | I. | Macculloch's Tree, Burgh,—*Frontispiece.* |  |
| ,, | II. | *Phyllites ardtunensis* sp. nov., | 85 |
| ,, | III. | Map showing the Distribution of Lava-Types and the Limit of Pneumatolysis, | 91 |
| ,, | IV. | Pillow-Lavas of Central Mull, | 134 |
| ,, | V. | Map showing Calderas, Major Intrusions, and Folds, | 165 |
| ,, | VI. | Map showing Ring-Dykes, | 307 |

# TERTIARY AND POST-TERTIARY GEOLOGY
## OF
# MULL, LOCH ALINE, AND OBAN.

## CHAPTER I.

### GENERAL INTRODUCTION.

IT may safely be maintained that Mull includes the most complicated igneous centre as yet accorded detailed examination anywhere in the world. This centre lies within Sheet 44 of the one-inch Map of Scotland. It is of Tertiary date, and affords the chief subject matter of the present volume.

The geographical limit of the area herein described is mainly fixed by the coast-lines of Mull and its attendant isles such as Ulva and Staffa; but eastwards it is extended to include the Loch Aline district of Morven and the Oban district of Lorne, in so far as the two fall inside Sheet 44. Thus to comprehend the accounts which follow it is essential for the reader to have a copy of Sheet 44 of the Geological Survey one-inch Map of Scotland. The remaining sheets which are to some extent involved are 43, 51 and 52. Sheet 44 is published, while the others will appear shortly.

Further introductory remarks may be grouped under three headings :—

>(1) An explanation is offered of the Plan of the Memoir, so that a reader may be able to find his way to topics of particular interest to himself without losing sight of their significance in the general story.
>
>(2) The Petrology of the district is briefly considered as a whole, with special regard to magma-types, their associations, and their time-relations. In this connexion, the Geological Survey analyses are gathered together, so as to assist in reference and comparison.
>
>(3) An Itinerary is sketched with a view to meeting the requirements of geologists who have an opportunity of visiting the exposures for themselves.

### PLAN OF MEMOIR.

Chapter II. is devoted to the progress of research. The results won by the Geological Survey and associated workers have been many and varied, but they leave undiminished the respect long entertained for the discoveries of men like Macculloch, the Duke of Argyll, Forbes, Gardner, Judd and Geikie. Even their dissensions have been of assistance in pointing the way for further enquiry. The Geological Survey entered a field where a wonderful amount of knowledge had already been gained, and where there was little further to be hoped for except as a result of detailed mapping. To summarize this accumulated knowledge from a historical standpoint is the main object of Chapter II.; but a glance is also taken at progress achieved during the course of the recent survey. In the body of the text, the usual convention is adopted of attaching the initials of specific authors to various sections. One of the main contributors, Dr. Clough, died before the Memoir was written, and others have left the Geological Survey. In such cases the initials of those, to whom descriptions should in justice be attributed, have been inserted between brackets.

Apart from this account of the development of research, and a final chapter devoted to economics, the rest of the memoir is essentially a geological history. Sequence in time has been adopted as the principal guide in the presentation of events. Naturally, however, many exceptions have had to be admitted in following out this general plan, partly on account of lack of knowledge, and partly because a too strict adherence would have led to many embarassing repetitions. It should also be pointed out that every account of a particular group of rocks is concluded with a petrological discussion. In most cases the field-relations and petrology are dealt with in a single chapter; in a few instances the petrology stands in a chapter by itself.

Chapters III. and IV. are devoted to the Tertiary sediments of the district; the former deals with field-relations, and the latter with fossils, more particularly the contents of the famous leaf-beds of Ardtun. The sediments are treated together, since, all told, they are of relatively insignificant bulk. They are taken at the beginning of the story because some of them are the earliest Tertiary rocks known in Mull. They tell of an arid climate that witnessed the upheaval and silicification of the chalk, and soon afterwards gave place to moist warm-temperate conditions. It was during this latter stage that volcanic activity awoke. The Ardtun leaves, and others from Carsaig, occur a little above the base of the volcanic pile. They have been examined by Professor Seward and Mr. Holttum, who agree with Mr. Gardner in assigning them to the Eocene. Regarding the period at which the Mull igneous centre became extinct, all that is certain is that the whole complex was profoundly eroded before the onset of the Quaternary Ice-Age.

The nature of the sediments, and the weathering of successive lava-flows, alike bespeak a terrestrial growth of the volcanic accumulation preserved to us in Mull.

Chapters V.-X. deal with the basalt and mugearite lavas still extant in the region. It is natural to consider these lavas in advance of other igneous rocks of the district, because, as a group, they constitute the first large-scale manifestation of igneous activity. There must be many intrusions in Mull of the same general date as these early lavas, and in fact some have been identified, but the great majority of the intrusions are obviously later. One may imagine that immense masses of lava, now stripped away by erosion, were poured out at the surface during the introduction of the later intrusions. It is certain at any rate that tremendous explosions took place at the Mull centre after some of the major intrusions had consolidated in the positions in which we now find them.

Of the six chapters allotted to the lavas, the first, Chapter V., is written as an introduction to the others, and the last, Chapter X., is concerned with petrology. Readers at all familiar with the subject, will turn instinctively to the descriptions of the columnar lavas of Staffa and its neighbourhood, and of Macculloch's submerged, but upright, tree as illustrated in the frontispiece. Both these picturesque topics are discussed in the earlier portion of Chapter VII. devoted to the lavas of Sheet 43. Still, such fascinating details have only a subsidiary place in the general story of the lavas of Mull. One finds one's self faced with the gradual accumulation of a great mass of olivine-rich basalts (Plateau Types of Plate III., p. 91), attaining locally 3000 feet in thickness and followed by a comparable development of olivine-poor basalts (Central Types). The terms Plateau and Central Types are introduced for local convenience, and their petrological significance is defined in Chapter X. During the outpouring of part at least of the Central Types, the lava-pile seems to have had the form of a gentle Kilauean dome, with a central caldera, measuring five or six miles across, and repeatedly renewed by subsidence (the south-eastern caldera of Plates III. and V., pp. 91 ,165). This caldera often filled with water during periods of quiescence, and the lavas, which entered the resultant crater-lake have been apt to assume pillow-structure (Plate IV. and Figs. 18, 21, pp .133, 151).

Another feature of outstanding interest is the degree to which the basalt-lavas (and to a less extent the later intrusions) have undergone pneumatolytic change round about the centre of activity. A line has been laid down on Plate III. to surround a district 16 miles in diameter, in which the olivine of the basalt-lavas is entirely decomposed. Outside this line, a very large proportion of the olivine is still fresh. Other changes such as the development of albite and epidote have also resulted from the pneumatolysis. For a discussion of the subject the reader should turn to Chapters V., VIII., IX., and X., where in addition an explanation is offered of the frequent failure of trap-featuring within the pneumatolytic area.

The question of the type of eruption responsible for the Mull lavas has already been touched upon, and may now be examined somewhat more fully. One thing is certain : the lavas were poured out with a minimum of explosive activity. The conditions of their formation must have differed considerably from those giving rise to strato-volcanoes of Vesuvian type. Beyond this it is doubtful whether the comparative absence of ash, and the extreme regularity of the lava-pile, lead to any definite conclusion. Such features characterise alike the lava-plains of fissure-eruptions, and the lava-domes of certain central volcanoes (Kilauean Type). Sir Archibald Geikie has done Hebridean geology a great service in insisting upon comparison with Iceland. In Iceland, Thoroddsen[1] points out that

[1] Thoroddsen's results are summarised by E. B. Bailey in 'Iceland—a Stepping-Stone,' *Geol. Mag.*, 1919, p. 466.

every type of volcanic activity can be recognized; and he enumerates, among others, 87 examples of fissure-eruption, and 16 of Kilauean lava-domes. The evidence which the pillow-lavas afford of a caldera in Central Mull during a considerable part of the lava-period (as one may for brevity call it) points strongly to the existence for the time being of a Kilauean lava-dome. In keeping with this interpretation, it may be mentioned that a certain small proportion of the basic intrusions concentrated in Central Mull are referable to the lava-period.

But let us consider the matter a little further. Sir Archibald Geikie has cited the profusion of dykes characteristic of Mull (Fig. 60, p. 357) as an argument in favour of the fissure-origin of the lavas of the island. He was well aware that many of the dykes—it is probably safe to say a large majority of them—are of later date than any of the lavas at present preserved. Still, he was able to demonstrate in Skye, and the same holds good in Mull, that a fair proportion are of relatively early date; and it is possible that some of them may even be as early as the lavas taken as a whole. The analogy of Iceland led Sir Archibald Geikie to infer that many of the Mull dykes are the consolidated feeders of fissure-eruptions, and that some among them have been the main source of supply of the Mull lavas still visible to-day.

The relationships of the Mull dykes will be returned to later on (pp. 10, 48). Meanwhile, the present writer may explain that he regards Mull as a volcanic centre of unusual longevity and complexity, with almost every conceivable type of eruption represented in its lavas, its agglomerates, and its intrusions; he thinks that this centre has repeatedly served as a focus of fissure-eruptions; but he is doubtful whether the lavas still spared by erosion are not, in the main, the products of a central volcano, an idea always linked with the name of Professor Judd (pp. 45, 49).

The view that fissure and central eruptions may have taken place in one and the same region within a comparatively brief period is in keeping with Thoroddsen's Icelandic observations; for on more than one occasion he refers to fissures cutting across lava-domes.

Chapter XI. supplies an account of all basic intrusions which have been assigned on local evidence to the same period as the basaltic lavas; and also of certain other dolerites and gabbros which do not readily find a place in the scheme of the Memoir.

Chapter XII. describes the granophyres of Glas Bheinn and Derrynaculen (Plate V., p. 165). These masses are the earliest large intrusions exposed by erosion in Mull. They seem to have risen in dyke-fashion along the fissure that bounds the caldera in which the pillow-lavas occur, until, on approaching the surface, they expanded laterally with irresistible force. At any rate, outside their outcrops, there has been developed a most wonderful series of arcuate folds concentric with the pillow-lava caldera. These folds are illustrated in Plate V., and Figs. 25 and 35 (pp. 174, 237), and are considered in some detail in Chapter XIII., along with other features of the Tertiary tectonics of Mull. The field-evidence proves conclusively that they are of very early date, compared, that is, with most of the intrusions of the Mull centre ; and it is largely on this account that a genetic connection between themselves and the Glas Bheinn and Derrynaculen Granophyres has been suggested. It may be pointed out, however, that there is some doubt whether the Glas Bheinn Granophyre, as now exposed, is not rather earlier than the folding. The granophyre has certainly been broken up by the first paroxysmal explosions of the Mull centre, dealt with in Chapters XV. and XVI., and it is not quite clear in the field whether the agglomerate resulting from these explosions is not affected by the folding. In Chapter XV. it is suggested that the balance of evidence favours the view that the agglomerate was showered down into already formed synclines ; in which case the time-sequence may be interpreted thus :—Glas Bheinn Granophyre accompanied by folding and followed by explosions. If, on the other hand, it is eventually found that the agglomerate has been involved in the folding, this suggested sequence will have to be a little modified and may perhaps read :—Glas Bheinn Granophyre consolidating in its upper portions and followed by a renewal of intrusion of similar magma accompanied by folding and explosion. A close association of intrusion, folding, and explosion is probable enough in view of Omori's account of the phenomena attending the upheaval of the New Mountain of Usu San in Japan.[1]

A break in the story must be noted at this point, for Chapter XIV. intervenes between the discussion of the folds and that of the agglomerates. It deals with an interesting set of alkaline intrusions of intermediate composition grouped together for petrographical convenience, in the absence of any very satisfactory chronological data. In some instances, these intrusions are accompanied by explosion-phenomena, and it is thought that certain of them may belong to the maximum period of explosive activity which supplies the main topic of the two succeeding chapters.

It has already been explained that these two, Chapters XV. and XVI., include an account of the first paroxysmal outbursts of the Mull centre following upon, or accompanying, the development of arcuate folding. Chapter XV. is devoted to the surface-agglomerates of these outbursts. Chapter XVI. describes the corresponding vents ; but it goes much further, and emphasizes a point, first established by Mr. Wilson, that central explosions of great importance were repeated at a much later period in the history of Mull, notably after the intrusion of the Ben Buie Gabbro of Chapter XXII.

Another feature of general interest dealt with at this juncture is the evidence for the most of Mull's explosive activity being connected with acid magma. Still another is the frequency of gneissic debris in the vents surrounding the calderas of Fig. 29 (p. 201).

Chapter XVII. introduces three important felsite-intrusions (Plate V., p. 165). Of these, the Beinn Mheadhon Felsite is of slightly later date than the first paroxysmal eruptions, whereas the Torness Felsite is probably rather earlier, while the Creag na h'Iolaire Felsite

---

[1] E. B. Bailey, 'The New Mountain of Usu San,' *Geol. Mag.*, 1912, p. 248.

cannot be very precisely dated. The Beinn Mheadon Felsite is the most interesting of the three, and furnishes an example of what appears to be a laccolithic swelling on a dyke, or sheet, inclined outwards from the Mull centre. It is also the largest of the many composite intrusions (p. 32) known in the district.

Chapter XVIII. discusses two large intrusions of augite-diorite balancing one another on either side of the north-west axis of symmetry through Loch Bà (Plate V., p. 165), first recognized by Mr. Wright. The augite-diorites are taken at this stage because they seem to be earlier than any of the cone-sheets of Mull; but all that is firmly established is that they are earlier than the Late Basic Cone-Sheets of Chapter XXVIII.

Here the reader has come to a stage in the history of Mull from which it is profitable to look forward and also to look back. The igneous centre from henceforth assumes a new character marked by an intermittent intrusion of numberless cone-sheets. Considering the profusion in which cone-sheets are represented in Mull, it is very strange how seldom they are met with at other eruptive centres investigated by geologists. At the present time cone-sheets are scarcely known outside Mull, except in Ardnamurchan, and Skye; and it was in Skye that they first obtained due recognition, receiving from Dr. Harker the title *inclined sheets*, now gradually giving place to the synonym *cone-sheets*. For a definition of the term, and a discussion of its application in Mull, the reader is referred to Chapters XIX., XXI., and XXVIII. of the present Memoir. To appreciate in some degree the distribution and aggregate bulk of the local cone-sheets, he must turn to the one-inch Map, Sheet 44, where cone-sheets are represented as groups rather than as individuals, and are lettered aI, bI, and tI, on a basis of composition combined, as far as is convenient, with relative age. The form taken by cone-sheet intrusions has suggested to various workers, notably Mr. Anderson, the application of magmatic pressure near the apex of the cone (p. 11).

Looking back over the events at the Mull centre, prior to the introduction of the first cone-sheets, one may epitomize the whole as follows :—At first, the outstanding event is the quiet outpouring of basalt-lava, olivine-rich to begin with, olivine-poor or olivine-free to end with. Thereafter, the story is of an acid magma that was either intruded as great subterranean masses, sometimes accompanied by marked disturbance of adjacent rocks, or else extruded at the surface by virtue of explosions.

Certain features, for instance, the development of one or two calderas, the location of granophyre intrusions, and the arcuate distribution of folds, indicate the early establishment of a ring-tendency (Plate V., p. 165). This ring-tendency finds much fuller expression at intervals during the succeeding period (Plate VI., p. 307).

Ring-fractures are more widely known in igneous geology than cone-fractures. It is not intended here to anticipate the few remarks offered in Chapter II. regarding their recognition in Mull. The reader who wishes to follow the subject farther afield may, however, turn to descriptions of Glen Coe,[1] Ben Nevis,[2] and Iceland[3]; and also, as Mr. Wright points out, to the Irish Geological one-inch Map (Sheets 59, 60, 70, and 71) of Slieve Gullion. It was in Iceland that the word *ring-fracture* (*Kreisbrüche*) was introduced by Thoroddsen. Mr. Anderson furnishes a dynamical discussion of the phenomenon (p. 11).

A very beautiful feature in the story of the intermittent injection of cone-sheets and ring-dykes is the distribution of these intrusions with reference to twin centres situated near Beinn Chàisgidle and the head of Loch Bà respectively (one-inch Map, Sheet 44, and Plate VI., p. 307). These twin centres are lettered $C_1$ and $C_2$ in Fig. 58 (p. 338); and, though they seem to have functioned to some extent simultaneously, there is no doubt that activity about $C_1$ tended to be replaced as time went on by activity about $C_2$. Thus $C_1$ is the main centre for the Early Acid and Basic Cone-Sheets of Chapters XIX. and XXI., and it also seems to have had a share in controlling the earliest of the Late Basic Cone-Sheets of Chapter XXVIII.; but presently activity shifted, and a large proportion of the Late Basic Cone-Sheets is grouped about $C_2$. In the same way, $C_1$ is the principal centre for the ring-dykes of Chapter XXIX., referable to the earlier part of the Late Basic Cone-Sheet period; while $C_2$ is the centre of the Loch Bà Felsite of Chapter XXXII., a ring-dyke later than all the cone-sheets of Mull.

The story of magmatic change after the appearance of cone-sheets in Mull is for the most part a repetition of what occurred in the pre-cone-sheet period. There is, it is true, a well-marked tendency for the earliest cone-sheets to be of acid or intermediate composition; but the resultant Early Acid Cone-Sheets may perhaps be regarded as a magmatic heritage from pre-cone-sheet days, and they are, after all, of subordinate bulk. The main magmatic succession is, as Dr. Clough clearly established, from olivine-rich magma responsible for the Early Basic Cone-Sheets of Chapter XXI., and the great olivine-gabbros of Chapter XXII., to olivine-poor or olivine-free magmas represented by the Late Basic Cone-Sheets of Chapter XXVIII., and the quartz-gabbros, granophyres, and felsites of Chapters XXIX.–XXXII. How the various petrological types of dykes described in Chapter XXXIV. fit into this scheme has only been very partially determined; but one point is beyond question: some among them, the latest of Mull's igneous manifestations, clearly demonstrate a return of olivine-rich magma in the final stage of activity.

Chapter XIX. furnishes an account of all the intermediate and acid cone-sheets of Mull (*see* one-inch Map, Sheet 44); most of them belong to an early stage of the cone-sheet period. As a body, they share with every other suite of minor intermediate and acid

---

[1] E. B. Bailey in 'Summary of Progress for 1905,' *Mem. Geol. Surv.*, 1906, p. 96; C. T. Clough, H. B. Maufe, and E. B. Bailey, 'The Cauldron-Subsidence of Glen Coe,' *Quart. Journ. Geol. Soc.*, vol. lxv., 1909, p. 611; 'The Geology of Ben Nevis and Glen Coe,' *Mem. Geol. Surv.*, 1916, Chap. VIII.

[2] H. B. Maufe *in* 'Summary of Progress for 1909,' *Mem. Geol. Surv.* 1910, p. 80; 'The Geology of Ben Nevis and Glen Coe,' Chap. X.

[3] T. Thoroddsen, summarized by E. B. Bailey, *in* 'Iceland—a Stepping-Stone,' *Geol. Mag.*, 1919, p. 466.

intrusions in the island, irrespective of age or habit, a marked tendency to composite association with basic magma. Again and again in such cases, one meets with intrusions which have intermediate or acid interiors, and relatively thin basic marginal layers; and there is no development of a chilled edge except at the exterior contacts of the two basic layers with country-rock (Chapters XVII., XXV., and XXXIV.; p. 32). In the present Memoir, the name composite intrusion is reserved for such intimate associations, although it has been used in a rather wider sense in the past to cover all complex acid and basic intrusions without insisting upon an absence of chilling at mutual contacts.

Chapter XX. deals with the Loch Uisg Granophyre and Gabbro. Both probably belong to some phase of the lengthy period characterized by Early Acid and Early Basic Cone-Sheets. The flat top of the granophyre, cutting across folded and highly baked lavas, is one of the most easily appreciated phenomena of Mull geology (Fig. 34, p. 231).

Chapter XXI. gives an account of the Early Basic Cone-Sheets.

Chapter XXII. describes the three chief olivine-gabbros of Mull (Plate V., p. 165). The Ben Buie Gabbro was intruded at a late stage of the Early Basic Cone-Sheet period; the Corra-bheinn Gabbro actually at the close; the Beinn Bheag Gabbro is less precisely dated. In form, the gabbros are great steep rather irregular masses. All three can be seen to break through vent-agglomerates, and yet to be themselves broken by a return of explosive activity at some date succeeding their own consolidation.

Chapters XXIII.-XXVII. cover almost all the sills and sheets of the district, except the cone-sheets to which reference has been so often made. It is thought that the majority of these intrusions date from some stage in the development of the Late Basic Cone-Sheets of Chapter XXVIII.; but there is very little evidence upon which to base a judgment. Apart from their unsatisfactory age-relations, the sills and sheets are in many ways noteworthy. Thus, one may allude to their relative concentration in South-West Mull, where they mark out a special field of intrusion (Chapters XXIII.-XXVI., Fig. 42, p. 258), with a sub-area in the neighbourhood of Loch Scridain, characterized by abundant pitchstone (Chapter XXIII.), and singularly interesting xenoliths often carrying sapphire (Chapter XXIV.). Sills and sheets other than those of the South-West Field are dealt with in Chapter XXVII.

The reader must realize that the intrusions considered above do not correspond in field-relations, or petrology (Chapters XXV.-XXVII.), with what Dr. Harker styles the "Great Group of Sills" in Tertiary Hebridean geology. As explained in some detail in Chapters V. and VI., the authors of the present memoir follow Sir Archibald Geikie in interpreting most of Dr. Harker's "Great Group" as the more solid portions of lava-flows.

Chapter XXVIII. treats of Late Basic Cone-Sheets. The time-relations of these cone-sheets to the ring-dykes of subsequent chapters are discussed in some detail, and evidence is advanced for the view that the centre of cone-sheet activity migrated during the period of the intrusion from $C_1$ to $C_2$ (Fig. 58, p. 338).

Chapter XXIX. starts with a definition of two important terms *ring-dyke* and *screen*. It then passes to the ring-dykes of Glen More and Beinn Chàisgidle (Plate VI., p. 307). This is the first region from which a numerous suite of concentric ring-dykes has been described. It is also a region of great complexity, for the ring-dykes are in many cases much cut up by cone-sheets. Accordingly an unusually full treatment is offered based upon Figs. 52 and 53 (pp. 308, 312), which reproduce with slight reduction the field-maps of two selected parts of the district. An interesting feature established for the ring-dykes is their occasional upward splitting, due to the magma concerned having found more than one ring-fissure available at the time of its intrusion. Another feature, and one of prime importance, is the clear evidence of gravitational differentation *in situ* afforded in several instances by these intrusions. Chapter XXX. is devoted to this aspect of the subject (Figs. 54-56, pp. 322-326). It is not surprising that the ring-dykes concerned should show such a tendency, for they are often more than 100 yards wide, and exposed on valley sides for over 1000 feet in a vertical sense, while their margins are unchilled, and their texture indicative of slow-cooling; moreover, their magma is of quartz-gabbro composition, and, as such, is particularly prone to allow of a migration of an acid residuum during the progress of crystallization.

Chapter XXXI. deals with the Glen Cannel Granophyre (Plate VI., p. 307), a great central intrusion with a dome-shaped roof largely stripped from it by erosion. In places the granophyre has been cut through for a thousand feet, but no floor is exposed. The granophyre has been intruded at the more north-westerly of the twin-centres of activity.

Chapter XXXII. discusses three very important ring-dykes, one of them clearly intruded with reference to the same north-westerly centre. These intrusions, the Knock and Bienn a' Ghràig Granophyres, and the Loch Bà Felsite (Plate VI., p. 307), furnished Mr. Wright with a first insight into the ring-structure of Mull. Later Mr. Richey completed the mapping of the Loch Bà Felsite, as shown in Fig. 58 (p. 338), and found in it the most perfect example of a ring-dyke known to science. All three are of relatively late date in the igneous history of Mull, in fact the Loch Bà Felsite may be claimed as the latest of all Mull intrusions except some of the north-west dykes of Chapter XXXIV. The Loch Bà Felsite almost certainly surrounds a cauldron-subsidence, and during its uprise its magma seems locally to have exploded. The two associated granophyres also furnish much material for thought. Nowhere in Mull is the form of major intrusions displayed to better advantage, and nowhere can one realize so graphically the part played by a screen of country-rock, than in the hills overlooking Loch na Keal. It is particularly noteworthy that a great intrusion, like the Beinn a' Ghràig Granophyre, can replace a large mass consisting of consolidated lavas and cone-sheets without upsetting the arrangement of the adjoining country-rock. Since there is evidence that in Beinn Fhada, at any rate, the roof of the intrusion has not been forced up, it seems fairly clear that the floor has sunk down, either *en masse*,[1] or piece-meal.[2]

[1] *Cf.* C. T. Clough, H. B. Maufe, and E. B. Bailey, 'The Cauldron-Subsidence of Glen Coe,' *Quart. Journ. Geol. Soc.*, vol. lxv., 1909, p. 669.
[2] *Cf.* R. A. Daly, 'The Mechanics of Igneous Intrusion,' 3rd paper, *Am. Journ. Sci.*, Ser. 4, vol. xxvi., 1908, p. 17.

Chapter XXXIII. is mainly of petrological interest and discusses certain hybrid complexes found at the margin of the Beinn a' Ghràig Granophyre.

Chapter XXXIV. supplies an account of all the normal dykes of Mull (Fig. 60, p. 357). It is pointed out that these dykes are of many different ages : some of them are the latest products of igneous activity represented in the district ; while others are of relatively early date, and quite possibly include examples contemporaneous with the lavas of Chapters V.-X. Most of the Mull dykes form part of a north-westerly swarm, rather more than 10 miles wide and well over 100 miles in length. The existence of the Mull Swarm seems to have been first definitely realized by Mr. Maufe, largely as a result of his experience in Sheet 36 of the one-inch Map (*cf.* Fig. 60). The swarm can be interpreted in exactly the same manner as the analagous Etive Dyke-Swarm[1] of Old Red Sandstone Age. Individual dykes of the Mull Swarm are regarded as a response to regional tension, or at any rate a relief of pressure, which opened, or helped to open, fissures directed north-west and south-east. The tension was in the great majority of cases relieved for the time being by very moderate movement, and accordingly individual dykes are narrow. The tension was, however, renewed again and again, and this accounts for the extraordinary number of the dykes and their wide disparity of age. Naturally, in the opening of a fissure, advantage would be taken of any particular structural weakness which the district presented ; and one can claim that it was the frequent subterranean weakness of the Mull centre, with the magma of its pipes (annular or otherwise) often in a molten condition and perhaps under excess of pressure, that determined the location, though not the direction of the Mull Swarm.

In Chapters V.-X., attention is drawn to the widespread pneumatolysis of the lavas found about the Mull centre. In later chapters it is pointed out that pneumatolytic changes can often be detected affecting the intrusions of the same central district. In fact, few of the more central intrusions of Mull, except such massive examples as the Ben Buie Gabbro, have retained their olivine fresh. Even the latest of the dykes of Chapter XXXIV. have suffered markedly in this respect, though not through quite so extensive an area as that shown for the lavas in Plate III. (p. 91). Thus one can recognize the presence of the subterranean molten plug of our theory not only in the localization of the dykes, but also in their alteration.

Another feature of interest dealt with in Chapter XXXIV. is the occasional evidence of explosions having occurred along the course of Tertiary dykes of the district. Naturally, this evidence greatly strengthens Sir Archibald Geikie's comparison between the Tertiary dykes of the Hebrides and those which have been responsible for fissure-eruptions during historic times in Iceland.

Chapter XXXV. treats of certain camptonite-dykes in and near Mull. It is uncertain whether these intrusions are of Tertiary or New Red Sandstone Age.

Chapter XXXVI. turns from the igneous history of Mull, and directs attention to what many regard as the most striking lesson

[1] C. T. Clough, H. B. Maufe, and E. B. Bailey, *op. cit.*, p. 674.

the district has to offer, namely the profound erosion which it has suffered during Tertiary times. Various episodes and features of this erosion are passed in review. One of the latest of them, reached not long before the onset of glacial conditions, has left clear traces of marine attack at levels lying between 100 and 160 feet above present high water.

Chapter XXXVII. passes on to the Glacial and subsequent history of the area. It is shown that, during the maximum glaciation, Mull maintained a sanctuary uninvaded by mainland-ice (Figs. 64 and 65, pp. 393, 395); that, during the valley-glaciation, several glaciers reached down into the sea of the period until after the elevation of adjacent 100-foot and 75-foot beaches; that, during the period of the Post-Glacial submergence responsible for the 25-foot raised beach of the district, Mull lay half included in the South-West Highlands, where this last-mentioned beach is very prominently developed, and half in the North-West Highlands, where it is much less clearly marked; and, finally, that men with Azilian culture occupied one of the caves of this 25-foot beach, when probably the sea had scarcely forsaken it.

Chapter XXXVIII. deals with the economic aspect of the region. Its most interesting feature is a series of lignite-analyses by Dr. W. Pollard.

An Appendix supplies a general Bibliography.

E. B. B. (as Editor).

## Dynamics of Cone-Sheets and Ring-Dykes.

A dynamical theory of the formation of cone-sheets and ring-dykes may be developed on the following lines. One may disregard the probable existence of a broad lava-cone and other irregularities, and suppose that, speaking very generally, the surface of the ground, at the periods at which these features were developed, was horizontal. Underneath this horizontal surface, at a depth of several miles, was a magma reservoir. Its shape may have been roughly that of a paraboloid of revolution.

One may suppose that at first the magma had a specific gravity equal to that of the surrounding rocks, and that it was under a pressure just high enough to have raised it to surface-level if there had been an outlet. Overlooking the fact that the rocks themselves were not quite homogeneous, one may further suppose that they were under the same horizontal and vertical pressures at every point as would obtain in a liquid with the same surface and the same specific gravity. While such was the case no fractures could arise.

*Cone-Sheets.* — If, however, these conditions were modified by an increase of pressure in the magma-basin, the pressure-system in the crust would have superimposed on it a system of tensions, acting across surfaces which near the basin were roughly conical. The fine firm lines in Fig. 1 are intended to show the intersection of these surfaces with the plane of the diagram. A superimposed system of pressures would also act across surfaces which cut the former orthogonally and are indicated in section by the fine broken lines. The superimposed tensions, together with the increased pressure of the magma, might cause a series of fractures to develop, along which the magma would intrude. The opening fractures would follow the fine firm lines of Fig. 1, and thus may have originated the cone-sheets of Mull. It will be seen from the diagram that if the surface were denuded to a certain depth, the cone-sheets exposed might be expected to be steeper in the central parts of the area of intrusion. This is actually the case.

*Ring-Dykes.*—If the conditions were reversed, and the pressure of the magma fell below that which was at first assumed, the originally "hydrostatic" pressure in the crust would be modified in a different way. Superimposed pressures would

12     Chapter 1.—General Introduction.

act across the surfaces whose trace is shown by the fine firm lines of Fig. 1, and superimposed tensions across those which are indicated by the broken lines. It seems likely that, in this case, surfaces of fracture would originate inclined at an angle to the surfaces across which there were maximum superimposed tensions, as in the case of normal faults.[1] The angle may have been about 20° or 30° Such surfaces of fracture correspond not to tension-cracks, but in theory at least, more nearly to planes of maximum shearing-stress. They deviate from the directions across which this stress is an absolute maximum, owing to certain considerations of friction. An attempt to show the trace of such surfaces, taking the angle mentioned as about 25°, has been made in the diagram. It can easily be seen that the theory explains the tendency to an outward slope, which is perhaps a feature of ring-dykes. If the fractures formed curves that were closed in cross-section, the rock inside them might tend to become detached, and to sink down into the magma. The gap between the subsiding mass and the stationary walls would widen with the subsidence, and this perhaps explains, in part, the greater width of ring-dykes when compared with cone-sheets. It is uncertain whether, in the majority of cases, the

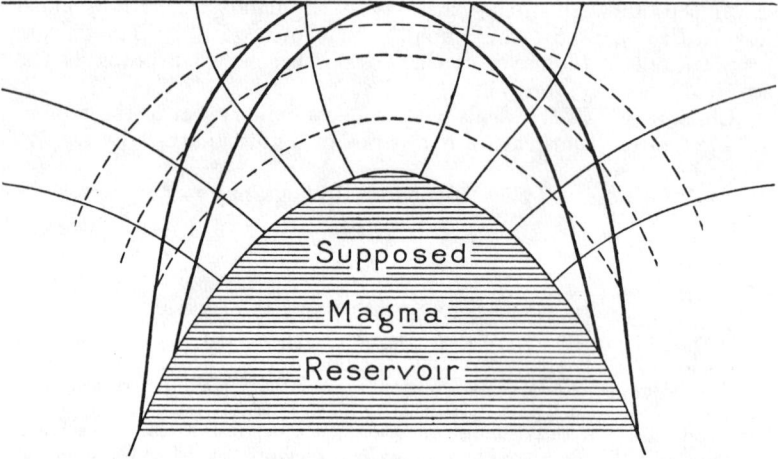

FIG. 1.—Stress-Diagram to show supposed mode of formation of Cone-Sheets and Ring-Dykes. For explanation of lines, see text.

ring-fractures continued upwards to the surface of the ground. When they did so, they must have given rise to circular depressions. Such a subsidence of the surface is known to have happened at Glencoe, and it appears to have happened in Mull in the case of the supposed south-eastern caldera.

The calculations necessary to fix the course of the surfaces of maximum and minimum tension on any particular assumption (as to the shape of the underground cavity, etc.), are extremely elaborate. They have not been fully carried out, but there can be no doubt that the general result will be roughly as shown in the diagram. The case is similar in certain respects, though not very closely, to that treated by Hertz,[2] who dealt with the stresses set up by the impinging of curved surfaces in elastic solids, or by French,[3] who showed that pressure exerted by a small steel ball on glass can lead to the formation of cone-fractures.     E. M. A.

[1] E. M. Anderson, 'The Dynamics of Faulting,' *Trans. Geol. Soc. Edin.*, vol. viii., 1905, p. 387.

[2] H. R. Hertz, 'On the Contact of Rigid Elastic Solids, and on Hardness,' *Miscellaneous Papers*. 1896, p. 163.

[3] J. W. French 'Percussion Figures in Isotropic Solids,' *Nature*, vol. civ., 1919-20, p. 312.

## MULL PETROLOGY.

In the body of the Memoir, the grouping of the petrological material is in large measure determined by the field-relations of the various rocks described. In the present chapter, a general survey of Mull petrology is attempted, in which special attention is paid to the composition of the Mull magma at various stages of its history. In this task, we are able to base our views on an exceptionally complete series of analyses. To Mr. E. G. Radley we are indebted for thirty-three rock-analyses from the Mull district, and to Mr. F. R. Ennos for five. Moreover, we have been able to draw upon published analyses of Mull types collected in other parts of the British Tertiary province, and in this connexion we quote freely from Dr. Harker's Memoirs for the Geological Survey, including ten analyses by Dr. W. Pollard.

In dealing with analysed material, we have carefully distinguished between what is typical, and what is exceptional. We have recognized certain compositional characteristics as belonging to what we term *magma-types*, illustrated in Figs. 2-4 and Tables I.-VII. The conception of magma-type is based upon composition alone. In this, it differs from the conception of rock-type which takes into account texture as well as composition. Thus a basalt and a gabbro may belong to one magma-type though admittedly representatives of different rock-types. The reader is warned, however, that the compositional definition of a magma-type is just as vague as that of a rock-type. All one can do is to state an *average composition* around which actual rocks seem to group themselves.

In the following pages, we have selected certain Mull magma-types for special discussion. A very large proportion of Mull rocks range themselves approximately in a magma-series graphically expressed in Fig. 2. Every stage in this series is copiously represented. Two other magma-types, selected as highly typical, are illustrated in Fig. 3. In addition, the extreme of alkaline variation reached by the Mull magma is shown in Fig. 4; in this last case the magma-types are of very little importance from the point of view of bulk.

After introducing the reader to the scope of magmatic variation illustrated in Mull, attention will be directed to certain intrusions which through differentiation *in situ* have produced a very large proportion of the magma-types illustrated in Fig. 2. Following upon this a general discussion of Mull differentiation is attempted in the light of observed magmatic sequences. Admittedly the treatment is speculative, but it is thought that it will help the reader to an appreciation of many of the observed facts of Mull petrology.

The account of magma-types and differentiation outlined above is illustrated by tables of analyses (I.-VIII.). These include most of the complete analyses of Mull material. There are other analyses, of minerals, xenoliths, and sediment, and these are grouped in Table IX. by themselves. They are printed in this chapter merely for convenience of reference.

## Chapter I.—General Introduction.

### NORMAL MULL MAGMA-SERIES OF FIG. 2.

*Plateau Magma-Type of Fig.* 2. This magma-type provides the bulk of the Plateau Basalt Lavas (Chapter X.), most of the Early Basic Cone-Sheets (Chapter XXI.), and many dykes including examples belonging to the latest phase (Chapter XXXIV.).

The magma-type is known with either a basaltic or doleritic crystallization (attaining to gabbroic in certain massive cone-sheets).

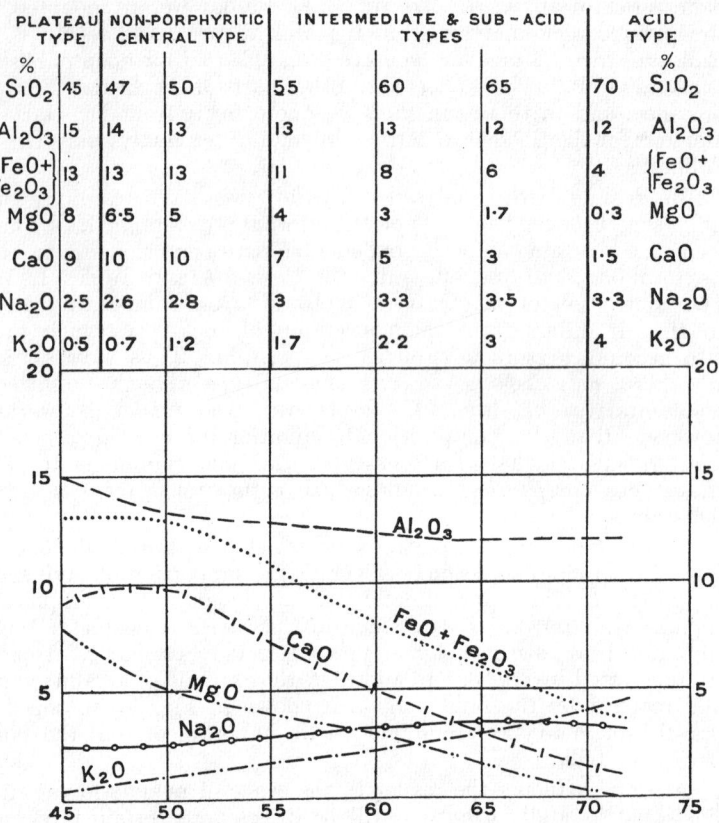

FIG. 2.—Variation-Diagram: Normal Mull Magma-Series.

Its essential minerals are olivine, augite, zoned basic plagioclase and iron ore, with a certain amount of interstitial remainder; analcite and natrolite are often present in this remainder where the olivine of the rock is fresh, but elsewhere serpentine seems to take their place. An approach to Non-Porphyritic Central Magma Type is marked in some doleritic representatives by a little residual quartz (Chapter XXI.).

A purple colour is characteristic of the augite of the rocks here considered. It is found in the augite of all the analysed specimens quoted in Table I., except E, which last, apart from its magnesia-content, corresponds with the basic end of the Mull Non-Porpyritic Central Magma. Though well-nigh universal, the purple

TABLE I. : PLATEAU MAGMA-TYPE OF FIG. 2.

|   | A | B | I. | II. | III. | C | D | E |   |
|---|---|---|---|---|---|---|---|---|---|
| $SiO_2$ | 43·94 | 45·24 | 45·37 | 45·48 | 45·52 | 46·46 | 46·61 | 47·64 | $SiO_2$ |
| $TiO_2$ | 2·45 | 2·26 | 2·87 | 3·48 | 2·85 | 2·07 | 1·81 | 1·27 | $TiO_2$ |
| $Al_2O_3$ | 14·03 | 15·63 | 15·16 | 15·66 | 14·30 | 15·48 | 15·32 | 14·15 | $Al_2O_3$ |
| $Cr_2O_3$ | tr. | tr. | ... | ... | ... | 0·02 | tr. | 0·01 | $Cr_2O_3$ |
| $V_2O_3$ | ... | ... | ... | ... | ... | 0·05 | ... | 0·06 | $V_2O_3$ |
| $Fe_2O_3$ | 1·95 | 5·56 | 3·38 | 3·64 | 3·43 | 3·63 | 3·49 | 5·18 | $Fe_2O_3$ |
| FeO | 11·65 | 7·19 | 11·58 | 10·56 | 9·00 | 10·23 | 7·71 | 7·96 | FeO |
| MnO | 0·32 | 0·23 | 0·31 | 0·20 | 0·19 | 0·48 | 0·13 | 0·33 | MnO |
| (Co,Ni)O | nt. fd. | tr. | nt. fd. | ... | ... | 0·02 | tr. | tr. | (Co,Ni)O |
| MgO | 10·46 | 7·82 | 6·72 | 6·99 | 10·65 | 6·80 | 8·66 | 7·38 | MgO |
| CaO | 8·99 | 9·38 | 8·11 | 8·24 | 9·54 | 9·05 | 10·08 | 11·71 | CaO |
| (Ba,Sr)O | nt. fd. | ... | nt. fd. | .. | ... | 0·02 | ... | nt. fd. | (Ba,Sr)O |
| $Na_2O$ | 2·68 | 2·01 | 2·90 | 2·68 | 2·21 | 3·01 | 2·43 | 2·38 | $Na_2O$ |
| $K_2O$ | 0·33 | 0 72 | 0·44 | 0·49 | 0·42 | 0·68 | 0·67 | 0·71 | $K_2O$ |
| $Li_2O$ | nt. fd. | ... | nt. fd. | nt. fd. | nt. fd. | ? tr. | ... | ... | $Li_2O$ |
| $H_2O+105°$ | 2·31 | 2·21 | 1·96 | 1·52 | 1·53 | 1·43 | 2·07 | 1·44 | $H_2O+105°$ |
| $H_2O$ at 105° | 0·85 | 1·12 | 1·18 | 0·93 | 0·70 | 0·89 | 1·10 | 0·19 | $H_2O$ at 105° |
| $P_2O_5$ | 0·20 | 0·20 | 0·29 | 0·26 | 0·23 | 0·30 | tr. | 0·09 | $P_2O_5$ |
| $CO_2$ | 0·16 | 0·49 | ... | 0·21 | 0·15 | nt. fd. | tr. | ... | $CO_2$ |
| $FeS_2$ | 0·04 | ... | ... | ... | ... | ... | ... | ... | $FeS_2$ |
| $Fe_7S_8$ | 0·06 | ... | ... | ... | ... | ... | ... | ... | $Fe_7S_8$ |
| $\frac{1}{2}S$ | ... | ... | ... | ... | ... | 0·08 | ... | ... | $\frac{1}{2}S$ |
| S | ... | ... | ... | nt. fd. | nt. fd. | ... | ... | 0·03 | S |
|   | 100·42 | 100 06 | 100·27 | 100·34 | 100·72 | 100·70 | 100·08 | 100·53 | ... |
| Spec. grav. | ... | 2·85 | 2·95 | 2·93 | 2·99 | ... | 2·87 | ... | ... |

A. (14174, Lab. No. 339). Dyke. Slac nan Sgarbh; Jura. CRINANITE, quoted from J. S. Flett, *Geology of Knapdale, Jura & North Kintyre*, 1911, p. 118. Anal. E. G. Radley.

B. (7854, Lab. No. 22). Sill. Ben Lee; Skye. DOLERITE, quoted from A Harker, *Tertiary Igneous Rocks of Skye*, 1904, p. 248. Anal. W. Pollard.

I. (15995, Lab. No. 390). Lava. Stream at pit, ⅛ m. N.N.E. of Pennycross House, 750 ft. above sea; Mull. BASALT, p. 137. Anal. E. G. Radley.

II. (19070, Lab. No. 460). Lava. Raised-beach cliff, 200 yds. W. of Lochaline Pier, between first two walls crossing raised beach west of Pier; Morven. BASALT, p. 136. Annal. F.R. Ennos.

III. (19071, Lab. No. 461). Lava. Thick flow on ashy material, little port, east side of Rudha Dearg, 1 m. west of Lochaline; Morven. BASALT, p. 137. Anal. F. R. Ennos.

C. (11727, Lab. No. 184). Lava. Orval; Rum. BASALT, quoted from A. Harker, *Geology of the Small Isles of Inverness-shire*, 1908, p. 57. Anal. W. Pollard.

D. (8185, Lab. No. 21). Lava. Drynoch; Skye. BASALT, quoted from W. Pollard, *Summary of Progress of Geol. Surv. for* 1899, p. 174; see also A. Harker, *Tertiary Igneous Rocks of Skye*, p. 31. Anal. W. Pollard.

E. (8062, Lab. No. 79). Cone-Sheet. Cuillins; Skye. DOLERITE, emended analysis quoted from A. Harker, *Tertiary Igneous Rocks of Skye*, p. 370. Anal. W. Pollard.

tint of the augite is more marked in some examples than in others; and it is difficult to connect this difference with the bulk-analyses of the rocks. Thus in A, B, I., and C (Table I.), the purple colour is very pronounced ; in II. and III., it is less conspicuous ; and, in E, it is wanting.

A marked feature of the crystal-development is that the augite does not occur as phenocrysts, nor does it build minute perfect prisms such as occur commonly in the ground-mass of British Carboniferous and Continental Tertiary basalts of similar basicity. These two related characteristics of the augite of the basalts of the British Teriary province were long ago pointed out by Professor Hull[1] and Sir Jethro Teall.[2] Our Mull experience agrees with Professor Hull's generalization, so far as basalts of Plateau Type are concerned, though small phenocrysts of augite are not uncommon in basalts of Central Types. It is noteworthy, however, that the ophitic augites of the Mull Plateau Type often completed their growth well within the crystallization-period of the associated felspar (p. 138).

Dr. Flett has grouped the analcite (or natrolite) bearing doleritic varieties of the Mull Plateau Type under the title *crinanite* (type-specimen A, Table I).

The analyses quoted in Table I. are taken from various West Highland localities, but in every case the rocks analysed can be matched in Mull, and may in this sense be regarded as representative of the Mull Plateau Magma-Type. The main chemical features of the type are summarized in Fig. 2 ; but it may be pointed out, in addition, that high $TiO_2$, about 2·5 per cent. is characteristic, and that a considerable range of MgO, from about 7-10 per cent. may occur without affecting the alkali-content.

As explained in Chapter XXXIV., there is, very occasionally, a definite approach to camptonite among undoubted Tertiary representatives of Mull Plateau Magma-Type. This emphasizes the importance of the comparison Dr. Flett has drawn between the crinanite type of dolerites and the camptonites.[3] The camptonites show a slight fall in $SiO_2$ and $Al_2O_3$, and an increase in CaO and $K_2O$. Two analysed Scottish camptonites, quoted by Dr. Flett, have 1·41 and 1·93 per cent., respectively, of $K_2O$. It must be borne in mind that, whether of Tertiary age or no, the camptonites of Scotland are of relatively insignificant bulk (Chapter XXXV.). Associated more extreme types, such as monchiquites and nepheline-ouachitites, are rarer still.[4]

Attention may be directed in this connexion to definitely alkaline segregation-veins developed in a small proportion of the Plateau Type of basalt-lava (Chapter X.).

*Non-Porpyyritic Central Magma-Type of Fig. 2.* This magma-type first appeared in the south-west part of the Mull district where it furnished wonderfully columnar lavas of Staffa Type (Chapters VII. and X.). It recurred in greater volume, and variety, as an important constituent of the Central Group of lavas (Chapters

---

[1] E. Hull, 'On the Microscopic Structure of the Limerick Carboniferous Trap-Rocks (Melaphyres),' *Geol. Mag.*, 1873, p. 160.

[2] J. J. H. Teall, ' British Petrography,' 1888, pp. 187, 246, 247.

[3] For analyses *see* E. G. Radley quoted by J. S. Flett in 'Geology of the Country near Oban and Dalmally,' *Mem. Geol. Surv.*, 1908, p. 126.

[4] J. S. Flett, *in* 'Geology of Colonsay and Oronsay with part of the Ross of Mull,' *Mem. Geol. Surv.*, 1911, pp. 41-46, 90, 91.

V. and X.). It supplied many of the sills (Chapters XXIII.-XXVII.), most of the Late Basic Cone-Sheets (Chapter XXVIII),

TABLE II.—Non-Porphyritic Central Magma-Type of Fig. 2.

| | Tholeiite Salen Type | Basalt Staffa Type | | | Basalt Compact Central Type | | Tholeiite Brunton Type | | Quartz-Dolerite and Tholeiite Talaidh Type | | |
|---|---|---|---|---|---|---|---|---|---|---|---|
| | I. | II. | III. | A | IV. | V. | VI. | VII. | VIII. | IX. | |
| $SiO_2$ | 47·35 | 47·80 | 49·76 | 52·13 | 50·54 | 53·78 | 51·53 | 51·63 | 52·16 | 53·97 | $SiO_2$ |
| $TiO_2$ | 1·75 | .... | 0·94 | .... | 2·80 | 2·28 | 1·57 | 2·00 | 3·25 | 1·24 | $TiO_2$ |
| $Al_2O_3$ | 13·90 | 14·80 | 14·42 | 14·87 | 12·86 | 12·69 | 11·05 | 11·77 | 11·95 | 14·65 | $Al_2O_3$ |
| $Fe_2O_3$ | 5·87 | .... | 3·95 | .... | 4·13 | 3·44 | 2·73 | 3·23 | 4·86 | 3·62 | $Fe_2O_3$ |
| FeO | 8·96 | 13·08 | 7·77 | 11·40 | 8·75 | 8·94 | 10·98 | 10·47 | 9·92 | 6·32 | FeO |
| MnO | 0·23 | 0·09 | 0·20 | 0·32 | 0·32 | 0·53 | 0·45 | 0·35 | 0·18 | 0·30 | MnO |
| (Co, Ni)O | nt. fd. | .... | nt. fd. | .... | 0·06 | nt. fd. | nt. fd. | 0·04 | .... | nt. fd. | (Co, Ni)O |
| MgO | 5·97 | 6·84 | 5·30 | 6·46 | 4·63 | 2·58 | 5·21 | 5·02 | 3·77 | 4·49 | MgO |
| CaO | 10·65 | 12·89 | 10·22 | 10·56 | 8·74 | 6·36 | 9·68 | 9·34 | 7·14 | 7·98 | CaO |
| BaO | .... | .... | 0·04 | .... | nt. fd. | 0·09 | nt. fd. | 0·03 | .... | 0·04 | BaO |
| $Na_2O$ | 2·73 | 2·48 | 2·49 | 2·60 | 2·89 | 2·74 | 3·48 | 2·90 | 2·36 | 2·54 | $Na_2O$ |
| $K_2O$ | 0·54 | 0·86 | 1·83 | 0·69 | 1·43 | 2·27 | 0·86 | 0·91 | 1·74 | 1·52 | $K_2O$ |
| $Li_2O$ | .... | .... | tr. | .... | nt. fd. | nt. fd. | tr. | nt. fd. | .... | tr. | $Li_2O$ |
| $H_2O+105°$ | 1·16 | } 1·41 | 1·03 | } 1·19 | 2·25 | 2·19 | 1·26 | 1·40 | 1·95 | 0·94 | $H_2O+105°$ |
| $H_2O$ at 105° | 1·04 | | 2·04 | | 0·17 | 1·19 | 0·71 | 0·68 | 0·56 | 1·92 | $H_2O$ at 105° |
| $P_2O_5$ | 0·24 | .... | 0·21 | .... | 0·34 | 0·55 | 0·22 | 0·29 | 0·24 | 0·27 | $P_2O_5$ |
| $CO_2$ | 0·32 | .... | 0·06 | .... | 0·33 | 0·08 | 0·08 | 0·11 | 0·18 | 0·51 | $CO_2$ |
| $FeS_2$ | .... | .... | 0·04 | .... | nt. fd. | 0·42 | 0·26 | 0·08 | .... | 0·09 | $FeS_2$ |
| S | 0·23 | .... | .... | .... | .... | .... | .... | .... | 0·18 | .... | S |
| | 100·94 | 100·25 | 100·30 | 100.22 | 100·24 | 100·13 | 100·07 | 100·27 | 100·44 | 100·40 | |
| Spec. grav. | 2·96 | .... | 2·72 | .... | 2·90 | 2·68 | 2·93 | 2·95 | 2·91 | 2·83 | |

I. (16808, Lab. No. 407). Dyke. Shore ¼ mile S.S.E. of Kintallen, and 2½ miles N.N.W. of Salen; Mull. THOLEIITE, Salen Type, p. 371. Anal. F. R. Ennos.

II. Lava. Staffa. BASALT, Staffa Type. Anal. A. Streng, quoted from Pogg. Ann. vol. xc., 1853, p. 114.

III. (20581, Lab. No. 669). Lava, which encloses Macculloch's Tree. Rudha na h-Uamha; Mull. BASALT, Staffa Type, p. 145. Anal. E. G. Radley.

A. Lava. Giant's Causeway; Ireland. BASALT, Staffa Type. Anal. A. Streng, quoted from Pogg. Ann., vol. xc., 1853, p. 114.

IV. (18474, Lab. No. 448). Lava. Monadh Beag, stream-junction 1000 yards N. of Ishriff; Mull. BASALT, Compact Central Type, p. 149. Anal. E. G. Radley.

V. (14824, Lab. No. 369). Lava. 1 mile N.E. of Loch Bà, 2 miles E. of Gruline House; Mull. BASALT, Compact Central Type, p. 149. Anal. E. G. Radley.

VI. (16810, Lab. No. 406). Dyke. Shore, ¼ mile E. of Arla, and 5½ miles S.E. of Tobermory; Mull. THOLEIITE, Brunton Type, p. 372. Anal. E. G. Radley.

VII. (16809, Lab. No. 411). Dyke. Shore ¼ mile N. of Kintallen, and 3 miles N.N.W. of Salen; Mull. THOLEIITE, Brunton Type, p. 372. Anal. E. G. Radley.

VIII. (18467, Lab. No. 444). Cone-Sheet. 70 yards S. of summit, Cruachan Dearg; Mull. QUARTZ-DOLERITE, Talaidh Type, p. 301. Anal. F. R. Ennos.

IX. (17170, Lab. No. 432). Basic Margin, Composite Sill. Rudh 'a 'Chromain; Mull. THOLEIITE, Talaidh Type, pp. 285, 286. Anal. E. G. Radley.

several of the ring-dykes (Chapter XXIX.), and a large proportion of the normal dykes (Chapter XXXIV.).

The magma-type is known as a partially devitrified glass. (Two silica-determinations by Mr. E. G. Radley quoted in Chapter XXVIII. for variolites of Cruachan Dearg Type give 50·66 and 53·65 per cent.), and with every grade of crystallization from this onwards to gabroic. The names which naturally apply to the various products of crystallization are compact or fine-grained basalt (lavas), variolites (pillow-lavas and minor intrusions), tholeiite (defined in Chapter XXV.—minor intrusions), quartz-dolerite (cone-sheets and ring-dykes), and quartz-gabbro (ring-dykes).

The essential minerals are augite, plagioclase felspar, and magnetite, with an acid residuum in which crystallization reveals quartz and alkali-felspar. Towards the basic end, olivine appears as a minor constitutent.

In non-variolitic lavas of the suite, the augite manifests a very strong tendency to granular crystallization in the ground-mass, and sometimes furnishes a few small phenocrysts. In intrusions, it favours a columnar or branching (cervicorn) form (Fig. 50, p. 302), which gives place, however, to an ophitic habit in relatively basic representatives, and in the earlier crystal-groupings of even the more acid rocks. The colour of the augite is usually brownish, but is often tinged with purple especially in olivine-bearing varieties.

A feature of the more acid and more crystallized members of the group is the obvious freedom of migration of the acid residuum, and the extent to which chemical action has proceeded between this residuum and the early-formed crystals. The conditions of Central Mull were clearly favourable to such chemical exchange both before and after complete consolidation, and it is fairly certain that auto-pneumatolysis and general pneumatolysis have often co-operated in accelerating the process.

The analyses of Table II. are, with the exception of A, all taken from Mull localities. The Giant's Causeway basalt (A) is included because it is a good example of the Staffa Type (Chapter X.). The main features of the magma-type as a whole are sufficiently summarized in Fig. 2; in addition, it may be noted that, as compared with the Plateau Magma-Type, there is a distinct, though not universal, fall in $TiO_2$.

*Intermediate to Sub-Acid Magma-Type of Fig.* 2. This magma-type is represented more particularly among the augite-diorites (Chapter XVIII.), Early Intermediate and Acid Cone-Sheets (Chapter XIX.), and Loch Scridain sills (Chapter XXIII.).

Its consolidation in these various occurrences ranges from glassy to dioritic. The products may be styled glassy andesite—pitchstone—and stony andesite (in both cases sills), craignurite (defined Chapter XIX.—cone-sheets), and augite-diorite (large masses).

The essential minerals are augite, plagioclase, and magnetite, with a very considerable residuum of glass or devitrification-products consisting largely of alkali-felspar and quartz. Olivine is almost always absent. Augite is brownish, and typically columnar or acicular; but may also be ophitic or cervicorn. Enstatite-augite is an essential of the inninmorites (andesite-type defined in Chapter XXV.), and a common accessory of many craignurites. Hypersthene is often met with in the leidleites (andesite-type defined in Chapter XXV.).

## Magma-Types.

More or less variolitic structure is often met with among the andesites.

TABLE III.—INTERMEDIATE TO SUBACID MAGMA-TYPE OF FIG. 2.

|  | Craignurite (basic) I. | Leidleite II. | Leidleite III. | Inninmorite IV. | Inninmorite V. | Craignurite (acid) VI. |  |
|---|---|---|---|---|---|---|---|
| $SiO_2$ | 55·82 | 59·21 | 61·69 | 62·37 | 64·13 | 66·27 | $SiO_2$ |
| $TiO_2$ | 1·62 | 1·06 | 1·00 | 1·06 | 1·19 | 0·87 | $TiO_2$ |
| $Al_2O_3$ | 11·47 | 14·06 | 14·43 | 12·04 | 13·15 | 11·92 | $Al_2O_3$ |
| $Fe_2O_3$ | 3·68 | 2·66 | 1·23 | 1·87 | 1·08 | 3·09 | $Fe_2O_3$ |
| FeO | 7·66 | 4·87 | 5·86 | 5·81 | 6·31 | 3·18 | FeO |
| MnO | 0·40 | 0·24 | 0·30 | 0·24 | 0·27 | 0·31 | MnO |
| (Co,Ni)O | 0·04 | nt. fd. | nt. fd. | nt. fd. | nt. fd. | nt. fd. | (Co,Ni)O |
| MgO | 4·08 | 3·71 | 2·81 | 0·97 | 1·08 | 1·44 | MgO |
| CaO | 7·88 | 5·95 | 4·97 | 3·51 | 3·62 | 3·30 | CaO |
| BaO | 0·03 | 0·03 | 0·04 | 0·07 | 0·09 | nt. fd. | BaO |
| $Na_2O$ | 2·58 | 2·06 | 3·20 | 3·47 | 3·64 | 2·89 | $Na_2O$ |
| $K_2O$ | 2·00 | 2·83 | 1·72 | 2·34 | 2·32 | 4·03 | $K_2O$ |
| $Li_2O$ | tr. | nt. fd. | nt. fd. | nt fd. | nt. fd. | tr. | $Li_2O$ |
| $H_2O+105°$ | 1·88 | 1·49 | 2·32 | 5·54 | 2·71 | 1·51 | $H_2O+105°$ |
| $H_2O$ at 105° | 0·66 | 2·06 | 0·25 | 0·44 | 0·36 | 0·78 | $H_2O$ at 105° |
| $P_2O_5$ | 0·23 | 0·20 | 0·24 | 0·30 | 0·31 | 0·17 | $P_2O_5$ |
| $Co_2$ | 0·08 | ... | ... | ... | ... | 0·53 | $Co_2$ |
| $FeS_2$ | 0·09 | nt. fd. | nt. fd. | nt. fd. | nt. fd. | nt. fd. | $FeS_2$ |
| Cl | ... | nt. fd. | 0·02 | ... | ... | ... | Cl |
|  | 100·18 | 100·43 | 100·08 | 100·03 | 100·26 | 100·29 |  |
| Spec. grav. | 2·88 | 2·61 | 2·64 | 2·50 | 2·57 | 2·65 |  |

I. (16800, Lab. No. 412). Cone-Sheet. Allt an Dubh-choire, 1220 yards above junction with Scallastle River; Mull. CRAIGNURITE (basic), p. 226. Anal. E. G. Radley.

II. (15997, Lab. No. 385). Sill, stony margin of III. ¼ mile N.N.E. of Mullach Glac an t-Sneachda (Loc. 1, Fig. 43, p. 261); Mull. Stony LEIDLEITE, p. 281. Quoted from E. M. Anderson and E. G. Radley, *Quart Journ Geol Soc.*, vol. lxxi, p. 212. Anal. E. G. Radley.

III. (15996, Lab. No. 384). Sill, glassy centre of II. Glassy LEIDLEITE, or LEIDLEITE-PITCHSTONE, p. 281. Quoted from E. M. Anderson and E. G. Radley, ibid, p. 212. Anal. E. G. Radley.

IV. (15989, Lab. No. 386). Sheet. Near head of stream from Tòm a' Choilich, ½ mile S.W. of Pennyghael (Loc. 4, Fig. 43, p. 261); Mull. Exceptionally glassy INNINMORITE, or INNINMORITE-PITCHSTONE, p. 283. Quoted from E. M. Anderson and E. G. Radley, *ibid*, p. 212. Anal. E. G. Radley.

V. (15990, Lab. No. 387). Sheet. $\frac{3}{16}$ mile S.W. of Trig. Station on Beinn an Lochain (Loc. 5, Fig. 43, p. 261); Mull. Fairly glassy INNINMORITE or INNINMORITE-PITCHSTONE, p. 284. Quoted from E. M. Anderson and E G. Radley, *ibid*, p. 212. Anal. E. G. Radley.

VI. (16802, Lab. No. 413). Cone-Sheet. Allt an Dubh-choire, 630 yards above junction with Scallastle River; Mull. CRAIGNURITE (acid), p. 225. Anal. E. G. Radley.

The early minerals of the whole group usually show marked interaction with the acid base, except where this latter has

20   Chapter I.—General Introduction.

consolidated as a glass. The effect is especially marked where there has been relative migration of crystals and residuum (Chapter XXX.).

The main chemical characteristics of the magma-type are summarized in Fig. 2. Comparison of the analyses of Table III. leaves it uncertain what chemical peculiarity determines the distinction

TABLE IV.—ACID MAGMA-TYPE OF FIG. 2.

|  | I. | II. | III. | IV. | V. |  |
|---|---|---|---|---|---|---|
| $SiO_2$ | 70.70 | 71.30 | 72.66 | 73.12 | 73.32 | $SiO_2$ |
| $TiO_2$ | 1.27 | 0.58 | 0.34 | 0.39 | 0.51 | $TiO_2$ |
| $Al_2O_3$ | 11.78 | 11.24 | 12.00 | 12.44 | 12.25 | $Al_2O_3$ |
| $Fe_2O_3$ | 1.32 | 1.80 | 2.03 | 2.09 | 2.77 | $Fe_2O_3$ |
| FeO | 3.45 | 2.84 | 2.04 | 1.65 | 2.20 | FeO |
| MnO | 0.07 | 0.31 | 0.18 | 0.17 | 0.12 | MnO |
| (Co,Ni)O | ... | nt. fd. | nt. fd. | nt. fd. | nt. fd. | (Co,Ni)O |
| MgO | 0.53 | 0.61 | 0.07 | 0.14 | 0.11 | MgO |
| CaO | 1.30 | 1.56 | 1.25 | 0.88 | 1.65 | CaO |
| BaO | ... | 0.07 | 0.12 | nt. fd. | 0.09 | BaO |
| $Na_2O$ | 2.48 | 3.44 | 3.26 | 3.90 | 3.92 | $Na_2O$ |
| $K_2O$ | 4.71 | 4.66 | 5.26 | 4.67 | 2.34 | $K_2O$ |
| $Li_2O$ | ... | ? tr. | nt. fd. | nt. fd. | nt. fd. | $Li_2O$ |
| $H_2O + 105°$ | 1.14 | 1.04 | 0.47 | 0.24 | 0.35 | $H_2O + 105°$ |
| $H_2O$ at 105° | 0.50 | 0.39 | 0.22 | 0.25 | 0.35 | $H_2O$ at 105° |
| $P_2O_5$ | 0.26 | 0.22 | 0.04 | 0.09 | 0.10 | $P_2O_5$ |
| $CO_2$ | 0.51 | ... | 0.24 | 0.05 | 0.06 | $CO_2$ |
| $FeS_2$ | ... | nt. fd. | nt. fd. | nt. fd. | nt. fd. | $FeS_2$ |
| S | 0.08 | ... | ... | ... | ... | S |
|  | 100.10 | 100.06 | 100.18 | 100.08 | 100.14 |  |
| Spec. grav. | 2.58 | 2.53 | 2.61 | 2.57 | 2.66 |  |

I. (18464, Lab. No. 443). Sill. S. of Coire Buidhe, between 800 and 900 foot contours, about ½ mile N. of Carsaig; Mull. FELSITE allied to inninmorite, p. 286. Anal. F. R. Ennos.

II. (16803, Lab. No. 394). Cone-Sheet. Craignure Bay, shore 50 yards N.N.W. of U.F.C. Manse; Mull. Fine GRANOPHYRE allied to craignurite, p. 226. Anal. E. G. Radley.

III. (14825, Lab. No. 370). Loch Bà Ring-Dyke. Top of deep gorge. ¾ mile N. of E. of Summit of Beinn a' Ghràig; Mull. RHYOLITE, p. 347. Anal. E. G. Radley.

IV. (14843, Lab. 372). Beinn a' Ghràig Ring-Dyke. Benmore Lodge, Loch Bà; Mull. GRANOPHYRE, p. 347. Anal. E. G. Radley.

V. (14841, Lab. No. 371). Knock Ring-Dyke. Beinn Bheag, ¼ mile S. of Knock; Mull. GRANOPHYRE, p. 349. Anal. E. G. Radley.

of leidleite from inninmorite. A very interesting point is the extent to which the more glassy representatives retain their magmatic water. This feature is discussed later on with additional partial analyses (Table XI., p. 263). The fall of $TiO_2$, which began to manifest itself in the Non-Porphyritic Central Magma-Types, is now pronounced.

*Acid Magma Type of Fig. 2.* The acid magma-type supplied

the early granophyres (Chapter XII.), the rhyolites and explosion-phenomena of Chapters XV. and XVI., three important felsite-intrusions (Chapter XVII.), cone-sheets (Chapter XIX.), the Loch Uisg Granophyre (Chapter XX.), a fair number of sills (Chapter XXV.), many ring-dykes (Chapters XXIX. and XXXII.), the Glen Cannel Granophyre (Chapter XXXI.), and a fair number of dykes (Chapter XXXIV.).

The range of texture is complete from glass to coarsely granophyric. Thus there are rhyolites, or acid pitchstones, felsites, fine granophyres, and coarse granophyres. Much of the Loch Bà Ring-Dyke of Chapter XXXII. is rhyolite, while the Knock, Beinn a' Ghràig, and Glen Cannel Granophyres of Chapters XXXI. and XXXII. afford excellent examples of coarse granophyre.

The predominant minerals are alkali-felspar and quartz ; and a tendency to granophyric structure is very pronounced. In most of the analysed specimens of Table IV., there is an appreciable proportion of soda-lime felspar, but this fails in the Beinn a' Ghràig (Anal. IV.) and Glen Cannel Granophyres. Augite is the common ferromagnesian mineral, and is generally some shade of pale brown, but in the Beinn a' Ghràig and Glen Cannel Granophyres it is a green pleochroic aegerine-augite. The brown-type of augite is not infrequently accompanied by enstatite-augite. Pyrogenetic hornblende occurs along with augite in the Glas Bheinn Granophyre (Chapter XII.) ; but hornblende- or biotite-granophyres are very rare in Mull, except as a product of contamination (Chapter XXXIII.). Dr. Harker[1] has pointed out that, among the major acid intrusions of the British Tertiary province, biotite is the characteristic ferromagnesian mineral of the more acid types (75 to 77 per cent. $SiO_2$), whereas hornblende and augite are more abundant in the less acid types (70 to 72 per cent. $SiO_2$) ; and that the quartz and felspar of these less acid types show a special tendency to granophyric intergrowth. The Mull occurrences fall into place in this generalization.

The chemical characteristics summarized in Fig. 2 include a strongly marked tendency to predominance of $K_2O$ over $Na_2O$. A low content of MgO is also striking. Reference to the analyses of Table IV. shows that the drop of $TiO_2$, noted above in connexion with increasing $SiO_2$, is still continued. Thus, whereas, in the Plateau Type (Table I.) with about 45 per cent. $SiO_2$, it is common to find 2·5 per cent. $TiO_2$, in the Acid Type with over 70 per cent. $SiO_2$, there is generally only about 0·5 per cent. $TiO_2$.

OTHER MAGMA-TYPES.

*Allivalite—Eucrite Magma-Series of Fig. 3.* This magma-series, so far as Mull is concerned, has only been recognized in bulk in Ben Buie where it furnishes the Ben Buie Eucrite, or Gabbro as the rock is generally styled in the course of the Memoir (Chapter XXII.). Anal. I., Table V., is from Ben Buie. For comparison Anal. B is borrowed from Dr. Harker's description of a eucrite from Rum, which differs somewhat from the Ben Buie rock in containing a fair proportion of hypersthene. Anal. A represents Dr. Harker's typical

[1] A. Harker, 'Tertiary Igneous Rocks of Skye,' *Mem. Geol. Surv.*, 1904, p. 153.

allivalite from Rum. There are several small masses of such allivalite in association with the Ben Buie Eucrite, and the greater part of the intrusion is definitely more allivalitic in composition than the analysed specimen I.

If one compares the mineral development of the Ben Buie Magma

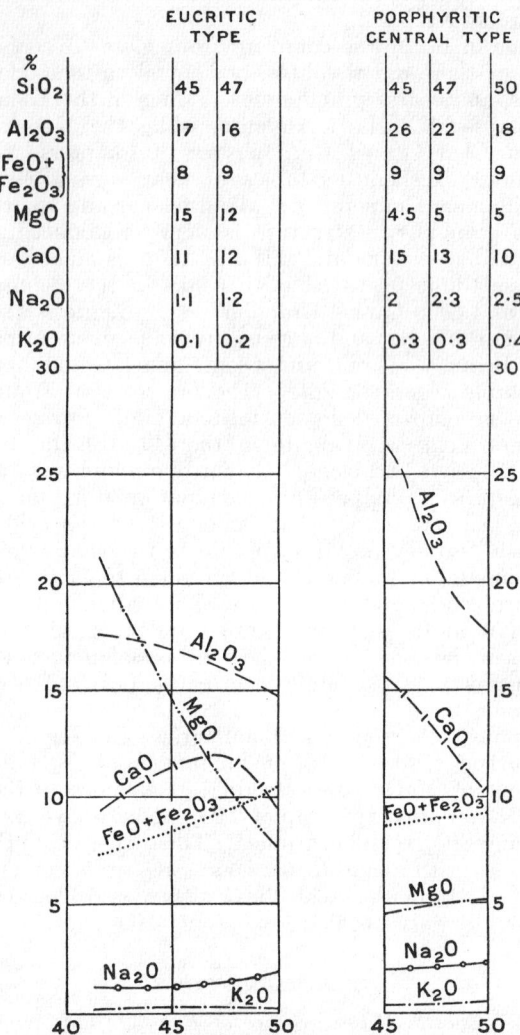

Fig. 3.—Variation-Diagrams: Allivalite-Eucrite Magma-Series; and Porphyritic Central Magma-Type.

with that of the Plateau Magma, one notes a marked falling off of iron-ores and alkali-residuum (alkaline zoning of felspars, interstitial analcite, etc.), and a disappearance of the purple tinge from the augite. At the more basic end of the series there is also a decided relative reduction in the amount of augite.

In a chemical comparison of the two magmas, the most striking features are a great fall in $TiO_2$, total iron, and alkalies, and a corresponding rise in MgO; there is also a distinct upward tendency in $Al_2O_3$ and CaO.

*Porphyritic Central Magma-Type of Fig. 3.* In Mull's history the appearance of this type on a tolerably large scale follows closely

TABLE V.—ALLIVALITE—EUCRITE MAGMA-SERIES OF FIG. 3.

|  | Allivalite | Eucrite | | |
|---|---|---|---|---|
|  | A | I. | B |  |
| $SiO_2$ | 42·20 | 46·66 | 48·05 | $SiO_2$ |
| $TiO_2$ | 0·09 | 0·47 | 0·49 | $TiO_2$ |
| $Al_2O_3$ | 17·56 | 16·71 | 15·35 | $Al_2O_3$ |
| $Cr_2O_3$ | 0·06 | ... | 0·14 | $Cr_2O_3$ |
| $Fe_2O_3$ | 1·20 | 2·69 | 1·86 | $Fe_2O_3$ |
| FeO | 6·33 | 5·87 | 7·53 | FeO |
| MnO | 0·18 | 0·12 | 0·28 | MnO |
| (Co,Ni)O | 0·13 | ... | 0·11 | (Co,Ni)O |
| CuO | 0·04 | ... | 0·05 | CuO |
| MgO | 20·38 | 12·36 | 12·53 | MgO |
| CaO | 9·61 | 12·57 | 11·02 | CaO |
| $Na_2O$ | 1·11 | 1·16 | 1·26 | $Na_2O$ |
| $K_2O$ | 0·11 | 0·27 | 0·19 | $K_2O$ |
| $Li_2O$ | ... | nt. fd. | ... | $Li_2O$ |
| $H_2O+105°$ | 1·13 | 1·24 | 0·45 | $H_2O+105°$ |
| $H_2O$ at 105° | 0·06 | 0·13 | 0·15 | $H_2O$ at 105° |
| $P_2O_5$ | ... | 0·13 | ... | $P_2O_5$ |
| $CO_2$ | tr. | 0·18 | 0·44 | $CO_2$ |
| S | 0·02 | nt. fd. | 0·20 | S |
|  | 100·21 | 100·56 | 100·10 |  |
| Spec. grav. | 2·96 | 2·97 | 2·95 |  |

A. (10464, Lab. No. 120). Major Intrusion. Allival; Rum. ALLIVALITE, quoted from A. Harker, *Geology of the Small Isles of Inverness-shire*, 1908, p. 80. Anal. W. Pollard.

I. (19069, Lab. No. 459). Major Intrusion. About 500 yards S.E. of summit of Ben Buie; Mull. EUCRITE, p. 250. Anal. F. R. Ennos.

B. (10471, Lab. No. 123). Major Intrusion. Allt Mòr na h-Uamha; Rum. EUCRITE, quoted from A. Harker, *Geology of the Small Isles of Inverness-shire*, 1908, p. 98. Anal. W. Pollard.

upon that of the Plateau Type on two widely separated occasions. Thus, lavas of the Porphyritic Central Type succeed lavas of Plateau Type (Chapter X.), and at a much later time cone-sheets of Porphyritic Central Type succeed cone-sheets of Plateau Type (Chapters XXI. and XXVIII.). On both these occasions there is a marked association of Porphyritic and Non-Porphyritic Central Types with perhaps a tendency for the Porphyritic to antedate the Non-Porphyritic. Among the sills (Chapters XXV. and XXVI.), there

24     Chapter I.—General Introduction.

is possibly a similar association to be noted. The Porphyritic Central Type is also represented among the dolerites and gabbros of Chapter XI. and the dykes of Chapter XXXIV.

TABLE VI.—PORPHYRITIC CENTRAL MAGMA-TYPE OF FIG. 3.

|  | Dolerite | Gabbro | | | Basalt | | | |
|---|---|---|---|---|---|---|---|---|
|  | I. | A | B | II. | III. | IV. | V. |  |
| $SiO_2$ | 45·54 | 46·39 | 47·28 | 48·34 | 47·24 | 47·49 | 48·51 | $SiO_2$ |
| $TiO_2$ | 1·06 | 0·26 | 0·28 | 0·95 | 1·46 | 0·93 | 1·46 | $TiO_2$ |
| $Al_2O_3$ | 23·39 | 26·34 | 21·11 | 20·10 | 18·55 | 21·46 | 19·44 | $Al_2O_3$ |
| $Cr_2O_3$ | ... | tr. | ... | ... | ... | ... | ... | $Cr_2O_3$ |
| $Fe_2O_3$ | 1·98 | 2·02 | 3·52 | 1·97 | 6·02 | 1·72 | 5·66 | $Fe_2O_3$ |
| FeO | 6·98 | 3·15 | 3·91 | 6·62 | 4·06 | 4·80 | 4·00 | FeO |
| MnO | 0·27 | 0·14 | 0·15 | 0·32 | 0·31 | 0·15 | 0·23 | MnO |
| (Co,Ni)O | ... | ... | ... | nt. fd. | 0·05 | 0·04 | 0·04 | (Co,Ni)O |
| MgO | 4·60 | 4·82 | 8·06 | 5·49 | 5·24 | 4·59 | 5·12 | MgO |
| CaO | 11·82 | 15·29 | 13·42 | 13·16 | 11·72 | 13·24 | 12·03 | CaO |
| BaO | ... | ... | ... | 0·10 | nt. fd | nt. fd. | nt. fd. | BaO |
| $Na_2O$ | 2·50 | 1·63 | 1·52 | 1·66 | 2·42 | 2·17 | 2·53 | $Na_2O$ |
| $K_2O$ | 0·44 | 0·20 | 0·29 | 0·98 | 0·15 | 0·42 | 0·25 | $K_2O$ |
| $Li_2O$ | ... | ... | ... | nt. fd. | nt. fd. | nt. fd. | nt. fd. | $Li_2O$ |
| $H_2O+105°$ | 0·72 | 0·48 | 0·53 | 0·44 | 2·24 | 2·54 | 0·48 | $H_2O+105°$ |
| $H_2O$ at 105° | 0·62 | 0·10 | 0·13 | 0·02 | 0·21 | 0·17 | 0·04 | $H_2O$ at 105° |
| $P_2O_5$ | 0·13 | tr. | tr. | 0·04 | 0·26 | 0·43 | 0·16 | $P_2O_5$ |
| $CO_2$ | ... | ... | ... | 0·11 | 0·19 | 0·08 | 0·09 | $CO_2$ |
| $FeS_2$ | ... | ... | ... | nt. fd. | nt. fd. | nt. fd. | nt. fd. | $FeS_2$ |
|  | 100·05 | 100·82 | 100·20 | 100·30 | 100·12 | 100·23 | 100·04 |  |
| Spec. grav. | 2·85 | 2·85 | 2·90 | 2·93 | 2·85 | 2·82 | 2·93 |  |

I. (15994, Lab. No. 389). Sill. Hillside between two streams S. of Coire Buidhe; Mull. SMALL-FELSPAR DOLERITE, p. 285. Anal. E. G. Radley.

A. (8043, Lab. No. 18). Major Intrusion, Cuillins. Sligachan River; Skye. OLIVINE-GABBRO, quoted from Harker, *Tertiary Igneous Rocks of Skye*, 1904, p. 103. Anal. W. Pollard.

B. (8194, Lab. No. 19). Major Intrusion, Cuillins. Coir 'a 'Mhadaidh; Skye. OLIVINE-GABBRO, quoted from Harker, *ibid.* p. 103. Anal. W. Pollard.

II. (14846, Lab. No. 373). Major Intrusion, Beinn na Duatharach. ⅝ mile N.N.W. of summit of B. na Duatharach; Mull. OLIVINE-GABBRO. Anal. E. G. Radley.

III. (18469, Lab. No. 445). Lava. ½ mile S.S.W. of Derrynaculen; Mull. BASALT, Porphyritic Central Type, p. 148. Anal. E. G. Radley.

IV. (18472, Lab. No. 447). Pillow-lava. ¼ mile slightly E. of S. of cairn on Cruach Choireadail; Mull. BASALT, Porphyritic Central Type, p. 150. Anal. E. G. Radley.

V. (18471, Lab No. 446). Lava. ⅜ mile N.E. of cairn on Cruach Doire nan Guilean, west side of a little lochan; Mull. BASALT, Porphyritic Central Type, p. 148. Anal. E. G. Radley.

The magma-type is always porphyritic in its finer crystallizations, with small or large phenocrysts of basic plagioclase felspar. The base may be partially vitreous, and often variolitic in chilled portions of pillow-lavas and minor intrusions ; elsewhere it is anything from basaltic to gabbroic in texture. In the gabbros the porphyritic tendencies of the felspar may be lost sight of, or at any rate much obscured.

The essential minerals are basic plagioclase, augite, and iron-ore ; olivine is often a minor constituent.

The outstanding feature of the type is the abundant early separation of basic plagioclase felspar. In quenched specimens, including many with variolitic tendencies, basic felspar may figure as the only phenocryst. In more thoroughly basaltic types, it is often associated with olivine and augite in small amounts. Where augite occurs as phenocrysts, its colour is brownish and the crystallization of the augite of the ground is typically granular ; in such cases, olivine is often absent. With increase of olivine, the augite is apt to assume the purple tint and ophitic habit so characteristic of the Plateau Type.

In a chemical comparison with the Plateau-Magma Type (Fig. 2 and Table I.), one notes a slight general increase in $SiO_2$, accompanied by a marked increase in $Al_2O_3$ and CaO. Alkalies remain approximately steady, while $TiO_2$, Fe, and MgO are all reduced.

In comparison with the Non-Porphyritic Central Magma-Type (Fig. 2 and Table II.), there is substantial agreement as regards MgO. The Porphyritic Type is, however, sometimes poorer in $SiO_2$, and is distinguished throughout by its abundant $Al_2O_3$, and, except towards the acid end, by its marked preponderance in CaO. Other differences are a reduction in $K_2O$, $TiO_2$, and Fe.

In comparison with the Eucritic Magma Type (Fig. 3 and Table V.), the Porphyritic-Central Type has a less extended range on the basic side. With equal silica, the porphyritic type is richer in $TiO_2$, $Al_2O_3$, CaO, $K_2O$, and $Na_2O$ ; iron remains roughly constant ; and MgO has fallen heavily.

The chemical peculiarity of the Porphyritic Central Magma-Type viewed in relation to all other associated magma-types is the relatively great concentration of $Al_2O_3$ and CaO; this peculiarity finds mineralogical expression in an abundance of basic felspar.

*Mugearite Magma-Type of Fig. 4.* This magma-type is very subordinate in Mull, except in the form of lavas among the Plateau Basalts, where it occurs more particularly at a high level in the Ben More succession (Chapters V. and X.).

The crystallization of the Mull mugearite-lavas is trachytic.

The essential minerals recognized are oligoclase, olivine, augite, and magnetite. Orthoclase is probably an important constituent, but has not been identified.

The mugearites of Mull are only represented by Anal. I., Table VII., which lies at about the acid extreme of the group. The limiting silica-percentages have not yet been chosen for mugearite, but it seems probable that they will agree approximately with 49 and 56 suggested on Fig. 4.

Compared with rocks of Non-Porphyritic Central Type, the mugearites are rich in alumina and alkalies and poor in lime and magnesia. These features find mineralogical expression in a notable relative reduction of basic felspar and augite.

## Chapter I.—General Introduction.

*Syenite, Trachyte, and Bostonite Magma-Type of Fig. 3.* There is much uncertainty regarding the exact period, or periods, at which this magma-type manifested its presence at the Mull centre. Various occurrences are grouped together in Chapter XIV. They range in crystallization from trachytic to fine-grained syenitic. Trachytic and bostonitic tuffs are also known (Chapters XIV. and XVI.).

The outstanding minerals are the alkali-felspars which make up almost all the rock in the bostonites—not analysed. In the syenite (Anal. II., Table VII.), aegerine-augite is the main ferro-

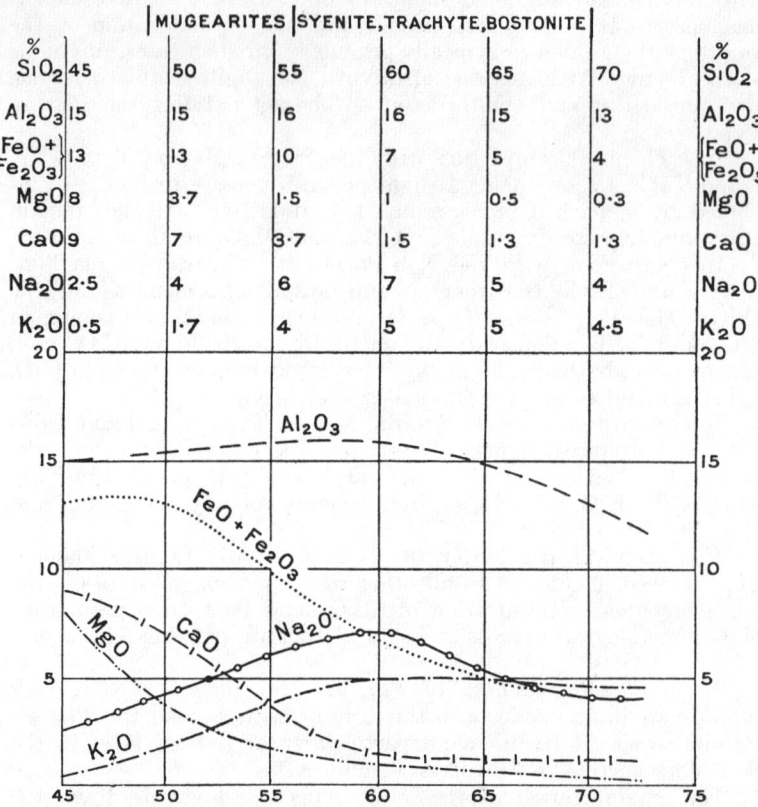

Fig. 4.—Variation-Diagram : Mull Alkaline Magma-Series.

magnesian mineral, with a little aegerine, enstatite-augite (pseudomorphed), and amphiboles; the latter are mostly referable to arfvedsonite, but varieties allied to riebeckite and barkevikite are also met with. In the Ardnacross Trachyte (Anal. III.), biotite predominates, but it is accompanied by aegerine-augite, and some decomposed amphibole. In the Bràigh a' Choire Mhòir Trachyte (Anal. III.), aegerine-augite and an ally of riebeckite are present in about equal proportions. In both these trachytes, there is reason to suspect the presence of nepheline.

## Magma-Types.

If one compare this magma-type with the Intermediate to Subacid Magma-Type of Fig. 2, one notes a great increase of $Al_2O_3$,

TABLE VII.—ALKALINE MAGMA-SERIES OF FIG. 4.

|  | Mugearite | | | | Syenite | Trachyte | | |
|---|---|---|---|---|---|---|---|---|
|  | A | B | C | I. | II. | III. | IV. | |
| $SiO_2$ | 49·24 | 49·92 | 50·70 | 55·76 | 58·81 | 60·13 | 63·12 | $SiO_2$ |
| $TiO_2$ | 1·84 | 2·04 | 1·89 | 1·78 | 0·76 | 0·73 | 0·51 | $TiO_2$ |
| $Al_2O_3$ | 15·84 | 12·83 | 14·60 | 16·55 | 14·81 | 16·53 | 15·44 | $Al_2O_3$ |
| $Cr_2O_3$ | tr. | tr. | .. | ... | ... | ... | ... | $Cr_2O_3$ |
| $V_2O_3$ | ... | 0·04 | ... | ... | ... | ... | ... | $V_2O_3$ |
| $Fe_2O_3$ | 6·09 | 6·96 | 5·23 | 3·10 | 4·58 | 2·86 | 1·73 | $Fe_2O_3$ |
| FeO | 7·18 | 6·21 | 7·68 | 6·02 | 4·21 | 2·55 | 3·53 | FeO |
| MnO | 0·29 | 0·52 | 0·42 | 0·22 | 0·27 | 0·46 | 0·27 | MnO |
| (Co,Ni)O | tr. | 0·03 | tr. | nt. fd. | nt. fd. | nt. fd. | nt. fd. | (Co,Ni)O |
| MgO | 3·02 | 3·78 | 4·15 | 1·08 | 0·80 | 1·20 | 0·62 | MgO |
| CaO | 5·26 | 7·25 | 7·20 | 3·23 | 2·33 | 1·61 | 1·31 | CaO |
| BaO | 0·09 | 0·09 | 0·08 | 0·07 | 0·03 | 0·11 | nt. fd. | BaO |
| SrO | tr. | tr. | tr. | ... | ... | ... | ... | SrO |
| $Na_2O$ | 5·21 | 3·72 | 3·71 | 6·28 | 5·60 | 8·06 | 5·81 | $Na_2O$ |
| $K_2O$ | 2·10 | 1·73 | 1·33 | 3·87 | 4·96 | 3·99 | 5·36 | $K_2O$ |
| $Li_2O$ | ... | tr. | ? tr. | tr. | nt. fd. | tr. | nt. fd. | $Li_2O$ |
| $H_2O+105°$ | 1·61 | 1·05 | 1·15 | 0·95 | 0·82 | 0·97 | 0·44 | $H_2O+105°$ |
| $H_2O$ at 105° | 1·08 | 3·58 | 2·08 | 0·80 | 2·00 | 0·55 | 0·14 | $H_2O$ at 105° |
| $P_2O_5$ | 1·47 | 0·45 | 0·49 | 0·40 | 0·20 | 0·57 | 0·25 | $P_2O_5$ |
| $CO_2$ | ... | nt. fd. | nt. fd. | 0·03 | ... | ... | 1·89 | $CO_2$ |
| $FeS_2$ | ... | ... | ... | nt. fd. | nt. fd. | nt. fd. | nt. fd. | $FeS_2$ |
| S | 0·03 | ? tr. | nt. fd. | ... | ... | ... | ... | S |
| F | 0·18 | ... | ... | ... | ... | ... | ... | F |
|  | 100·46* | 100·20 | 100·71 | 100·14 | 100·18 | 100·32 | 100·42 | |
| Spec. grav | 2·79 | ... | ... | 2·67 | 2·64 | 2·51 | 2·89 | |

A. (8732, Lab. No. 80). Sill. Druim na Crìche; Skye. MUGEARITE, quoted from A. Harker, *Tertiary Igneous Rocks of Skye*, 1904, p. 263. Anal. W. Pollard. *The total is 100·53 − 0·07 (oxygen equivalent of 0·18 fluorine).

B. (11731, Lab. No. 179). Sill. Eilean a' Bhaird; Canna. DOLERITIC MUGEARITE, quoted from A. Harker, *Geology of the Small Isles of Inverness-shire*, 1908, p. 130. Anal. W. Pollard.

C. (11732, Lab. No. 186). Sill. Pass S. of Fionn-Chrò; Rum. DOLERITIC MUGEARITE, quoted from A. Harker, *Geology of the Small Isles of Inverness-shire*, p. 130. Anal. E. G. Radley.

I. (20582, Lab. No. 670). Lava of Ben More horizon. Below road, 290 yards E. of Kinloch Hotel; Mull. MUGEARITE, p. 144. Anal. E. G. Radley.

II. (15991, Lab. No, 388). Intrusion. Gamhnach Mhòr, Carsaig Bay; Mull. SYENITE, p. 189. Anal. E. G. Radley.

III. (15753, Lab. No. 393). Plug in Ardnacross Vent. Shore ¼ mile W. of Rudh' an t-Sean-Chaisteil; between Salen and Tobermory; Mull. TRACHYTE, p. 192. Anal. E. G. Radley.

IV. (14821a, Lab. No. 368). Plug in Vent. Bràigh a' Choir' Mhòir, W. of Salen; Mull. TRACHYTE, p. 191. Anal. E. G. Radley.

$Na_2O$, and $K_2O$, and a fall in MgO and CaO. Comparison between Tables III. and VII. shows furthermore a marked fall in the $TiO_2$ of the alkali-rocks.

The chemical characteristics of the Intermediate Alkaline Magma-Type finds mineralogical expression in an abundance of alkali-felspar, and also in the alkali-content of the associated ferro-magnesian minerals.

DIFFERENTIATION-COLUMN ; GLEN MORE RING-DYKE.

In Chapter XXX., a detailed account is given of certain ring-dykes which furnish examples of gravitational differentiation on a large scale. The type-sections are provided by the Glen More Ring-Dyke, two exposures of which have supplied the material for Figs. 54-56 (pp. 322-6), and the analyses of Table VIII. It is obvious that these analyses fit with wonderful precision the values given in Fig. 2 for the Normal Magma-Series of Mull. It should be clearly understood that the curves of Fig. 2 were drawn to meet the requirements of Tables I.-IV., and that they had taken their present form before any comparison was made between them and Table VIII.

The agreement between the curves of Fig. 2 and the analyses of Table VIII., is, of course, not complete. Some of the differences appear to be mere irregularities, as for instance, the relatively high lime and low iron of Anal. II. Others, though smaller, seem to be of a more reliable nature. Thus :—

MgO decreases somewhat more steeply from the basic to the intermediate range of Table VIII. than in Fig. 2.

$Na_2O$ is slightly higher throughout, the difference amounting to about 1 per cent. at intermediate compositions.

$K_2O$ is slightly higher at intermediate and subacid compositions. At the latter, the excess may reach almost 1 per cent.

Another difference between the rocks of Table VIII. and those of the normal magma-series of Tables II.-IV. is shown in the relative concentration of $P_2O_5$ at intermediate compositions in Table VIII. In Tables II.-IV., there is a tendency for $P_2O_5$ to decrease fairly regularly from the basic end of the series to the acid. The condition of affairs as regards $P_2O_5$ in Table VIII. is probably not accidental (p.   ), but it would be idle to discuss in the present state our knowledge why it is not reproduced in Tables II.-IV.

Most petrologists will probaly agree that the same general conditions as determined the differentiation illustrated in Table VIII. are sufficient to account for the development of the greater part of the magma-series illustrated in Fig. 2. These conditions, as explained in Chapter XXX., are the early crystallization of olivine, augite, and basic felspar, leaving behind a partial magma of alkali-felspar and quartz endowed with high fluidity and low temperature of consolidation ; the migration of the acid residuum, impelled by gravity, has led to differentiation, though it is possible to show that the process has been in some measure hindered by chemical exchange between solid and liquid.

It would not be proper to conclude this section without noting the essential agreement between ourselves and Dr. Bowen [1] as regards

---

[1] N. L. Bowen, 'The Later Stages of the Evolution of the Igneous Rocks,' *Journ. Geol.*, 1915, vol. xxiii., Supplement, pp. 1-91 ; and 'The Reaction Principle in Petrogenesis,' *Journ. Geol.*, vol. xxx., 1922, pp. 177-198.

the efficacy of differentiation through crystallization and also the

TABLE VIII. — DIFFERENTIATION - COLUMN OF GLEN MORE RING - DYKE AS EXPOSED IN CRUACH CHOIREADAIL AND COIR' AN T-SAILEIN, 2½ MILES APART.

|  | Cruach Choireadail, Fig. 54, p. 322. | | | | Coir' an t-Sailein. | | |
|---|---|---|---|---|---|---|---|
|  | I. | II. | III. | IV. | V. | VI. |  |
| $SiO_2$ | 49·90 | 51·32 | 56·22 | 68·12 | 50·04 | 57·18 | $SiO_2$ |
| $TiO_2$ | 2·56 | 0·98 | 2·74 | 1·26 | 2·56 | 3·25 | $TiO_2$ |
| $Al_2O_3$ | 12·70 | 13·96 | 12·45 | 13·08 | 13·32 | 10·75 | $Al_2O_3$ |
| $Fe_2O_3$ | 4·20 | 2·48 | 3·09 | 1·02 | 4·71 | 4·96 | $Fe_2O_3$ |
| FeO | 7·88 | 7·10 | 7·58 | 3·26 | 8·07 | 6·24 | FeO |
| MnO | 0·36 | 0·34 | 0·43 | 0·39 | 0·33 | 0·32 | MnO |
| (Co, Ni)O | nt. fd. | nt. fd. | nt. fd. | nt. fd. | nt. fd. | nt. fd. | (Co, Ni)O |
| MgO | 5·88 | 5·78 | 2·78 | 0·71 | 5·01 | 2·15 | MgO |
| CaO | 10·39 | 11·51 | 5·93 | 1·81 | 10·02 | 5·73 | CaO |
| BaO | nt. fd, | nt. fd. | 0·04 | 0·04 | nt. fd. | 0·06 | BaO |
| $Na_2O$ | 2·86 | 3·50 | 3·82 | 4·15 | 3·28 | 4·62 | $Na_2O$ |
| $K_2O$ | 0·95 | 1·16 | 2·67 | 4·47 | 1·08 | 2·67 | $K_2O$ |
| $Li_2O$ | nt. fd. | nt. fd. | nt. fd. | nt. fd. | tr. | tr. | $Li_2O$ |
| $H_2O + 105°$ | 1·65 | 1·27 | 1·35 | 1·16 | 1·45 | 1·31 | $H_2O + 105°$ |
| $H_2O$ at 105° | 0·67 | 0·36 | 0·44 | 0·40 | 0·27 | 0·33 | $H_2O$ at 105° |
| $P_2O_5$ | 0·20 | 0·24 | 0·50 | 0·22 | 0·28 | 0·46 | $P_2O_5$ |
| $CO_2$ | 0·09 | 0·09 | 0·05 | 0·06 | 0·08 | 0·08 | $CO_2$ |
| $FeS_2$ | nt. fd. | nt. fd. | nt. fd. | nt. fd. | nt. fd. | nt. fd. | $FeS_2$ |
|  | 100·29 | 100·09 | 100·09 | 100·15 | 100·50 | 100·11 |  |
| Spec. grav. | 2·95 | 2·91 | 2·77 | 2·55 | 2·97 | 2·71 |  |

I.–IV. Ascending Sequence, Cruach Choireadail.

I. (18463, Lab. No. 442). Glen More Ring-Dyke. ½ mile W.N.W. of Creag na h-Iolaire, Glen More; Mull. QUARTZ-GABBRO, p. 328. Anal. E. G. Radley.
II. (18462, Lab. No. 441). Glen More Ring-Dyke. Fully ½ mile S.S.E. of cairn on Cruach Choireadail; Mull. QUARTZ-GABBRO, p. 328. Anal. E. G. Radley.
III. (18461, Lab. No. 440). Glen More Ring-Dyke. Fully ¼ mile S.S.E. of cairn on Cruach Choireadail; Mull. Allied to CRAIGNURITE, p. 328. Anal. E. G. Radley.
IV. (18460, Lab. No. 439). Glen More Ring-Dyke. 130 yards E.S.E. of cairn on Cruach Choireadail; Mull. Allied to acid CRAIGNURITE, p. 328. Anal. E. G. Radley.

V. and VI. Ascending Sequence Coir'an t-Sailein.

V. (18455, Lab. No. 437). Glen More Ring-Dyke. Rather more than ½ mile W.N.W. of Corra-bheinn; Mull. QUARTZ-GABBRO, p. 328. Anal. E. G. Radley.
VI. (18456, Lab. No. 438). Glen More Ring-Dyke. Fully ¼ mile W.S.W. of Corra-bheinn; Mull. Allied to CRAIGNURITE, p. 328. Anal. E. G. Radley.

importance of such reactions as are apt to occur between early crystals and residual magma in the course of the process.

## Chapter I.—General Introduction.

### FURTHER COMMENTS ON MULL DIFFERENTIATION.

In the preceding section a sketch was offered of the main process of differentiation which is regarded as responsible for the greater part of the magma-series illustrated in Fig. 2. A few words may be added concerning the place of the Plateau Magma-Type in this scheme.

There are two reasons, both inconclusive, for assuming the Plateau Magma-Type as the parental stock from which the others have been derived :—

(1) It is of great bulk and early manifestation.

(2) It is possible to arrange most of the Mull magmatic time-sequence into two great cycles, each beginning with the Plateau Magma-Type.

The second of these two statements may be elaborated as follows ;—

*Cycle I.* starts with the Plateau Basalt Lavas and leads through the Central Lavas (of both Porphyritic and Non-Porphyritic Magma-Types) to manifestations of intermediate and acid magma. These latter include the early granophyres of Chapter XII., the main explosions of Chapters XV.-XVI., the felsites of Chapter XVII., the Early Acid Cone-Sheets of Chapter XIX., and the Loch Uisg Granophyre of Chapter XX.

*Cycle II.* starts with the bulk of the Early Basic Cone-Sheets of Plateau Magma-Type (Chapter XXI.), and continues, after the eucritic episode of Ben Buie (Chapter XXII.), with Porphyritic Early Basic Cone-Sheets of Porphyritic Central Magma-Type (Chapter XXII.), and Late Basic Cone-Sheets of predominantly Non-Porphyritic Central Magma-Type (Chapter XXVIII.). During the Late Basic Cone-Sheet period an extensive series of ring-dykes (Chapters XXIX.-XXXII.) was intruded ranging from Non-Porphyritic Central to Acid Magma-Type ; among these, the acid predominates towards the close.

In commenting upon the above time-scale, it must be pointed out that locally, in the Staffa district, the Mull lavas begin with Non-Porphyritic Central Magma-Type. These particular lavas have been neglected in the statement given above because they are of relatively small volume. In addition to this, the Early Acid Cone-Sheets do not seem to be wholly earlier than their Early Basic Associates so that there appears to have been an overlap of the two cycles. It may be added for the sake of completeness, though it does not affect the argument, that a recurrence of Plateau Type of Magma is evidenced at the close of Cycle II. by certain very late basic dykes (Chapter XXXIV.) cutting the Loch Bà Felsite of Chapter XXXII.

If it be admitted that the Plateau Magma-Type holds a parental position in Mull petrology, a difficulty is at once manifest. In its most typical representatives, analcite and natrolite, rather than quartz, seem to be the last products of consolidation. How then was the passage brought about from the Plateau Magma-Type to the unstable Non-Porphyritic Central Type ?

A possible answer is that diopside, spinel, and silica might result during crystallization as an alternative to aluminous augite, or that a magnesian olivine, magnetite, and silica might develop instead of a ferriferous olivine ; but Mull does not seem to supply evidence bearing upon such matters. Possibly the change from the one magma-type to the other was due in part to assimilation, as Professor Daly[1] has argued in comparable cases. There is again no direct evidence bearing upon this point ; all that can be said is that if assimilation has been of importance in modifying the Mull Magma it must have been accomplished at a high temperature under conditions admitting of complete admixture of melted sediment and original magma. There is no inherent impossibility in this conception. At Tràigh Bhàn na Sgurra, a thin sill of Non-Porphyritic Central Magma-Type has melted the gneiss forming its roof through a thickness amounting locally to 4 or 5 feet (p. 266); and, in many other sills of the same class, more or less completely fused xenoliths of sedimentary origin occur surrounded by crystalline products of interaction (Chapter XXIV. ; for instances of assimilation, *see* Chapter XXXIII.). Under the more extreme temperature conditions that prevailed in the probably deeper-seated reservoir that supplied the plateau-lavas, it is easy to imagine modification of magma by a limited but complete absorption of country-rock.

If it is tempting to see in the transition from Plateau Magma-Type to Non-Porphyritic Central Magma-Type a suggestion of an absorption of siliceous country-rock, it is equally tempting to suppose that the Porphyritic Central Type of Fig. 3 might have been initiated through an absorption of argillaceous schists or gneisses. The growth of anorthite through interaction of basic magma and argillaceous xenoliths is dealt with in some detail in Chapter XXIV.

The place of the Eucritic Magma-Type of Fig. 3 in the differentiation-scheme of Mull is too hypothetical to deserve discussion.

The Mugearite Magma-Type of Fig. 4 can be derived from the Plateau Magma-Type by elimination, through crystallization, of a considerable proportion of olivine, augite, and basic felspar. It has already been pointed out that the augite of the Plateau Magma-Type, though commonly ophitic, often *concludes* its crystallization at a relatively early date. Examples are cited (p. 138) where augite-felspar complexes lie embedded in a felspar-olivine base.

The Trachytic Magma-Type of Fig. 4 has much in common with the felspathic segregations of the Plateau Type of basalt-lavas (p. 138), and basalt-dykes (pp. 361, 369).

### EXPLANATION OF MAGMA–SEQUENCE.

*In general Complex*. If we accept the Central, Intermediate, and Acid Magma-Types as derivatives of the Plateau Magma-Type, there still remains the problem why they manifested themselves during the two great cycles of activity in the order stated, from basic to acid.

Dr. Bowen would probably suggest, as he has done in similar cases,[2] that the sequence of magma-types has been controlled by the progresssive variation in composition of the residual magma

---

[1] R. A. Daly, 'Igneous Rocks and their Origin,' 1914.

[2] N. L. Bowen, 'The Later Stages of the Evolution of the Igneous Rocks,' *Journ. Geol.*, vol. xxiii., 1915, Supplement, p. 81.

left liquid towards the summit of a great reservoir during long-continued crystallization. This view has much in common with Dr. Harker's[1] suggestion that "the earliest erupted volcanic rock will represent more or less the undivided magma of the province." In interpreting the Mull record we adopt the ideas expressed above with two reservations :—

(1) We accept the Plateau Magma-Type as parental, neglecting the Staffa-Type for reasons already explained.

(2) We think that the main reservoir was charged on three occasions, rather than one, with parental magma. Thus we account for what we style the two major cycles of activity, and also the final recurrence of the Plateau Magma-Type as represented among the dykes that cut the Loch Bà Felsite.

While we employ the theory of progressive differentiation in accounting for the Mull magmatic cycles, the reader should remember that there is an alternative suggestion applied by Mr. Barrow[2] in regard to the Pre-Tertiary plutonic intrusions in the Highlands, and by Dr. Harker[3] in dealing with the Tertiary plutonic rocks of Skye. According to this interpretation, the order of decreasing basicity among the plutonic intrusions is determined by remelting from the base upwards of a density-stratified, and completely consolidated, magma. These two authors were concerned with instances where ultrabasic rocks have functioned as liquids rather than as mere crystalline precipitates, and it is probably impossible to avoid some species of remelt-hypothesis to meet the difficulty thus encountered. In another direction, a special type of remelt-hypothesis, which we shall now attempt to elaborate, is of assistance in co-ordinating certain peculiarities of composite minor intrusions.

*Composite intrusions.* As already explained, the term composite intrusion is employed in this Memoir in a more restricted sense than in the past. There are three conditions in our definition :—

(1) Composite intrusions are composed of parts of recognizably different composition.

(2) As a complex they chill exteriorly against country-rock.

(3) The component parts do not chill against one another—if they do the intrusion is *multiple*, not *composite*.

Such composite intrusions are exceedingly numerous in Mull (p. 8). Probably quarter at least of the acid to intermediate minor intrusions of the island are composite. Their habitual arrangement in such cases is acid interior with relatively thin basic marginal layers, chilled exteriorly. Always the acid has followed the basic; and Professor Judd and Dr. Harker have given such good descriptions of composite intrusions showing this character in the Hebrides generally that there can be no doubt of the definiteness of the phenomenon.

To account for the magmatic sequence in such cases Dr. Bowen invokes the same principle of progressive differentiation as is

[1] A. Harker, 'Natural History of Igneous Rocks,' 1909, p. 329.
[2] G. Barrow, Discussion of Garabal Hill, *Quart. Journ. Geol. Soc.*, vol. xlviii., 1892, p. 121.
[3] A. Harker, 'Natural History of Igneous Rocks,' 1909, p. 330.
[4] N. L. Bowen, 'The Later Stages of the Evolution of the Igneous Rocks,' *Journ. Geol.*, vol. xxiii., 1915, Supplement, pp. 13, 14.

outlined in the previous section. He pictures a fissure as having served for the passage of magma during so considerable a period that the composition of the upper part of the reservoir of supply has been radically altered by progressive crystallization. We are inclined to think that this hypothesis demands too long a period of infilling for the very numerous minor intrusions concerned. In fact we agree with Dr. Harker[1] that the differentiation of the magmas involved was already effected before the date of their intrusion.

The conception of prior differentiation lends itself to two alternative explanations of the observed magmatic sequence, basic to acid. Some may hold that the sequence is controlled by the remelting of a completely consolidated stratified magma (*see* previous section). Our difficulty in accepting this hypothesis is that the phenomenon to be explained is too regular. The impression we have gained is that acid and basic magma must have been available in very many cases at one and the same time, and that it is some inherent physical difference that has given precedence to the basic magma in the invasion of cold fissures. Our own suggestion as regards magma-sequence in composite intrusions is a variant of the remelt-hypothesis, in that it involves remelting of a special type as a preliminary condition. Stated fully it as is follows :—

(1) Magmatic differentiation, though controlled by sinking of crystals, may, under certain circumstances, give rise to liquid differentiates through remelting of fallen crystals at lower levels. The necessary condition is that the magma-basin is kept hot below, while it is cooling above. A concentration of heavy remelted matter at relatively low levels, would, in itself, tend to check thermal convection-currents; while upward diffusion would merely lead to further precipitation of solid phases at high levels where support is lacking.

(2) A fissure communicating with a reservoir in which the liquid contents are stratified, may often cut the wall of the reservoir obliquely, and thus offer itself as a channel of intrusion for both acid and basic magma. It may be expected in such a case that hot basic magma would proceed much more readily along a cold fissure than relatively cool acid magma. Once basic magma has prepared the way, acid magma can easily follow; field-experience teaches that acid magma under hot subterranean conditions is extremely mobile; whereas under the conditions which generally attend the progress of lava-flows it is notoriously viscous.

OTHER TOPICS.

Additional subjects of petrological interest include :—Pneumatolysis (Chapters V., X., XXXIV., etc.) ; Hybridization (Chapter XXXIII.) ; Xenoliths of the Loch Scridain Sills (Chapter XXIV.) ; and Devitrification of Pitchstone (Chapter XXIII.).

Table IX. gathers together analyses, other than bulk-analyses of igneous rocks, made from material collected in the Mull district. These analyses are mostly of minerals, and, taken with those already to be found in Dr. Harker's Geological Survey Memoirs on Skye and the Small Isles, they furnish a fairly complete record of chemical research in regard to Hebridean minerals of Tertiary date. For analyses of the Mull coals, or lignites, the reader is referred to Chapter XXXVIII. dealing with the economic side of the subject.

H.H.T., E.B.B.

[1] A. Harker, 'Natural History of Igneous Rocks,' 1909, p. 344.

## TABLE IX.

| | Phenocrysts | | Amygdale—Minerals | | | | | | | | | | | | Mud-Stone | Xenoliths and enclosed Spinel | | |
|---|---|---|---|---|---|---|---|---|---|---|---|---|---|---|---|---|---|---|
| | | | Outside Pneumatolysis Limit | | | | | | | Inside Pneumatolysis Limit | | | Inside Contact-Zone | | | | | |
| | I. | II. | III. | IV. | V. | VI. | VII. | VIII. | IX. | X. | XI. | XII. | XIIa. | XIII. | XIV. | XV. | XVI. | XVII. | XVIII. |
| $SiO_2$ | 49·72 | 50·80 | 53·74 | 53·41 | 52·95 | 53·39 | 48·91 | 46·51 | 46·62 | 46·75 | 46·21 | 46·10 | 45·9 | 38·69 | 37·66 | 39·26 | 49·74 | 38·67 | 0·77 |
| $TiO_2$ | 0·85 | | 0·82 | | | | | | | | | nt. fd. | | 0·12 | 0·12 | 1·33 | 1·49 | 0·90 | 0·50 |
| $Al_2O_3$ | 0·90 | 31·54 | | 1·76 | 1·71 | | 0·11 | 2·40 | 3·90 | 24·82 | 27·00 | 25·05 | 26·0 | 28·54 | 21·84 | 27·34 | 34·99 | 37·27 | 60·84 |
| $Fe_2O_3$ | 1·72 | | | | | | | 1·14 | 0·66 | | | | | 6·97 | 4·07 | 1·57 | 1·53 | 2·78 | 4·26 |
| FeO | 27·77 | | | | | | 2·97 | 1·85 | 1·08 | | | 0·55 | | 0·22 | 0·34 | 17·18 | 0·34 | 5·07 | 24·00 |
| MnO | 0·98 | | | | | | 2·27 | | | | | nt. fd. | | 0·29 | 0·53 | 0·04 | 0·15 | 0·17 | 0·15 |
| (Co, Ni)O | nt. fd. | | | | | | | | | | | nt. fd. | | | nt. fd. | | nt. fd. | nt. fd. | nt. fd. |
| MgO | 12·69 | 12·83 | | | | | | 0·47 | | | | 0·32 | | 0·49 | 0·45 | 0·34 | 0·66 | 2·35 | 9·37 |
| CaO | 3·80 | | 31·19 | 31·69 | 31·48 | 33·41 | 0·56 | 33·40 | 33·98 | 14·20 | 13·45 | 14·17 | 14·3 | 23·78 | 33·06 | 0·74 | 0·88 | 4·65 | 0·36 |
| BaO | | | | | | | 40·39 | | | | | | | | | | | | |
| $Na_2O$ | 0·23 | 3·96 | 9·94 | 8·11 | 8·04 | 8·80 | 0·22 | 0·36 | 0·89 | 0·89 | | tr. | | tr. | 1·17 | nt. fd. | nt. fd. | | |
| $K_2O$ | 0·12 | tr. | nt. fd. | 2·42 | 2·42 | | 1·16 | 1·45 | 0·57 | | | | | 0·03 | 0·75 | 1·08 | 3·76 | 3·63 | |
| $Li_2O$ | tr. | | | | | | | | | | | nt. fd. | | tr. | nt. fd. | 2·05 | 1·72 | 3·01 | |
| $H_2O$ +105° | 1·27 | 0·52 | 3·38 | 3·66 | 4·07 | 4·46 | 4·17 | 12·61 | 12·11 | 13·64 | 13·78 | 13·78 | 13·8 | 0·99 | 0·20 | 7·20 | tr. | 1·63 | 0·14 |
| $H_2O$ at 105° | 0·08 | | 0·6 | | | | | | | | | 0·13 | | 0·09 | nt. fd. | 1·44 | 3·44 | 0·22 | nt. fd. |
| $P_2O_5$ | | | | | | | | | | | | | | | | 0·34 | 0·61 | | |
| $CO_2$ | | | | | | | | | | | | | | | | | 0·86 | | |
| $FeS_2$ | | | | | | | | | | | | | | | | nt. fd. | 0·04 | | |
| | 100·13 | 99·65 | 99·67 | 101·05 | 100·67 | 100·06 | 100·76 | 100·19 | 99·81 | 98·80 | 100·44 | 100·13 | 100·0 | 100·21 | 100·19 | 100·04 | 100·21 | 100·35 | 100·39 |
| Spec. grav. | 3·44 | 2·72 | | | | | 2·605 | | 2·423 | | | 2·285 | | 3·488 | 3·61 | | 2·65 | 2·91 | 3·837 |

I. Uniaxial Augite. II. Labradorite. III.–VI. Pectolite. VII. Xonolite. VIII, IX. Tobermorite.[1] X.–XII. Scolecite. XIII. Pink Epidote. XIV. Garnet. XV. Basal Mudstone (altered). XVI. Uncontaminated argillaceous xenolith. XVII. Contaminated argillaceous xenolith. XVIII. Dark-green Spinel.

[1] In British Museum Students' Index, Tobermorite is listed as a synonym of Gyrolite.

## Other Petrological Topics. 35

### EXPLANATION OF TABLE IX.

#### PHENOCRYST IN INNINMORITE (CHAP. XXV.)

I. UNIAXIAL AUGITE, Lab. No. 402, separated from V of Table III. Anal. E. G. Radley, quoted from A. F. Hallimond, *Min. Mag.*, vol. xvii., 1914, p. 99.

#### PHENOCRYST IN BASALT (CHAP. XXXIV.)

II. LABRADORITE (? from N.W. Dyke, p. 358). S.E. of Tobermory ; Mull. Collected by J. W. Judd, *Quart. Journ. Geol. Soc.*, vol. xlii, 1886, p. 71. Anal. T. H. Holland, *Min. Mag.*, vol. viii., 1889, p. 156.

#### AMYGDALE-MINERALS,* COLLECTED OUTSIDE LIMIT OF PNEUMATOLYSIS OF PL. III. (CHAPS. V. AND X.)

III. PECTOLITE. Dearg Sgeir, N.W. of mouth of Loch Scridain ; Mull. Quoted from M. F. Heddle, *Trans. Geol. Soc. Glasgow*, vol. ix., 1892, p. 244. Anal. Stuart Thomson.

IV. PECTOLITE. Same locality, M. F. Heddle, *op. cit.*, p. 245. Anal. Stuart Thomson.

V. PECTOLITE. Same locality, M. F. Heddle, *op. cit.*, p. 245. Anal. Stuart Thomson.

VI. PECTOLITE. Same locality, M. F. Heddle, *op. cit.*, p. 245. Anal. C. Robertson.

VII. XONOTLITE. Kilfinichen, Loch Scridain ; Mull. Anal. M. F. Heddle, quoted from *Min. Mag.*, vol. v., 1884, p. 4.

VIII. TOBERMORITE. North of Tobermory Pier ; Mull. Anal. M. F. Heddle, quoted from *Min. Mag.*, vol. iv., 1882, p. 119.

IX. TOBERMORITE. Near Lighthouse, N. of Tobermory ; Mull. Anal. M. F. Heddle, quoted from *Min. Mag.*, vol. iv., 1882, p. 120.

X. SCOLECITE. Staffa. Quoted from M. F. Heddle, *Mineralogy of Scotland*, 1901, vol. ii., p. 106. Anal. Fuchs & Gehlen, *Schweigger's Journal*, vol. xviii., 1816, p. 13.

#### AMYGDALE-MINERALS, COLLECTED WITHIN LIMIT OF PNEUMATOLYSIS OF PL. III., BUT NOT CONTACT-ALTERED (CHAP. X.)

XI. SCOLECITE. Near Beinn na Croise, S.E. of Loch Scridain (locality quoted from Heddle's *Mineralogy*, p. 107) ; Mull. Anal. A. J. Scott, *Ed. New. Phil. Journ.*, vol. liii., 1852, p. 282.

XII. SCOLECITE, Lab. No. 403. North-east slope of An Gearna ; Mull. Quoted from W. F. P. M'Lintock, *Trans. Roy. Soc. Edin.*, vol. li., 1917, p. 5. Anal. E. G. Radley.

XII*a*. SCOLECITE, calculated for formula $CaAl_2Si_3O_{10} + 3H_2O$.

#### AMYGDALE-MINERALS, COLLECTED WITHIN CONTACT-ZONE OF BEINN A' GHRÀIG GRANOPHYRE (CHAP. X.)

XIII. Pink EPIDOTE, Lab. No. 404. N.E. slope of An Gearna ; Mull. Quoted from W. F. P. M'Lintock, *op. cit.*, p. 6. Anal. E. G. Radley.

XIV. GARNET, Lab. No. 416. N.E. slope of An Gearna ; Mull. Quoted from W. F. P. M'Lintock, *op cit.*, p. 8. Anal. E. G. Radley.

#### BASAL MUDSTONE COLLECTED WITHIN LIMIT OF PNEUMATOLYSIS OF PL. III.

XV. MUDSTONE, Lab. No. 395. At lava-base, 1 mile S. of entrance to Loch Don ; Mull. Anal. E. G. Radley, pp. 58, 59.

#### LOCH SCRIDAIN XENOLITHS (CHAP. XXIV.)

XVI. SILLIMINITE-BUCHITE, Lab. No. 431, with 0·48 per cent. of sapphire. Uncontaminated portion of Xenolith in IX., Table II, Nuns' Pass, Carsaig ; Mull. Quoted from H. H. Thomas, *Quart. Journ. Geol. Soc.*, vol. lxxviii., 1922, p. 236. Anal. E. G. Radley.

XVII. (16612, Lab. No. 396), CONTAMINATED XENOLITH, with basic plagioclase, mullite, sapphire (1·45 per cent.), and spinel, in same exposure as XVI. Quoted from H. H. Thomas, *op. cit.*, p. 236. Anal. E. G. Radley.

XVIII. Dark-green SPINEL, Lab. No. 397, between magnesian pleonaste and ferrous hercynite. Separated from XVII. Quoted from H. H. Thomas, *op. cit.*, p. 247. Anal. E. G. Radley.

## Chapter I.—General Introduction.

### Itinerary.

Geological visitors to Mull must be prepared for bad weather, and also for difficulty in obtaining quarters. The average annual rainfall at Glencannel, in the centre of the mountains (Sheet 44), is about 117 inches; at Gruline, near the mouth of Loch Bà (Sheet 44), 79 inches; at Ulva, near the mouth of Loch na Keal (Sheet 43), 65 inches; at Quinish, in northern Mull (Sheet 52), 56 inches; at Oban, on the mainland east of Mull (Sheet 44), 53 inches; and at Iona, west of Mull (Sheet 43), 47 inches.

May falls in the centre of a comparatively dry period; and, with June, has the further advantage of being a rather easy month in which to obtain accommodation. On the other hand, travelling facilities are better in the tourist-season, roughly speaking July and August, when it is possible, for instance, to visit Staffa by steamer.

The regular ports of call in Mull and district are Craignure, Lochaline (Morven), Salen, Croggan, and Carsaig, all in Sheet 44; Iona Ferry, Bunessan and Staffa, Sheet 43; and Tobermory, Sheet 52. Most of the roads, with the exception of the Glen More road connecting Loch Spleve and Loch Scridain, are traversed by postal gigs or motors on steamer-days.

The mountainous district, where the complicated geology lies, is largely given over to deer-stalking, and strangers are not welcome off the beaten track during August and September.

There are but few hotels or inns, but on the other hand there are several farms or cottages ready to take in one or two visitors if arrangements are made in advance. In the following statement all that is attempted is a brief summary of the geological features that can be reached on foot, or with the help of a bicycle, from the main centres of population. If a party of geologists plans an excursion to the island, they will be well-advised to take a tent, and camp alongside conveniently situated houses.

### Sheet 44.

Salen is taken first in this itinerary as the main village within Sheet 44. Other places are dealt with in clockwise rotation.

*Salen.*—Village with Hotel.

(*a*) Sound of Mull, east to Garmony Point: contrast of Lavas of Area 9 (Chap. VIII.), lying inside Pneumatolysis Limit of Pl. III. (Chap. V.), and those of Area 4a (Chap. VI.), outside same; Dykes (p. 359); Bàn Eileanan exposure of Craignure Anticline (Chap. XIII.).

(*b*) Hills between Sound of Mull and Glen Forsa: Agglomerate-Vents (p. 206); Beinn Mheadhon Felsite (Chap. XVII); Early Acid and Basic Cone-Sheets (Chaps. XIX. and XXI.).

(*c*) Glen Forsa: Fan and eskers (Chap. XXXVII.); Killbeg Ring-Dyke (p. 343); Loch Bà Felsite Ring-Dyke and faulting of Early Basic Cone-Sheets (pp. 339, 238); Gaodhail River section of Augite-Diorite and Late Basic Cone-Sheets (Chaps. XVIII. and XXVIII.).

(*d*) Loch Bà: Knock and Beinn a' Ghràig Granophyres separated by Screen; Caps on Beinn a' Ghràig Granophyre; Late Basic Cone-Sheets cut by Granophyre; Loch Bà Felsite Ring-Dyke—all in Chap. XXXII. (Pl. VI., etc); Hybrids (Chap. XXXIII.); Glen Cannel Granophyre (Chap. XXXI.).

(*e*) Ben More Massif: Amygdaloidal lavas of Pneumatolytic Zone both without and within Contact-Alteration Zone (pp. 128, 142, 152); Ben More Mugearite (Chap. VIII.); Sills (Chap. XXVI.).

(*f*) Loch na Keal, South Shore; Contrast of Lavas of Area 8 (Chap. VIII.), inside Pneumatolysis Limit, and those Area 2 (Chap. VI.), outside same; Sills (Chap. XXVI.); Dykes, altered within Pneumatolysis Limit (Chap. XXXIV.); Glaciated Pot-Holes (Chap. XXXVII.).

(*g*) Loch na Keal, North Shore: Lavas of Area 3 (Chap. VI.); Dykes (Chap. XXXIV.).

(*h*) Sound of Mull, north-westwards: Bràigh a' Choire Mhòir trachytic vent (Chap. XIV.): Lavas of Area 3 (Chap. VI.); Dykes (Chap. XXXIV.); See also Ardnacross Trachyte of Sheet 52.

*Lochaline.*—Village with Inn (in Morven).

(*a*) East and west of Loch Aline: Lavas of Area 4 (Chap. VI.); Dykes (Chap. XXXIV.).

(*b*) West of Loch Aline: Inninmore Fault (Fig. 26, p. 182); Inninmorite Sheet-and-Dyke-Complex (p. 293); Camptonite-Dykes (Chap. XXXV.).

*Craignure.*—Hamlet with Inn.

(*a*) Sound of Mull, south-east to Duart Bay: Craignure Anticline and Early Acid Cone-Sheets (Chaps. XIII. & XIX.).

(*b*) Dùn da Ghaoithe Range: Big-Felspar and other Basalt-Lavas of Area 5 (Chap. VIII.); Agglomerates of Coire Mòr Syncline (Chap. XV.); Early Acid and Basic Cone-Sheets (Chaps. XIX. & XXI.).

(*c*) Sound of Mull north-west to Garmony Point: Craignure Anticline (p. 176). Composite Cone-Sheets (pp. 223, 228).

*Lochdonhead.*—Village.

(*a*) North of Loch Don: Sand-Moraine in relation to Raised Beaches (p. 405).

(*b*) Peninsula south of Loch Don: Lavas of Area 5 (Chap. VIII.); Marginal Tilt, Duart Bay Syncline, Loch Don Anticline, Coire Mòr Syncline, Loch Spelve Anticline (Pl. V, and Fig. 25, Chap. XIII.); relation of Cone-Sheets to Loch Don and Loch Spelve Anticlines (pp. 222, 236); Port Donain Wrench-Fault (p. 183).

(*c*) Road towards and into Glen More: Roadside sections of Early Basic Cone-Sheets, west of Ardachoil (Chap. XXI.), and Vent-Agglomerate, opposite Gleann Sleibhte-coire (Chap. XVI.); Pillow-Lavas, Coire Gorm and Beinn Fhada, and beside track leading to Glen Forsa (Pl. IV. and Fig. 18, Chap. IX.).

(*d*) Sgùrr Dearg: Big- and Small-Felspar Dolerites (Fig. 22, Chap. XI.); relations of Vent-Agglomerates to Early Gabbros, Granophyres, etc. (Fig. 30, Chap. XVI.); relations of Early Basic Cone-Sheets to Loch Spelve Anticline and Vent-Agglomerates (Fig. 35, Chap. XXI.).

(*e*) Gleann Lirein: Coire Mòr Syncline and Beinn Mheadhon Felsite (Figs. 30 & 35, and Chaps. VIII., XIII., XV., and XVII.).

*Oban, Kilninver, Clachan Bridge, etc.*—Plenty of accommodation (Mainland of Lorne).

Dykes of Mull Swarm (Chap. XXXIV.); Explosion-Vents on dykes (pp. 364-5); See also *Staffa* (Sheet 43).

*Croggan.*—Hamlet.

(*a*) South-east coast to Port a' Ghlinne: Lavas of Area 6 (Chap. VIII.); Dykes (Chap. XXXIV.); Explosion-Vents on Dykes (pp. 365-6).

(*b*) Inland: Gentle arcuate folds (Pl. V., Chap. XIII.).

*Loch Buie.*—Village.

(*a*) South of Loch Uisg: Lava-succession in Area 6 from Plateau to Central Group (Pl. III., Chaps. V. & VIII.); Arcuate folds cut by Loch Uisg Granophyre (Fig. 34, and Chaps. XIII. & XX.); flat top and contact-altered roof of Loch Uisg Granophyre (p. 153 and Chap. XX.); Loch Uisg Gabbro (Chap. XX.).

(*b*) Kinlochspelve: Barachandroman outcrop of Surface-Agglomerate (pp. 197-8); Moraines of shelly clay (p. 406).

(c) North of Loch Uisg and Loch Spelve and east of Gleann a' Chaiginn Mhòir: Relation of Ben Buie Gabbro to Early Basic Cone-Sheets (Chap. XXI.); Agglomerate-Vents (p. 207); Glas Bheinn Granophyre and its assimilation of sandstone (Chap. XII.); moraines (p. 399).

(d) North of Loch Buie: Ben Buie Gabbro (Fig. 38, Chap. XXII.); dual relation of Ben Buie Gabbro to Vent-Agglomerates (pp. 199, 247); relations of Ben Buie Gabbro to Acid and Basic Cone-Sheets (pp. 222, 245); banded granulitic suite of Loch Fuaran (p. 252).

(e) West of Loch Buie: Lava-succession in Area 7 from Plateau to Central Group (Pl. III., Chaps. V. & VIII.); coast-section of Cone-Sheets (p. 239); Folds and Faults (Pl. V. and pp. 175, 178).

*Carsaig.*—A few houses.

Desert-Sandstone (p. 57); Lavas of Area 1 with examples of double-tier Columnar Flows (Chap. VI.); intercalated Coals and Sediments (pp. 63, 64); Loch Scridain Sills with pitchstone and xenoliths, including Rudh' a' Chromain Sill (Figs. 42-48, Chaps. XXIII.–XXV.); Syenite and Bostonite (Chap. XIV.).

*Kinlochscridain.*—Inn.

(a) Westwards both sides of Loch Scridain: Lavas of Areas 7 and 8 (Chap. VIII.), inside Limit of Pneumatolysis, contrasting with those of Areas 1 and 2 (Chap. VI.), outside same; Loch Scridain Sills with pitchstones and xenoliths (Fig. 42-48, Chaps. XXIII.–XXV.); Macculloch's and other Trees and associated Columnar Lavas (Frontispiece, and pp. 108, 113; *tide* must be at least *half out*). Dr. Heddle's tachylyte selvage (p. 265); Pre-Glacial Beach-Notch (p. 387). Most of these are more easily approached from *Bunessan* or *Iona* (Sheet 43).

(b) Ben More Massif: see *Salen* (e).

(c) Glen More road from Inn to Craig Cottage: Mugearite lava (p. 124); Plateau and Central Lavas of Areas 7 and 8 (Chap. VIII.); Derrynaculen Granophyre (Chap. XII.); relations of Early and Late Basic Cone-Sheets to Ben Buie, Corrabheinn, and Glen More Gabbros (Chaps. XXI., XXII., XXVIII., XXIX). Gravitational differentiation (Figs. 54-56, Chap. XXX.); Pillow-Lavas (p. 134).

(d) South-east of Inn: Bostonite (pp. 187-8). See also *Loch Buie* (c & d).

*Glen More.*—A road runs along the through-valley known as Glen More connecting the north end of Loch Spelve with Kinlochscridain. There are cottages very well situated for the geology at Craig and Ishriff. It may be possible to obtain quarters, or at any rate it is easy to tent at one or other situation. The sections can also be reached by walking, cycling, or driving from Lochdonhead or Kinlochscridain.

(a) Near Craig: See *Kinlochscridain* (c).

(b) Near Ishriff: Ring-Dykes of Figs. 52 and 53 (Chap. XXIX); Pillow-Lavas pp. 134, 312); Beinn Bheag Gabbro (Fig. 37, Chap. XXII.).

## Sheet 43

*Bunessan.*—Village with Hotel.

(a) North of Loch Scridain, by boat from one of the coastal cottages: see under *Kinlochscridain* (a).

(b) South of Loch Scridain: Ardtun Leaf-Beds (Chaps. III. and IV.). Prof. Cole's tachylyte-selvage (p. 265). Loch Scridain Sills with pitchstones and xenoliths (Chaps. XXIII-XXV). Lavas with double-tier columnar jointing (pp. 108, 145).

(c) South and west of village: camptonite-dykes (Chap. XXXV.).

*Iona.*—Village with two hotels. Visitors can reach by motor-boat exposures listed under *Bunessan* and *Staffa*.

*Staffa.*—Uninhabited. Visited by Tourist-Steamer from *Oban*, by sailing-boat from *Gometra*, or by motor-boat from *Iona* (Chap. VII.).

*Ulva and Gometra.*—Islands with Inn.

(*a*) Staffa, see above.

(*b*) The islands themselves : Pipe-Amygdales and Segregation-Veins in Lavas (pp. 114, 138) ; Preglacial Beach-Platform and Sea-Cave (pp. 388, 390).

(*c*) North of Loch Tuath : Pipe-Amygdales and Auto-Intrusion phenomena in Lavas (pp. 113-4) ; Preglacial Beach-Notch (p. 388).

*Gribun.*—Village on main road between Salen and Kinlochscridain.

Desert-Sand and Silicified Chalk (pp. 54-6). It is possible, though wearisome, to reach Macculloch's Tree, etc., listed under *Kinlochscridain* (*a*) by descending cliff near Stac Glas Bun an Uisge ; Sills (Chap. XXIII–XXVI) ; Preglacial Beach-Notch, (p. 387).

## Sheet 51

*Dervaig.*—Village with Hotel.

Lavas (Chap. VII.) ; 'S Airde Beinn Plug (Chap. XI.) ; Sills (Chap. XXVII) ; Dykes (Chap. XXXIV.) ; Preglacial Beach-Notch (p. 388).

## Sheet 52

*Tobermory.*—Village with Hotels.

Ardnacross Trachyte-Plug (p. 186) ; see also *Dervaig* (Sheet 51).

E.B.B. (as Editor).

# CHAPTER II.

## HISTORY OF RESEARCH.

In the following pages an epitome is attempted of the long series of researches that have been concerned with the Tertiary igneous rocks of Mull. Other subjects, such as glacial phenomena, raised beaches, and diatomite, are sufficiently covered by references in the Bibliography at the close of the volume.

Staffa, interpreted, means the *Island of Staves*, so that it is certain that the columnar structure of its basalt attracted attention from early Norse adventurers; not until 1772, however, did its fame penetrate to the scientific world. In that year Sir Joseph Banks, Dr. Solander, Archbishop Troil, and others called at Staffa on their way to Iceland, and it is noteworthy that they did not learn of its existence till they reached the Sound of Mull. Banks' descriptions, sketches, and measurements, as published by Pennant in 1774, are of value to this day. Though he does not enter into theoretical discussions, he calls the rock " Basaltes, very much resembling the *Giant's* Causeway in Ireland," (1774, p. 310),[1] and he compares some of its appearances with those of lava (1774, p. 307). In Troil's account, 1780, we find the following very significant statement :—" The stratum beneath the pillars here mentioned, is evidently *tuffa*, which had been heated by fire, and seems to be interlarded, as it were, with small bits of basalt ; and the red [? bed] or stratum above the pillars, in which large pieces of pillars are sometimes found irregularly thrown together, . . . . is evidently nothing else but lava."

In 1773, before Sir Joseph Banks' account of Staffa appeared, Dr. Samuel Johnson touched at the headland of Ardtun near the mouth of Loch Scridain in Mull, and he tells us how the columns of this shore were pointed out to him as no less deserving of notice than those of Staffa itself. When it is remembered that Desmarest published his paper on the volcanic origin of columnar basalt in 1774, and Hutton his *Theory of the Earth* in 1785, it will be realized at what an opportune time the news of Staffa and Ardtun was circulated among geologists.

In 1788, a vulcanist in the person of Abraham Mills visited both the localities mentioned above. He had no hesitation in speaking of their rocks as lavas. He also furnishes an account of the Ardtun sediments (1790, p. 79), though not their leaf-beds ; and he described the thin coal-seam that crops out on the north-east side of the Ardtun peninsula (1790, p. 83).

Mills was followed by another vulcanist Faujas Saint Fond, who published in 1797 an account of his important travels through

[1] Dated references in this chapter refer to Bibliography, p.

England and Scotland. He compares the columns of Staffa with others he was familiar with in Vivarais (1799, English edition, Vol. II., p. 47). He suggests that where they are irregular they betoken rapid cooling (1799, p. 57), an idea which is elaborated in Chapter VI. of the present Memoir. He notes how the craggy tracts of Mull are "composed of different currents of basaltic lava" (1799, p. 109). He draws attention to the red weathered tops of some of these lavas, but wrongly ascribes them to calcining (1799, p. 111). He records an emergence of fossiliferous limestone from beneath lava on the east coast (1799, p. 123). Altogether, he supplies a wonderfully-modern pioneer-account of the geology of Mull. Saint Fond also describes the lavas and conglomerates of Oban, though naturally he attempts no age-comparisons between them and the basalts of Mull.

It is interesting to find Saint Fond speculating on the explosions which might have carried up from below the many granite-boulders in Mull that we now know have been transported by ice from the mainland.

Between 1794 and 1799, Robert Jameson undertook an extensive exploration of the Scottish Isles. He had been initiated at Edinburgh into the simplicities of Wernerian doctrine, and his memoir, published in 1800, is a good representative of its kind. When he tells us of Mull, he smiles at the enthusiasm of his predecessors, and explains that their word *lava* means nothing else than *basalt*. At the same time he notes that the island furnishes examples of basalt veining basalt, a phenomenon unknown to foreign mineralogists. He also points out that Mull is predominantly composed of trap,. with primary strata restricted to the Ross and Gribun, and with minor occurrences of sandstone and limestone, sometimes fossiliferous. He describes the Ardtun coal-seam examined by Mills, and adds an account of a more important seam outcropping on Beinn an Aoinidh (1800, pp. 214, 221). These coals are interbedded among the basalts, and are dealt with in Chapter III. of the present Memoir. A notice of a piece of black pitchstone-porphyry (1800, p. 213) on the shore of Loch Scridain furnishes an introduction to a subject that figures largely in Chapter XXIII. of this Memoir.

John Macculloch was selected in 1811 to undertake a series of geological investigations in Scotland ; and in 1819 he published his famous monograph on the Western Isles. He tells us the circumstances under which the idea of writing a book, rather than a series of papers, came into his mind. He had been at work since five in the morning, and had visited and explored Staffa, and now at midnight he sat watching at the helm while his men slept at their oars, and the birds on the motionless sea ; all the time, the mountains of Mull and the walls of Iona shared his vigil in the red glow of the mid-summer twilight.

Macculloch had to deal with a region that lent itself to broad generalization and he adopted the following classification : Primary, Secondary, and Overlying. For him the base of the Old Red Sandstone furnished the lower limit of the Secondary Formation ; while the Tertiary lavas and intrusions were more or less unstratified representatives of the Overlying. Macculloch paid sufficient

## Chapter II.—History of Research.

attention to fossils to recognize the existence of Lias in the Post-Old Red portion of his Secondary, but he never elaborated in this direction.

It so happens that this simple classification served Macculloch's purpose admirably in most of the islands. Difficulties, however, were encountered in comparing Mull with Kerrera and the Mainland of Lorne. Macculloch referred the Lorne traps to the same great Overlying Formation as constitutes the bulk of Mull. In this he was mistaken, for the Lorne lavas are of Old Red Sandstone age. At the same time, he came very near to realizing the distinction, for he classed the conglomerates of Kerrera, Oban, and Appin (Benderloch) with the Old Red Sandstone, although they contain pebbles of "basalt, greenstone, amygdaloids, cavernous trap, and clinkstone or compact felspar"; and from this observation he deduced the former existence of Primary traps, of which the Glencoe assemblage is "in all probability the remains" (1819, Vol. II., pp. 117, 121, 122).

At the time he wrote his account of the Western Isles, Macculloch was very reticent in regard to the origin of trap. His leaning towards some species of igneous theory is however clearly indicated: he advanced the hypothesis that stratified traps might have resulted from melting *in situ* of sediments (1819, Vol. I., pp. 357-360, 378), and pointed out that the recurrent traps and conglomerates of Canna are very suggestive of submarine volcanic action (1819, Vol. I., pp. 419-458); he also produced evidence to suggest that traps, like lavas, are in many cases augite-, rather than hornblende-, bearing (1819, Vol. I., p. 381); and he cites instances of vesicular traps to confound those who claimed that traps, in contrast to lavas, are always amygdaloidal (1819, Vol. I., p. 458).

Perhaps his clearest statement of approximately this date is to be found in his *Geological Classification of Rocks*, 1821, p. 594, a book which he regarded as a supplement to his *Western Isles*. Macculloch points out here that differences of opinion exist regarding the classification of the products of recent and extinct volcanoes. "But it will be found that the chief confusion has arisen from prejudices respecting the trap rocks, which some of these observers have thought to attribute to an aqueous origin. . . . . That more correct theory of the trap rocks, which now begins so generally to prevail, will hereafter remove many of these obscurities."

It is curious to find Macculloch speaking of obscurities, since his own obscurities are proverbial. But, as Sir Archibald Geikie puts it, "so laborious a collection of facts, and so courageous a resolution to avoid theorising upon them, gave to his volumes an almost unique character. His descriptions were at once adopted as part of the familiar literature of geology. His sections and sketches were reproduced in endless treatises and text-books."

Space forbids more than brief mention of other contributions by Macculloch to our knowledge of Mull geology.

(1) He shows by maps and sections the main distributional and structural relationships of Gneiss, Secondaries, and Trap, in both Mull and Morven (1819, vol. iii., pp. 27, 63).

(2) Like Jameson, he found the trap of Mull 'traversed by great veins of basalt, which also cross the strata beneath' (1819, vol. i., p. 567). This was in

keeping with his experience elsewhere (pp. 238, 384, 385, 394); in fact, he tells us that further work would probably show, within the Overlying Formation of the Hebrides, great masses of trap of two dates feeding earlier and later veins. There is not much in this conception comparable with the idea of successive lava-streams, which so readily presented itself to Saint Fond, and even to Macculloch himself in the special case of Canna, but still it marks an important stage in the realization of a time-scale for the Hebridean complexes.

(3) He figured and described the tree which appears as the frontispiece of this Memoir. He recognized the tree as coniferous and regarded it as occurring in a vein of conglomerate, apparently thinking that both tree and conglomerate might be occupying a fissure in a trap older than themselves (1819, vol. i., pp. 197, 568). It is interesting to recall that the exact locality of Macculloch's tree was for many years lost sight of until it was brought to the Duke of Argyll's notice by Mr. Bell of Tavool.

(4) He drew attention to the poor development of trap-featuring in parts of Mull (1819, vol. i., pp. 570, 576). This subject is debated in Chap. V. of the present Memoir.

(5) In his map of Mull, he partially separated the Beinn a' Ghràig Granophyre from the general trap. At the same time he regarded the junction of trap and granophyre on the slopes overlooking Loch na Keal as affording evidence of transition from the one rock-type to the other (1819, vol. i., pp. 577, 578).

(6) He noticed a metamorphosed xenolith on the north shore of Loch Scridain as a 'mass of the primary strata' entangled in trap (1819, vol. i., p. 569). The Loch Scridain Xenoliths furnish the subject matter of Chap. XXIV. of the present Memoir.

(7) Finally, he recognized that the Overlying Trap must be extremely old before its 'exterior strata could have been shaped into distinct mountains by the abrasion of their edges or the loss of extensive portions; before the separation of Staffa from Mull, for example, could have taken place' (1819, vol. i., p. 575).

Macculloch's *Western Isles* was quickly followed, probably in 1820, by Ami Boué's *Essai*, which remains to this day our only text-book of Scottish geology. It was complete up to date, and was enriched by many new observations, coupled with a frank recognition of volcanic phenomena as such. Boué's wide experience in Central Scotland leads to a suggestion (1829, p. 146) that the Lorne lavas are of Old Red Sandstone Age; but further on in the book (1820, p. 227) he concludes, with considerable hesitation, that the lavas of the Western Isles and Lorne, are probably all of them later that the *Gryphea*-limestone, and that some, perhaps all, are like their neighbours of Antrim, later than the Chalk. This question of time-classification was destined to remain open for many years to come.

One of Ami Boué's most interesting observations relates to the boles, or iron-clays, to use Macculloch's name for them. Boué noted boles among the lavas of Mull and elsewhere in the Hebrides (1820, pp. 237, 246); he compared them with red clays beneath the lavas of Mezen and Cantal, and, interpreting them as fine ashes, drew the pregnant and quite novel deduction that some of the Hebridean basalts flowed from subaerial vents.

Boué pays considerable attention to the basaltic veins of the Hebrides (1820, p. 272). He does not follow Hutton in regarding them as injected from below, but rather as fed from the superincumbent lavas. In this he agrees closely with Macculloch.

In 1850, one of the most important finds of Scottish geology was announced by the Duke of Argyll. Leaves, in a beautiful state of preservation, had been discovered a few years previously at Ardtun by Mr M'Quarrie, of Bunessan, who brought them to the notice of the Duke of Argyll. The Duke, realizing their scientific importance, made a thorough investigation of the beds in which they occur, and entrusted his material to Professor Edward Forbes for determination. Professor Forbes declared the leaves Tertiary, and tentatively ascribed them to the Miocene. Later work by Mr. Gardner (1887), and after a long interval by Professor Seward and Mr. Holltum, has modified Professor Forbes' suggestion by substituting Eocene for Miocene; for further information regarding the history of research on this extremely important subject, the reader may consult Chapters III. and IV. of the present Memoir. Suffice it here to say that the Duke of Argyll drew the following deductions from the Ardtun leaves, and from chalk-flint which he recognized in an associated conglomerate :—

The Ardtun lavas above the sedimentary layer are undoubtedly of Tertiary date, comparable with those of the Antrim coast and the products of a subaerial volcano.

Strangely enough, at the same time as the Duke of Argyll demonstrated the existence of Tertiary lavas at Ardtun, he suggested a Jurassic date for the mass of Hebridean lavas, including most of those constituting Mull. In this, he was influenced by Professor Forbes, who claimed that some of the Skye basalts, correctly interpreted by Macculloch as injections into Secondary sediments, might in reality be interbedded lavas.

Sir Archibald Geikie in his early days followed the lead given by the Duke of Argyll and Professor Forbes. In the map accompanying his paper on the *Chronology of the Trap-Rocks of Scotland* (1861, p. 654), he groups Hebridean and Lorne lavas alike as Jurassic, with the exception of the Ardtun exposures. In his *Scenery of Scotland* (1865), he shows the Hebridean lavas as doubtfully Jurassic, and the Lorne lavas as of Old Red Sandstone Age. Two years later, 1867, he has righted matters completely : he claims the Mull lavas, with their occasional interbedded coals, as Tertiary ; notices a fresh occurrence of flint-conglomerate among them near Carsaig ; and points out that the Ardtun lavas lie near the base of the series, which, all told, amounts to more than 3000 feet in thickness, and consists of basalt below and pale lavas above (pale weathering basalts and mugearites, Chapter VIII. of present Memoir). He also draws attention to alteration of the lavas of Ben More near the syenite (granophyre) of the district.

Throughout his writings Sir Archibald Geikie has laid particular stress on the importance and extent of the north-west dykes accompanying the Hebridean lavas. In his British Association address of 1867, he argues that many of them may have acted as feeders of lava-streams (1868, p. 53).

In 1871, Professor Zirkel greatly advanced our petrological knowledge of Mull, more particularly by describing olivine-gabbro which he had collected in the complicated central region of the island. It is at this stage that the microscope was adopted as an auxiliary of the hammer by Hebridean geologists.

In 1872, Scrope, in the second edition of his *Volcanoes*, pointed out that the double-tier type of jointing, so clearly seen in Staffa, is exactly comparable with what he had long ago recorded in the case of columnar lavas of Auvergne. He attributed the upper irregular columns to heat-radiation from the lava-top, and the lower regular columns to heat-conduction from the lava-base. Scrope's point is more clearly stated by Professor Judd (1874, p. 225) than it is in the original, and it is to Judd that we are indebted for the reference.

Professor Judd would always be remembered by students of Mull's Tertiary rocks, were it only for his zoning of the local Jurassic, and his discovery of the unconformable Upper Cretaceous of Gribun and Morven. In addition to this, his paper of 1874 records the first serious attack made on the region of central complication, and it also establishes as a general proposition that the Tertiary eruptions were terrestial.

Before entering upon detail, the reader may be reminded that Professor Judd reached the general conclusion that Mull, Ardnamurchan, Rum, and Skye are the basal wrecks of great central volcanoes. His history of Mull itself included the following stages:—

> (1) Major Acid Phase: the centre supplied abundant acid lavas and accompanying intrusions.
> (2) Pause: erosion reduced the acid volcano to a wreck.
> (3) Major Basic Phase: the centre revived, furnishing floods of basic lava and accompanying intrusions; the mobile basalt-flows spread far beyond the remnant of their acid predecessors.
> (4) Extinction, decay, and erosion of central volcano.
> (5) Puy Phase: minor sporadic outbursts of acid and basic material found their way to the surface at scattered foci.

The above propositions contain much that is erroneous; but they all have a basis in observation, and should be borne in mind in reading the critical summary given below.

Professor Judd's maps and sections show several very important features hitherto left more or less unrecorded:—

> (1) **Difference of character of Mull Tertiary lavas as centre of activity is approached.** We recognize this difference by enclosing Judd's peculiar central lavas within a line termed the Limit of Pneumatolysis (Pl. III., p. 91). Judd called his peculiar lavas 'felstones.' In this and several other respects, Judd's statement of the case was wrong; but our discussion of his 1890 paper will show that it contained the germ of a great truth.
> 
> (2) **Agglomerates and Breccias of Central Mull,** often with gneiss-fragments. Judd's work in this direction is in danger of being discredited through his local mistakes. In Ben More, he misinterpreted some lavas particularly rich in zeolites as basaltic agglomerates (1874, pp. 240, 247), and others with a fissile structure as stratified tuffs. At the same time many of his agglomerate-outcrops are correct, and his insistence upon their importance is a great advance upon the passing references of previous observers.
> 
> (3) **Profusion of Acid and Basic Intrusions in Central Mull,** the latter occurring more particularly in a horseshoe outcrop open towards the north-west, and reaching the surface in innumerable branches inclined towards the centre of the complex.

It is well to understand that Judd at this stage mainly relied, in matters petrological, upon his reading of Zirkel's recent paper. He grouped his rocks into two great series; Acid, ranging in texture from 'granite' to 'felstone,' and Basic ranging from gabbro to basalt and tachylyte. His determination of 'felstone' was mostly based upon field-appearances, and, as already hinted, we find him continually misinterpreting altered basalt as 'felstone.' Add to this the preconception that a simple succession from Acid to Basic, or *vice versa*, was involved, and the *rationale* of many of his theories is supplied :—

> (1) Naturally he failed to determine an age-relationship between his basalts and 'felstones' on any consideration of superposition. He wrongly interpreted this difficulty by supposing that the basalts abut against the eroded edges of the 'felstones.'
> 
> (2) He realized a very striking phenomenon of Central Mull, namely the repeated cutting of other rocks, including some important acid intrusions, by basic sheets ; and on this he founded his law of basic follows acid. As stated above, however, Prof. Judd always admitted a minor final stage of sporadic eruptions in which there might be a recrudescence of acid magma (1874, p. 272).

We have already outlined some of the theoretical consequences of his too general deduction that basic follows acid. It is, however, worth while pointing out that it did not prevent him from realizing the following very important facts :—

> (1) The 'granites' are often clearly intrusive into the 'felstones.'
> (2) The Beinn a' Ghràig 'Granite' is an intrusion with caps of 'felstone' lava (1874, p. 246).
> (3) Mull is a region of marked central subsidence (1874, p. 256).

Away from the district of complication, Professor Judd gave a very good account of the picturesque 'S Airde Beinn plug in the country west of Tobermory. In fact, it furnished him with a type-example of puy-eruption.

In conclusion, it may be recalled that Judd's conception of Mull as a great central volcano is at the present time regaining adherents (*See* Chapter I. of the present Memoir) ; and further, that no one, not even Sir Archibald Geikie or Macculloch, has emphasized more successfully than Judd the vast erosion that has succeeded the extinction of Hebridean vulcanicity.

In 1880, Sir Archibald Geikie published a most valuable account of impressions gained the previous year during a traverse of the lava-fields of western America, already ascribed by Richthofen to fissure-eruptions. Geikie claimed that the Hebridean plateaux must be relics of a Brito-Icelandic lava-field fed from fissures, the dyke-contents of which are now revealed by erosion. We shall postpone comment on this conception until we consider Sir Archibald Geikie's classic of 1888 based upon explorations in large measure inspired by his visit to America.

Meanwhile, in 1883, Professors Judd and Cole described several occurrences of basalt-glass in the Western Isles : they demonstrated by analysis that true basalt-glass does occur, though they significantly remark that it is generally restricted to narrow selvages ; and they quoted two analyses by Sir Jethro Teall giving 46·68 and 47·46 per cent. of $SiO_2$ for glasses collected by themselves in Mull.

In 1886 Judd continued his account of the basalt-glasses by publishing a comprehensive survey of all varieties of crystallization of basic magma from basalt to gabbro. The paper is especially noteworthy for its announcement of the conception of **petrographical provinces** (1886, p. 54), with special reference to the Brito-Icelandic province of Post-Cretaceous date.

There can be little doubt that, in accounting for dissimilarities of microstructure in rocks of more or less similar texture, Judd laid too great stress on contrasts of tranquility and flow, and too little on differences of chemical composition. Thus it is now recognized that what we term the Plateau Type of basalt is particularly prone to *ophitic*, and the Central Non-Porphyritic Type to *granular*, crystallization of the contained augite. In another connexion, Judd's assumption that iron avoids crystallizing as magnetite under plutonic conditions (1886, p. 79) leads one to suppose that he was comparing iron-poor gabbros, or eucrites, such as figure in the Cuillins and Ben Buie, with iron-rich basalts and dolerites (*cf.* Tables I., V., VI., pp. 15, 23, 24 of present Memoir). In conclusion, it may be pointed out that many interesting subjects such as **the action of surrounding magma upon crystals, and the action of steam and other gases at the surface** are broached in this important paper; while **schillerization**, already referred by Judd in 1885 to solvent action under pressure, is dealt with once again. In this latter connexion it is noteworthy that salite-structure is in our own accounts that follow often attributed to action of residual acid magma on early-formed augite.

In 1888, Professor Cole returned to the subject of tachylyte, and described a well-known occurrence at Ardtun where a basalt-sill occurs with glassy selvages. His analysis shows 53·03 per cent. $SiO_2$, but it is so high in $Al_2O_3$ that it seems to stand in need of confirmation.

The same year, Professor Kendall gave an extensive synopsis of investigations of chilled edges of dykes in Mull. He found more or less definite tachylytic selvages in many cases; but, at the same time, he pointed out that the degree of marginal chilling of an intrusion is sometimes materially affected by the conductivity of the country-rock.

The great feature of 1888, however, was the appearance of Sir Archibald Geikie's comprehensive account of British Tertiary volcanic action. In this work, Geikie constantly acknowledges discoveries by colleagues on the Geological Survey, notably Dr. C. T. Clough's observations on the dykes of Cowal; he also points out that during two visits to Mull he received much assistance from Mr. H. M. Cadell, while in matters petrological he was largely guided by Dr. F. Hatch.

Tertiary dykes are described with particular care, and one realizes how truly Geikie tells us that they are to his mind " by far the most wonderful feature in the history of volcanic action in Britain." When he mentions Arran, Mull, Eigg, and Skye as furnishing illustrative districts for gregareous north-west dykes (1888, p. 33) he comes very near formulating the conception of a Dyke-Swarm discussed already in Chapter I. In his treatment of the dynamics of dykes and fissure-eruptions, he makes constant references to a paper by Professor Hopkins (1888, pp. 71, 74, 110).

Geikie clearly showed that the major intrusions of Mull are later than such lavas as have escaped erosion; but he was of opinion that they had been in large measure located by "the larger or more closely clustered vents of the plateau-period" (1888, p. 183); and he was ready to believe that possibly they might themselves have supplied lava-streams now denuded away. As regards the great bulk of the lavas of the region, he was certain that they had resulted from scattered vents along the course of dyke-fissures. In referring the reader back to the discussion of this subject in Chapter I., it is well to emphasize two elements of the problem disclosed by recent work:—

(1) The pillow-structure of many of the lavas found within the central region of Mull suggests a central crater-lake.

(2) The satellitic grouping of the Mull Swarm of dykes makes it clear that the Mull Centre was already marked out *before* the Mull Swarm had developed to any appreciable extent.

Sir Archibald Geikie's comparison of the Hebridean gabbros with Gilbert's laccoliths (1888, p. 143) is not helpful in Mull; but his account of the acid intrusions near the mouth of Loch Bà is excellent, and marks a considerable advance upon Judd and Macculloch. He not only traces the course of the Beinn a' Ghràig and Knock Granophyres (he takes no notice of the separating screen of Chapter XXXII.), but he also recognizes the Loch Bà Felsite as an individual dyke continuing for miles (1888, pp. 152-157). He gives a sketch (1888, Fig. 43), showing in masterly fashion the granophyre with a conspicuous capping of lava, and the course taken by the associated felsite across hill and dale. He did not realize any arcuate tendency in the outcrops of either granophyre or felsite; but he confirmed Judd's observations of intrusive contacts between granophyre and lavas—the latter now correctly interpreted as altered basalts,—and he emphasized the extraordinary absence of all signs of disturbance.

Sir Archibald Geikie successfully demonstrated that the acid intrusions of Mull are later than the basic lavas; but when he claims a comparatively simple basic to acid sequence among the intrusions themselves, he overstates his case very much as Professor Judd had already done, only in the reverse direction. As a matter of fact, the earliest major intrusions of Mull are acid (Chapter XII.), and in spite of all the complications of the district it so happens that it is much more *obvious* in the field that basic intrusions follow acid, than *vice versa;* but Geikie covers such instances of this relationship as he noticed by referring them to a late recurrence of basalt-veins and dykes (1888, pp. 145, 158). As explained in Chapter I., the magma-sequence in Mull, in the most generalized view possible, cannot be accommodated in less than *two* cycles from basic to acid, and even then one has to admit a basic recrudescence at the end of the second cycle.

Two very useful features of Sir Archibald Geikie's 1888 account of the Mull lavas are his further definition of the Pale Group of Ben More and his description of the gradual changes one meets with as one approaches the plutonic masses: in the Pale Group he notes the occurrence of fissile lavas, the mugearites of the present Memoir; and among the altered basalts he records an incapacity to weather either spheroidally or to a loam, and comments on a development of epidote both in amygdales and in interlacing veins (1888, p. 138).

The introduction of the term granophyre into Hebridean geology furnishes an interesting petrological feature of the paper.

In 1889, Professor Judd published a reply to Sir Archibald Geikie, and instituted a comparison between Mull and the Hawaian volcanoes.

In 1890, Professor Judd returns in earnest to the subject of the 'felstones.' In some respects the results of his further study are disappointing. His discrimination between lavas and intrusions remains inadequate, and his reading of the lava-sequence is still inverted. His petrology, too, is often inaccurate. To show how difficult all this makes it to do justice to his results, the reader is warned that the two analyses Judd furnishes of early propylite-lavas from Mull must be interpreted as analyses of intrusions later than any lavas preserved in the island; and that when he speaks of hornblende-propylites he means, so far as Mull is concerned, intermediate or acid intrusions with pseudomorphs after acicular or columnar augite. Attention is directed to these blemishes mainly to clear the ground for recognition of an important generalization which emerges from the confused statement of his observations. He found throughout his 'felstone'-area, now called the 'propylite'-area, widespread changes which he attributed to s o l f a t a r i c a c t i o n; these changes result in a decomposition of felspars and ferromagnesian minerals, and a development of epidote, chlorite, magnetite, etc. He also recognized definite c o n t a c t-a l t e r a t i o n of much more restricted geological scope, and often characterized by crystallization of granular aggregates of augite and magnetite, and of scales of biotite. A considerable portion of the present Memoir is concerned with the elaboration of Judd's dual conception of metamorphism outlined above. It has stood the test of minute enquiry, and the only important detail that we should like to criticize at the present juncture is the over-emphasis Judd lays upon the presence of sulphides in his propylitic area. Our experience is that sulphides are rare in this position.

Later publications by Professor Judd and Sir Archibald Geikie have added very little that is new in regard to Mull geology. Professor Judd, however, has been shown to be wrong in his interpretation of a well-exposed gabbro-granophyre junction in Skye; and this has reacted unfavourably upon the general reception of his views. One feels, in going over the old controversy, that Sir Archibald Geikie proved himself much more capable than Judd of escaping from the influence of initial mistakes. Both authors had great ideas to present to us; but in Judd's case, the great ideas were too often shrouded behind errors, which others could easily detect, and he himself seemed unable to disperse.

In 1895, Dr. Heddle wrote an account of a tachylyte-selvage which is quoted from on p. 265 of the present Memoir.

In 1899, Dr. Currie reported upon a particularly interesting amygdale-assemblage characteristic of some of the lavas of Maol nan Damh, Ben More, and contrasted it with what he was familiar with in 'Torosay,' meaning thereby the western portion of Gribun Peninsula. Currie comments on the abundance of scolecite, epidote, and a green mineral which he calls celadonite, though it is later spoken of by Dr. M'Lintock and ourselves, without critical determination, as chlorite. To account for an abundant development of this peculiar amygdale-assemblage, Mr. Currie invokes pneumatolysis very much as Judd had done before him.

In 1901, the posthumous appearance of Dr. Heddle's volumes on the *Mineralogy of Scotland*, edited by Mr. J. G. Goodchild, brought together a number of scattered observations regarding the amygdale-minerals, etc., of the Mull archipelago. We supply (p. 425) a species-index of this work, so far as it is concerned with the subject-matter of the present Memoir, and this enables us to omit from our Bibliography many of the detailed references it contains. Ten of the analyses of Table IX. (p. 34) are to be found in Heddle's pages, and the brief account of the amygdale-assemblages of the Mull plateau-region (p. 140) is based almost altogether on his statements.

In 1915, Dr. M'Lintock carried Mr. Currie's work a step further. He identified albite, previously mistaken for heulandite ; and he showed that the Maol nan Damh amygdale-assemblage undergoes marked change in the vicinity of the Beinn a' Ghràig Granophyre. Among the new minerals set up, garnet and hornblende are conspicuous. Dr. M'Lintock's research into these matters has been so thorough that we have drawn freely upon his accounts in preparing Chapter X. of the present Memoir. Only in one matter of detail, do we differ. Dr. M'Lintock thinks that the Maol nan Damh amygdale-assemblage was developed through auto-pneumatolysis of the lavas containing it, whereas we shall give reasons later for following Mr. Currie and Professor Judd in ascribing it to pneumatolysis, or solfataric action, belonging rather to the volcano than to its individual lavas. At the same time, auto-pneumatolysis is a pronounced feature in some departments of Mull geology ; and its separation from general pneumatolysis is in many cases a matter in which no final opinion can be expressed.

Dr. M'Lintock worked alongside members of the Geological Survey and acknowledges help from Mr. J. E. Richey in the field, supplemented by chemical analyses by Mr. E. G. Radley. Before sketching the history of the recent survey upon which the present Memoir is based, it is but proper to acknowledge the assistance derived from Dr. A. Harker's *Tertiary Igneous Rocks of Skye* published by the Geological Survey in 1904. There are two main features in which the Skye Memoir has contributed to lighten the task of those who have laboured in Mull ; the first of these is the recognition of Inclined Sheets, or Cone-Sheets, as we now style them ; the second is the clear and accurate statement of petrology, which we have found in large measure as applicable to Mull as it is to Skye.

### GEOLOGICAL SURVEY.

The mapping of Mull was commenced for the Geological Survey in 1902, when the small corner of the island projecting into Sheet 36 of the one-inch Map was examined. Further work was postponed until 1907, from which date yearly progress was made until the outbreak of war, 1914. In 1920 the survey of the whole island was completed. The corresponding *Summaries of Progress* contain a brief contemporary record of this work, accompanied by chemical analyses. Moreover, initials of authors are inserted in the body of the present Memoir, so that all that is required here is a very brief statement in regard to certain selected pivotal dates.

*Cone-Sheets.* The fact that Mull reproduces the Inclined Sheet system of the Cuillins was realized by Mr. Wright and myself when we accompanied Dr. Harker to Skye in 1909 (1910, pp. 31, 34). That the basic cone-sheets (inclined sheets) of Mull could be separated into an earlier suite preceding the Corra-bheinn Gabbro and a later suite following the same was subsequently clearly shown by Dr. Clough (1913, p. 44).

*Magma-Sequence.* Dr. Clough recognized in the early basic Cone-Sheets and the Ben Buie and Corra-bheinn Gabbros an olivine-rich assemblage followed by an olivine-poor or olivine-free assemblage, represented by Late Basic Cone-Sheets and Coir' an t-Sailein Gabbro, the latter merging into granophyre (1913, pp. 44, 45). In regard to an earlier stage of the igneous centre, he suspected that what we now term the Porphyritic Central Types of lava succeed the Plateau Types (1912, p. 34).

*Ring-Dykes.* On analogy with Glen Coe and Ben Nevis, Mr. Wright introduced the idea of ring-structures in the Loch Bà district, where he was the first to emphasize the importance of *screens* in igneous tectonics (1910, pp. 32, 33; 1911, pp. 31-35). Mr. Richey completed the tracing of the Loch Bà Felsite, showing beyond question that it is a ring-dyke (1914, p. 49; 1915, p. 36); and the writer demonstrated that ring-dykes are a dominant feature of the Glen More district (1914, p. 51).

*Axis of Symmetry.* Mr. Wright realized a north-west axis of symmetry in Mull geology (1911, p. 38). Mr. Richey found that two centres of ring-dykes lie on this axis, and Mr. Wilson, in preparing a one-inch reduction showing outcrops of cone-sheets, found the same to hold good in relation to these intrusions (1915, p. 36).

*Gravitational Differentiation.* The writer advanced the idea that gravitational differentiation is exhibited by the contents of certain ring-dykes observed by Dr. Clough and himself (1914, p. 51). At first the interpretation adopted was one of gravitational separation in a liquid emulsion, but after consultation with Dr. Thomas and Mr. Hallimond this has been exchanged for gravitational separation of early crystals and residual magma.

*Pillow-Lavas.* Pillow-lavas were independently recognized in Central Mull by Dr. Clough and the writer in 1913 (1914, pp. 47, 50). When they were found to be common, they were interpreted as a record of a crater-lake (1915, pp. 39, 40).

*Vent-Agglomerates.* It was early realized by the writer that Judd's agglomerates have two contrasted relationships : in the one case they are more or less conformable to the lavas ; in the other, they are very transgressive and come into contact with disintegregating gneiss and Tertiary plutonic rocks (1910, p. 30). At first it was thought that the transgressive breccia was a superficial accumulation, but this view was already much weakened when Dr. Clough found breccia both earlier and later than the Corra-bheinn Gabbro (1914, p. 51); and was finally replaced by a vent-interpretation when Mr. Wilson got the same relationship on a large scale in the case of the Ben Buie Gabbro (1915, p. 38).

*Arcuate Folding.* The strong folding of south-east Mull was the first feature of interest to attract the writer's attention (1908, pp. 67, 68). Its acuate character made it the subject of a series of maps (1910, p. 28; 1913, p. 46; in colaboration with Mr. Wilson 1914, p. 44). Its interpretation as a result of the intrusion of granophyres along the old caldera edge was suggested (1915, p. 41).

*Dykes.* As explained in Chapter I., Mr. Maufe had arrived at his conception of the dyke-swarm before he visited Mull. The theoretical interpretation adopted for Mull is the same as that previously advanced by the writer for the Etive Swarm of Old Red Sandstone age.

*Loch Scridain Pitchstones.* Pitchstones, their sheath-and-core structure, and their association with olivine-free dolerites in southern Mull were reported on by Mr. Lightfoot (1911, p. 30). They have been made the subject of a special paper by Mr. E. M. Anderson and Mr. E. G. Radley (1917), in which the rôle played by water as an inhibiter of crystallization is discussed. The same paper includes preliminary definitions of two new rock-types, leidleite and inninmorite, by Dr. Thomas and myself. An essential constituent of inninmorite is a uniaxial augite described in detail by Mr. Hallimond (1914).

*Loch Scridian Xenoliths.* These xenoliths, as already noted attracted the attention of Macculloch. Later on, Compton and Rose recorded occurrences of graphite (1821, p. 374, and 1851, p. 102). Little more was done until Mr. Cunningham Craig, Dr. Clough, and Dr. Flett described a very interesting xenolithic sill (with cordierite, etc.) at Tràigh Bhàn na Sgurra (1911, *Geology of Colonsay etc.*, p. 92). Mr. Anderson described other xenoliths (1912, p. 34), and eventually he sent in material in which Dr. Thomas recognized sapphire (1913, pp. 48, 66). A little later, Mr. Wilson and Mr. Tait discovered the Rudh' a' Chromain locality for this mineral (1913, pp. 48, 66). Many other interesting xenolith-localities are now known, mainly as a result of work by Mr. Anderson and Dr. Clough. The whole subject has been investigated by Dr. Thomas, who has published a detailed account (1922).[1]

*Tertiary Desert.* The view that the Tertiary period in Mull was ushered in under desert conditions first occurred to the present writer in 1920 (1921, p. 36; 1924).

*Petrology.* For a brief summary of the petrological section of Chapter I. by Dr. Thomas and myself, see 1923, p. 113.

E.B.B.

[1] See footnote, p. 268.

# CHAPTER III.

## TERTIARY SEDIMENTS OF MULL AND LOCH ALINE.

### INTRODUCTION.

As regards bulk, sediments play an almost insignificant rôle in the Tertiary accumulations of the Mull district. At the same time, they furnish particularly interesting information concerning the climatic conditions which attended and preceded the outpouring of the Mull lavas ; and they also date with some degree of precision the earlier stages of the volcanic history.

The band of sediment which includes the Ardtun Leaf-Beds, at the mouth of Loch Scridain (Sheet 43), has been familiar to all students of geology since its description by the Duke of Argyll and Professor Edward Forbes in 1851, and its fuller treatment by Mr. Starkie Gardner in 1887. Its field-occurences will be dealt with in the present chapter ; but its palæobotanical contents, more especially as represented in material collected by Mr. Tait for the Geological Survey, are treated separately by Professor A. C. Seward and Mr. R. E. Holttum in Chapter IV. The general geological conclusions arrived at by these two authors may be summarized as follows : The leaves of the Ardtun Leaf-Beds seem to have fallen into the waters of a still lake ; the climate was temperate, though perhaps warmer than is met with in most of the British Isles to-day ; the date was Eocene, possibly Lower Eocene.

Chapter IV., although mainly concerned with the fossil plants of Ardtun, deals also with beetles from the same locality and with additional plants from near Carsaig.

The Leaf-Beds of Ardtun form part of a thin sedimentary intercalation near the base of the lavas in South-West Mull. In the same district, impersistent seams of lignite, or inferior coal, are met with sporadically at various levels in the Plateau Basalts. These presumably accumulated under the warm-temperate conditions responsible for the plants of the leaf-beds. To the same genial climate may be referred the repeated weathering of basalt-lava to red soil, to which attention is frequently directed in succeeding descriptions of the Plateau Basalts. In keeping with this evidence of interbasaltic weathering and development of bole, is the recurrence of a singularly wide-spread mudstone actually at the base of the Tertiary lavas. So far as exposures afford a chance of judging, there are comparatively few places in Ardnamurchan, the Loch Aline district, and Mull where such a basal mudstone, a few feet thick, is not to be found. Its appearance very strongly suggests the lateritic decay of basic igneous rock. It is difficult to regard it as a product of decay of lava in mass, since nowhere is there any suggestion of a pre-existing lava-flow. The four-fold hypothesis here advanced is:—that the basal mudstone is the result of lateritic decay of a wide-spread basaltic ash; that this ash was the initial product of vulcanicity in the district; that the climate of the time was warm and moist; and that a long interval of volcanic repose followed the initial explosion. Dr. Lee was the first for the Geological Survey to suggest an ash-origin for the basal mudstone. He also points out in the sequel that the deposit seems to have undergone subsequent alteration within the Pneumatolysis Limit of Plate III. The analysed specimen (XV. p. 34) comes from the area of alteration, and this perhaps accounts for the marked preponderance of FeO over $Fe_2O_3$.

The evidence of moist conditions during, and immediately preceding, the accumulation of the Plateau Basalts is further shown by the occasional occurrence of conglomerates with rounded pebbles of flint (sometimes, it is thought, including silicified chalk). Such conglomerates had long been known associated with the Leaf-Beds of Ardtun, and on what is very probably the same horizon near Carsaig Arches of the south coast of Mull. Other earlier examples, ante-dating the Basal Mudstone, have been met with during the course of the Survey in the east corner of Mull.

Passing back in time, one seems to enter upon the record of a desert climate. Locally the bottom Tertiary accumulation consists of a few feet of sand in which what are taken to be wind-rounded grains are prominently represented. It is pointed out in the Memoir dealing with the Mesozoic rocks of the district that desert conditions probably prevailed along neighbouring shores during the accumulation of a large portion of the Upper Cretaceous marine deposits of the districts. It is therefore very difficult to say whether the Tertiary sands here dealt with may not be re-assorted Upper Cretaceous sands. The evidence which most strongly supports the view that the desert climate continued after the upheaval of the chalk is supplied, not so much by the sands themselves, as by a prevalence of silicification attributable to the period. This evidence may be summarized as follows:—

(1) The chalk was elevated as a limestone for it has weathered in carious fashion with fissures everywhere.

(2) Its cavities are filled-in locally with sand including many wind-rounded grains which cannot be matched in Cretaceous or other sediments of the immediate vicinity; silicification has replaced the chalk by some cherty substance, and this latter often serves as matrix for quartz-grains of the fissures.

(3) Locally, the chalk has disintegrated into fragments which are enclosed in sand; here again silification of the chalk has been instituted, and the sand has been involved to some extent, so that each fragment of silicified chalk is surrounded by a cherty halo enclosing quartz-grains.

(4) The quartz-grains of the sand, where not enclosed in chert, show a pronounced tendency to develop crystalline facets which, when examined in sunlight under a lens, are seen to glitter brightly.

The silicification-phenomena outlined above are exceptional in degree. They recall accounts given by Rogers and others of surface-quartzites, etc., from South Africa. S. Passage in *Die Kalahari* accepts silicification as characteristic of a desert climate, and offers a theoretical explanation based upon the tendency for solutions under desert conditions to dissipate by evaporation rather than by drainage.

In accordance with the above, it is here suggested that the desert climate of late Cretaceous times continued after the elevation of the chalk. In the absence of direct evidence, the emergence of the chalk is taken as marking the beginning of Tertiary times. It is fairly certain (from analogy with Morven and Antrim) that this elevation happened at some date later than the Upper Chalk zone of *Belemnitella mucronata*. Possibly the elevation actually occurred near the end, rather than at the end, of Cretaceous times; in which case the desert climate may belong wholly to the Cretaceous. Be this as it may, there are obviously good grounds for accepting Professor Judd's suggestion that Tertiary follows Cretaceous, in the Hebridean record, without any striking time-interval. This suggestion has, since Judd wrote, been much strengthened by Gardner's dating of the Ardtun leaves as Lower Eocene—a claim which Professor Seward and Mr. Holttum tentatively accept.

Before passing on to the more detailed discussion of the various exposures, one may notice a very difficult question that arises in connexion with the silicification of the chalk. It is not clear what was the source of the silica. It seems very improbable that much of it has been derived from the chalk itself; Professor Rupert Jones supplied Professor Judd with descriptions of foraminifera and *Inoceramus*-prisms, easily recognizable in thin slices of silicified chalk from Mull and Morven, so that it is clear that to some extent the chalk has been pseudomorphed rather than its silica concentrated; moreover, fragments of silicified chalk embedded in sandstone and surrounded by a halo of cemented grains (which betokens silicification after enclosure in the sand) are not flattened. Nor is it at all probable that the silica has been derived from solution of the Tertiary sand, since in one case silicified chalk is to be seen with its cavities infilled by a mudstone of later date than the sand. Probably the main source of silica is the underlying sediments, and the solutions concerned have been brought up by capillarity.

In the preceding remarks, the familiar Ardtun Leaf-Beds have been taken as a starting point. Now that the general nature of the evidence has been given in outline, it is convenient to group the detailed consideration of the exposures in such a way that the earlier parts of the story receive first attention.

## Desert and Associated Deposits.

*Gribun (Sheet* 43). The Gribun sections are the only ones in Mull which show the chalk as a deposit and not merely *remanié*. There are four exposures all told, and in three of them the chalk is only seen in land-slips; but this does not prevent their use as indications of the nature of the geology of the district. In all four exposures, the chalk is silicified to a hard white rock, preserving a curiously chalky aspect, and very occassionally showing fossils to the naked eye.

Two of the landslip-exposures occur just north of the straggling village of Gribun near the ruins marked Clachandhu on the one-inch Map. One is in broken-up landslip-material between the two first streams shown north of Clachandhu. The other is a coherent landslip serving as the low coastal cliff of a raised-beach along which runs the road, just where it passes the most northerly cottage of Gribun. This latter exposure is particularly serviceable in reading the post-elevation history of the chalk. One can examine the silicified chalk in detail. It is 6 ft. thick and rests on a fine white sandstone,[1] 10 ft., and this in turn on fossiliferous greensand. On close examination it is seen that the silicified chalk is traversed in all directions by sand, often completely cemented by a cherty matrix of smoother fracture than that replacing the chalk itself. The majority of the sand-grains are angular or subangular, but a considerable proportion are thoroughly well-rounded. A slice (22090), cut across chalk and sandstone, shows much of the chalk represented by clear cryptocrystalline silica, but with white opaque patches. The matrix of the quartz-grains is white and opaque resembling the material of these patches. The hardness of this material is taken as proving its siliceous composition. Rounded quartz-grains of 0·4 mm. diameter are well represented in the sandstone.

The next exposure occurs in Allt na Teangaidh above Balmeanach Farm, and has the advantage of furnishing a complete section between the silicified chalk and the basalt-lavas. The chalk is 12 ft. thick, and shows the same type of phenomena as just described. It is overlain by 3 ft. of sandstone followed by 12 ft. of dark red-brown mudstone, that as usual forms the floor for the basalt-lavas. A slice (20807) of the sandstone overlying the chalk shows a number of beautifully rounded quartz-grains about 0·2 or 0·3 mm. in diameter. These are distributed at random among numerous angular grains of distinctly smaller dimensions. Much of the slide has a white opaque matrix recalling the siliceous cement mentioned above in (22090). More locally, quartzite has been developed through the addition of quartz to the original clastic grains. The line between added and original quartz can sometimes be clearly traced by dirt. There is a little mica, albite, and zircon.

The last exposure of silicified chalk is four miles south-west of Allt na Teangaidh. Here several big blocks of the chalk lie at the base of the landslip that forms the southern part of The Wilderness, exposed on the coastal cliff on the south side of Aird na h-Iolaire. In one of these blocks, the silicified chalk has dark red mudstone penetrating its cavities.

Between Allt na Teangaidh and the Wilderness there is another very interesting exposure of part of the basal Tertiary sediments. It is afforded by a little waterfall some yards above a gorge, that reaches the sea close to a rock known as Caisteal Sloc nam Ban. The waterfall shows Triassic pebbly sandstone overlain by 5 ft. of greenish thin-bedded (? Rhaetic) sandstone with films of black shale near the base, and these in turn by 2-6 ft. of *remanié* chalk (silicified), and 2 ft. of white sandstone. Above the sandstone there is a gap which may well hide the basal mudstone. The *remanié* chalk has a matrix of sand, and the quartz-grains of this matrix, as well as those of the overlying sandstone, show wind-rounding masked to some extent by growth of secondary facets. Some of the quartz figured originally as small pebbles rather than grains, and in this case partial facets are particularly conspicuous under a lens. The fragments of chalk have not suffered deformation during silicification. A slice from the overlying sandstone (20808) shows many well rounded quartz-grains, in places enclosed in a matrix of cryptocrystalline silica, but elsewhere with outgrowths of quartz; this added quartz is in some cases easy, and in others difficult, to distinguish from the original grains. A few patches of cryptocrystalline silica suggest pseudomorphs after small fragments of chalk.

The two first lavas seen above the Tertiary sediments just described are somewhat columnar basalts. The higher of the two is a doleritic basalt extending for some distance along the cliff. It has a vesicular upper portion, which has weathered on top to a bright-red earth, but a little lower down shows cavities filled in with 'millet-seed' quartz-sand. It is suggested that in this case sand, fashioned in the

---

[1] This white Upper Cretaceous sandstone, seen here and at Allt na Teangaidh, is of finer texture than the Tertiary sandstone of the district. Examined with a lens it shows much less rounding of its constituent grains, except in the case of its very few larger grains which at once attract attention for their smooth more or less spherical, or ovoid, form. It is pointed out in the companion Memoir that this sandstone has probably resulted through desert sand blowing into the Upper Cretaceous sea. It is therefore open to anyone to suggest that the Tertiary sands described in the text are merely *remanié* of Cretaceous desert sand, coarser in texture than that exposed at Gribun, though comparable with what is met with in Morven. On the other hand, as already pointed out, such an interpretation leaves the Tertiary silicification-phenomena unexplained. *Apropos* of the possible derivation of the round grains of the Tertiary of Gribun, it is well to emphasize the fact that the grains of the local Triassic, Rhaetic, and Liassic sandstones show no conspicuous rounding at all.

## Desert Deposits.

desert period, has travelled into position during the subsequent warm-temperate climate that witnessed the outpouring of the lavas.

One last exposure of the Gribun district remains for comment. It lies between two faults shown on the one-inch Map at the Gribun School between Balnahard and Balmeanach. It is almost certain that the rocks concerned are altered Trias sandstone and conglomerate; but they have undergone a silicification which renders them extremely like the chalk and immediately associated Tertiary sandstone. It is therefore not improbable that they represent a local silicification of Trias sandstone and conglomerate performed during the Tertiary desert stage. The Triassic sediments, where unaltered in this neighbourhood, have a calcareous matrix.

The silicification has kept fairly constantly to one horizon—the uppermost few feet exposed. What intervenes between the zone of silicified rock and the Tertiary lavas is hidden. The junction with typical Triassic sediments below is, however, well-exposed.
E.B.B.

*Carsaig* (*Sheet* 44). It is in the Carsaig district that the Lower Tertiary sediments attain their greatest development, in thickness as well as in variety. Some of the beds described here as Tertiary were considered by Prof. Judd to be Chalk, but detailed examination along the outcrop supports the view that the Carsaig Chalk is all *remanié* and should be ascribed to the base of the Tertiary. The observed masses of chalk do not form bedded layers, but are dispersed without order in a sandstone matrix.

The best exposure is in the stream above Feorlin Cottage, 50 yds. above the left affluent. Underlying the basalt-lavas is a purple mudstone, 3 ft. thick, which rests on a bed made up of chalk-fragments embedded in a black sandy matrix. This is underlain by a massive white pebbly sandstone with red staining, 20 ft. thick. Five feet from the top of this sandstone, is a lenticular bed, 2 ft. thick and some 80 ft. long, made up of white flints. The base of the section consists of Cenomanian glauconitic sandstone passing down to concretionary limestone. The lateral variation of these beds is rapid and considerable. In the affluent 50 yds. below the highest bed seen is a white sandstone. This is underlain by a 2 ft. layer of chalk-fragments resting on a 5 ft. sill, which caps the massive white sandstone referred to above.
G.W.L.

The recognition in 1920 of what appears to be wind-rounding as a feature of the quartz-grains of the Tertiary sandstone of Gribun led to a re-examination of the white sandstone of the main stream above Feorlin Cottage. A pocket-lens shows beautiful rounding of many of the quartz-grains of the sandstone. Another point noticed was a white cherty halo surrounding included chalk-fragments and serving as matrix to adjacent sand-grains. From this latter phenomenon it is clear that silicification of the chalk occurred after the latter was embedded in the sand; and the fact that it proceeded without a flattening of the fragments shows that silica was precipitated to take the place of the lime removed. A slice (20809) was cut from a hand-specimen of sandstone enclosing chalk. Only silicified sandstone is shown with quartz-grains bound together by cryptocrystalline silica. Perhaps half the grains show marked rounding, and oval sections are common measuring 0·3 mm. by 0·2 mm.
E.B.B.

The amount of chalk-fragments dimishes eastwards from the stream above Feorlin Cottage: at the waterfall Eas Mheanain, 400 yds. to the east, there is no individual layer of chalk-debris, but angular fragments are irregularly scattered throughout the mass of the white pebbly sandstone. Still farther east, at Eas na Dabhaich above Pennycross House, no chalk occurs in the white pebbly sandstone, which is here a little glauconitic. It merges downwards into glauconitic sandstone with crushed lamellibranchs. Both the lamellibranchs and the glauconite are probably *remanié* Cenomanian.

The sandstone with chalk can be traced on the western side of the bay, as far as the waterfall at the Nuns' Pass Quarries, where it dies out.

In the gully above Àird Ghlas the following section shows a type of sedimentation different from that on the east side of the bay:—

|  | Ft. |
|---|---|
| Basalt-lava | |
| Soft shaly sandstone with fragments of flints | 1 |
| Greenish sandstone with flints and lumps of chalk | 2 |
| Purple mudstone | 1 |
| Greenish sandstone with a few lumps of chalk and flints | 5 |
| Gap | 1 |
| Cenomanian | — |

From some inaccessible place a little south-west of this point, large fragments of chalk up to a foot across have fallen on to the shore. There is no difficulty in believing they may come from *remanié* masses in the Tertiary as farther east.

The intercalated mudstone in the Àird Ghlas section is not of appreciable horizontal extent. The sandstone is apparently always greenish on this side of the bay, instead of white as on the other side. The rock to which the term chalk is applied here has much the appearance of chalk, but is silicified and consequently very hard. Notes on its microstructure, by the late Rupert Jones, are appended to Prof. Judd's Memoir.[1]

G. W. L.

*Tobermory* (*Sheet* 52).—In the road-section above the distillery at Tobermory, grey fossiliferous Liassic shales are followed immediately by one or two feet of red-brown Tertiary mudstone occurring at the base of the lavas. Some 70 yds. farther north, however, in a section at the back of the houses that stand below the cliff, three feet of olive-green sandstone interpose between the Liassic shales and the Tertiary mudstone. This sandstone has abundant rounded quartz-grains, some with secondary facets. White mica is also conspicuous.

A mile and a half south-east of the town, an exposure of soft white sandstone occurs beside a path that skirts the south-west side of the lake in the grounds of Aros House. The sandstone is largely composed of beautifully rounded quartz-grains with very subordinate white mica. There is no section connecting the exposure with those of neighbouring basalt-lavas, but the probability is that the sandstone is of Tertiary date, and that it lies at the base of the lavas and is bounded on the north-east by a fault.

E.B.B., G.V.W.

## Flint-Conglomerate, Etc., Eastern Mull

In the year 1909, when the basal flint-conglomerate of Eastern Mull was mapped, it was judged to consist of normal flint *and* silicified chalk, both occurring as water-worn pebbles. If this view is correct, it supplies clear proof that the period of silicification ante-dates the conglomerates.

*Craignure District* (*Sheet* 44). No Tertiary sediments were noticed at the base of the lavas in the disturbed coast-sections near the entrance to the Sound of Mull, whether at Bàn Eileanan, Craignure Bay, or Duart Bay.

Inland, however, a good exposure of flint-conglomerate with abundant mudstone-matrix occurs above the road close to a deer-fence half a mile north-west of Torosay Castle. Three hundred yards farther south, the lavas are resting on a foot of flint-conglomerate and this upon a flinty bed, with grey limestone beneath, referable to the Upper Cretaceous.

Some way south of this, a fault throws back the base of the lavas 600 yards to the west. A flint-conglomerate has been noted at intervals separating the lavas from the Lower Lias. Four hundred yards south-west of Upper Achnacroish Farm, the Tertiary basement-bed is 30 ft. thick, and consists of sandstone with only occasional flint-pebbles.

*Port Donain Peninsula* (*Sheet* 44).—Round the Loch Don Anticline, no basement Tertiary sediments are to be seen. On the east coast, however, there is a capital section of them on the shore half a mile east of Auchnacraig Farm. The Cenomanian is overlain by a sandstone containing water-rounded white flints, up to a foot in diameter, along with numerous broken lamellibranchs, in both cases *remanié* of the Cretaceous. This bed is lenticular, varying in thickness up to 6 or 7 ft. South of a well-marked gully the sequence is a little different. The flints form a distinct conglomerate-band, 2 ft. thick, above a Cenomanian sandstone with *Exogyra* in place. The conglomerate has a sandstone-matrix. At both localities, the flint-conglomerate horizon is overlain by a foot or two of dark mudstone which is represented by Analysis XV. (p. 34). This mudstone occasionally contains fragments of flint and black shale, the former being numerous in the lower 8 inches as seen south of the gully. The top is decidedly sandy and gritty at this point, while it becomes shaly north of the gully. No plant-remains were noticed. A very small patch of flint-conglomerate occurs at Port Donain, but none is seen farther south.

[1] J. W. Judd, 'Secondary Rocks of Scotland,' *Quart. Journ. Geol. Soc.*, vol. xxxiv, 1878, p. 739.

## THE BASAL MUDSTONE.

The most persistent Tertiary sediment of the Mull district is a muddy, often unbedded, rock of fine texture which breaks up into small irregular fragments. It is met with again and again at the actual base of the lavas, so that there can be little doubt as to its stratigraphical individuality. The bed is certainly absent in a few fully exposed undisturbed sections, but still its relative persistence is remarkable when taken in conjunction with the small thickness of the deposit which very rarely reaches 10 feet. As indicated already, the mudstone has probably in the main originated from lateritic weathering of basaltic ash. Sometimes a slight admixture with normal sediment can be detected. E.B.B., G.W.L.

The usual colour of the deposit is a purplish brown, or deep red, evidently determined among other things by an abundance of ferric oxide. Within the area included by the Limit of Pneumatolysis (Plate III., p. 91), this red tint fails ; and the mudstone as seen south of Craignure, and in the Port Donain Peninsula is dark, almost black. Taken in connexion with certain abnormal characters exhibited by the underlying Mesozoic sediments and overlying Tertiary lavas, the local colour in this case seems to be due to pneumatolytic change. At any rate, the analysis of a dark mudstone from the east coast (XV., p. 34) shows 17·18 per cent. FeO as against 1·57 per cent. $Fe_2O_3$, which is probably very different from what would be found in the deep-red mudstone. G.W.L.

Another local variation of colour is one that may be ascribed to original conditions, rather than pneumatolysis. In the coastal sections of the Croggan Peninsula, the basal mudstone is very pale green or buff in colour, and is probably more aluminous, and correspondingly less ferruginous, in composition than usual. The occurrence has an additional interest, as it is suggested later on that a similar variety of mudstone is the source of the sapphire-bearing xenoliths described in Chapter XXIV. E.B.B., G.V.W.

Most of the exposures in Mull have been mentioned already, but a brief recapitulation is given below before proceeding to an account of the Loch Aline evidence :—

*Gribun (Sheet* 43).—Two exposures have been described above. At Allt na Teangaidh 12 ft. of mudstone intervene between the desert Tertiary sand, below, and the basalt-lavas, above. At the south end of The Wilderness, the mudstone may be seen choking the cavities of a land-slipped block of silicified chalk.

An additional exposure occurs at the roadside in a great coherent landslip north-east of Balmeanach Farm. The locality is easy to find as it is situated at the sharp bend of the main road where it comes nearest to the farm-house.

*Carsaig (Sheet* 44).—Two exposures have been recorded above. In the stream above Feorlin Cottage, the basal mudstone, 3 ft. thick, occurs in its normal position immediately below the lavas. Underneath the mudstone itself, is the desert Tertiary sand.

South of this, above Àird Ghlas, a 1-ft. mudstone for a short distance can be traced separated by 3 ft. of sandstone from the overlying lavas  This section is unique in showing what appears to be the basal mudstone separated by other sediment from the lavas ; but there is, of course, no difficulty in accounting for such a local peculiarity. Below the mudstone there are 5 or 6 ft. of Tertiary sandstone. G. W. L.

Nothing was seen of basal mudstone in the disturbed section of An Coileim south of Glenbyre Farm.

*Croggan Peninsula (Sheet* 44).—In the An Garradh section, at a point indicated on the one-inch Map, there are about 20 ft. of pale buff-coloured marl (17398) above Middle Lias sandstone. This marl is regarded as an aluminous development of the basal mudstone.

G.V.W.

Mudstone is definitely absent near the mouth of a small stream south-east of Beinn na Sròine, where the only Tertiary sediment is a very thin sandstone to be described at the foot of the page.

Farther north-east, at Port na Muice Duibhe, there is a section at the base of the lavas, much complicated by dykes and sills, but at the same time showing some 20 ft. of pale-green brittle shale or mudstone. This mudstone differs from the ordinary type in being distinctly bedded. It also at one point carries white concretions (15867).

*Port Donain Peninsula (Sheet 44).*—Dark brittle mudstone is met with as the only Tertiary sediment at the base of the lavas south of Port nam Marbh.

North of Port na Tairbeirt, the coast-sections, as already noted, show a foot or two of similar mudstone above flint-conglomerate.

*Craignure District (Sheet 44).*—The only known occurrence of mudstone, in the comparatively few available exposures of this district, occurs above the road at the deer-fence half a mile north-west of Torosay Castle. The rock has been referred to above as flint-conglomerate with abundant mudstone-matrix.

E.B.B.

*Tobermory (Sheet 52).*—The list of Mull exposures is completed by recalling those at Tobermory, where, in the town, as already mentioned, one or two feet of mudstone is locally the only Tertiary sediment at the base of the lavas, while elsewhere it is accompanied by three feet of underlying sandstone with round wind-worn grains.

G.V.W.

*Loch Aline, Morven (Sheet 44).*—The basal mudstone occurs in several sections on the two sides of Loch Aline. The following exposures may be enumerated :— shore, north of pier on west side of Caolas na h-Airde ; cliff, north of Achadh Forsa, 6 ft. thick ; left bank of burn that flows into Rannoch River above Achranich, near Fountainhead ; waterfalls from Allt Leacach southwards, 5 ft. thick ; shore, west of Am Mìodar, over 4 ft. thick ; faulted inlier in Allt na Samhnachain.

The mudstone in these sections always lies at the base of the lavas and is probably the only Tertiary sediment represented. It is often underlain by a white sandstone with some rounded quartz-grains, but this is regarded as of Upper Cretaceous age.

The mudstone is of normal character, muddy, unbedded, and breaking into small irregular fragments. Its predominant colour is purplish brown, sometimes red. No pebbles are present, and clastic elements are represented by small grains of quartz visible with the aid of a lens. At one or two places, namely near the Fountainhead above the pier south of Achranich, it contains small masses of impure coal or charred vegetable remains, which led to search for coal, as testified by two shallow day-levels. The same mudstone crops out on the left bank of Allt na Socaich. The valley here is sometimes called the Coal Glen from the fact that lignite was once obtained from it at a spot long since completely hidden by scree-materials. From analogy with the sequence at Beinn Iadain, farther north, Prof. Judd inferred that the lignite belonged to a sandstone-series overlying the chalk ; but such a series is not known to be represented in the immediate vicinity of the Coal Glen, while on the other hand the mudstone near by contains small quantities of a coaly matter approaching lignite, so that it seems more probable that the lignite of the Coal Glen was obtained from the mudstone.

G.W.L.

*Ardnamurchan (Sheet 52).*—Although Ardnamurchan falls outside the scope of the present memoir, it is proper perhaps to note that a basal mudstone comparable in type and thickness with that of the Mull and Loch Aline district has been found in several exposures.

## OTHER BASAL SEDIMENTS.

The preceding accounts of the Tertiary sediments beneath the lavas can be completed by two further notes :—

*Croggan Peninsula (Sheet 44).*—The only Tertiary sediment at the mouth of the small stream that enters the sea south-east of Beinn na Sròine is a very thin sandstone containing numerous fragments of lava-form rocks (15854-8), and a few of sandstone, in a matrix often rich in quartz-grains. The rock-fragments are occasionally water-rounded, and a small proportion of the sand grains seem wind-rounded.

The lava-enclosures are too weathered for accurate determination, but several of them look like trachyte.                                                                   E.B.B.

*Inninmore, Morven* (Sheet 44).—A path leads to Inninmore Cottage at the angle of the bay of that name. A mile west of the cottage, and 170 yds. west of a prominent fault, this path takes a sharp bend. Just at this point, there is a small exposure of a greenish micaceous sandstone immediately overlying Triassic cornstone, and overlain by a sill. The relations of the sandstone to the underlying Trias are by no means clear; in fact, the sandstone was taken to be part of the Trias until search for fossils led to the unexpected find of a fossil leaf. This is not well-preserved, or specifically determinable, yet it is an undoubted Dicotyledon stated by Mr. Clement Reid to resemble oak in its primary venation. The position of the sandstone is below the basalts; but its age relative to that of the Loch Aline mudstone cannot be ascertained, the mudstone being absent here.    G.W.L.

LEAF-BEDS AND ASSOCIATED GRAVELS OF ARDTUN (SHEET 43).

The famous Leaf-Beds of Ardtun are included in a belt of sediment outcropping, as shown on the one-inch Map, for about a mile close to the coast of the Ardtun Peninsula between Loch na Làthaich and Loch Scridain. The Duke of Argyll distinguished three leaf-beds in the sequence: a top leaf-bed, styled by him the 'first'; a mid or 'second'; a bottom or 'third.' Their positions are indicated in the descriptions given below. The whole series was mapped for the Geological Survey by Mr. Bosworth, but the detailed researches of Mr. Starkie Gardner[1] so fully cover the subject, that the following account of field-relations is drawn mainly from his descriptions. For further information the reader should refer to the original paper which is illustrated by six text-figures.

Mr. Tait has made a large collection of plants for the Geological Survey from the leaf-beds. He also obtained a few beetle and molluscan remains. The plants and beetles are discussed in the following chapter. The molluscs have been found by Dr. Lee "not sufficiently well preserved to permit identification yet it seems that they include a *Cyrena*, beside gasteropods." They occurred actually in a leaf-bed.

The best exposure is in a ravine indicated by a note on the one-inch Map. Gardner measured the section when cleared by quarrying; and it still remains in satisfactory condition. His measurements for the east side of the ravine are:—

|  | Ft. | In. |
|---|---|---|
| Rudely columnar basalt cut by intrusive sheet | — | |
| Sandstone, more or less fissile | 8 | 0 |
| Indurated gravel | 7 | 0 |
| Hard bed, with *Onoclea* | 1 | 0 |
| Black Leaf-Bed (Mid Leaf-Bed) | 2 | 4 |
| Indurated gravelly sand | 2 | 0 |
| Carbonaceous rubble | 1 | 0 |
| Amorphous basalt | — | |

The columnar basalt on top is the lower part of a lava; the amorphous basalt at bottom is the upper part of another lava, which downwards becomes columnar. The position of the Duke's top leaf-bed is at the top of the sedimentary sequence, but the 8-foot sandstone exposed in Gardner's excavation did not prove fossiliferous, and was separated from the overlying basalt merely by a

[1] J. S. Starkie Gardner, 'On the Leaf-Beds and Gravels of Ardtun, Carsaig, etc., in Mull,' *Quart. Journ. Geol. Soc.*, vol. xliii., 1887, p. 270.

parting of carbonaceous rubble. The bottom leaf-bed was not found on this side of the ravine ; but on the west side it is 1 foot thick, a pale buff or cream-coloured laminated sandstone resting on the amorphous basalt.

Westwards the sediments thin somewhat, and as seen in a fine section in another ravine 120 yards from the main locality, are only 14 feet thick.

Eastwards they thicken for a space, and 80 yards from the ravine taken as our starting point, they provided Starkie Gardner with the following measurements :—

|  | Ft. | In. |
|---|---|---|
| Columnar basalt | — | |
| Position of Top Leaf-Bed (grassed over) | 2 | 0 |
| Gravel | 25-40 | 0 |
| Black Mid Leaf-Bed | 2 | 6 |
| Gravel, about | 7 | 0 |
| Grey clay, with faint leaf-impressions | 2 | 0 |
| Laminated sandstone, with leaf-impressions | 0 | 6 |
| Fine limestone, with rare but beautiful leaves | 0 | 3 |
| Clay, with layer upon layer of beautiful leaves at base | 1 | 0 |
| Clunch with rootlets | 0 | 7 |
| Amorphous basalt, becoming columnar at base | — | |

The lower 4 feet or so may be grouped as the Bottom Leaf-Bed. Gardner describes the stratum at the base of the 1 foot clay as the most interesting of the whole series : "This consists for an inch or two of layer upon layer of leaves in the most perfect preservation, and retaining almost the colour of the dead leaves themselves. One of the most striking, as well as most abundant, is *Ginkgo,* of large size and purple colour. Still more conspicuous is the large *Platanites hebridicus,* Forbes, one leaf exposed measuring full 15½ inches in length and 10½ inches in breadth. Many other kinds of leaves appeared to be almost equally fine, and the characteristic dicotyledonous trees of this locality possessed at that period relatively large foliage. In the same bed were coniferous branches like the living *Taxodium* (*Glyptostrobus*) *heterophyllum* and *Cephalotaxus.* Unfortuately every effort to remove and preserve these specimens has failed. There are rush-like stems, from 1 to 3 inches in diameter towards the base, but the beds are almost destitute of monocotyledons, and no trace of Ferns or even *Equiseta* has been seen in them. This lowest leaf-bed passes into a thin seam of coal in one direction, and rests upon 6 to 7 inches of whitish, clunchy, and concretionary clay, with rootlets, and with softer clay filling in the rough surface of the underlying basalt."

Eastwards the sediments thin, but there is no ground for Mr. Gardner's statement that they fail altogether.

A few words may now be said regarding the contents of the gravels and sands. The sands abound in quartz-grains with only a comparatively small proportion of rounded individuals. The most noteworthy pebbles of the gravels, or conglomerates, are water-worn chalk-flints (first recorded by the Duke of Argyll) and porphyritic lavas. The former seem to the writer to resemble genuine flints rather than the silicified chalk of Gribun, etc. ; but Professor Cole[1] thinks that the latter type may be well repre-

[1] G. A. J. Cole, 'Note on the Gravel of Ardtun,' *Quart. Journ. Geol. Soc.,* vol. xliii., 1887, p. 276.

sented. He, in fact, suggests that silicification may have occurred after inclusion in the gravels; but it is fairly certain that these well-worn pebbles were hard at the time they were transported. The porphyritic lavas represented among the pebbles have long been recognized as distinct in type from the neighbouring basalt-flows. They have generally been assumed to be of Tertiary age, but this is by no means certain. The suggestion is here advanced that they may be porphyritic augite-andesites, and allied types, of Lower Old Red Sandstone age (20764-6, 20768-9). Another noteworthy type is of less common occurrence (20767), but has been met with by Professor Cole as well as ourselves. It is an intrusive type consisting mostly of zoned alkali-felspars, often with perthitic interiors; the ground is microgranophyric with subordinate quartz; the ferromagnesians are mainly decomposed, but include a little pale biotite. The rock has been styled a sadinophyre or sanidine-felsite by Professor Cole. As in the case of the lava-pebbles it does not seem safe to assume that it is of Tertiary age, although it vaguely recalls certain abnormal Tertiary intrusions.

The whole deposit, gravels, sands, and leaf-beds, may well be interpreted as a fluvio-lacustrine series, with a root-bed locally developed at the base.                                E. B. B.

## SEDIMENTS OF MALCOLM'S POINT (SHEETS 36 AND 44).

Mr. Starkie Gardner has described what he very reasonably regards as a reappearance of the Ardtun sediments on the south coast of Mull at Malcolm's Point west of Carsaig. The point falls within Sheet 36 of the one-inch Map, and has been dealt with by Dr. Harker in the Geological Memoir on that Sheet. It is also engraved on the margin of Sheet 44. Dr. Harker's account is reproduced below :—

"The stratified deposits rise from sea-level at Carsaig Arches and may be followed continuously in the lower part of the cliff to Malcolm's Point and some half a mile beyond. Their position is thus very near the base of the basalt succession. Near the Arches they are only a few feet thick, consisting of bedded basaltic tuff with sandy material. The deposit rests on basaltic lava, and is covered by a dolerite sill. Followed past Malcolm's Point, the bedded group thickens, and there is more non-volcanic material mingled with the basaltic *débris*. At about 500 yards beyond the Point the thickness is some 12 feet, and there are isolated rolled pebbles of flint, which become more numerous, until the deposit may be described as a conglomerate. The last good section, near the limit of the map [*i.e.* Sheet 36], shows 15 feet of conglomerate, composed of rolled flint pebbles up to 4 or 6 inches in diameter, and passing up into a bedded basaltic tuff, surmounted by the same dolerite sill. Below the conglomerate is about 30 feet of amygdaloidal basalt."

Near Carsaig Arches, Mr. Gardner noted 2 feet of impure sand, with indistinct vegetable-markings at the base of the deposit at one place, and finer material with a thin band of lignite at another.

What may well be a continuation of the same outcrop is seen farther east at the foot of the cliff a third of a mile south-west of the Nuns' Pass (Sheet 44). The section reads as follows :—

|  | Ft. |
|---|---|
| Lava . . . . . . . . . | — |
| Sill . . . . . . . . . | 2 |
| Sandstone . . . . . . . . | 3 |
| Sill . . . . . . . . . | 8 |
| Highly vesicular basalt, probably lava . . . | — |

The sandstone is black and pebbly and to some extent made of basalt-fragments. A layer of flint-pebbles occurs near its base.

### Sediments East of Carsaig (Sheet 44).

An addition to our knowledge of the Mull leaf-beds was made during the survey of the district. Fifty yards west of An Dùnan, about one mile east of Carsaig Bay, at the landward end of a little gully, there is exposed below high-water mark a lenticular bed composed of black shale, gritty sandstone, and a 3-inch coal seam, intercalated between basalts (*see* below). From the black shale, Mr. Tait obtained a small suite of plants rather badly preserved. They are discussed in the following chapter.  G.W.L.

### Lignites of South-West Mull.

A local association of very thin seams of lignite, or inferior coal, has been noticed above in connexion with the sediments described from Ardtun, Malcolm's Point, and east of Carsaig. Recurrences of the same type are rather characteristic of this south-west district of Mull, and have long attracted attention alike on account of the hopes of mineral wealth that they have given rise to and also because of their genuine scientific interest. As regards their economic value, Sir Archibald Geikie has rightly said "they seem to be always lenticular patches"; moreover they are of inferior quality (Chapter XXXVIII.).

*Beinn an Aonidh* (Sheet 44).—The most important seam of lignite, or coal, occurs near the south coast of Mull, 40 ft. below a porphyritic lava, which serves as a convenient index in correlating the separate exposures. The coal is seen at the top of the cliff at Dearg Bhealach above Tràigh Cadh' an Easa', about 400 ft. above sea-level, and perhaps not very much more than this figure above the base of the lavas. The outcrop is shown on the 1-inch Map, and details will be given in Chap. XXXVIII. on Economics. At present it may be enough to mention that it can be traced as far inland as Àiridh Mhic Cribhain, and that the same coal occurs on the western flank of Beinn an Aonidh, where in one place it is reported to have been 3 ft. in thickness. It is associated with carbonaceous shale, but is very irregular, and in addition often burnt by intrusions. Some graphitic sediment also occurs in one of the streams which drain the north-west slope of Beinn an Aonidh; but the mode of occurrence is doubtful, and the material obtained may have been xenolithic.  E.M.A.

*Carsaig* (*Sheet* 44).—A coal-seam from this locality is referred to in the writings of Sir Archibald Geikie. It is exposed about 70 ft. up in the coastal cliff, below Sgùrr Mhòr immediately east of the important fault shown on the one-inch Map. The coal is here 18 inches thick, brownish-black in colour, dirty, soft, and friable. It is associated with several feet of sandy shale. An intrusion of columnar basalt cuts across the seam obliquely. The extent of the exposure is not more than 40 yds.; but about 200 yds. farther east, coal is seen on the raised-beach platform, unfortunately much broken by intrusions. Possibly the planty sediments mentioned above belong to the same horizon, but this point is not, as yet, definitely established.  G.V.W.

*Shiaba* (*Sheet* 43).—Another coal, sometimes worked for home-use by the Shiaba shepherd, is seen near the bottom of a waterfall called Eas Dubh, half a mile east of

Shiaba Cottage near the southern margin of Sheet 43. It has for the most part a dull brown colour, and seems of a parroty nature. In one place it is two feet thick but diminishes to 6 inches within a few yards. It dips steeply east and cannot be traced far.

*Gowanbrae and Ardtun, Bunessan (Sheet* 43).—In his description of the Ardtun leaf-beds, the Duke of Argyll refers to two outcrops of coal in the vicinity. One of these is shown on the one-inch Map at Gowanbrae. It was opened up in shallow pits, apparently on the Duke's instructions, some time in the seventies. The trials must have proved unsatisfactory, as no attempt was made to carry the work further. Around the pits, which are still visible, there are pieces of dark shale with fragments of wood and traces of leaves, but no coal.

The other outcrop is clearly visible where shown on the one-inch Map close to the shore of Loch Scridain, south-east of Tòrr Mòr.[1] The exposure was recently opened up, and from 6 inches to 1 ft. 2 in. of coal were seen resting on a reddish-brown clay containing recognizable fragments of basalt. Most of the seam was very inferior with no more than 6 inches of bright coal. A columnar basalt-lava serves as roof, and has its under surface apparently chilled, and also corrugated into subparallel waves.                                                                                                                    (C.T.C.)

*Macculloch's Tree (Sheet* 43).—A thin seam, or streak, of lignite occurs in places at the base of the lava which envelopes Macculloch's well-known upright tree (p. 111).

A very few other instances of the same kind are mentioned in the literature of the subject, but were not located during the geological survey of the district.                                               E. B. B.

## OTHER SEDIMENTARY INTERCALATIONS.

The Ardtun, Malcolm's Point, and Carsaig outcrops, described already, stand almost alone in supplying evidence of non-volcanic detritus having entered the district during the accumulation of the lavas. Three other cases are known ; two of them from south-west Mull are sufficiently described elsewhere (pp. 56, 115) ; the third is represented by a minute outcrop of reddish yellow sandstone, from one to two feet thick, intercalated between basalt-flows on the north shore of the Sound of Mull, south-east of Loch Aline and a third of a mile west of the site of Ardtornish House (Sheet 44). It is quite barren of fossil-remains, and its position is low in the basalt-series, perhaps between the first and second flows. G. W. L.

In the descriptions of the basalt-lavas which follow, reference will be constantly made to thin layers of red bole often met with between flows in the district outside the Limit of Pneumatolysis of Plate III., p. 91. Inside this limit, such red volcanic muds soon fail for one reason or another. Attention may be directed, however, to a good example of a sedimentary intercalation unusually high up in the sequence. It occurs a little above the Ben More mugearite-zone, about half a mile due south of Ben More summit (Sheet 44). It consists of 8 to 9 ft. of sediment, partly breccioidal, partly a carbonaceous shale with obscure plant-remains. Another sedimentary intercalation is known in the same general district exposed in a stream east of Beinn nan Gobhar. It seems to overlie a bostonite shown on the one-inch Map, and its nature will be discussed more precisely in this connexion (Chapter XIV.). Its thickness is about 5 ft.                                                                                                                               E. M. A.

---

[1] The Tòrr Mòr referred to is the small hill of that name between two and three miles north-east of Bunessan, and not, as has been erroneously stated, 5 miles east of Ardtun.

E

## Chapter III.—Tertiary Sediments.

Very little sediment accumulated in the south-east caldera of Plate III. (p. 91), although there is good reason to believe that it was often occupied by a lake during pauses in the accumulation of the Central Type of lavas (Chapter V.). There is, however, an interesting exposure of volcanic sand and mud which may be referred with some confidence to this period. It occurs on the north-east face of Beinn Bheag, north of Beinn Talaidh, and is lettered i on the one-inch Map (Sheet 44). It is greatly cut by the Late Basic Cone-sheets of Chapter XXVIII., but can be seen to be flatly bedded and to lie upon the chilled top of a small-felspar dolerite, which in character strongly recalls pillow-lavas well represented in the same general district. Much of the deposit consists of recognizable mineral-debris, felspar and augite, some of it remarkably fresh and of a character such as the pillow-lavas, or corresponding tuffs, might readily supply (18653, 18667).

E. B. B.

## CHAPTER IV.

### TERTIARY PLANTS FROM MULL.[1]

By A. C. SEWARD AND R. E. HOLTTUM.[2]

WITH A DESCRIPTION OF A NEW BEETLE.

By T. D. A. COCKERELL.

#### INTRODUCTION.

THE specimens which form the subject of the present Report were sent to one of us for examination from the Museum of the Geological Survey, Edinburgh. They were collected by Mr. Tait from two localities (i) the Ardtun leaf-beds at Bunessan, the locality from which Mr. Starkie Gardner collected numerous specimens (p. 61), and from (ii) a new locality at Carsaig discovered by Mr. Tait : the latter beds occur near the base of the volcanic succession and are believed to be approximately on the Ardtun horizon (p. 64). The great majority of the fossils are impressions of leaves usually far from complete and unfortunately without any carbonaceous films which could be examined microscopically. Our attention has been mainly concentrated on the Geological Survey collection, but use has also been made of the more satisfactory material collected by Mr. Starkie Gardner and now in the British Museum. Though it is seventy years since Tertiary plants were discovered in the Island of Mull the flora as a whole has never been thoroughly investigated. We hope that in the near future it may be possible to undertake a more thorough examination of all the available material from the Island. Our immediate object is to determine as far as we can the specimens submitted to us, to form an opinion on the botanical character of the vegetation and of the geological age of the Mull leaf-beds.

For many years past the investigation of the older Tertiary floras of Britain has been neglected while on the other hand the researches of the late Mr. Clement Reid, and more recently those of Mrs. Reid, have demonstrated the possibilities of palæobotanical work when the material consists largely of fruits and seeds.

The literature on Tertiary plants is both voluminous and scattered, and unfortunately many of the published generic names have been adopted without any evidence that systematists could accept

---

[1] M.S. received March 11, 1921 (Editor).

[2] It is only fair to state that Mr. Holttum has done the more laborious and difficult part of the work ; he is responsible for the determination and description of the Dicotyledons and he has also assisted me in the description of the other fossils.

(A. C. Seward).

as satisfactory. Many leaf impressions, however perfect, cannot be identified with confidence, and it is even asserted that in the absence of fruits and seeds or flowers the palæobotanist's task is foredoomed to failure. There would seem to be two alternatives ; either wholly to neglect fossil angiospermous leaves or with the assistance of expert systematists to endeavour to steer a middle course between the over-confidence of the enthusiast, who cannot resist the temptation of naming specimens which are indeterminable, and the extreme caution of the botanist, who declines to commit himself to definite opinions which cannot be supported by evidence such as he is accustomed to demand from recent plants. In dealing with Tertiary plants it is easy to be destructive and to throw doubt on the conclusions of other authors. Mere destructive criticism is of little value from the point of view of definite progress. Our aim is to discard material that in our opinion cannot be determined with reasonable confidence, and to satisfy ourselves that the opinions expressed are based on evidence that would not be considered inadequate by botanists possessing a considerable knowledge of the taxonomy of recent plants.

Our thanks are due to Dr. Lee and Mr. Bailey of Edinburgh for references to literature, to Dr. Kitchin of the Jermyn Street Museum, to members of the Botanical and Geological Departments of the British Museum, particularly to Dr. Rendle and Mr. W. N. Edwards. We are also indebted to Mrs. Reid for examining and reporting upon some specimens submitted to her.

In 1851 the Duke of Argyll communicated a paper to the Geological Society on the Geology of the Ardtun leaf-beds, with a note by Professor Forbes on the fossil plants.[1] Of the numerous fossils obtained a few of the "most perfect impressions of plants, mostly of leaves only" were selected for illustration, and Forbes with commendable caution stated that "without much more data than such impressions, however perfect, afford, anything like a specific diagnosis satisfactory to botanists may not be constructed." He stated that the plants undoubtedly indicated a Tertiary and probably a Miocene age. The Duke of Argyll from an examination of the manner of occurrence of the leaves in his middle, or second, leaf-bed concluded that they must have been shed "autumn after autumn into the smooth still waters of some shallow lake, on whose muddy bottom they were accumulated, one above the other, fully expanded and at perfect rest." He drew attention to the presence of only small twigs and the absence of trunks or large branches.

In 1870 Dr. Grieve and Mr. Mahony exhibited a series of fossils from the leaf-beds of Mull at a meeting of the Natural History Society of Glasgow. The leaf-beds were assigned, for reasons not specified in the Report, to the Miocene period.

A Committee of the British Association appointed to collect and report on Tertiary plants from the north of Ireland issued four Reports in 1879, 1880, 1881, and 1883. In the third Report (drawn up by W. H. Bailey) it is stated that "by the identification of these plant remains we are enabled to fix the period in which they lived as being lower Miocene. . . . they also afford strong evidence of

[1] Argyll (1851). For references cited in this chapter *see* Special Bibliography, p. 89.

being contemporaneous with other volcanic districts such as those of the Island of Mull on the west coast of Scotland and of north Greenland, where mid-European plants such as these once flourished." The few illustrations of the plants given are not very satisfactory, but Gardner[1] has figured in his *Eocene Flora* some good specimens of Conifers with cones from the north of Ireland. It is to be noted that Gardner[2] subsequently gave reasons for referring the plant remains to a Lower Eocene age.

In 1881 Mr. W. E. Koch contributed notes *On Mull and its Leaf-beds* to the Geological Society of Glasgow. He described a section through the middle leaf-bed " on the north-east side of a glen leading to the sea." His description agrees fairly well with that later given by Gardner except that he found no break in the fossiliferous succession between the middle leaf-bed and a brown soil below, which latter contained roots and branches, one of them "5 inches across and another 2 inches across, and $\frac{1}{2}$ in. thick. This bed," he added, "rests on eruptive matter and here I believe we are on the site of the old forest." The discovery of a local basal root-bed was confirmed by Gardner ; but its presence hardly warrants Mr. Koch in dispensing with the swampy lake in which the Duke of Argyll pictured the accumulation of fallen leaves.

The Committee appointed by the British Association to report on the fossil plants of the Tertiary and Secondary beds of the United Kingdom issued Reports in 1885 and 1886 without any reference to the Mull Flora. The second Report contains a paragraph remarking on the inadequacy, from a botanical point of view, of leaf-impressions alone for the identification of genera and species of plants.

Several references are made to the Mull plants in a Monograph on the British Eocene Flora by Baron Ettingshausen and Mr. Starkie Gardner. In the first part, by both authors,[3] a description is given of the Fern mentioned by Forbes as *Filicites hebridicus*: this is transferred to the genus *Onoclea*. The type-specimen, illustrated both by Forbes and by the later authors, is in the Jermyn Street Museum. In the second part, by Gardner,[4] several Gymnosperms are described and illustrated. Gardner states that the Mull flora includes types that are also met with in the English sub-tropical Middle Eocene Flora such as *Podocarpus*, the widely distributed southern hemisphere Conifer, together with species recorded by Heer from Greenland in rocks regarded by him as Miocene. Gardner in 1887 [5] gave an account of the lava sheets of Mull and their relation to the leaf-beds, based on personal exploration and quarrying operations. In the latter contribution he discusses the geological age of the Ardtun leaf-beds and points out that Heer's pronouncement on the Miocene age of Scottish, Greenland, and other plant-bearing beds had been generally accepted as an authorative decision, but adds that in his opinion the evidence clearly indicates an early Eocene horizon.

[1] Gardner (1886).
[2] Gardner (1887).
[3] Ettingshausen and Gardner (1882).
[4] Gardner (1886).
[5] Gardner (1887).

# Chapter IV.—Tertiary Plants and Insects.

The following lists give (A) the plants recorded by previous authors with the addition of the names adopted in the present paper and (B) the plants described by us :—

### A.

| Forbes, 1851. | A.C.S. and R.E.H. |
|---|---|
| *Taxites* (?) *Campbelli* Forbes. | *Elatocladus Campbelli* (Forbes). |
| *Filicites* (?) *hebridicus* Forbes. | *Onoclea hebridica* (Forbes). |
| *Equisetum Campbelli* Forbes. | *Equisetum Campbelli* Forbes. |
| *Rhamnites* (?) *multinervatus* Forbes. | |
| *R. major* Forbes. | |
| *R. lanceolatus* Forbes. | |
| *Platanites hebridicus* Forbes. | *Platanus hebridica* Forbes. |
| *Alnites* (?) *MacQuarrii* Forbes. | |
| Gardner, 1886. | |
| *Cryptomeria Sternbergii* Goepp. | *Pagiophyllum Sternbergi* (Goepp.). |
| *Ginkgo adiantoides* (Ung.). | *Ginkgo adiantoides* (Ung.). |
| *Podocarpus eocaenica* Ung. (recorded but not figured from Mull). | |
| *Podocarpus Campbelli* Gard. | *Podocarpus Campbelli* Gard. |
| Gardner, 1887. [1] | |
| *Sequoia Langsdorfii* Heer. | *Sequoiites* (?) *Langsdorfi* (Brongn.). |
| *Glyptostrobus europæus* Heer. | |
| *Podocarpus borealis* Gard. | |
| *Quercites greenlandicus* Heer. | *Quercus greenlandica* Heer. |
| *Boehmeria antiqua* Gard. | |
| *Grewia crenulata* Heer. | |

### B.

PTERIDOPHYTA.
    Equisetales. *Equisetum Campbelli* Forbes.
    Filicales. *Onoclea hebridica* (Forbes).

GYMNOSPERMÆ.
    Ginkgoales. *Ginkgo adiantoides* (Ung.).
    Coniferales.
        Abietineæ. *Pinites* sp.*
        Cupressineæ. *Cupressites MacHenryi* Bail.*
        Sequoiineæ. *Sequoiites* (?) *Langsdorfi* (Brongn.).*
        Podocarpineæ. *Podocarpus Campbelli*. Gard.
        Araucarineæ (?). *Pagiophyllum Sternbergi* (Goepp.).
    Coniferales incertæ sedis.
        *Elatocladus Campbelli* (Forbes).
    Coniferous Wood.
        A. *Cupressinoxylon* sp.
        B. Wood in Lava.

ANGIOSPERMÆ. Dicotyledones.
    Betulaceæ. *Corylites hebridica* sp. nov.
    Dicotyledonous Wood (? Betulaceæ).
    Fagaceæ. *Quercus greenlandica* (Heer).
    Platanaceæ. *Platanus hebridica* (Forbes).
    Dicotyledones incertæ sedis.
        *Phyllites platania* (Heer).
        *Phyllites ardtunensis* sp. nov.
        *Phyllites* spp.

PLANTÆ INCERTÆ SEDIS.
    Specimen A.*
    Specimen B.*
    Specimen C.
    Specimen D.
    Specimen E.*
    Specimen F.

[1] This list includes only the species additional to those recorded by Forbes.
* The species with an asterisk are from Carsaig ; the others from Ardtun.

DESCRIPTION OF SPECIMENS.

INSECTA.

Mr. Starkie Gardner figured an elytron of a beetle and the "hind wing of a Cercopid insect"[1] which was found in association with some of the Mull plants. The few specimens of elytra among the fossils submitted to us were shown to Professor Cockerell of the University of Colorado who kindly contributed the following description of an elytron which he makes the type of a new species, the first Tertiary insect from Scotland to receive a name.[2]

"*Carabites scoticus* sp. nov. Elytron 5 mm. long and 2 mm. wide, the apex obtuse; inner basal corner rectangular; margins very slightly convex except at apex and outer base; ten striæ, not counting the inner absolutely marginal one; striæ weakly and closely punctate, but the general effect sharp; outermost stria marginal except near base; third and fourth striæ (counting from inner margin) joining a considerable distance from apex, with a short appendiculation beyond; seventh stria ending before the sixth or eighth. Eocene; Isle of Mull. Much smaller than the elytron figured by Gardner,[3] and abundantly distinct from the Eocene beetles described from the south of England. It resembles such genera as *Anchomenus* in the present fauna of Britain, but is different and probably represents an extinct generic type. It is placed in the genus *Carabites* in the absence of more complete material. In the same collection are two other elytra, too imperfect to describe. One is at least very close to the above. The other is only about 3 mm. long, weakly striate; apparently a weevil.

"For permission to examine these interesting specimens I am indebted to Professor Seward."

PTERIDOPHYTA.

EQUISETALES.

*Equisetum Campbelli* Forbes. The largest specimen is an impression of an aerial shoot 12·5 cm. long with internodes 1–1·5 cm. in length (Fig. 5A). The leaf-sheaths with acuminate segments (Fig. 5B) extend almost the whole length of the internode. Imperfectly preserved pieces of adventitious roots are seen near the base of the shoot. There is no evidence of branching. It is clearly impossible to give a complete diagnosis even of the vegetative features: in size the specimen shown in Fig. 5A agrees with shoots of the recent species *Equisetum maximum*, particularly with fertile shoots in which the leaf-sheaths are relatively large and there are no branches.

An examination of the original specimen figured by Forbes[4] convinced us of its specific identity with the specimen shown in Fig. 5. Forbes's figure is misleading as the actual specimen shows very

[1] Gardner (1887), Pl. XIII, Figs. 8, 9.
[2] Cockerell (1921), p. 22, Fig. 28.
[3] *Ibid.* Fig. 8.
[4] Forbes in Argyll (1851), Pl. III., Fig. 6.

clearly a portion of a leaf-sheath. Fig. 5c represents an elongate, oval impression of an unexpanded part of a shoot, probably a fertile shoot, of the same type. The surface is covered by overlapping segments of crowded leaf-sheaths.

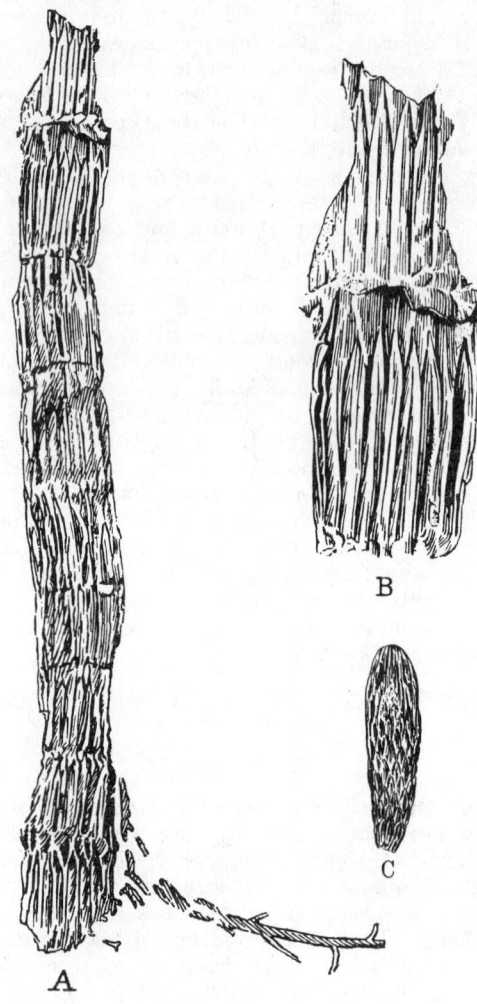

FIG. 5.—*Equisetum Campbelli* Forbes.
    A. Nat. size. [T. 2905 E.].
    B. Leaf-sheaths. ×2 [T. 2905 E.].
    C. Unexpanded shoot. Nat. size. [T. 2887 E.].

The specimen reproduced in Fig. 6 may or may not belong to the species represented in Fig. 5. An imperfectly preserved slender stem is seen in close association with a slightly crushed spherical body which may be a tuber similar to the subterranean tubers of certain recent species.

## Pteridophyta.

### FILICALES.

*Onoclea hebridica* (Forbes). The fragmentary impressions of this fern do not enable us to add anything to the description of the Mull specimens previously published,[1] but through the courtesy of Mr. W. N. Edwards of the Geological Department of the British Museum we are able to give the accompanying figure (Fig. 7) of some spores which he obtained from fertile fronds collected by Mr Gardner in the Island of Mull.

These spores agree closely with those of the recent species *Onoclea sensibilis* and support the generic identification based on sterile

FIG. 6.—*Equisetum* sp.   Nat. size.   [2900 E.].

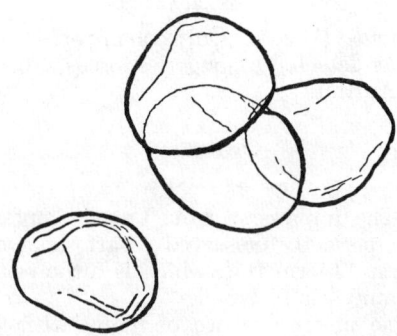

FIG. 7.—*Onoclea hebridica* (Forbes).   Spores.   × 384.   [*British Mus.* V. 14848A.].

fronds. The genus *Onoclea*, represented by a single species, occurs in eastern North America from Newfoundland to Florida and extends to Saskatchewan and Nebraska; it also occurs in Japan and north-east Asia. Like many other plants it has a discontinuous range and the fossil records show that the genus had a wider and more continuous geographical distribution in the early Tertiary period. Christ[2] points out that *Onoclea* affords a good example of the similarity between the flora of east Asia and that of eastern North America which was first recognised as an interesting phytogeographical problem by Asa Gray.

Sterile fronds of *Onoclea* were described by Forbes and later

---

[1] Forbes in Argyll (1851), Pl. II., Figs. 2A, 2B; Ettingshausen & Gardner (1882), p. 68, Pl. XIII., Figs. 5, 6.

[2] Christ (1897), p. 284, and (1910), p. 340.

## Chapter IV.—Tertiary Plants.

by Gardner from Mull. The same species is recorded by Heer[1] from Greenland in beds assigned by him to the Miocene; by Newberry[2] from Dakota, and by Knowlton[3] from Montana, in beds belonging to the Fort Union and other series, which are believed to be Lower Eocene in age. It has also been found in other North American localities.[4] The only previous record of the association of sterile and fertile fronds is that by Knowlton.[5] Lesquereux[6] in his *Tertiary Flora* described specimens of *Caulinites fecundus* from Erie, Colorado, which Knowlton[7] refers to the genus *Onoclea*. It is possible that Knowlton is correct, but the figures are not convincing and moreover there are no sterile leaves. Hollick and Berry both figure fertile fronds as *Onoclea inquirenda* Holl. from the Upper Cretaceous (the former[8] from Long Island and Martha's Vineyard, the latter from South Carolina[9] and Maryland[10]), but in the absence of sterile fronds the identification as *Onoclea* cannot be considered certain.

Newberry, in his original remarks on the fossil leaves, notes the wide range of variation in the living species and states that he is unable to separate the fossil from the recent fern.

### GYMNOSPERMÆ.

#### GINKGOALES.

*Ginkgo adiantoides* (Ung.). Only one imperfect fragment of a leaf of this species was detected among the fossils submitted to us from the leaf-beds of Ardtun.[11]

#### CONIFERALES.

#### ABIETINEÆ.

*Pinites* sp. The impression from Carsaig represented in Fig. 8 is probably an imperfectly preserved dwarf shoot of a five-needled Pine. The longest filiform leaf, which is incomplete, is 4 cm. long and not more than 0·4 mm. broad.

It is impossible in the absence of more satisfactory material to identify the specimen with confidence, but we are inclined to regard it as evidence, though not amounting to proof, of the occurrence

---

[1] Heer (1868), p. 86, Pl. I., Fig. 16; (1869), Pl. XL., Fig. 6; (1883), Pl. LXX., Fig. 6 [Specimens obtained in Greenland in 1921 by the authors of this paper confirm the accuracy of Heer's determination].

[2] Newberry (1898), p. 8, Pl. XXIII., Fig. 3; Pl. XXIV., Figs. 1-5.

[3] Knowlton (1902), p. 705, Pl. XXVI.

[4] I take this opportunity of correcting a careless mistake In the *Hooker Lecture*, published in vol. xlvi. of the Linnean Society's Journal, 1922, leaves of *Onoclea*, indistinguishable from those of the recent plant, are said to have been recorded from Upper Cretaceous rocks in various parts of North America (*loc. cit.*, p. 222). The American specimens are from the Fort Union, the Lance, and Paskapoo formations, all of which are believed to be Lower Eocene in age. I am indebted to Mr. Knowlton and Mr. Hollick for drawing my attention to this unfortunate mistake A. C. S.

[5] *loc cit.*

[6] Lesquereux (1878), p. 101, Pl. XIV., Figs. 1-3.

[7] Knowlton (1898), p. 153.

[8] Hollick (1906), p. 32, Pl. I., Figs. 1-7.

[9] Berry (1914), p. 14, Pl. II., Figs. 7, 8.

[10] Berry (1916) p. 764, Pl. LI., Figs. 1, 2.

[11] Gardner (1886), p. 99, Pl. XXV.

## Gymnospermæ.

in the Mull flora of a species of *Pinus* similar in its foliage to the existing species *Pinus Strobus*. Gardner makes no mention in his monograph of the occurrence of five-needled Pines. A similar foliar spur is figured by Saporta[1] as *Pinus palæostrobus* Ett. from the Eocene plant beds of Aix, a species originally described by Ettingshausen[2] from Häring in the Tyrol.

### CUPRESSINEÆ.

*Cupressites MacHenryi* Bail. The impression of a small piece of branched shoot from Carsaig shown in Fig. 9 appears to be identical with the Irish species first figured by Bailey[3] and afterwards more fully described and more adequately illustrated by Gardner from much better material, including cones, from the Irish beds. Gardner[4] substitutes for Bailey's designation the name *Cupressus Pritchardi* on the ground that blocks of a Cupressineous wood from

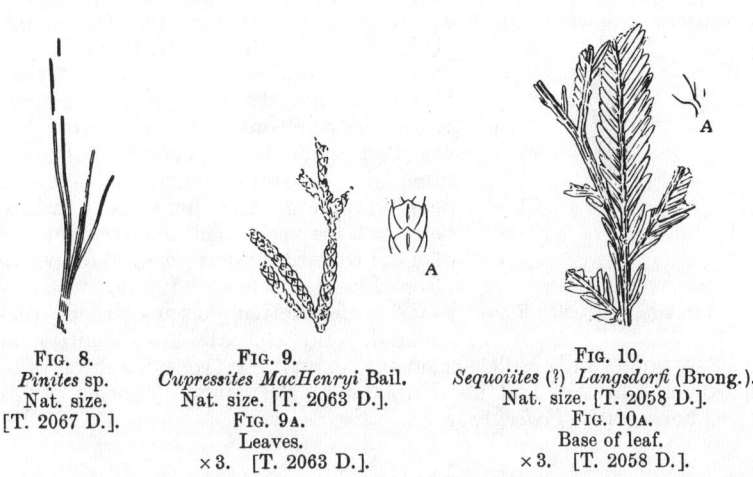

FIG. 8.
*Pinites* sp.
Nat. size.
[T. 2067 D.].

FIG. 9.
*Cupressites MacHenryi* Bail.
Nat. size. [T. 2063 D.].
FIG. 9A.
Leaves.
× 3. [T. 2063 D.].

FIG. 10.
*Sequoiites* (?) *Langsdorfi* (Brong.).
Nat. size. [T. 2058 D.].
FIG. 10A.
Base of leaf.
× 3. [T. 2058 D.].

Lough Neagh, which Goeppert named *Pinites Pritchardi*[5] and Kraus later referred to *Cupressinoxylon*, were from the tree which bore the type of shoot figured by Bailey as *C. MacHenryi*. This assumption may or may not be correct, and we prefer therefore to retain Bailey's specific name.

The leaves of the Mull specimen are small and scale-like and are arranged in decussate pairs (Fig. 9A). This type of shoot has not previously been recorded from Mull.

### SEQUOIINEÆ.

*Sequoiites* (?) *Langsdorfi* (Brongn.). Among the specimens sent to us there are a few from Carsaig which appear to be identical

---

[1] Saporta (1873), p. 98, Pl. I., Fig. 17.
[2] Ettingshausen (1855), p. 35, Pl. VI., Figs. 22, 23.
[3] Bailey (1869), Pl. XV., Fig. 5.
[4] Gardner (1886), p. 82, Pl. XVI., Figs. 8, 9 ; Pl. XVIII., Fig. 1 ; Pl. XIX.
[5] Seward (1919), p. 305.

with the small twig compared by Gardner with those from Greenland described by Heer as *Sequoia Langsdorfii*.[1] The best of the Mull specimens is shown in Fig. 10. Fig. 10A illustrates the clearest example we have noticed of the manner of attachment of a leaf by a petiole-like constricted base. We do not think, however, that this is sufficiently clear to indicate whether it represents the real constriction seen at the base of a leaf of *Taxus* or the apparent constriction in *Sequoia sempervirens* due to the twisting of the leaf as described by Gardner.[2] We consider therefore that we have not sufficient evidence to assign these specimens with certainty to any particular genus, though we believe them to be probably identical with the specimen figured by Gardner in 1887.

PODOCARPINEÆ.

*Podocarpus Campbelli* Gard. A few rather poor specimens of leaves are most probably of the same species as the better preserved examples collected by Gardner and named by him *Podocarpus Campbelli*.[3] Gardner regards it as significant that similar leaves have not been found in the Arctic floras described by Heer. More recently Schindehütte[4] has recorded what he considers to be the same species from the Lower Miocene near Homberg, the flora as a whole being sub-tropical. This may be evidence of a southward migration of the species after Eocene times. In no instance have reproductive organs been discovered, and the absolute identity of the German and Mull specimens cannot be considered certain. Gardner, however, had no doubt as to the generic identity of his specimens with *Podocarpus*.

FIG. 11.
*Pagiophyllum Sternbergi* (Goepp.).
Nat. size. [T. 2910 E.].

? ARAUCARINEÆ.

*Pagiopyllum Sternbergi* (Goepp.). The piece of vegetative shoot represented in Fig. 11 appears to be identical with a rather larger specimen figured by Gardner[5] from Mull as *Cryptomeria Sternbergii* (Goepp.). Under this name Gardner also figured several good specimens of shoots and cones from Ireland,[6] and it was the close resemblance of the fossil cones to those of the recent Japanese conifer *Cryptomeria japonica* that led to the inclusion of the Irish specimens in the genus *Cryptomeria*. Unfortunately no cones of the *Cryptomeria* type, or of any other type, have been recorded from Mull, but only a few pieces of sterile branches. The majority of the Irish fossils are characterized by leaves which seem to be identical in their almost straight form and in arrangement with those of *Cryptomeria*

---

[1] Gardner (1887), p. 289, Pl. XIII., Fig. 1.
[2] Gardner (1886), p. 101.
[3] *Ibid.*, p. 97, Pl. XXVI.
[4] Schindehütte (1907)), p. 15, Pl. 1, Fig. 3.
[5] Gardner (1886), Pl. X., Fig. 3.
[6] *Ibid.*, Pl. X., Figs. 10-13; Pls. XX., XXI.

*japonica*. On the other hand, in the Mull impressions the leaves are rather stouter and relatively shorter and distinctly falcate as in twigs of the recent *Araucaria excelsa* and allied species. In the absence of cones it would be somewhat rash to assign the fossil fragments to *Araucaria*, though there is little doubt that the genus existed in the Eocene floras of western Europe. We prefer to adopt provisionally the non-committal generic name *Pagiophyllum*. In our opinion the Mull specimens are identical with the species instituted by Goeppert as *Araucarites Sternbergii* and by some authors subsequently transferred to the genus *Doliostrobus*.[1]

### Coniferales Incertæ Sedis.

*Elatocladus Campbelli* (Forb.). The Mull specimen on which Forbes[2] founded this species under the name *Taxites* (?) *Campbelli* is part of a vegetative shoot with lateral branches having spirally disposed linear leaves in two ranks. Forbes's illustration adequately represents the specimen, which is now in the Jermyn Street Museum, and shows the decurrent and twisted leaf-bases. Gardner[3] at first regarded this species as *Sequoia Langsdorfii*, but after examining additional material he reverted to the opinion of Forbes and proposed the generic name *Taxus*.[4] Reference has already been made to a small twig figured by Gardner as *Sequoia Langsdorfii*[5] and compared by him with Heer's Greenland specimens. Fig. 10 illustrates what we believe to be the *Sequoia* type. The Mull collection also includes four indistinct specimens from Ardtun which appear to be identical with *Taxites Campbelli*. We are not, however, convinced that Gardner has definitely proved this to be an undoubted species of *Taxus*, and in the absence of stronger evidence we prefer to adopt Halle's provisional generic term *Elatocladus*.[6]

### Coniferous Wood.

A. *Cupressinoxylon* sp. Some pieces of lignitised wood from the upright tree described and figured by Macculloch[7] (see Frontispiece, and pp. 43, 111 of this Memoir), and subsequently mentioned by Gardner,[8] were submitted to us for examination. They present the appearance of charred wood and are very friable in texture; the tissues are relatively soft and not petrified. By suitable treatment it was possible to prepare sections with a razor. Though the preservation is far from perfect the anatomical features are sufficiently well shown to enable us to refer the tree to the comprehensive genus *Cupressinoxylon*.[9] This generic type includes members of the Cupressineæ, Podocarpineæ, and some other recent Conifers. It is clearly impossible to assign the wood with any confidence to a

[1] *See* Seward (1919), p. 267.
[2] Forbes (1851), p. 103, Pl. II., Figs. 1A, 1B.
[3] Gardner (1886), p. 41, Pl. X., Figs. 1, 1A.
[4] *Ibid.*, p. 101, Pl. XXVII., Figs. 1-3.
[5] Gardner (1887), p. 289, Pl. XIII., Fig. 1.
[6] Halle (1913) p. 82. *See also* Seward (1919), pp. 352, 417, 429.
[7] Macculloch (1819), Pl. XXI. Fig. 1.
[8] Gardner (1887), p. 283, footnote.
[9] Seward (1919), p. 186.

particular type of vegetative shoot. The trunk may be that of the tree which bore the foliage shoots identified as *Cupressites MacHenryi* or it is possible that the vegetative shoots were those named *Sequoiites* (?) *Langsdorfi*. The anatomical characters of imperfectly preserved specimens do not afford any satisfactory means of distinguishing between wood of the *Sequoia* type and that of *Cupressinoxylon*. The wood is clearly not that of a Pine or, in all probability, of any other member of the Abietineæ ; it is certainly not Araucarian and it is not the wood of a *Taxus*.

The annual rings are well marked, but owing to the crushed condition of the tissues it is impossible to give a detailed description of the spring and summer wood. There are no resin-canals. The frequent occurrence of dark brown material, presumably resin, filling the cavities of cells in vertical rows, with an occasional transverse wall preserved, indicates the presence of resiniferous xylem-parenchyma which is not confined to any particular region. There is usually a single row of circular bordered pits on the radial walls of the tracheids ; rarely the pits are in contact and slightly flattened. Rims of Sanio are occasionally preserved. The medullary rays are numerous and uniseriate, and deep. One or two pits occur in the field (*i.e.* the area bounded by the radial walls of a ray cells and the vertical walls of a tracheid as seen in a radial section) which are either oval or circular : in the oval pits the major axis is approximately radial (horizontal). There are no pits in the tangential or transverse walls of the ray cells.

B. Fragments of Coniferous wood enclosed in lava from river mouth, S.E. of Tavool House.

The occurrence of these fragments is dealt with on p. 113. Their preservation is much less satisfactory and we cannot with any confidence refer them to a genus. The annual rings are numerous and narrow, the narrower late summer tracheids, which are very sharply contrasted from the larger spring tracheids, being confined to a very narrow zone in each ring. There are no resin-canals and no indication of resiniferous parenchyma. In the absence of such characters as the pitting of the tracheids and the pitting of the medullary-ray cells it is impossible to attempt a diagnosis of the material.

## Angiospermæ : Dicotyledones.

### Betulaceæ.

*Corylites hebridica* sp. nov. A careful examination of the type of Forbes's species *Alnites MacQuarrii* in the Jermyn Street Museum shows that, with the exception of the slightly cordate base, the leaf is incomplete and the apparent teeth on the margin are the result of tearing of the lamina. There is one [1] imperfect leaf in the present collection which is probably identical with *Alnites MacQuarrii* as figured by Forbes, but in the absence of more adequate material we cannot make any definite suggestion as to the true nature of this species. The specimen is not, we believe, specifically identical with

[1] Specimen No. T. $297_2$ E.

the majority of leaves figured by later authors under the name *Corylus MacQuarrii*, though it was taken by Heer[1] as the type of this species and other writers have so regarded it. Moreover the recently collected leaves which we believe to be examples of the *Corylites* type, and which are undoubtedly identical with *Corylites MacQuarrii* as figured by Gardner,[2] are not in our opinion specifically identical with the torn specimen figured by Forbes. We give expression to this view by adopting the name *Corylites hebridica* for the leaves sent to us for examination. The characters of these leaves vary within fairly wide limits, but the following account may be taken as typical of the majority. The base is more or less cordate and rather narrow, the lamina increasing gradually and evenly in width up to just over half way towards the apex, after which it tapers gradually with a slight inward curve to the tip. The smallest leaves are hardly more than 5 cm. long by 3 cm. at their

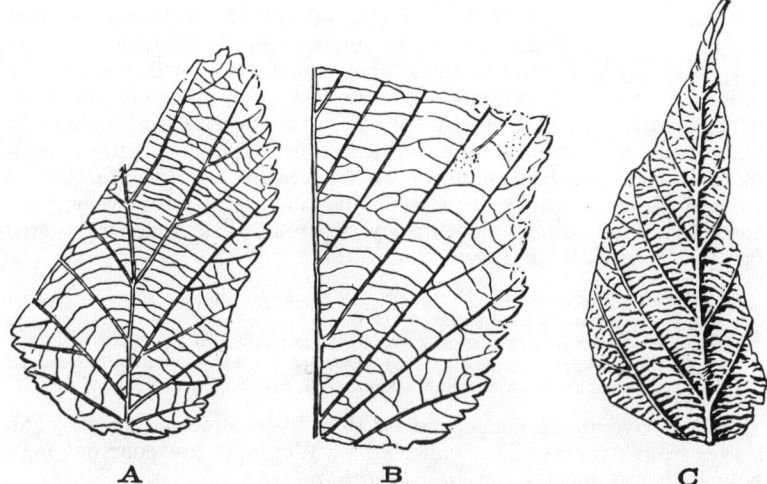

FIG. 12.—*Corylites hebridica* sp. nov. Nat. size. [T. 2968 E, T. 2996 E, T. 2997 E.].

greatest width, while the largest leaf in the British Museum collection is almost 13 cm. long by 10 cm. broad. One leaf which is 10 cm. long is only about 4 cm. broad; in all cases the length is greater than the breadth. The secondary veins are numerous, 10 to 14, and almost straight; the lower ones are closer and at a larger angle to the midrib; on their lower side they bear strong tertiary veins which pass into marginal teeth. The teeth are broad and short, with usually a sharp point; those at the ends of the secondary veins are generally larger than the others and often somewhat more prominent though never so prominent as they normally are in most recent species of *Corylus*.

The specimens represented in Figs. 12A and 12B show the marginal teeth; Fig. 12A shows part of a very slightly cordate base. Fig. 12c represents the acuminate apex of a young leaf.

[1] Heer (1868), p. 104, Pl. IX., Fig 1.
[2] Gardner (1887), Pl. XV., Fig. 3.

## Chapter IV.—Tertiary Plants.

The question of referring these leaves to a recent genus is a very difficult one. Gardner[1] remarks that "the total absence everywhere in the Eocene of anything like nuts, and their abundance in the Pliocene, renders it difficult to believe that the genus *Corylus* was actually in existence during early Tertiary periods." Heer's figures[2] of Hazel nuts are certainly not convincing. The fossil shown in Fig. 13 may possibly be a nut split open, but it is not well enough preserved to be identified with certainty.

On leaf characters alone the recent genera *Carpinus, Betula,* and *Alnus* cannot confidently be separated.[3] Many of the *Corylites*

FIG. 13.
(?) Nut split open.
× 1½. [T. 2938 E.].

leaves are very much like those of *Carpinus Betulus*, though the largest and broadest leaves are far nearer the *Corylus* type. The recent species, the leaves of which appear to agree most nearly in form and in range of variation with these fossils, is *Corylus rostrata*, though a striking point of difference is the smaller number of secondary veins. *Corylus Avellana* has on the average much broader leaves and the characters of the edge and apex are different.

A comparison of these leaves with other fossils is also difficult. One of us has carefully examined a number of Heer's figured specimens in the British Museum of what he called *Corylus MacQuarrii* from various localities in Greenland. Nearly all are very poor specimens and often inaccurately figured. The following agree fairly closely with the leaves from Mull :—

*Corylus MacQuarrii.* Heer, *Flor. Foss. Arct.* vol. v., Pt. I. (Grinnell Land.); especially Pl. VI., Figs. 4 and 6.
*C. MacQuarrii.* Heer, *Flor. Foss. Alaskana* (1869), Pl. IV.
*C. MacQuarrii.* Heer, *Flor. Foss. Arct.* vol. i. All the leaves referred to this type except those from the Mackenzie River which are very doubtful.

The above are all considered by Heer to be Miocene in age. The leaves from Atanikerdluk[4] collected by Whymper are poor but may belong to the species *Corylites hebridica*.

*Corylus* leaves from Menat (Oligocene or Eocene) described by Laurent[5] do not agree very closely with the Mull specimens: that shown on his Plate VI., Fig. 6 is perhaps the nearest. They have fewer secondary veins, like the recent species of *Corylus*, and show other differences. One cannot be certain that these leaves are not a variety of *C. hebridica*, but they certainly differ notably from the usual Mull type. Menzel's leaves from the Posener Ton[6] (Upper Miocene) are somewhat like the smaller, narrower leaves from Mull but they appear to agree more closely with *Carpinus* than with *C. hebridica*.

There are various references of leaves by American authors to *C. MacQuarrii*, but none of them are very satisfactory. One of

---
[1] Gardner (1887), p. 290.
[2] Heer (1868), Pl. IX., Fig. 5; (1874), Pl. III. Fig. 10.
[3] *See* Reimann (1917), pp. 21-25.
[4] Heer (1869), p. 469, Pl. XLIV., Fig. 11A; Pl. XLV., Fig. 6B.
[5] Laurent (1912), pp. 79-83, Pl. VI. Figs. 5, 6; Pl. VIII., Figs. 3, 4, 5; Pl. IX., Fig. 1.
[6] Menzel (1910), Pls. XII.-XV.

*Angiospermæ : Dicotyledones.*

Lesquereux's figures [1] may be of the same type as the Mull leaves. Newberry's leaves,[2] which he calls *C. MacQuarrii*, are less like the Mull type than those referred to *C. americana fossilis* and *C. rostrata fossilis*. Ward gives some good figures of leaves referred to *Corylus*, but his one leaf of *C. MacQuarrii* [3] is hardly to be distinguished from his *C. americana fossilis* which closely resembles one of Laurent's figures [4] which he calls *C. MacQuarrii*.

The comparative study of these American leaves is not easy; so few examples are given from each locality that it is impossible to get a general idea of the range of form of each type. In general it may be said that none of the figures mentioned shows a really close agreement with the Mull *Corylites*. However the experience gained by examining specimens of one recent species from different localities indicates that it is not impossible that some of the American fossil leaves may be varieties of the species found at Mull. But one cannot say exactly how much latitude should be allowed in this matter.

Though we cannot assert, from a botanical point of view, that the Mull leaves are proved to belong to the genus *Corylus*, we think that they agree more closely with that genus than with any other and may reasonably be included in *Corylites*.

### *Dicotyledonous Wood (? Betulaceæ).*

We have examined sections of portions of a prostrate trunk which occurs at Uamh Mhic Cuill, Rudha na h-Uamha, enclosed in the same lava as the *Cupressinoxylon* described above (*see* also p. 112).

The wood appears to be most nearly related to that of members of the Betulaceæ, particularly *Betula*, but there are certain differences. It agrees with *Betula* in the radial groups of vessels, the oblique ends of the vessels being crossed by numerous fine scalariform bars, and in the number and size of the medullary rays some of which are uniseriate and some as much as four cells in width. No aggregate rays are present in the sections examined. Vessels appear to be evenly distributed throughout the transverse section. No annual rings have been detected, but only a small part of the section shows well-preserved structure. Points of difference from *Betula* are the presence of irregular large scalariform or rounded bordered pits on the wall of contact of one vessel with another, instead of the closely packed very small pits of *Betula*, and a peculiar ray character which is present to a small extent in *Betula alba*, but more marked in species of *Fagus*. The broader rays of the fossil wood consist of narrow cells, radially elongated, with one or more rays of larger cells, not radially elongated above and below them. The cells of the uniseriate rays are also of this latter kind, and it frequently happens that two broader rays are connected vertically by a plate of the larger cells. In the Birch there are indications

---

[1] Lesquereux (1883), Pl. XLIX., Fig. 4.
[2] Newberry (1898), Pl. XXXII. Fig. 5; Pl. XLVIII., Fig. 4.
[3] Ward (1887), Pl. XIII., Fig 7.
[4] Laurent (1912), Pl., VI. Fig. 6.

of this but the disparity of size between the two kinds of cells is not so great.[1]

It is hoped to publish a more detailed description of this wood in the near future. At present we refrain from giving it a name.

FAGACEÆ.

*Quercus greenlandica* Heer. Gardner[2] in his paper of 1887 refers to a "new and very rare leaf at Mull" which "is perhaps to be identified with the *Quercus greenlandica* Heer, from Atanikerdluk, some of the specimens of which seem, however, to have been placed in *Castanea Ungeri* by that author, though they do not resemble the Miocene chestnuts of either Europe or America." Among the leaf impressions in the red-brown matrix there are many which appear to us to be identical with Heer's *Q. greenlandica*[3] and with Gardner's figured leaf. These impressions are all fragmentary and of varying size. The largest was probably about 10 cm. long with at least 14 secondary veins on either side, and the smallest is about 5 cm. long with 10 pairs of secondary veins.

There appears to be no distinction between *Q. greenlandica* and *Fagus castaneæfolia* Ung. as figured by Heer in the *Flor. Foss. Arct.* Vol. I., Plate XLV., and the latter in his *Flor. Alaskana*[4] is identified with *Castanea Ungeri*. In Vol. VII., p. 85 of the *Flor. Foss. Arct.* he includes the leaves figured in Vol. I. of the same work (Plate XLVI., Figs. 1-3) as *Fagus dentata* Ung. with *Castanea Ungeri*.

It is extremely difficult to separate these leaves into different species, especially when one takes into account the close resemblance between leaves of some existing species of *Quercus* and those of species of *Castanea*, and the small amount of difference between the leaves of different species of the latter genus. Thus it is possible that Heer is justified in separating the leaves called by him *Castanea Ungeri*, though some of those referred to *Fagus castaneæfolia* are indistinguishable from *Q. greenlandica*. He records but does not figure flowers and fruit of *C. Ungeri* from Atanikerdluk.[5]

There is one reference of leaves found in America to this species by Newberry[6] from the Miocene of Alaska, but this seems to us very doubtful, the leaves being long and narrow, with teeth more like those usually found in *Castanea*.

As several authors have remarked, it is impossible to decide definitely on leaf characters alone, whether a doubtful leaf belongs to the genus *Quercus* or to *Castanea*. The very blunt and prominent teeth of *Q. greenlandica* resemble very closely those of *Q. Prinus* and other oaks and cannot be matched in any Castanea: though not absolutely certain, it would seem very probable that we are here dealing with a true *Quercus*.

---

[1] In the examination of this wood we have received valuable assistance from Mr J. Line, of Emmanuel College, Cambridge.
[2] Gardner (1887), p. 291, Pl. XIV. Fig. 2.
[3] Heer (1868), p. 108, Pl. VIII. Fig. 8; Pl. X., Figs. 3, 4; Pl. IX., Fig. 4; Pl. XLVII., Fig. 1.
[4] Heer (1869²), p. 32.
[5] Heer (1883), p. 84.
[6] Newberry (1898), p. 75, Pl. LI., Fig. 3; Pl. LIV., Figs. 1, 2.

# Angiospermæ : Dicotyledones.

## PLATANACEÆ.

*Platanus hebridica* (Forbes). The only pubished figures of leaves of this species are those of Forbes,[1] both of which show incomplete specimens, so that it is doubtful whether any of the edge is shown in either. He remarks that "this leaf is one of the most abundant and characteristic of all those found in Ardtun." Gardner gives a further description and records the presence of numerous male flowers and detached anthers, some of which he illustrates;[2] also one cluster of female flowers.

The leaves vary considerably in size and shape. Gardner records one 37 cm. in length. In the present collection we have only three undoubted specimens, all imperfect; the smallest of these, which shows the edge of the lamina most clearly is represented in Fig. 14.

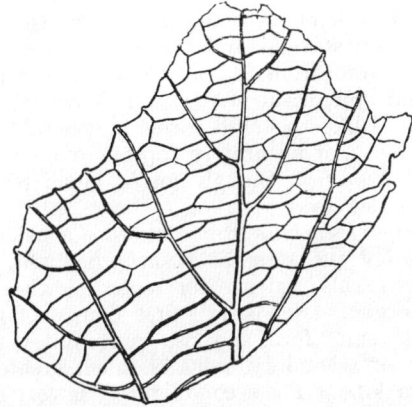

FIG. 14.—*Platanus hebridica* (Forbes). Nat. size. [T. 3022 E.]

Of the leaves in the British Museum many are nearly 20 cm. in length, and are all imperfect. They are typically three-lobed, the lateral lobes being usually rather narrow and ending in sharp points. There occur also large leaves in which these lateral lobes are almost suppressed. The edge is nearly smooth, or with small projections at the ends of the main veins (Fig. 14). There is nothing to be seen of the large prominent recurved teeth of recent species. The two main lateral veins branch from the midrib at the base of the leaf (often not opposite each other) and usually curve inwards slightly. They bear numerous smaller veins on their lower sides, which end in the small marginal teeth. There are only a few other secondary veins branching from the midrib. The leaf lamina may be decurrent as much as 2 cm. below the junction of the main lateral veins, forming a cuneate base, but in the broader, strongly lobed leaves it is less decurrent and the base is almost square to the petiole. Heer[3] pronounced these leaves to be probably *Platanus aceroides* Goepp. though he had only seen Forbes's figures and not the actual

[1] Forbes (1851), Pl. III., Fig. 5; Pl. IV. Fig. 1.
[2] Gardner (1887), Pl. XIII., Figs. 13-15.
[3] Heer (1859), pp. 313, 314.

84       Chapter IV.—Tertiary Plants.

specimens : this identification was used by him as an argument in favour of the Miocene age of the leaf-beds. Goeppert's leaves from Schossnitz,[1] including the original of *P. aceroides*, which he divided into several forms, are probably all one species and are much more akin to the recent *P. occidentalis* than to the Mull leaves.

Gardner[2] says of the *Platanus* leaves " they (*i.e.* the Mull species) have been collected at Atanikerdluk, but of much smaller size, by Whymper, Colomb, and others, and being wholly different from any Miocene form, should bear the name given by Forbes." Heer's Atanikerdluk leaves[3] may be identical with *P. hebridica*, but they are too imperfect to be determined with certainty. The same remark applies to those from Iceland.[4] The Mackenzie River leaves[5] are still less satisfactory. We have examined the specimens from Disko Island[6] called by Heer *P. Guillelmæ* Goepp. : these are all very poor impressions, the best being shown on Plate XLVII., Fig. 1 (even in this specimen the edge is very indistinct). The veins are very slender and straight and the general appearance of these leaves is very different from that of those from Mull. The leaves from Hare Island (off Disko I.) (Lower Miocene) referred to *P. aceroides*[7] are more like the Mull forms in general appearance, but the large sharp teeth, if accurately represented, are very different from those of *P. hebridica*, though not so much recurved as those on the Schossnitz leaves.

Newberry's large *P. nobilis* from Dakota is considered by Knowlton to be probably the same as *P. hebridica* and he quotes this supposed identity in his evidence for regarding the American Fort Union beds as Eocene.[8] Newberry's first figure[9] of *P. nobilis* shows a leaf of much the same form as those from Mull, with similar edge, but the number of secondary veins is much greater and the leaf has a small extra lobe. The second figure[10] is very much more like the Mull leaves, as is also that of Ward[11] from the Laramie beds. It is quite likely therefore that both the American and the Scottish leaves belong to the same species ; they are certainly closely related. Perhaps also *P. Raynoldsii* var. *integrifolia* Lesq.[12] from Golden, Colorado, and Black Buttes, Wyoming, both localities regarded as Lower Eocene, may be included in the same group. Lesquereux's *P. aceroides* and *P. Guillelmæ*,[13] which are hardly to be distinguished from each other, as he says, are both from his third group considered to be Lower or Middle Miocene. Both these agree far better with Goeppert's Schossnitz leaves than with *P. hebridica*. Knowlton's *P. aceroides latifolia*[14] from the Raton formation

[1] Gœppert (1855), Pls. IX., X., XI.
[2] Gardner (1887), p. 290.
[3] Heer (1868), Pl. XII., Figs., 1-8 ; Pl. XLVII., Fig. 3.
[4] *Ibid*, Pl. XXXII.
[5] *Ibid*, Pl. XXI., Fig. 17B ; Pl. XXIII., Figs. 2B, 4.
[6] Heer (1869), p. 473, Pls. XLVII., XLVIII., XLIX., Figs. 4B, C, D.
[7] Heer (1883), p. 96, Pl. XC.
[8] Knowlton (1909), p. 225.
[9] Newberry (1898), Pl. XXXIV.
[10] *Ibid*,. Pl. L.
[11] Ward (1887), Pl. XVI.
[12] Lesquereux (1878), p. 185, Pl. XXVI., Figs. 4, 5 ; Pl. XXVII., Figs. 1-3.
[13] *Ibid*, pp. 183-185, Pl. XXV.
[14] Knowlton (1917), p. 321, Pls. XCII., XCIII., XCIV.

PLATE II.—*Phyllites ardtunensis* sp. nov. ⅔ Nat. size. [British Museum.]

(Eocene) is very like the Mull leaves in form and venation as also in size.

Lesquereux's *P. primæva*[1] from the Cretaceous Dakota group is a very interesting species. Its leaves vary greatly in form from the variety *grandidentata*, hardly to be distinguished from Heer's *P. aceroides* from Hare Island,[2] to a variety with almost entire leaves. One leaf is recorded as 17 cm. long by 20 cm. in width, but those figured are all much smaller. Lesquereux considers this species to be essentially the same type as *P. Guillelmæ* and *P. occidentalis* and we think he is very likely correct. This species is perhaps the ancestor of the types widely spread in Miocene and recent times, and possibly also of the large leaved type including the American *P. nobilis* and the British *P. hebridica*. This latter type would seem to have disappeared early, as it is not recorded later than the Eocene.

Upper Cretaceous leaves from Kunstadt described by Krasser[3] are very like the less markedly lobed leaves of *Platanus hebridica*; there is close agreement in venation, margin, and base, and many of the Kunstadt leaves are large. Krasser describes eight leaf forms, admitting that it is possible that all belong to one botanical species.

### Dicotyledones Incertæ Sedis.

*Phyllites platania* (Heer). One leaf impression in the recently obtained collection, though showing no edge, is very similar in venation and size to some large leaves obtained by Gardner. These leaves agree closely with those called by Heer *Quercus platania* from Atanikerdluk[4] though the veins of the latter in the specimens examined in the British Museum are finer and less strongly marked. Schimper[5] remarks that the reference of these leaves to *Quercus* is "fort contestable," and with this we agree though we are unable to suggest their true position. A complete leaf from Ardtun in the British Museum measures 18 cm. by 9 cm. and a larger one, which is incomplete, is 13 cm. in width. Dawson[6] records this species from the Laramie of Bow River (Canada) and figures a comparatively small complete leaf. He notes that the leaves appear to be variable, and that *Q. platania* was "evidently one of the most magnificent of the Laramie species in point of size."

*Phyllites ardtunensis* sp. nov. The British Museum collection of Mull plants includes numerous well-preserved impressions of leaves of the type figured by Gardner on Plate XIV. (Fig. 3) of his paper published in 1887, though most of them are larger than his figure. The specimen reproduced in Plate II. (British Museum coll.) is 15 cm. long. Among the fossils sent to us there are two very imperfect specimens which belong to this type. The leaves

---

[1] Lesquereux (1892), p. 72, Pls. VIII.-X.
[2] Heer (1883), Pl. XC.
[3] Krasser (1896), pp. 138-144, Pls. XII.-XV.
[4] Heer (1868), p. 109, Pl. XI., Fig. 5; Pl. XLVI., Figs. 8, 9; (1869), p. 472, Pl. XLVI., Fig. 5; Pl. LV., Fig. 3; (1883), p. 91, Pl. LXVIII., Fig. 1.
[5] Schimper (1870), II., p. 657.
[6] Dawson (1889), p. 72, Pl. XI.

are ovate, usually with a broad, almost straight base at right-angle.
to the petiole. At the apex the lamina tapers off evenly to a point.
On the edge are numerous small, not quite evenly placed, strongly
recurved narrow teeth. The secondary veins are numerous and
strongly marked. They are more crowded at the base where they
are almost straight and nearly at right-angles to the midrib; higher
up they are distinctly curved. Near the margin of the leaf the
secondary veins usually fork in a very characteristic fashion, sometimes twice or even three times, the ends of their branches entering
the marginal teeth. Of these leaves, Gardner remarks that they
seem as yet peculiar, and we can only repeat that statement. In
searching the records of both American and European Tertiary
floras we cannot find any leaves which at all closely resemble them.

Among fossil leaves that most nearly resemble them in shape
and in the arrangement of the secondary veins is Saporta's *Fagus
pristina*,[1] but the teeth are absent and there is no branching of
secondary veins. The texture was probably something like that of
leaves of *Fagus* or *Castanea* and the teeth of some species of the
latter genus are very like those of the fossil examples, but the general
form is very different. The branching of the veins is similar to
that often seen in *Ulmus effusa* and the recurved teeth of that species
sometimes resemble those of *Phyllites ardtunensis*, but are double
instead of single. The closest resemblance in leaf form and arrangement of the secondary veins that we have seen among recent leaves
is in the Chinese genus *Davidia* (Nyssaceæ); but the leaves of *Davidia*
are smaller.

*Phyllites* spp. The Survey collection includes leaves undoubtedly
specifically identical with those figured by Gardner (1887) in his
Figs. 3 and 4, Plate XVI., also larger leaves probably of the same
species as those named by Forbes [2] *Rhamnites major* and *R. multinervatus*; the latter may be an unexpanded leaf of *R. major*. Of
the former leaves, Gardner says,[3] "there are a number of simple
ovate leaves resembling those of the Bay and Laurel etc., as well
as the *Rhamnites* of Forbes; but it seems scarcely probable that
any of these are capable of generic identification from the leaves
alone." The description of the drawings of these leaves is "evergreen leaves, like those of the Myrtaceae."

We feel at present entirely unable even to suggest the probable
genera to which these leaves belong, and as in this case the leaves
are of no use for comparison of the Mull flora with those of other
localities we do not propose to describe the fossils in detail.

PLANTÆ INCERTÆ SEDIS.

*Specimen A.* The fragment represented in Fig. 15A is part of
a lamina characterised by numerous very oblique and rarely forked
veins; those on one side being rather more crowded than those
on the other. No definite midrib is recognisable, but it is not improbable that lower in the lamina the line where the convergent

[1] Saporta (1893), Pl. V., Fig. 5.
[2] Forbes (1851), Pl. III. Figs. 2, 3.
[3] Gardner (1887), p. 291.

## Incertæ Sedis.

veins meet was represented by a median strand. It is impossible to determine whether the fragment is a portion of a leaf or a phylloclade; the venation agrees closely with that of the phylloclades of the recent Conifer *Phyllocladus*. Ettingshausen[1] in an account of the Tertiary flora of Australia refers some specimens, much larger than ours and showing both axes and flattened appendages (probably phylloclades), to *Phyllocladus asplenioides* in which the phylloclades resemble but are not identical with the Mull fossil. Numerous examples of impressions of leaves or phylloclades, in

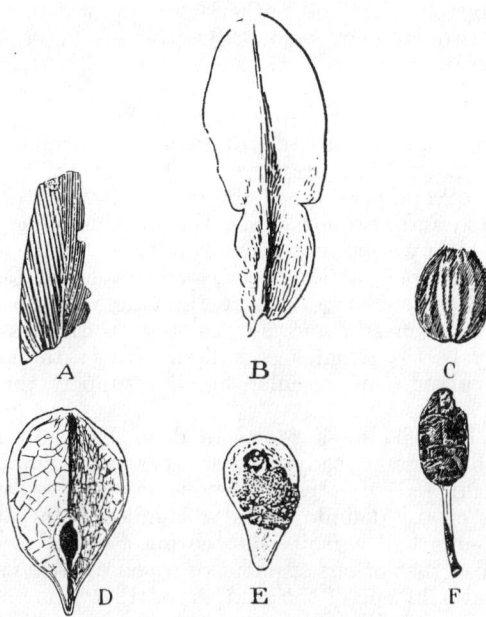

FIG. 15.—*Plantæ Incertæ Sedis.*

Specimen A. Part of a Lamina. ×1½. [T. 2055 D.].
Specimen B. Nat. size. [T. 2068 D.]
Specimen C. ×2. [T. 2893 E.].
Specimen D. ×3. [T. 2895 E.].
Specimen E. ×6. [T. 2045 D.].
Specimen F. Nat. size. [T. 2945 E.].

most cases detached and not borne on an axis, are figured by American authors[2] from Upper Cretaceous beds under the generic name *Thinnfeldia*, *Phyllocladus*, or *Protophyllocladus*.. We are inclined to think that the American fossils are true leaves and not flattened branches (phylloclades). The fragment shown in Fig. 15A is very similar to the species *Phyllocladus subintegrifolius* Lesq., which was afterwards transferred by Berry to his genus *Protophyllocladus*, and at an earlier date assigned by Newberry to *Thinnfeldia*, a genus characteristic of Jurassic and Rhaetic rocks.

[1] Ettingshausen (1886), p. 94, Pl. VIII., Figs. 28-31.
[2] For references to records of American species *see* Seward (1919), pp. 414, 415.

We propose to refrain from giving a name to the Mull specimen as it is too small to determine with any degree of confidence, but we are inclined to think it may be identical with, or at least closely allied to, *Protophyllocladus subintegrifolius* (Lesq.).

*Specimen B.* The specimen shown in Fig. 15B is an impression on black shale of a leaf-like organ 2 cm. long, with a thick prominent keel at the smaller end which dies out distally. It is characterized by a partial division of the lamina into a smaller proximal and a larger ovate distal end. No veins have been detected. It may be a bract or possibly a leaf of leathery texture which was partially divided into two laminar regions like the leaves of some recent species of *Citrus*.

*Specimen C.* The specimen shown in Fig. 15c appears to be a fruit or seed, ·5 cm. by ·4 cm. Its surface is covered by a film of carbonaceous matter irregularly cracked, with no apparent vascular bundles except possibly at the edge. It has irregular ridges and is in shape and size not unlike the fruit of *Carpinus Betulus*. The absence of any signs of vascular strands on the ridges or of any scar of attachment at the base seems to rule out *Carpinus* as a possibility. Mrs. Reid suggests that the ridges are merely due to the nature of the matrix and not to any special features of the fruit. It may be the remains of a fleshy fruit : the carbonaceous film and the absence of vascular bundles support this view.

*Specimen D.* The fossil represented in Fig. 15D resembles a scale-leaf with a smaller body attached near its pointed end. The surface is composed of a carbonaceous layer irregularly cracked with no trace of any definite vascular bundles. Mrs Reid suggests that it represents half a fruit pod bearing a single seed, the form being not unlike that of one segment of a pod of *Hedysarum* though the mode of attachment of the seed shows that it is not a member of the Leguminosæ. We think that this is a probable explanation of the specimen, but we do not feel able to determine its systematic position.

*Specimen E.* A small scale rounded at one end and bluntly pointed at the other is shown in Fig. 15E. Near the rounded end there is a definite circular scar as if some organ had been attached. A mass of black material covers part of the surface. We cannot make any definite statement as to the nature of the specimen.

*Specimen F.* The stalked object shown in Fig. 15F consists of a comparatively thick mass of carbonaceous material irregularly grooved. The size and shape suggest comparison with an old ' cone ' of an Alder, but as it is impossible to make out any definite scales this cannot be regarded as a satisfactory explanation of the specimen.

CONCLUSION.

The collection dealt with in this paper is by no means fully representative of the Mull flora. Some of the most interesting types, *e.g.* the leaf called by Gardner *Boehmeria antiqua*, are absent or represented by very poor specimens in the material which we have examined. A few very imperfect leaf fragments we have been unable to identify and have not described them.

The vegetation from the leaf-beds of Ardtun indicates that the climate was temperate like that of the regions in which at present summer deciduous forest is typically developed. The great majority of the leaves from Mull are those of deciduous trees such as *Platanus*, *Corylites*, and *Quercus* though the probable presence of such Conifers as *Podocarpus* and *Araucaria* points to a warmer climate than exists over most of the British Isles to-day. The possible presence of evergreen shrubs is another indication of a warm, fairly moist climate. The leaves were most probably preserved by falling into a still lake round the marshy borders of which *Onoclea* and *Equisetum* flourished.

We see no reason to dissent from Gardner's opinion, expressed in 1887, that the flora is Eocene ; it may well be, as Gardner believes Lower Eocene. A relationship with the Lower Tertiary floras of Greenland and other Arctic regions—regarded by Heer as Miocene, but by Gardner and other more recent writers as Eocene—is undoubtedly indicated by *Corylites, Quercus greenlandica*, and possibly by *Quercus Platania*. Relationship with the Laramie floras of North America is also suggested by *Onoclea, Platanus*, and possibly *Quercus Platania*.

The relationship to the sub-tropical Eocene floras of southern Europe is less marked : these according to Berry exhibit a closer resemblance to the Eocene floras of Carolina and Georgia.

## BIBLIOGRAPHY.

ARGYLL, DUKE OF (1851), On Tertiary Leaf-beds in the Isle of Mull. With a note on the Vegetable Remains from Ardtun Head by E. Forbes. *Quart. Journ. Geol. Soc.*, Vol. vii., p. 89.

BAILEY, W. H. (1869), Notice of Plant-remains from beds interstratified with the Basalt in the county of Antrim. *Quart. Journ. Geol. Soc.*, Vol. xxv., p. 357.

BERRY E. W. (1911), The Flora of the Raritan Formation. *Geol. Surv., New Jersey*, Bull iii.

—— (1914), The Upper Cretaceous and Eocene Floras of S. Carolina and Georgia. *U.S. Geol. Surv.*, Prof. Paper 84.

BERRY, E. W., and others (1916), Maryland Geological Survey : Upper Cretaceous.

CHRIST, H. (1897), Die Farnkräuter der Erde. Jena.

—— (1910), Die Geographie der Farne. Jena.

COCKERELL, T. D. A. (1921), Some British Fossil Insects. *Canadian Entomologist*, Vol. liii., p. 22.

DAWSON, SIR W. (1889), On Fossil Plants collected by Mr R. A. MacConnell on Mackenzie River, and by T. C. Weston on Bow River. *Trans. R. Soc., Canada*. Vol. vii., Sect. iv., p. 69.

ETTINGSHAUSEN, C. VON (1855), Die Tertiäre Flora von Häring im Tirol. *Abh. K.k.geol. Reichs.* Wien., Bd. ii., Abt. ii., No. 2.

—— (1886), Beiträge zur Kenntniss der Tertiärflora Australiens. *Denkschr. K. Akad. Wiss.* Wien., Bd. liii.

FORBES, E. (1851), *See* Duke of Argyll.

GARDNER, J. S., and C. VON ETTINGSHAUSEN (1882), Monograph of the British Eocene Flora. *Palæont. Soc.*, London.

GARDNER, J. S. (1886), *Ibid.*, Vol. ii.

—— (1887), On the leaf-beds and gravels of Ardtun, Carsaig, etc. in Mull, with notes by G. A. J. COLE, *Quart. Journ. Geol. Soc.*, Vol. xliii., p. 270.

## Chapter IV.—Tertiary Plants and Beetles.

GOEPPERT, H. R. (1855), Die Tertiäre Flora von Schossnitz in Schlesien, Görlitz.
HALLE, T. G. (1913), The Mesozoic Flora of Graham Land. *Wiss. Ergeb. Schwed. südpol. Exped.*, 1901-03, Bd. iii., Lief. 14.
HEER, O. (1859), Flora Tertiaria Helvetiæ, Bd. iii.
—— (1868), Flora Fossilis Arctica, Bd. i., Zürich.
—— (1869), Contributions to the fossil Flora of N. Greenland, being a description of plants collected by E. Whymper during the Summer of 1867. *Phil. Trans. R. Soc.* (Included in the *Flor. Foss. Arct.*, Bd. ii.).
—— (1869), Flora Fossilis Alaskana. (*Flor. Foss. Acad.*, Bd. ii.).
—— (1874), Nachträge zur Miocenen Flora Grönlands (*ibid.*, Bd. iii.).
—— (1883) Flor. Foss. Arct. Bd. vii.
HOLLICK, A. (1906), The Cretaceous Flora of southern New York and New England. *U.S. Geol. Surv.* Monograph 50.
KNOWLTON, F. H. (1898), Catalogue of the Cretaceous and Tertiary plants of North America. *U. S. Geol. Surv.*, Bull. 152.
—— (1902), Report on a small collection of fossil plants from the vicinity of Porcupine Buttes, Montana. *Bull. Torrey Bot. Club*, Vol. xxix.
—— (1909), On the stratigraphic relations and paleontology of the Hell Creek Beds, Ceratops Bed, and equivalents, and their reference to the Fort Union formation, *Proc. Washington. Acad. Sci.*, Vol. xi.
KNOWLTON, F. H., and W. T. LEE (1917), Geology and Paleontology of the Raton Mesa and other regions in Colorado and New Mexico., *U.S. Geol. Surv.* Prof. Paper 101.
KRASSER, F. (1896), Beiträge zur Kenntniss der Kreideflora von Kunstadt in Mähren, *Beit. Paleont; Geol. Österr.-Ungarn. und des Orients*, Bd. x., Heft. iii., Wien.
LAURENT, L. (1912), Flore fossile des Schistes de Menat, *Ann. Mus. d'hist. nat. Marseille*, Geol., Tome xiv.
LESQUEREUX, L. (1874), Contributions to the Fossil Flora of the Western Territories. Pt. i. The Cretaceous Flora, *Rep. U.S. Geol. Surv. Territ.*, Vol. vi.
—— (1878), *Ibid.*, Pt ii., The Tertiary Flora, *Rep. U.S. Geol. Surv. Territ.*, Vol. vii.
—— (1892), The Flora of the Dakota group, *U.S. Geol. Surv.* Monograph xvii.
MACCULLOCH, J. (1819) Description of the Western Islands of Scotland, London.
MENZEL, P. (1910), Pflanzenreste aus dem Posener Ton. *Jahrb. K. Preuss. Geol. Landesanst*, Bd. xxxi, Tl. i, Heft i.
NEWBERRY, J. S. (1895), The Flora of the Amboy Clays, *U.S. Geol. Surv.*, Monograph xxvi.
—— (1898), Later extinct floras of N. America, *U.S. Geol. Surv.*, Monograph xxxv.
REIMANN, H. (1917), Die Betulaceen und Ulmaceen des Schlesischen Tertiärs, *Jahrb. Preus. geol. Landesanst*, Bd. xxxviii., Tl. ii, Heft i.
SAPORTA, G.de (1873), Études sur la végétation du Sud-est de la France a l'époque Tertiare, *Ann. Sci. Nat.* [5] Tome xvii.
—— (1893), Révue des Travaux de Paleont. vég. parus en France dans le cours des ann. 1889-92, *Rév. Gén. Bot.*, Tome v.
SCHIMPER, W. P. (1870), Traité de Paléontologie Végétale. Tome ii.
SCHINDEHÜTTE, G. (1907), Die Tertiärflora des Basalttuffes von Eichelskopf be Homberg. *Abh. K. Preus. Geol. Landesanst*, N.F. Heft liv.
SEWARD, A.C. (1919), Fossil Plants, Vol. iv., Cambridge.
WARD, L. F. (1887), Types of the Laramie Flora, *U.S. Geol. Surv.*, Bull. 37.

A.C.S., R.E.H.

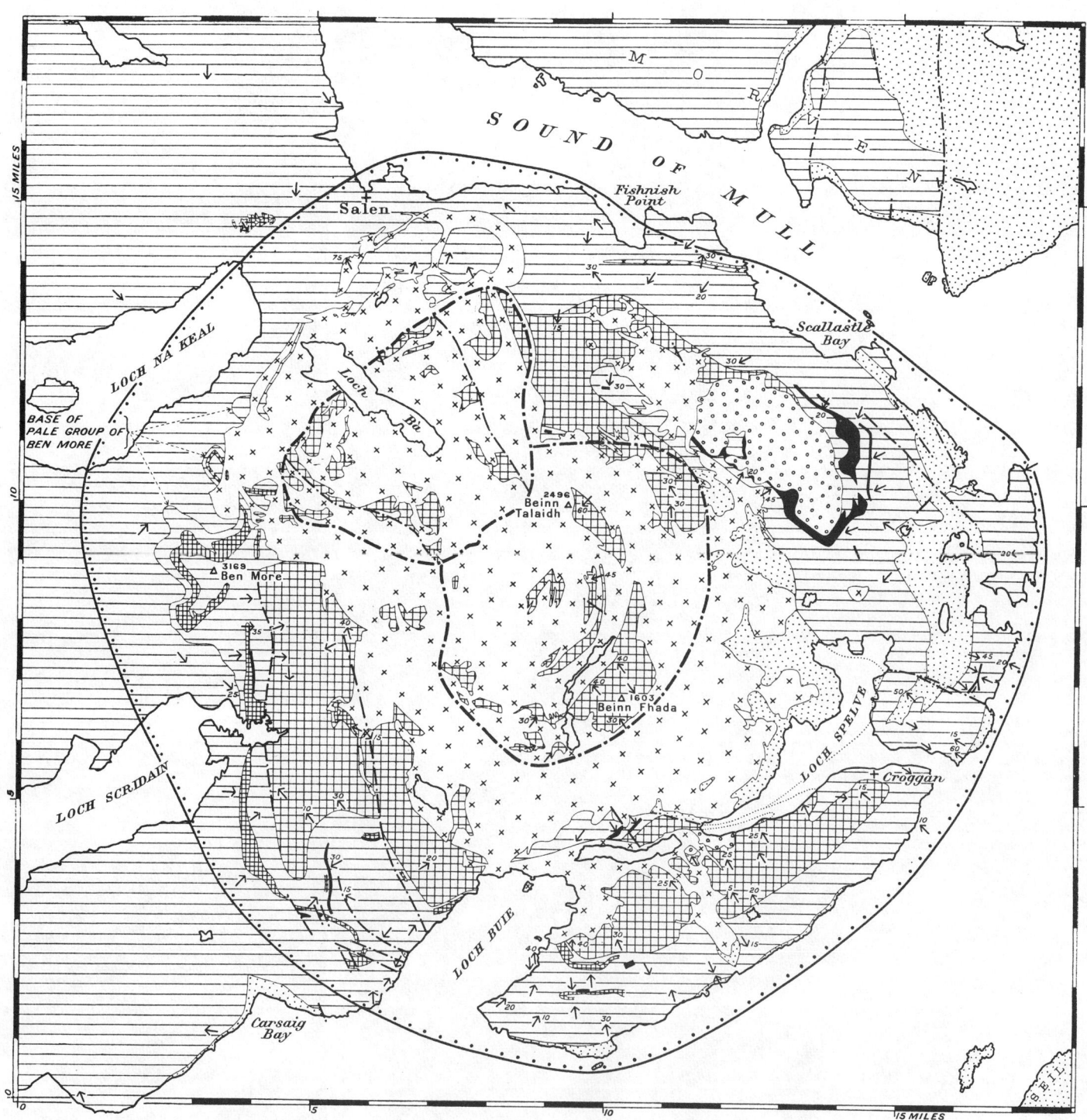

PLATE III. MAP OF LAVAS ABOUT MULL CENTRES.

# INDEX TO PLATE III.

### TERTIARY SURFACE ROCKS:—

#### Basalt-Lavas.

Agglomerate.    Common Central Types.    Big-Felspar Basalt.    Plateau Types.    Mugearite-Lavas.

### TERTIARY INTRUSIONS (including Vent-Agglomerates, but omitting Sheets and Dykes):—

### PRE-TERTIARY ROCKS:—

### SIGNS:—

Dips, in degrees.

—·—·— Margins of Main Calderas.

—·—·— Other Faults.

· · · · · Limit of Pneumatolysis.

### GENERALIZED SUCCESSION OF MULL LAVAS.

Central Group.

Pale Group of Ben More.

Plateau Group.

### DISTRICTS INDIVIDUALLY TREATED IN TEXT.

The subdivisions of sheet 44, are arranged as follows:—

1—4a. Outside Pneumatolysis Limit.

5—9. Between Pneumatolysis Limit and Central Calderas.

10. Central Calderas.

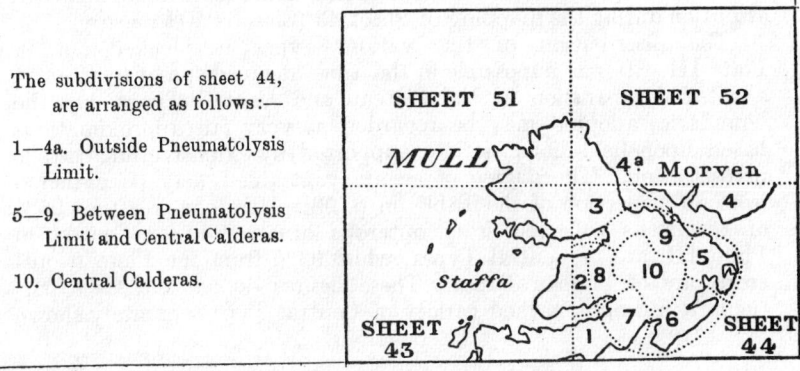

# CHAPTER V.
## BASALT AND MUGEARITE LAVAS: INTRODUCTION.
### Lava-Succession.

With the exception of a few relatively recent rhyolitic flows to be considered in Chapter XV., the lavas of Mull are basalts with a few mugearites. The latter (lettered W on the one-inch Map) are largely confined to a particular horizon, where, though subordinate in bulk, they constitute a distinctive unit within Sir Archibald Geikie's well-known Pale Group of Ben More.

The basalts fall into two main divisions, the one rich in olivine, the other poor. In general, both divisions are lettered B on the one-inch Map, though special types, as will be explained presently, are distinguished as B' and fB respectively. The olivine-rich basalts are classed together as Plateau Types since they constitute practically the whole of the lava-pile as preserved to us in the peripheral plateau regions of Mull except in the south-west corner of the island mostly included in Sheet 43 beyond the limits of Plate III. They also continue into the central mountainous region—and as a matter of fact reach to the summit of Ben More—, but in central Mull they are abundantly associated with olivine-poor varieties for which the designation Central Types has been reserved.

The petrology of the lavas is dealt with in Chapter X. It is sufficient here to mention that the mugearites are often characterized in the field by their pale weathering and marked fluxion; the Plateau Basalts by their olivine, fresh or decomposed as the case may be, and their non-porphyritic crystalline aspect; the Central Basalts by their porphyritic felspars or—where non-porphyritic—by their extremely compact texture. A minor, but interesting, feature of a few among the Plateau Basalt Lavas is the occurrence of contemporaneous segregation leading to inconspicuous pegmatitic veins. The phenomenon is widespread, and has received special attention during the mapping of Sheet 43 (Chapter VII., also p. 138).

The distribution of the various types is blocked out in Plate III. It was impossible in the time at our disposal to attempt a detailed separation of the Plateau and Central Types, but the boundaries adopted may be regarded as very fair approximations based upon field-diagnosis and supported by extensive microscopic investigation. The degree of accuracy attained may be gathered from an inspection of the Table on p. 93. Here one sees that 96% of specimens collected from outcrops ornamented as Central in Plate III. are of Central Types, while 93% from the Plateau outcrops are of Plateau Types. These figures do not include a particularly strongly marked variety of Central Type separately shown

## Lava-Succession.

in Plate III. as Big-Felspar Basalt (fB of one-inch Map). Several examples of this type have been mapped, and as they all fall distinctly within the Plateau outcrops they serve as a useful index of a definite, though restricted, interdigitation of Central and Plateau Types. Another example of this interdigitation is afforded by the prevalence of a compact, non-porphyritic, olivine-poor basalt, described in the sequel (p. 145) as Staffa Type, among the earlier, often columnar, lavas of south-west Mull (mostly restricted to Sheet 43).

TABLE X.—SYNOPSIS OF MULL BASALTS MICROSCOPICALLY EXAMINED FROM 1-INCH MAP, SHEET 44.

| District. | Ornament employed in Pl. III. | Number of Slides Examined. | | Number of Slides with— | |
|---|---|---|---|---|---|
| | | Plateau Types. | Central Types. | Olivine fresh. | Olivine all decomposed. |
| Area 1 | Plateau | 16 | 1 | 15 | 2 |
| ,, 2 | Plateau | 8 | 1 | 4 | 5 |
| ,, 3 | Plateau | 6 | Nil. | 5 | 1 |
| ,, 4 | Plateau | 8 | Nil. | 8 | Nil. |
| ,, 4a | Plateau | 4 | Nil. | 2 | 2 |
| ,, 5 | { Central | Nil. | 1 | Nil. | Nil. |
| | { Plateau | 60 | 1 | Nil. | 60 |
| ,, 6 | { Central | 2 | 36 | Nil. | 3 |
| | { Plateau | 38 | 6 | Nil. | 42 |
| ,, 7 | { Central | 2 | 21 | Nil. | 5 |
| | { Plateau | 17 | 1 | Nil. | 18 |
| ,, 8 | { Central | 1 | 5 | Nil. | 1 |
| | { Plateau | 107 | 5 | Nil. | 112 |
| ,, 9 | { Central | Nil. | 2 | Nil. | Nil. |
| | { Plateau | 52 | 8 | Nil. | 62 |
| ,, 10 | Central | Nil. | 74 | Nil. | 33 |
| Sheet 44 (*inclusive*). | { Central | 5 | 139 | ... | ... |
| | { Plateau | 316 | 23 | ... | ... |

Areas 1-4A lie outside the Pneumatolysis Limit. They supply 44 slides, showing olivine (or its pseudomorphs); and in 34 cases the mineral is fresh. Areas 5-10 lie inside the Pneumatolysis Limit. They supply 336 slides, showing pseudomorphs after olivine, and none in which fresh olivine has been noted.

None of the sliced specimens mentioned was collected with a view to determining the condition of the olivine.

The age-relationships of the various types so far alluded to can be summarized as follows :—

The main mugearite-development, that of Ben More, is quite clearly an episode in the latter half of the Plateau Basalt period

Where comparison is possible the big-felspar basalts are in some cases slightly older, in some cases slightly younger, than the Ben More Mugearites (Areas 6, 7, and 8 of Sheet 44, shown in Index to Plate III.). All that is definitely certain in Area 5 is that the big-felspar basalts are high up in the Plateau Basalt Group.

## 94  Chapter V.—Basalt and Mugearite Lavas: Introduction.

The main mass of Central Types succeeds the main mass of Plateau Types. For this statement the distribution of the two types in relation to observed dips and the form of the ground in Areas 6 and 7 (Plate III.; cf. also Plate V.) may be taken as sufficient justification. Equally significant perhaps is the fact that, except in south-west Mull where the Staffa sub-group is developed, one cannot point to direct superposition of Central Types on the Pre-Tertiary platform; in fact, close approach, or contact, of the two occurs only very exceptionally, and is almost certainly a result of faulting.

In Area 2, the Plateau Group has been subdivided by Sir Archibald Geikie into a Dark Group below, and a Pale Group above. Where this separation has been effected, the basalts of the Pale Group are lettered B' on the one-inch Map. The age-relations of these two subordinate groups is self-evident in the field.

It is appropriate to refer here to the conspicuous columnar jointing which, as hinted at above, characterizes several of the basalt-lavas in the lower part of the Plateau Group of south-west Mull. The feature is almost confined to Sheet 43, where Staffa furnishes a famous illustration. The columnar lavas of Sheet 43 continue some little distance into Area 1 of Sheet 44 and the contiguous portion of Sheet 36, and are there finely exposed in the cliffs west of Carsaig. An isolated occurrence is also known at Bloody Bay, Sheet 52. Most of these early columnar lavas, though referred to the Plateau Group on the score of geographical and chronological convenience, are petrologically of Central rather than of Plateau Type, and are treated under a special heading in Chapter X. Several interesting aspects of their wonderful jointing are dealt with more particularly in Chapter VII.

The Plateau Group as a whole must attain a thickness of some 3000 feet in Ben More, but it seems hardly possible that the group much exceeds 1500 feet along the shores of the Sound of Mull (Areas 5 and 9), or in the Croggan Peninsula (Area 6). The variation of thickness just alluded to cannot be established beyond doubt owing to difficulties of estimation; but within the limits of the Plateau Group the same kind of change is strongly suggested by other evidence. Thus it is quite probable that the Ben More and Derrynaculen mugearites are on approximately the same horizon, although at Ben More there are about 800 feet of Plateau Basalts above, while at Derrynaculen, 3 miles south-east of Ben More, there are scarcely any at all. The Derrynaculen development agrees closely with that near Glenbyre on Loch Buie.

The Central Group cannot be accurately measured, but locally it seems to be quite as thick as the Plateau Group at its maximum, while in the Coire Mòr Syncline (pp. 98, 121) it fails altogether.

### Pneumatolysis and Trap-Featuring.

One of the many interesting features of the Mull lavas is the difference of their condition inside and outside the line drawn on Plate III. as the Limit of Pneumatolysis—a difference in keeping with what Dr. Lee has already pointed out in connexion with the Mesozoic and Tertiary sediments (p. 59).

Outside the limit, one notices that the Plateau Types of basalt are apt to weather with rusty red surfaces ; some rot spheroidally to soft red loam ; while others may develop pustular surfaces, crowded with little rounded eminences left by a wearing away of intervening less resistant matrix. It is an easy matter, too, to secure specimens which show glistening specks of olivine when examined with a lens. Inside the limit, weathering yields sombre grey and brown surfaces, smooth except for veins and amygdales ; and fresh olivine is unprocurable. The results of microscopical examination tally with field-observation in this matter : of 336 olivine-basalts sliced from within the Pneumatolysis Limit, not one has retained a vestige of fresh olivine ; while, of forty-four specimens examined from outside the limit, thiry-four carry this susceptible mineral more or less intact.

Another important characteristic of the pneumatolytic core of Mull is a widespread development of albite and epidote filling vesicles and cracks. In many parts of the district included within the Pneumatolytic Limit, the lavas have minute cracks running in all directions and now filled with epidotic veins. In such localities the tracery of the veins, upstanding on weathered surfaces, arrests attention ; and so too does the resistance to weathering of amygdales which is a common feature of much of the central region of Mull.

Here one may notice a scenic distinction of the pneumatolitic core which happens to be the basis for very divergent interpretations. The lavas occurring within the Pneumatolytic Limit do not show step-, or trap-, featuring nearly so prominently as those outside. The change is not abrupt ; for instance, much of Area 9 is strongly terraced. The explanation adopted here is that subsequent folding, shattering, and pneumatolysis (not to mention interruption by intrusions, and the development of contact-aureoles) are quite sufficient to account for the relatively amorphous weathering of central Mull lavas. In a district where individual amygdales and the contents of cracks weather as prominences, it is not surprising to find that vesicular bands weather at approximately the same rate as solid basalt. Dr. Harker, however, attributes the scenic contrast to an absence of more massive basaltic bands in central Mull, which he accounts for by supposing that the massive bands, responsible for scarps in the surrounding trap-country, have been excluded from central Mull by previous contact-alteration of the lavas of that region ; and from this he deduces that the massive bands are sills. Dr. Harker happens to have described a part of the Carsaig cliffs in this sense, but his position is mainly based upon his wide experience of other Hebridean occurrences,[1] and is correspondingly difficult to criticize. At the same time the detailed examination of Mull has supplied the following data which have carried conviction to those employed on the research :—

(1) The normal massive bands of the trap-country are, as Sir Archibald Geikie and others have maintained, the central portions of lava-flows because they are usually inseparably associated with slag above and below, without any

---

[1] Dr. Harker's position is stated by himself in the Geol. Survey Memoirs on Skye and the Small Isles, and in 'The Sgùrr of Eigg ; some comments on Mr. Bailey's Paper,' *Geol. Mag.*, 1914, p. 306.

## Chapter V.—Basalt and Mugearite Lavas: Introduction.

chilling at the mutual contact.[1] Our experience is that not 1 per cent. of the sheets we interpret as lavas, other than pillow-lavas, shows the smooth, compact, manifestly-chilled bottom almost universally met with among intrusions of comparable size and texture; and that none of them has a smooth, compact, chilled top. Dr. Harker in some measure recognizes this peculiarity, and interprets it as an indication of high temperature affecting the country-rock during the supposed period of sill-intrusion. It is noteworthy, however, that the Early Basic Cone-Sheets of Chap. XXI, though commonly coarser in their interior crystallization than any of our lavas, everywhere show chilled margins. Another hypothesis connects the lack of chilling with the conductivity of the country-rock; but the objection, here, is that one can see innumerable transgressive intrusions chilling against the self-same country-rock that fails to chill the bands under discussion.

(2) In many parts of Central Mull, for instance on the slopes of Ben More, the same type of interbanding of slag and solid can be detected in the field. In such cases, both slag and solid have suffered pneumatolytic change, and the trap-featuring of the succession is materially reduced. The statement that the interbanding of slag and solid is comparable in the trap and central districts, apart from differences referable to pneumatolysis, is based on field-observation and microscopic study. The massive bands of the trap country *are* represented profusely in Central Mull *in an altered condition*.

It should be added that Dr. Harker, in claiming the massive bands of the trap-islands as sills, emphasizes the fact that they sometimes show transgressive relations. It must be realized, however, that the authors of the present Memoir have also found transgressive sills, locally in great abundance, for instance, in south-west Mull (Fig. 42, p. 258), and have been impressed with their difference from the normal massive bands of the trap-succession. In regard to the latter, it may be mentioned that transgressive relations are sometimes quoted which seem to be of the type ascribed in Chapter VII. to auto-intrusion of a lava into itself. Such auto-intrusion must be a common phenomenon from the very nature of the case.

Having informed the reader of the interesting discussions which have arisen out of the scenic contrast on the two sides of the Pneumatolytic Limit drawn on Plate III., we may now pass to another aspect of the question, namely, the date of the pneumatolysis. There can be little doubt that the conditions giving rise to the change appeared and reappeared at widely different epochs, and it is hardly to be imagined that the whole central district was subjected simultaneously to vapour-action. Here it is sufficient to consider two instructive occurrences. Area 9 furnishes good examples of shattered epidotized lavas, which are freely cut by Early Basic Cone-Sheets (bI of one-inch Map) subsequent to the epidotic veins. The sheets are, it is true, altered, and only their most massive representative, that of Beinn Chreagach Mhòr, retains any fresh olivine; but they are not veined (p. 239). On the other hand, Area 10 supplies instances where the Early Basic Cone-Sheets are themselves altered and veined in the most conspicuous manner (p. 236).

The pneumatolysis-area contains several intrusions of plutonic

---

[1] It is often possible to recognize the line of junction of two lavas, since the top of the underlying lava may be weathered, or there may be some mineralogical difference. In such cases, the slaggy top is generally found to be much thicker than the slaggy base, for which latter 6 inches is a common measurement. In other cases, it may be impossible, owing to the absence of chilling, to say where the one lava ends and the other begins.

rock, and these are surrounded by recognizable aureoles of contact-alteration (pp. 128,151). In a number of cases it can be shown that pneumatolysis preceded contact-alteration, and the reverse relation has not yet been proved. The subject has not been fully investigated, but it would be natural to expect that products of contact-alteration, owing to their compact texture, should be relatively resistant to pneumatolysis just as they are to present-day weathering.

The Mull amygdales have yielded much of the information at present available in regard to the relations of pneumatolysis and contact-alteration. Another interesting feature of the Mull amygdale-assemblages is the rich development of scolecite in the Ben More district.

### Conditions of Accumulation.

It is now convenient to touch upon the conditions which prevailed during the accumulation of the Mull lavas. The main conclusion of earlier workers, that the Mull basalts were poured out on dry land, has been accentuated by detailed study. The Cretaceous sea gave place to desert sands, and these to swamps and forests, amidst which great floods of lava made their appearance. One cannot speak of forests in connection with Mull geology, without thinking of the leaf-beds of Ardtun and Macculloch's upright tree often styled the 'fossil tree of Burgh' (*Frontispiece*). The latter is the most arresting single geological phenomenon in the island, a coniferous trunk forty feet high, submerged in lava, but still erect. All the same, two or three trees, rare leaf-beds and sporadic coal-seams, mostly restricted to south-west Mull, are not enough to furnish a complete picture of the conditions of the period. Of more general significance is the evidence of subaerial weathering afforded by reddened surfaces and thin red boles met with time after time in the Plateau Basalt succession; and there is also the well-nigh universal absence of sedimentary infillings of cavities in the lavas—only two cases of sandstone-filled cavities have been noted.

The red-weathered tops referred to above are abundantly displayed in many of the peripheral regions of Mull, but fail almost at once inside the Pneumatolysis Limit. Two explanations may be advanced, and both probably have an element of truth. The volcanic accumulation in the central region may have been more rapid than at the periphery, thus checking the accumulation of weathered products. Also the pneumatolysis of central Mull may have rendered the red tint of surface-weathering evanescent, as it certainly seems to have changed the colour of the basal mudstone (p. 59).

Though red tops are not available, there is nothing in the greater part of the central pneumatolysis area to suggest that aqueous instead of terrestrial conditions prevailed during the lava-period. Everything continues normal until in one restricted tract (outlined as the more south-easterly of two calderas in Plates III. and V.), evidence of an aqueous environment is abundantly displayed. It is afforded by widespread and repeated development of pillow-lavas within this caldera (Plate IV. and Fig. 18, pp. 134, 133).

G

## Chapter V.—Basalt and Mugearite Lavas: Introduction.

That pillow-structure is confined to lavas poured out into water is a very generally accepted thesis, and has been strongly supported of late years by Dr. Tempest Anderson's observations at Matavanu. What seems to clinch the matter in the case of the Mull examples is a marked superficial chilling of the individual pillows (Fig. 21, pp. 132, 151)—and the same phenomenon has been recorded from many other well-known localities. Clearly a fluid must have been responsible for the chilling of these ellipsoidal surfaces, a fluid which has left no other trace of its presence. Air scarcely seems competent to chill the surfaces of incandescent pillows. Water seems the only agent available.

It is proper to state that a definite approach to pillow-structure has sometimes been noted outside the caldera-area, especially in south-western Mull (Sheet 43). It shows itself as a somewhat lenticular flow-structure, generally with pipe-amygdales springing from the base of each flow-band; and it is regarded (p. 114) as a subaerial counterpart of pillow-structure.

Microscopic evidence supports the contrast between the south-eastern caldera and the surrounding country. It is well known that variolitic structure is a common microscopic associate of pillow-structure. In keeping with this, one finds that thirty-four per cent. of the basalt slides from Area 10 are variolitic as against two per cent. from the rest of Sheet 44. Another interesting microscopic peculiarity of the Mull pillow-lavas, as compared with their fellows, is the invasion of early-formed vesicles by froth (Fig. 21, p. 151).

The present-day restriction of pillow-structure to the south-eastern caldera of central Mull appears to be sufficiently definite to call for interpretation. The basalts of the region are all of Central Types, and as they approach closely on the west to outcrops of Pre-Tertiary rocks, this in itself makes it tolerably certain that they occupy a region of marked subsidence (Chapter XIII.). The question naturally arises whether this subsidence is merely responsible for the preservation of the pillow-lavas, or whether it may not equally have been responsible for their development.

In the first place, it might be claimed that the pillow-lava suite, which on the local evidence is some few thousand feet thick, is a downthrown remnant of a once wide-spread formation. But two circumstances make this hypothesis difficult :—

> (1) The lavas in question are shown by their relation to neighbouring intrusions to be of early date in Mull's igneous history. This, combined with their close petrological resemblance to other developments of Central Types in Mull, makes it probable that they belong to the same general period as these latter. Accordingly, if they mark a wide-spread and a long continued submergence of Mull as a whole, it is strange not to find their counterpart in the Coire Mòr and Loch Spelve Syncline where the Mull lavas are overlain by later agglomerate.
>
> (2) The pillow-lava suite is very little interrupted by sediment.

It is reasonable to suppose, therefore, that the pillow-lava suite of Central Mull is the local equivalent of much of the Central Group found round about; and that it accumulated in a frequently renewed crater-lake. If, as analogy would suggest, the crater was placed centrally upon a flat lava-dome, the entry of external drainage would be prohibited, and sedimentation would thus be reduced to a minimum. The picture that rises before one's eyes is of the caldera-sink of Kilauea with its movable floor[1]; only one must imagine a Kilauea with intermittent periods of repose during which the fires of Halemaumau give place to the waters of a lake.

The possible co-operation of normal fissure-eruptions in building up the lava-pile of Mull as a whole has already been discussed (p. 4).  E.B.B. (as Editor).

[1] J. D. Dana, 'Characteristics of Volcanoes,' 1890, p. 78.

## CHAPTER VI.

### BASALT AND MUGEARITE LAVAS.

#### OUTSIDE PNEUMATOLYSIS LIMIT.

##### SHEET 44.

FIVE Areas, 1-4a of Sheet 44 as outlined in the Index Map (p. 91), are considered in this chapter. They have much in common, and to avoid reiteration it may be stated that they all furnish good examples of trap-featuring, rusty present-day weathering, spheroidal exfoliation, and abundance of fresh olivine easily recognized in the field (*see* also Table X., p. 93). They also very commonly exhibit reddened tops due to contemporaneous weathering of Tertiary date, and most of the districts furnish good examples of red boles. A rather minor feature is a tendency to pimply surfaces as explained (p. 138).

The great majority of the lavas do not carry porphyritic felspars, but various exceptions will be specially noted in the sequel.

##### AREA 1, SHEET 44 : CARSAIG BAY TO LOCH SCRIDAIN (Index Map, p. 91).

What may be described from the scenic point of view as typical ' plateau country ' extends from Glen Leidle and Carsaig to the western boundary of the map. Trap-features are more pronounced than east of Glen Leidle, but at the same time they are not caused by every lava. Except as regards its southern margin, the region is not one of marked relief.

The sequence is entirely basaltic. Very few of the flows are porphyritic, but one of this type may be traced for about half a mile to the south of Creag an Fheidh. Another occurs on the southern coast west of Tràigh Cadh' an Easa', at a height of about forty feet above the coal to which reference has been made in Chapter III. It can be traced inland to Airdh Mhic Cribhain, and occurs again on the western slope of Beinn an Aoinidh, always in the same relation to the coal to which it forms an index.                E.M.A.

The district is noteworthy for the spectacular exposures afforded in its southern coastal cliff, ranging up to 1000 feet in height. The difficulty the geologist finds is that much of what he looks at he cannot reach to hammer. He has to walk, or row, along the base of the cliff, or to examine it from above. Still what is seen can be compared with more accessible sections elsewhere in Mull, and thus interpreted with considerable confidence. The view here taken is that the cliffs consist essentially of a succession of basic lava-flows, interrupted occasionally by well-defined sills of basic and intermediate composition. In Chapter V. it has already been pointed out that a very different interpretation of these cliffs has been advanced by Dr. Harker :[1] he recognizes as superficial flows only the more conspicuously amygdaloidal parts of the sequence, whereas the present authors regard these as often nothing more than slaggy tops and bottoms ; and he holds that what are treated here as the massive interiors of successive lava flows are in reality sills of a quite later date. His account refers only to that portion of the district which extends into one-inch Sheet 36. In comparing the two descriptions it should be noted that the sills described by Dr. Harker are additional to the sills of basic and intermediate composition which are recognized in Chapter XXIII. of the present Memoir. The former are called olivine-dolerite by Dr. Harker, while the sills about which no question arises are generally tholeiite or andesite. At the same time, this difference should not be overstressed, for it happens that the majority of the more conspicuously columnar sheets of south-west Mull, classed by Dr. Harker as sills and by ourselves as lavas, are olivine-basalts of markedly tholeiitic affinity (Staffa Type, Chapter X.).

No instance has been noted by the present writers where the debatable olivine-bearing sheets (whether one call them dolerite or basalt does not affect the question) have compact non-vesicular tops. On the other hand, they do exceptionally show bases of this type. For instance, an interesting example of a massive sheet with variable basal character is exposed south-east of Creachan Mòr. This sheet can be followed along the cliff-margin for over a quarter of a mile. Just west of the indentation of the cliff, which is determined by erosion along a fault running inland to Allt Cnoc nam Piob, it has the typical slaggy base of a lava ; but 600 yards farther west, to the south of the more westerly of two small lochans shown on the map, it has a compact, and to a certain extent chilled, bottom-layer. Occasional chilling of lava-bases is, as already stated, a widespread, though very subordinate, feature of Mull geology.

A particularly convincing reason, for interpreting as lavas most of the inaccessible massive sheets of the part of the Carsaig cliffs, described by Dr. Harker, lies in the following facts. Reference will be made in the next chapter to the frequency with which the non-vesicular part of a columnar lava is divided along a definite plane into a lower zone with massive regular vertical columns, and an upper zone in which the columns are curved and relatively thin. The type-locality for this double-tier type of jointing is Staffa, where observations, which will presently be recorded, leave no doubt that one is dealing with a lava (p. 111) ; and there are many other instances—in the Gribun and Ardtun peninsulas. Of the numerous examples from the Mull area which have been examined at close quarters, there has been only one where the point was doubtful. In agreement with this, one finds that Scrope and Judd, the former building on his experience of the Auvergne, have developed a theoretical explanation (pp. 44, 109) which would not be in accordance

[1] A. Harker *in* 'The Geology of the Seaboard of Mid Argyll'; *Mem. Geol. Surv.*, 1909, p. 79.

## Chapter VI.—Lavas outside Pneumatolysis Limit.

with subterranean conditions of cooling. Now the basic sheets of the cliffs in the part referred to are largely columnar, and in some cases they reproduce the Staffa appearances precisely. On looking east across a narrow gully from the top of the cliff immediately above Carsaig Arches one sees the following sequence :—

> Layer of curving columns ⎫
> Layer of regular columns ⎭
> Slaggy layer.
> Layer of curving columns ⎫
> Layer of regular columns ⎭
> Slaggy layer
> Layer of curving columns ⎫
> Layer of regular columns ⎭

The double columnar layers must be regarded as examples of the 'olivine-dolerite sheets' referred to, which there is thus strong reason to regard as interior or basal portions of lavas. The slaggy layers are probably in the main the tops of the two lower of the three lavas represented in this section—for experience teaches that slaggy bases are usually quite subordinate in thickness to slaggy tops. If this supposition be correct, the two lavas referred to are each formed of three layers of about equal thickness.   E.B.B., E.M.A.

The term 'columnar' might be not unreasonably applied to many of the plateau basalts of Mull in which a columnar appearance is caused by the intersection of more or less rectangular systems of joints. It is possible that these are not ordinary joint-planes, but were formed by contraction on cooling. If so, they are related in origin to the joints bounding polygonal columns. In the description of this area, however, the term columnar is restricted to lavas in which the columns are polygonal. Columnar basalts in this sense extend along the southern cliff westwards from Carsaig Arches into Sheet 43. Above the Arches, they reach upwards to nearly 1000 feet from the bottom of the lava-group. This appears to be a maximum development. As a general rule in the Mull area, they underlie the great mass of non-columnar, or not-polygonally-columnar, members ; and near Gribun (see p. 108) are confined within a much shorter distance from the base.

Judging from what is seen on the southern cliffs, fifty feet may be taken as an average thickness of an individual flow ;  One or two cases can be seen where a lava attains a hundred feet ; and one, at the bottom of the cliff at Tràigh Cadh' an Easa', appears to be over 180 feet in thickness.

A very constant feature in this district, as in others outside the Pneumatolysis Limit, is the more or less pronounced reddening by ferric oxide of the tops of the lavas. This reddening appears to have been produced on the weathered surface of each flow before the superposition of the next in order ; it affects only a small thickness of the lava, except where it has worked its way downwards along cracks and joint-planes. It is due to incipient production of laterite, checked before it proceeded to any great extent by the protection afforded by succeeding flows.   E.M.A.

Reference has been made in Chapter III. to certain intercalations of sediment and associated ash near the base of the lava-sequence. Of these, the deposit exposed in the cliffs between Carraig Mhòr and An Dùnan, east of Carsaig, is the most noteworthy. It varies from 10 to 20 ft. in thickness, and consists in the main of a breccia of pale green to buff-coloured fragments of fine-grained vesicular basalt. The interstices are filled with black sedimentary material, and the whole contains a good deal of calcium carbonate both as a filling to the vesicles and dispersed through the rock.    (B.L.)

Between the southern cliffs and Loch Scridain, conditions are not good for observing such intercalations, and in only one case has a thin band of tuff been detected.

In the southern part of the district, there is a very gentle rise of the lavas to the east, which, though hardly perceptible in the featuring, brings the base of the assemblage from below sea-level at Carsaig Arches to about 300 ft. above sea-level at Carsaig.

E.M.A.

### AREA 2, SHEET 44 : LOCH SCRIDAIN TO LOCH NA KEAL (INDEX MAP, p. 91).

The lavas on the north side of Loch Scridain are of the same general character as those on the south. Trap-featuring is rather more pronounced. Red-weathered tops are again found, and are sometimes overlain by a foot or more of red bole, in the deposition of which water may have played a part. An easily located example of such a bole occurs beneath a porphyritic flow on Aird Kilfinichen, a little above high-water mark on the south coast of this peninsula. Another place where a geologist is sure to find red tops and boles is among the exposures afforded by the many small streams descending into Gleann Doire Dhubhaig, on the north side of Coirc Bheinn.                                                                           E.M.A., J.E.R.

While the majority of the lavas of the area are non-porphyritic, a well-defined porphyritic zone, consisting sometimes of one, sometimes of two flows, is traceable on both sides of Kilfinichen Bay. The field-characters leave no doubt as to the contemporaneous nature of these porphyritic lavas ; for instance, the little stream east of Dùn Scobuill, just inside Sheet 43 across the border of Area 2, furnishes an excellent exposure of a thick slaggy top belonging to the upper of the two flows—the only one to extend so far west. The porphyritic elements are fairly abundant felspar-phenocrysts, usually rather more than a quarter of an inch long, but exceptionally reaching 3 inches. These Kilfinichen lavas have not been coloured on the one-inch Map (Sheets 43, 44) as Big-Felspar Basalts, since their porphyritic character is not sufficiently conspicuous. At the same time, they furnish an excellent example of the intercalation of porphyritic lavas in the lower part of the Plateau Group. Their outcrop has been traced from Aird Kilfinichen, where there are two flows of the type, for a mile up the north-east side of the valley a little above the road. They have been picked up again on the other side of the valley, and here the upper flow makes a conspicuous scarp and shelf above Seabank Villa. Beyond Tiroran, only this upper flow persists, and it has been traced almost a mile into Sheet 43. Its outcrop at its farthest west exposures crosses the Tiroran-Tavool Road.

104    Chapter VI.—Lavas outside Pneumatolysis Limit.

Attention may be directed to interesting examples of what seems to be auto-intrusion within a lava. The Abhuinn Bail' a' Mhuilinn flows over bare rock, and in dry weather exposes an exceptionally clear section for some distance above its junction with Allt a' Mhuchaidh and, again, at a point a little farther upstream. Amygdaloidal lava is exposed cut by irregular, but distinct, amygdaloidal intrusions (18507 by 18508, and 18509 by 18510) of similar type. The intrusions have chilled edges so that if, as seems possible, they are portions of the lava forced from below into an already consolidated crust, this latter must have been fairly cool at the time. Observations of the coastal cliffs of Sheet 43 have furnished many examples of various types of auto-intrusion described later (Chapter VII.).    E.M.A.

On Coirc Bheinn, the highest mountain of the district, there is an interesting outlier of lavas with scolecite, albite, and epidote in their amygdales. The cavities thus filled are larger than is general among the lavas of the neighbourhood. Scolecite-bearing lavas, on much the same geological position occur in the Ben More country (Area 8), where they have attracted considerable attention. The Coirc Bheinn examples seem to deserve further investigation, since they seem to fall outside the Pneumatolysis Limit of Plate III. (p. 91), whereas their Ben More analogues are well inside.

J.E.R.

Area 3, Sheet 44 : Glen Aros District, Loch na Keal to Sound of Mull (Index Map, p. 91).

Almost all the lavas are non-porphyritic olivine-rich basalts —thoroughly typical of the plateau country in general. An exception is a flow with conspicuous felspar-phenocrysts which forms a strong crag on the western top of Mcall na Caorach, and has been traced from this point in a north-westerly direction into Sheet 52. Another exception is a mugearite shown on the map at Tom a' Chrochaire, and in a small isolated crop a mile farther west. This mugearite is a very platy rock. Near the south-western end of its main exposure, it is seen resting irregularly, with slightly chilled base, on a red bole overlying amygdaloidal basalt ; while its top is probably in places brecciated—at any rate mugearite-breccia underlies the most westerly of the associated basalt-outliers. Even within the limits of its outcrop, the mugearite is seen to be impersistent ; and it is probably quite distinct from, and earlier than, the Ben More flow of similar type.

Along the coast north of Salen, a gentle dip in the south-east direction brings successive lava-flows to the surface at long intervals; and no better section is known in Mull for illustrating the thin vesicular base, the thick massive interior, and the thick slaggy top so characteristic of individual lava-streams. The evidence is particularly clear, since in a majority of cases the slaggy top of each flow is found to have been reddened, and very often to have been covered with red bole, before the outpouring of the next lava of the sequence. A few notes regarding this illuminating coast-section may be of service :—

South of Sgiath Ruadh, red bole is seen filtered down into cracks in reddened amygdaloidal lava.

At Port an Tobire, a reddened top is exposed in which the amygdales are grouped in bunches—quite a common arrangement.

For about a mile north of Rudha Àrd Ealasaid, the lavas are distinctly platy.

At the north end of the bay, 400 yards S.E. of Kintallen, there are good exposures of a reddened lava-top with overlying bole.

Opposite Kintallen, a banded lava occurs in which there are marked alternations of vesicular and solid layers. J.E.R.

Leaving the Sound of Mull one may glance for a moment at exposures at Rudha Mòr on Loch na Keal. Here, pimply-weathering basalt is seen—a type of basalt which, as already stated (p. 100), is rather characteristic of the plateau country as a whole, outside the Limit of Pneumatolysis ; and close by is an instance of contemporaneous pegmatitic veining such as is more fully described in Chapter VII. E.B.B.

AREA 4, SHEET 44 : LOCH ALINE, MORVEN (INDEX MAP, p. 91).

The Loch Aline district is more accessible from Oban and the south than any other part of the plateau region outside the Limit of Pneumatolysis. Fortunately it serves as a very good example of its kind. The rusty weathering cliffs rising above Inninmore Bay have been known to draw from the passing tourist the remark " there must be iron in those rocks." Moreover trap-featuring is displayed to special advantage through contrast arising from juxtaposition of lavas with gneisses along the wonderful Inninmore Fault (Fig. 26, p. 182).

Cliff- and shore-sections are easily reached from the steamer-pier. They show, as is usual, that the majority of the flows are massive in their centres and become amygdaloidal and brecciated towards their tops and bottoms. The amygdaloidal band at the top of any particular flow is always thicker than at the bottom, and is often marked by purple and red staining. There is a capital exposure of a brecciated base at the head of the little bay east of Rudha Dearg, three quarters of a mile west of the pier ; and a reddened scoriaceous top is well seen along the shore east of the same. Just at the entrance to Loch Aline, on the western shore about 200 yards south of the ferry, massive basalt with a foot of slightly vesicular rock at its base can be seen resting on a reddened scoriaceous top. Such examples might be multiplied indefinitely. In the precipitous northern face of Glais Bheinn, east of the loch, seven distinct thin red layers have been counted in the basalt-sequence.

A tendency to rude columnar jointing is quite common, but there is no example of a definitely columnar lava. Sometimes, the dominant structure is platy jointing parallel to the plane of flow.

As elsewhere in the plateau country outside the Pneumatolysis Limit, spheroidal weathering is widespread and conspicuous.

A point of great interest is an appearance suggestive of thinning of the lava-group northwards away from Mull. On the west side of Loch Aline, the lowest lavas seem to be dying out northwards,

106  *Chapter VI.—Lavas outside Pneumatolysis Limit.*

and to be overlapped by upper members of the sequence, which thus come down on to the floor of Cretaceous sediments. In Glais Bheinn, too, there would seem to be only some 500 ft. in the northern escarpment as opposed to 1100 or 1200 ft. in the southern slope—but just possibly here the northern sequence may be curtailed by a fault concealed beneath talus. (H.B.M.), G.W.L.

### Area 4a, Sheet 44 : Fishnish, Mull.

The Sound separates Fishnish from Area 4 in Morven, but the Pneumatolysis Limit divides it even more markedly from its geographical continuation (Area 9) in Mull. Only the lowest of the Plateau Basalts are met with ; and some of them are rather doleritic in texture. The north coast of the peninsula shows a succession of massive bands with slaggy tops and less-marked vesicular bases—typical lava-streams. The massive bands weather rusty at their surfaces, and on fracture show specks of fresh olivine.

Easily reached by road from Salen or Craignure, the district furnishes an excellent starting point from which to investigate the reality of the Pneumatolysis Limit. E.B.B.

# CHAPTER VII.

## BASALT AND MUGEARITE LAVAS.

### OUTSIDE PNEUMATOLYSIS LIMIT.

### SHEETS 43, 51, AND 52.

ONLY a brief anticipation of the separate memoirs dealing with the maps enumerated in the heading of this chapter is attempted in the following pages.

Speaking generally, one may say that cliffs and coast-sections afford numberless exposures of basalt-lavas with slaggy tops and subordinate slaggy bases. Sheet 43 is the best provided in the matter of cliff-sections, including those above the road leading along the south shore of Loch na Keal and southwards through Gribun.

Contemporaneous weathering of lava-tops is also well displayed in these three maps. Sometimes it has led merely to a rotted superficial layer, at other times it has been accompanied by accumulation of red bole.

Another characteristic of the district as a whole is the very widespread preservation of fresh olivine. As usual it can easily be recognized in hand-specimens with a lens, especially on a sunny day. It is scarcely necessary to add that rusty and spheroidal weathering are both prominent, and that pimple-weathering is not uncommon.

The whole assemblage of flows falls within the Plateau Group as defined in Chapter V., and lies outside the Limit of Pneumatolysis drawn on Plate III. It must be remembered, however, that non-porphyritic, compact, olivine-poor basalts, of Central rather than Plateau Type, are well-represented among the earlier, often columnar, lavas of Sheet 43 (pp. 30, 94). E.B.B., E.M.A., J.E.R., G.V.W., G.A.B.

### SHEET 43 : THE WESTERN COASTS OF MULL, WITH ULVA, THE TRESHNISH ISLES, AND STAFFA.

Trap-featuring is nowhere better displayed than in Sheet 43. Particularly good examples are afforded in the part of the district north of Loch Tuath, in Ulva and Gometra, and in Bearraich, the mountainous termination of the Gribun Peninsula. A curious capriciousness may be noticed in this respect. While the Gribun Peninsula as a whole is characteristically terraced, there is a marked poverty of such features in that part of the district which centres about Creag nam Fitheach. As instances of flows with porphyritic felspar, such as occasionally occur in the Plateau Group, mention may be made of four lavas exposed in and around Allt an Fhir-eòin draining into Loch Assapol.

108     *Chapter VII.—Lavas outside Pneumatolysis Limit.*

The Leaf-Beds of Ardtun are of cardinal importance to the student of Mull lavas, since they date the early part of the sequence. Along with the basal sediments of the district, and a few impersistent coal-seams, they have already been described in Chapter III., and their fossil-contents in Chapter IV.

*Columnar Basalts : Staffa and elsewhere.*—The geological phenomenon which has attracted most attention to Sheet 43 in the past, apart perhaps from the Leaf-Beds of Ardtun, is the beautiful columnar structure exhibited by a small proportion of the lavas. All the examples known are referable with fair certainty to a low position in the Plateau Group, with a maximum development in the south-east corner of the map. A geographical as well as a time limit has been recognized. The columnar lavas of the south-east corner of Sheet 43 continue for some distance eastwards into Sheets 36 and 44 along the Carsaig cliffs, but otherwise no typical columnar basalt is known in Mull, except a solitary occurrence low down in the lava-sequence at Bloody Bay (Sheet 52, p. 118). In the following rather disconnected paragraphs, attention is directed to the various exposures met with in Sheet 43 :—     E.B.B., E.M.A., J.E.R., G.A.B.

A continuation of the zone of columnar lavas which has been described in the cliffs above Carsaig Arches (Chap. VI.) follows the southern coast to the east of Eas Dubh, and reappears in the Ardtun Peninsula.

In the gully formed by the more easterly of the two streams, which are shown in the one-inch Map to descend the cliffs above Aoineadh Beag, a single flow is seen have the following section :—

|  | Ft. |
|---|---|
| Non-columnar zone, reddened at top | 60 |
| Zone with irregular curving columns | 60 |
| Zone with regular vertical columns | 10 |

The lavas immediately above and below the intercalation of sediment which contains the Ardtun Leaf-Beds at the mouth of Loch'Scridain are both strikingly columnar. The underlying member shows two zones of columns, with the upper zone merging into slag above.     E.M.A.

In the Gribun Peninsula north of Loch Scridain, one realizes very clearly the restriction to a low horizon of the few typical columnar basalts that occur. They are exposed on both sides of the peninsula, more especially in the main river east of Tavool House, where they extend for 400 yds. inland from the sea, and again in the coastal cliffs reaching from Carraig Mhic Thòmais to the Wilderness. In the continuation of the cliffs just mentioned, they are not visible much east of Caisteil Sloc nam Ban ; and, in keeping with this, are definitely seen to be absent from the Allt na Teangaidh section by the roadside above Balmeanach.     E.B.B.

North of the Inch Kenneth anticline (p. 172), two columnar flows extend along the coast from Ulva Ferry to Tòrr Mòr. They occur practically at the bottom of the pile of lavas constituting Ulva, and presumably lie very near the base of the lava-group considered as a whole.

The columnar flow of Fingal's Cave, which has rendered Staffa famous, is also almost certainly on a low horizon.     G.A.B.

A very fine columnar basalt forms the two islands of Réidh Eilean and Stac Mhic Mhurchaidh on the west of Iona, and is possibly to be regarded as the last exposure of the Mull Plateau Basalts towards the south-west. The texture is, however, more doleritic than usual ; and no definite proof has been obtained that the islands are not part of a sill. The columns, which appear to be fully a hundred feet in height in the case of Stac Mhic Mhurchaidh, are vertical and regular, and if they form part of a lava, this must have been of unusual thickness.     E.M.A.

Even in this brief notice, it is necessary to discuss the origin of certain features of columnar jointing. Many, but not all, of

## Sheet 43: Columnar Basalts.

the more conspicuously columnar lavas of Mull show three well defined zones : a lower zone of massive regular columns, a middle zone of closer-spaced wavy columns, an upper zone of slag. The phenomenon is illustrated at Staffa, the Wilderness, the river at Tavool, the Ardtun Peninsula, and the Carsaig cliffs. The line of demarcation between the two tiers of columns is, as a rule, very even and well-defined ; although it is obvious on inspection that both tiers consist of one rock, and are merely differentiated by their jointing. The line between the upper zone of columns and the slag is often irregular, and there are numerous clear instances of an insensible passage from the one into the other. Of these, special mention may be made of those afforded by Staffa and the columnar basalt underlying the Ardtun Leaf-Beds, and also of some half dozen flows accessibly exposed at the base of the Wilderness cliff above Aird na h-Iolaire. E.B.B., E.M.A., G.A.B.

Scrope and Judd, as already noticed (p. 44), give good reason to believe that the close-spaced wavy columns have resulted from rapid irregular cooling of the upper surface (which may safely be ascribed largely to convection) ; and that the more massive and straight columns have developed through slow regular cooling of the lower surface. Professor Iddings, adopting much the same standpoint, argues that the spacing of the columns is controlled by rate of cooling, and their direction by a tendency to grow always at right angles to a retreating isothermal surface corresponding with a particular temperature. It may seem, at first sight, that the rate of the cooling responsible for the two tiers of columns has been the same in both cases, since often the two tiers are of about the same depth. But it must be remembered that the upper tier is covered with a thick layer of slag, and that this has, of course, retarded the initiation of its cooling and consequent jointing.

Professor Iddings strongly supports his thesis by pointing to a very striking phenomenon sometimes observable at the approach of two systems of columns, where the columns, instead of coalescing, bend abruptly into rough parallelism. Professor Iddings[1] draws his illustrations from America, and Mr. James[2] from Australia ; Fig. 16 is a sketch by Mr. G. M. Sinclair from a photograph taken at Staffa, while the Frontispiece shows another, though more complicated, example furnished by the mainland of Mull. The explanation advanced by Professor Iddings is sufficiently illustrated in Fig. 16.

It must not be thought that final mutual avoidance is a general phenomenon of the approach of two systems of columns. It is the exception rather than the rule. The alternative is coalescence, resulting from an alignment of certain margins of the lesser columns and the disappearance of others. Here the weakness of neighbouring jointed basalt is evidently a controlling factor. Combined with this, is the relatively slow cooling from below. The form of the critical isothermal surface retreating rapidly and irregularly away from the top of the flow is in general little affected by loss of heat

[1] J. P. Iddings, 'Igneous Rocks,' 1909, vol. i., Figs. 13-17, pp. 320-325.
[2] A. V. G. James, 'Factors producing Columnar Structure in Lavas and its Occurrence near Melbourne, Australia,' *Journ. Geol.*, vol. xxviii., 1920, Fig 13, p. 467.

110     Chapter VII.—Lavas outside Pneumatolysis Limit.

downwards even after joining with its counterpart below. Thus columns of the upper tier approach exceedingly close to columns

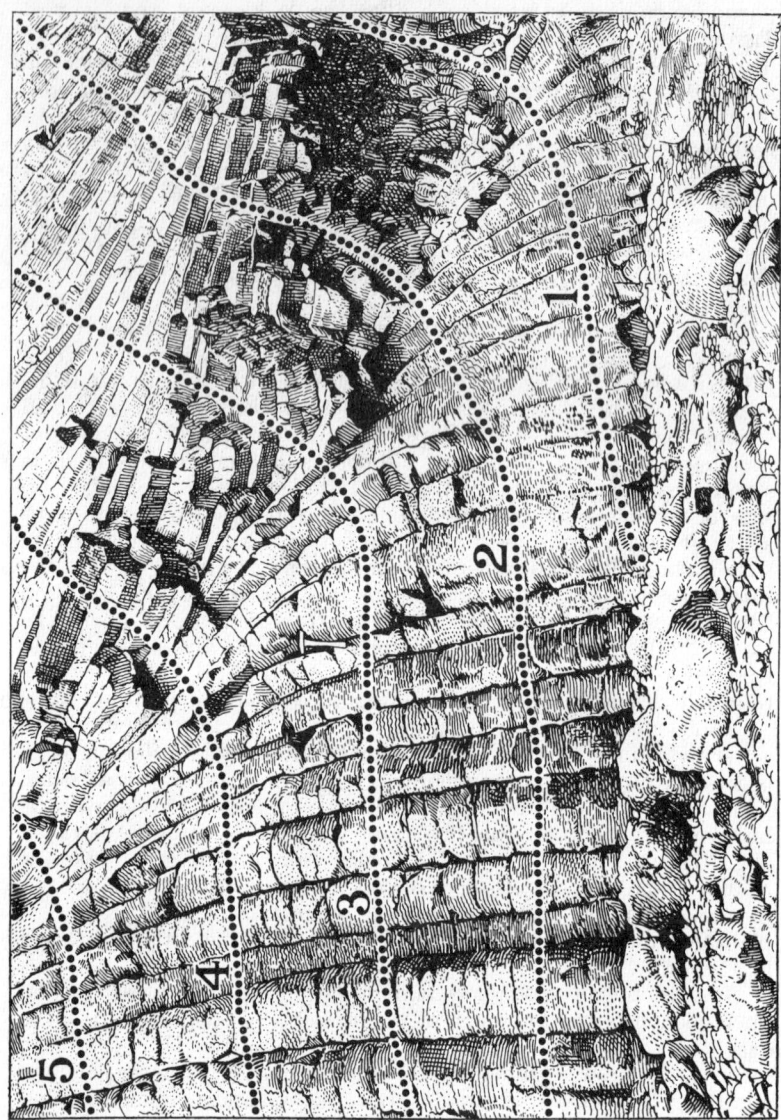

Fig. 16.—Columnar Jointing, North of Landing-Stage, Staffa. The dotted lines, 1 to 5, indicate inferred successive positions of the critical isotherm, at right angles to which the columns have developed. (Modern beach-gravel shown at bottom of sketch.)

of the lower tier without experiencing any tendency to deflection. The relative sluggishness of downward coloing is sufficiently evidenced by the course of the even boundary that commonly separates the two tiers of columns in lavas of this class, for the separation follows a plane at right angles to the columns of the lower tier.  E.B.B.

Staffa has been mentioned several times. Its main constituent is the Fingal's Cave Lava. This lava is underlain by red ash exposed from Fingal's Cave along the greater part of the west coast; while in Meallan Fulann it passes under succeeding basaltic flows. The Fingal's Cave Lava is generally divisible into a lower zone of massive regular columns, a middle zone of narrow wavy columns, and a top zone largely of slag. The base of the lava is most conveniently reached from above at Port an Fhasgaidh where it is seen to be vesicular and brecciated. The two columnar zones are typically exposed at Fingal's Cave; and the visitor, as he approaches the cave along the causeway, has plenty of opportunity for seeing that there is no line of separation in the basalt thus variously jointed. The slaggy zone occupies the east coast for some 300 yds. south from Goat Cave. An interesting feature of this upper portion is the occurrence of masses of slag, often brecciated, which have been carried forward by still liquid lava, now in large measure columnar. Towards the top of the flow, columnar structure fails altogether. Several instances can be examined of passage from columnar basalt to slag; and vesicular structure often manifests itself within the limits of a well-defined column. At Goat Cave, thin ash and carbonaceous sediment separate the Fingal's Cave Lava from the chilled columnar base of an overlying flow. The actual top is again seen at sea-level north of Meallan Fulann.  G.A.B.

*Macculloch's Tree.* Another famous columnar lava of the district is that which flowed round and over Macculloch's Tree, otherwise known as the 'Fossil Tree of Burgh' (*Frontispiece*) The cast of the tree-trunk stands in a recess of the coastal cliff, 150 yds. north of the waterfall of Allt Airidh nan Caisteal (this is the most southerly of two neighbouring waterfalls, about 70 yds. apart, a little south of Rudha na h-Uamha). A visitor will find it most convenient to approach by boat from the Bunssan side, or by motor-boat from Iona. If he comes on foot, he is well-advised to start from the Tavool side. In any case it is *essential* that he should time his arrival for a *falling half-tide*.

The cast of this tree is seen in vertical section and looks like a pipe, nearly 5 ft. broad and about 40 ft. high, and largely filled with breccia consisting of lumps of 'white trap' and charred wood. The lower part of the pipe encloses a partially silicified semi-cylinder of wood glittering with quartz-crystals. The diameter of the cylinder is about 3 ft., and outside is a hollow where soft black coniferous wood (p. 77), a couple of inches thick, may be dug out with a hammer. In Macculloch's day the wood continued as a coat around the silicified cylinder, but since then it has been ruthlessly stripped by curio-hunters until now it is no longer part of the spectacle—though the recollection of its former presence is a great stimulus to the observer. The pipe rises vertically in the heart of columnar basalt, and the columns as they approach it turn so as to meet it everywhere at right angles. The roots of the tree are not seen; but close at hand a carbonaceous mud, sometimes with a film of coal, is exposed beneath the lava ; and under this there follows a prominent bed of red ash.

In the continuation of the same lava, 230 yds. north of the upstanding tree, is a cave with two entrances. In the overhanging northern face of the main entrance, there lies a gently inclined dicotyledonous tree (p. 81), 10 ft. long and 6-8 inches thick, enclosed in a brecciated portion of the flow.

It is probable that the two trees mentioned are merely remnants of a forest. At any rate, the lower portion of the lava has characteristics which suggest that it consolidated in an unusual environment. Magnificent exposures of this lower part of the flow are afforded by the cliffs of Rudha na h-Uamha. It is seen to rest regularly on top of the carbonaceous layer and ash already mentioned ; but it is within itself wonderfully irregular, for it consists of a chaotic assemblage of columnar and brecciated basalt. The columnar basalt traverses the brecciated, or sweeps round great masses of it 40 or 50 ft. high. Everywhere the columns are at right angles to the junction-line of the two types, and thus most arresting combinations are produced, including rosettes and fringes.

What has been spoken of above as brecciated basalt is always more or less vesicular ; but its solidity varies greatly, for sometimes this part of the complex flow is merely a vesicular basalt with little or no brecciation, whereas elswhere it is an ash-like accumulation of fragments of different degrees of porosity, bound together by infiltrated calcite. One cannot strictly define the brecciated portion as basal. The columnar part of the complex is equally often found in contact with the underlying black mud, and in such a case shows distinct chilling—probably the mud was wet. There is, however, no chilled edge to the columnar part at the well-defined contacts of solid lava and breccia.

It is undeniable that the columnar basalt was in a sense intrusive into the brecciated. But the non-transgressive relation of the complex as a whole with reference to the underlying black mud—as seen for quarter of a mile—negatives the possibility that the intrusion took place under cover. The columnar basalt and the breccia together constitute a unit, and the breccia's character stamps the unit as superficial. The two trees, the one enveloped in solid, the other in breccia, afford confirmation of this view. Probably we have here a lava that poured into a forest, where its first advance gave rise to a series of explosions. The flow continued, and great masses of slag and breccia were caught up and involved in the irresistible flood.

To test this explanation a careful search was undertaken for other traces of the postulated forest. Apart from the two trees, only occasional fragments of wood could be found ; but in places the non-columnar basalt encloses short ribbons of what looks like a carbonaceous distillate.

Two other localities may be mentioned in connexion with this wonderful lava. Between the two waterfalls already mentioned as a guide to the upright tree, one may see the massive columns of the lower part of the lava in regular disposition and overlain —Staffa fashion—by narrow wavy columns. In the cliff just north of the next little stream (200 yds. south of Allt Airidh nan

Sheet 43 : *Tavool Tree and Auto-Intrusion.* 113

Caisteal), the top of the flow is very accessible, and the minor columns can be examined merging upwards into slag.

*Tavool Tree.*—An inconspicuous, but readily accessible, analogue of Macculloch's Tree is exposed near the north shore of Loch Scridain in the bed of a little river, or stream, that enters the sea 500 yds. south-east of Tavool House. There is a foot-bridge near the mouth of this stream, and a beautiful waterfall a little farther up. Half-way between these two landmarks, the bed of the stream for 15 yds. shows a mass consisting largely of fragments of coniferous wood (p. 78) ; some of these are several inches long, and are intimately mixed up with vesicular basalt, locally altered to 'white trap.' The basalt belongs to the basal part of a strikingly columnar flow, with the double-tier type of jointing well-displayed in the waterfall. As the columns approach the mass of vegetable matter, they seem to tilt up away from it ; but exposures are too restricted to give much information on this point.

The first exposure on the sea-shore, 400 yds. east of the mouth of the stream, shows a sill, with sapphire-bearing xenoliths (Chapter XXIV.), and cutting with chilled base a basaltic breccia made of angular lumps of vesicular basalt, among which are many pieces of black wood. The breccia is rather more like an ash than a lava. Its position, judged by neighbouring exposures, is just at the base of the columnar flow seen in the stream. E.B.B.

*Auto-Intrusion.*—In the descriptions just given of the upper part of the Fingal's Cave Lava, and also of the lower part of the lava enveloping Macculloch's Tree, it will be noted that complex relations of slag and solid are recorded which may fitly be ascribed to auto-intrusion. Still-fluid lava has evidently involved, entangled, and carried forward previously semi-consolidated portions. The numerous cliff-sections of the district frequently show phenomena of this kind. No finer example is known than that exposed in the cliffs of Little Colonsay at Port Nam Faochag. A thick mass of solid lava forming the cliffs east of the port suddenly gives place north-westwards to a banded complex of solid and slag looking at first sight like a succession of flows.

More accessible coast-sections showing auto-intrusions occur in Mull at Lòn Reudle, at the mouth of Loch Tuath. The exact positions of two important examples are noted on the one-inch Map. In the more westerly exposure, comparatively solid black lava is well seen making a cliff-face and projecting as a tongue 4 yds. long into neighbouring slag. The tongue has concentric zones of amygdales parallel with its margin. The slag also shows a certain amount of banding conforming with the margin of the tongue, but in this case the conformity seems to have been imposed, and the banding of the slag is notably disturbed and broken. The junction of the two rocks, solid and slag, is well-defined, though there is no sign of marginal chilling. The contrast of the two has been accentuated by subsequent weathering which has reddened the slag and left the solid black.

The more easterly section shows a banded face of slag and solid, each band horizontal and roughly 6 ft. thick. In one very conspicuous case, a solid band is interrupted by a narrow vertical belt of slag apparently in continuity with the horizontal bands of slag above and below. The exposures are a little too high up the cliff to be actually hammered, but the phenomenon is so conspicuous that it deserves special mention.  J.E.R.

*Pipe-Amygdales.*—A few words may now be offered regarding a particularly interesting type of flow-banding illustrated by several examples within Sheet 43. The flow-banding is lenticular, and is characterized by the zonal arrangement of the amygdales. Each band has a basal zone with pipe-amygdales, a middle zone free of amygdales, and an upper zone with normal more or less spherical amygdales closely grouped. There may often be half a dozen such flow-bands seen in superposition, and they have as yet only been noticed in the upper parts of flows. Easily located examples are to be found :—

(1) In the cliff, Dùn Bhuirg, Loch Scridain.
(2) At summit of the conspicuous landslip-hill (An Sithean), above the road 700 yds. N.E. of Balmeanach, Gribun.
(3) Sàilean Mòr, Gometra.
(4) Roadside, 300 yds. E. of Kilninian Church, N. of Loch Tuath.
(5) Shore, E. of Sloc an Neteogh, also north of Loch Tuath.

The last-named example is interesting on account of the diminutive thickness of the bands.  E.B.B., J.E.R., G.A.B.

The conditions which led to the flow-banding under consideration were probably strictly superficial. The layers are distinctly separated from one another. There is, it is true, no clearly marked chilling at the surface of separation ; but there is, on the other hand, no tendency for the round vesicles of the top of one band to extend upwards and mingle with the pipes of the next. Moreover, the contact seems in some cases definitely imperfect, for reddening can occasionally be observed (as for instance in the 4th example cited) along the junction-line of successive bands, suggesting that Tertiary soil-water found room to enter. It looks indeed as if each band were in a sense an individual surface-flow of miniature dimensions. A group of such bands indissolubly connected betokens intermittent advance of a lava, wave upon wave (p. 98).  J.E.R.

*Segregation-Veins.*—Another point of general interest shown by a small proportion of the lavas of the district is the occurrence of inconspicuous pegmatitic or aplitic veins consisting of augite and felspar. Such veins are best displayed among the lavas of Gometra, where, for instance, numerous examples may be examined 300 yds. south of the 503 ft. summit. Another easily found exposure is in the Dùn Bhuirg cliff, Loch Scridain, already mentioned for its exhibition of flow-banding and pipe-amygdales. In this latter case, the pegmatitic veins, or strings, occur in what is almost certainly the massive downward continuation of the flow-banded portion above. The crystallization of the veins is coarse enough to be decipherable with the naked eye ; and one can see minute augite-prisms embedded in a felspathic ground, or projecting into associated amygdales. The veins occur in lavas which retain their olivine fresh, and they do not seem to be accompanied by any notable signs of pneumatolysis. The microscopic appearances are dealt with in Chapter X.  E.B.B., G.A.B.

*Sand-Infillings.*—In the introductory remarks of Chapter V., attention was drawn to two occurrences of cavities in the Tertiary lavas of Mull filled in with quartz-sand. One of these is in the main lava-escarpment of the Gribun Peninsula, where the stream from Fionna Mhàm leads down to Caisteil Sloc nam Ban. Climbing up beside the stream, one finds that the second lava encountered is a distinctly columnar doleritic basalt, easily identified since it figures prominently in the cliffs. Its massive columnar portion merges upwards into slag with a bright-red weathered top. Not far below this weathered surface, there are miniature caves choked with bedded millet-seed sand.  E.B.B.

The other occurrence alluded to can be investigated in numerous exposures between Port Burg and Sloc an Neteogh, on the north shore of Loch Tuath. In this case again the bedding of the sandstone into the lava-cavities is quite clear, but the sand-grains are not wind-rounded.  J.E.R.

*Ulva Ferry.*—Mention has already been made of two columnar lavas seen near Ulva Ferry. Attention may now be directed to some other points of interest in this neighbourhood. At the ferry itself, only one columnar flow is exposed. It forms the Mull shore, and above it comes ash with a thick reddened top. The junction of the ash with the underlying columnar flow is concealed by raised beach; but the ash itself constitutes the lower part of the old coastal cliff above the road from near the School to Laggan Bay, and in the latter neighbourhood is well seen in stream-exposures. From this point on, it is cut out locally by a mugearite-plug constituting Na Torranan, but reappears in the road- and shore-sections of Camas an Lagain. Near the plug, the ash has what is, for Mull, a quite exceptional thickness, rougly about 30 ft. In Ulva, the same bed has been traced for some distance passing by Ulva House and the Church, but it is there only two or three feet thick, and is altered to red bole.

The Laggan ash is almost everywhere overlain by a platy mugearite. A good illustration of the relationship between the two is laid bare at the foot of the waterfall on the north shore of Camas an Lagain. The mugearite here rests with slaggy irregular base on the ash, and contains some exceptionally long pipe-amygdales springing up from its bottom layer. Individual pipes seem to measure some two or three feet. in length.  G.V.W.

## SHEET 51 : THE NORTH-WEST COAST OF MULL.

Trap-featuring is very general in that part of Mull included within the limits of Sheets 51, and is nowhere more impressed upon the observer than along the zig-zag course of the road leading from Dervaig eastwards towards Tobermory. There is also another and independent element in the scenery deserving attention in the half of the district lying north-east of Dervaig, for here a large proportion of the ridges and valleys are lineated north-west and south-

## Chapter VII.—Lavas outside Pneumatolysis Limit.

east. As will be shown in Chapter XXXIV., this is mainly due to the presence of a great belt of north-west dykes cutting the lavas of Quinish and Mishnish, and but sparingly represented in Mornish.

The most interesting single exposure of the district is to be found in the cliffs of Bloody Bay, at the eastern margin of the map, 500 yds. east-south-east of Ardmore Farm, and of easy access from above, or by boat from Tobermory. An old quarry is situated in the cliff-face at the top of a steep grassy slope leading down to the sea. It was opened in a sandstone for building the foundations and causeway of Rudha nan Gall Lighthouse (Sheet 52). The sandstone is red and carries local irony concretions. It is lying flat and shows bedding and false-bedding. Its grains are occasionally rounded. Its age is quite uncertain. The quarry-face is some 50 ft. high and 100 ft. long, and occasional exposures show that the

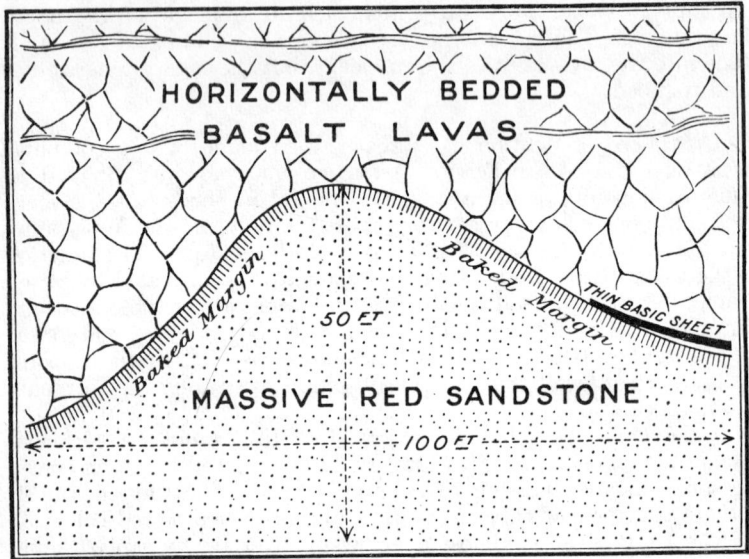

Fig. 17.—Tertiary Lavas overlying a Knob of Sandstone at Bloody Bay. Quoted from 'Summary of Progress for 1920,' p. 37.

sandstone continues at about the same level for 200 yds. farther north to where (and again south of the quarry) a north-north-west fault brings lavas down to the sea-shore. The interest of the quarry from the present point of view is the manner in which it shows an eroded knoll of sandstone projecting up into manifestly unconformable lavas (Fig. 17). There is no other instance known of obvious discordance at the base of the Mull Tertiary lavas. The sandstone where it comes in contact with the covering lava is notably hardened, and has developed a system of rude columnar joints extending inwards for a few inches. That one is not dealing with an intrusion-phenomenon seems clear, for the basalt concerned has no approach to a chilled edge and is associated with slaggy material.
G.V.W.

## Sheet 51 : North-West Coast of Mull.

For the rest, the geology of the lavas of Sheet 51 is a counterpart of much of Mull outside the Pneumatolysis Limit. There are no columnar lavas to attract attention, but the observer will find plenty to interest him in such matters as reddened tops. A good succession of basalt-flows with red weathered tops is seen in the ravined coastal cliff, north of the 200-ft. hill, east of Treshnish Point. Other good examples occur in the cliff above the road to Calgary pier, between the plantation and the stream that are shown on the one-inch Map. Another is afforded at the base of Dùn Leathan on the coast of Quinish. J.E.R., G.V.W.

A quite unusually thick red bole perhaps deserves special mention. It measures over 12 ft. in thickness, and is partly made up of lava rotted *in situ*. Though a conspicuous feature in the field owing to its dimensions and bright colour, it is not easy to locate on the map. The reader on consulting Sheet 51 will note that two streams are shown betewen Beinn Bhuidhe and Càrn Mòr. The more westerly of these two, 700 yds. north of where it crosses the map-margin, is joined on its west side by a tributary (omitted from the one-inch Map). The exposure of red bole occurs in a little gorge some 300 yds. up this tributary.

The thick red bole just referred to is overlain by a lava (20949) of very compact texture exactly reproducing the characters of the compact non-porphyritic lavas of Central Mull described in Chapter X. The lavas of Beinn Bhuidhe as a whole, it may be added, are persistently platy. Of course, platy joining is not uncommon among the Plateau Basalts, though in Sheet 51 as elsewhere it is subordinate to a rude blocky approach to columnar structure.

An example of a sparsely porphyritic lava forms a continuous scarp above the Dervaig-Salen road, $2\frac{1}{2}$ miles south-east from Dervaig. As a rule, it need scarcely be added, the lavas of the district are free from obvious felspar-phenocrysts.

Sometimes the surface of basalt-lavas in Sheet 51, as in other parts of Mull outside the Pneumatolysis Limit, weather with a multitude of pimply excrescenses. A good example may be examined below the road along Calgary Bay, at the foot of the cliff above the southern termination of the rocky foreshore half a mile south of Lainne Sgeir. Here solid pimply-weathering basalt may be seen giving place laterally to vesicular basalt with large druses containing conspicuous crystals of analcite.

Just at the margin of Sheet 51, south from Treshnish Point, the observer may study a good example of massive interbanding of solid and slag within what appears to be limits of a single flow. The slaggy layers themselves exhibit banding on a small scale. J.E.R.

Contemporaneous pegmatitic veining of the lavas has been noted occasionally in this area as elsewhere in Mull. The veins are little segregatory felspathic patches including minute augite-prisms, and are accompanied by similarly constituted linings to amygdales. Examples occur on the northern ridge of Beinn na Cille, about three-quarters of a mile from the summit; others again in a gorge of Allt Mòr, due east of Calgary Castle, where pegmatitic crystallization is seen in an amygdaloidal portion of a banded lava. G.A.B.

## Chapter VII.—Lavas outside Pneumatolysis Limit.

### SHEET 52 : TOBERMORY, MULL.

The geology of the Tobermory district is quite typical of its kind, and is largely covered by the general remarks introducing the present chapter. Trap-featuring is well-displayed, and the tendency to a north-west and south-east lineation of ridge and hollow—already commented on in the case of Sheet 51—is clearly discernable near Tobermory and also along the hollow of Loch Frisa.

The prevalence of non-porphyritic basalt may be emphasized by mention of an exceptional flow with large felspar-phenocrysts which has been followed from Area 3 of Sheet 44 across the map-boundary to the source of Allt Ardnacross in Coire Arla.    J.E.R.

Of columnar structure, only one really good example is known from the district. It occurs at the cliff-bottom 800 yds. east-south-east of the point where Glac Mhòr empties into Bloody Bay. The lava is of the triple-zoned variety. Its lower portion consists of massive vertical columns, its central portion of thinner wavy columns, and its top of typical slag with reddened surface. Its actual base is hidden by the sea.

Another, but less perfect, example of columnar structure is afforded by the Calve Island basalt, as must often have been noticed by passengers entering Tobermory Harbour by the main entrance. In agreement with what has been found in Sheet 43, both these columnar lavas are near the base of the lava-series.

A very convenient path connects Tobermory with the Rudha nan Gall Lighthouse, and mention may be made of a few points of interest which can be seen along its course. Starting from Tobermory, one soon meets with several exposures of a lava showing more or less lenticular banding, with pipe-amygdales springing from the bases of individual bands. The appearances are rather puzzling, but they recall certain occurrences in Sheet 43, already dealt with in the earlier part of this chapter. Of more exceptional nature, and much more likely to escape attention, is a small group of pipe-amygdales, each pipe about 3 ft. in height. This is seen in the cliff-wall of the path some 400 yds. beyond the exposure of a 30 ft. basalt dyke, which goes out to sea at Rudha na Leip. Continuing along the path, one presently turns a corner that brings the Lighthouse into view. A hundred yards farther on, there is a particularly striking example of pegmatitic segregation-veins within a lava. The lava itself has undergone pronounced weathering giving rise to a rusty loam. The veins consist, as usual, of felspar and augite.

The pegmatitic veins so well seen near the Lighthouse are merely a sample of what can be found here and there throughout the district. Another easily located example is met with at the roadside half-way between the Poorhouse and the bridge over Tobermory River.                                                                                                     G.V.W.

In describing Area 3 of Sheet 44 (Chapter VI.), the coast-section on the Sound of Mull was specially referred to as affording a convenient introduction to the characteristics of Mull lavas outside the Pneumatolysis Limit. Its continuation across the southern boundary of Sheet 52 is equally useful. Reddened tops are abundantly exhibited. Typical instances occur on the little peninsula projecting into the bay into which the stream drains that passes by Arla Farm; and again where the map shows an islet surrounded at high tide just north of the mouth of Allt Ardnacross. Other features of interest within this stretch of coast are the banding of some of the lavas, for instance at Rudha Gorm and also about 200 yds. south of the mouth of Allt Ardnacross. The banding consists of an alternation of thin layers of compact and amygdaloidal basalt.

J.E.R.

## CHAPTER VIII.

### BASALT AND MUGEARITE LAVAS.

#### Between Pneumatolysis Limit and Central Calderas.

##### SHEET 44.

FIVE areas, 5–9 of Sheet 44 as outlined in the Index Map (p. 91), are dealt with in this chapter. The lavas of these five areas are strikingly constrasted with those considered in the last two chapters, for they have not retained any of their olivine fresh (*cf.* Table X., p. 93). In harmony with this, they do not weather to rusty loam, or with spheroidal exfoliation, under present-day conditions; nor do they show reddened tops or red boles dating back to Tertiary times, except towards the outer boundary of the district considered as a whole. A minor character of the present-day weathering is the failure of pimply surfaces.

The development of epidotic amygdales and veins is another widespread characteristic of the lavas of Areas 5–9. Such epidotic veins often become very conspicuous as one approaches the Central Calderas; but they fail in the other direction some distance within the Limit of Pneumatolysis.

Central and Plateau Types of basalt-lava are strongly developed. The main features of their distribution can be gathered from Plate III. (p. 91). Much importance also attaches to the mugearite-zone of Ben More.

##### Area 5, Sheet 44: Scallastle Bay to Loch Spelve (Index Map, p. 91).

Visitors are likely to approach the district from Craignure where the Oban steamer is met by a ferry-boat. They will find everywhere that trap-featuring is but poorly defined. In part, this is referable to the folding and disturbance of the lavas (Chapter XIII.), and in part, no doubt, to pneumatolysis. The whole area lies within the Pneumatolysis Limit, and fresh olivine and rusty surfaces are nowhere to be seen.

Good coastal exposures of dark basalt-lavas can be examined between Duart Point and Port Donain across Loch Don. West from Grob a' Chuthaich, near the entrance to Loch Spelve, the lavas are often broken and weather pale-grey.

The base of the lava-series is seen at several places, and practically all the lavas of the area are referable to the Plateau Basalt Group—with intercalations of big-felspar basalt towards the summit of the pile.

*Auchnacraig Syncline.* A careful collection was undertaken by Mr. Tait of a lava-suite from the exposed base at Port Donain to the centre of the syncline at Auchnacraig (Fig. 25, p. 174). Of the slides (18121-18162), (18136) may be discarded as an intrusion, and the remainder are Plateau Basalt Types, save for (18161-2) representing the big-felspar basalts shown on the map and section near the top of the sequence. This Auchnacraig outcrop of big-felspar basalts was first noticed by Mr. Tait. Epidote is a common mineral of the district, both in amygdales and veins.

*Coire Mòr Syncline* (Plates III. and V., pp. 91, 165). Another important suite (18822-18834) collected by Mr. Tait extends along Abhuinn Lìrein upstream from the Jurassic outcrop. Of these, two (18832-3) come from the big-felspar basalt exposure in line with the head of Abhuinn Barr Chailleach. The remainder are for the most part clearly referable to Plateau Basalt Types, with a mugearitic variety (18830). It is noteworthy that two variolitic slides occur, of which (18826) is very reminiscent of the variolites of the south-east Caldera (Chapter IX.), though (18829) is quite distinct.

The big-felspar basalts of the Coire Mòr Syncline are noteworthy, since their outcrops have proved of great importance in elucidating the structure. They are easy to trace in two bands on the east side of the syncline and in Abhuinn Lìrein. On the south-west side, their outcrops are greatly interrupted by intrusions (*cf.* one-inch Map). It is the close conformity of the agglomerate of the district to the big-felspar basalt outcrops that leads to the conclusion that the agglomerate is here an accumulation on top of the basalt-lavas, and not the contents of a vent (Chapter XV.). This lends significance to the general absence of the Central Group of Mull lavas in the type-locality of the Coire Mòr Syncline.

The most striking outcrop of big-felspar basalt is that afforded by Maol nan Uan. That one is dealing with lavas, and not with an intrusion, is well shown by the contemporaneously weathered slaggy top of the same mass as exposed in the Scallastle River. The most westerly exposure of big-felspar basalt in Gleann Lìrein may also be mentioned as thoroughly vesicular.

*Various.* A point which cannot be passed over in silence is that, of eleven big-felspar basalt slides examined, three (14226, Maol nan Uan; 18161-2, Auchnacraig) have a distinctly variolitic base. Possibly this implies pillow-structure which was overlooked in the field, but perhaps more probably it is a mere coincidence.

Mention has been already made of a mugearitic flow from Abhuinn Lìrein. An easily located example of a mugearite-lava (15555), low down in the sequence, occurs at Ardchoirk, north of Loch Don. It rises from among raised-beach deposits, and has not been followed far, and accordingly is not distinguished on the map.

An interesting occurrence of pegmatitic veining or segregation, perhaps the first noticed in Mull, can be examined on the shore 200 yds. north-north-west of the school at Scallastle. The veins are quite unusually coarse (15547); and they traverse a green doleritic basalt (15546), which is interpreted as a lava for it seems to be linked to a well-marked vesicular top (15545). An interesting microscopic feature of the top is its variolitic structure. Probably,

122 *Chapter VIII.—Lavas between Pneumatolysis Limit & Calderas.*

if the frothy portions of flows in Mull had been freely sliced, many of them would show variolitic characters; in the case of pillow-lavas one finds such structures commonly developed in compact specimens.

E.B.B.

### Area 6, Sheet 44 : Croggan Peninsula
(Index Map, p. 91).

For convenience, a small tract of lavas north of Loch Uisg will be considered in this section, along with the peninsula lying between Loch Spelve and Loch Buie.

The whole district is characterized by a lack of fresh olivine and consequently of rusty weathering tints. Good coastal exposures extend from Croggan Pier by way of the Firth of Lorne to where granophyre appears near the head of Loch Buie. The usual associations of solid and slag are repeatedly seen; and good instances of pipe-amygdales springing from the base of a lava can be examined in the bay east of Rudha Mhàirtein, and again as one approaches the stream that flows from Coire na Caise.

Coastal exposures belong almost exclusively to Plateau Types. Central Types have only been encountered on the south shore of Loch Buie in the immediate neighbourhood of the Loch Uisg Granophyre. The coast-line of Loch Spelve, where one might expect to meet the Central Types in force, yields practically no exposure of any sort.

The Central Types, though not seen on the coasts, are abundantly exposed in the low rocky hills of the interior of the peninsula. Their outcrop is so definitely in relation to well-marked dips away from the intimately associated Plateau Types that their superposition can be safely inferred. The Central Types here referred to are characterized in the field by their small felspar-phenocrysts or— where non-porphyritic—their extremely compact texture.

A long suite of specimens (18901-18942) was collected by Mr. Tait to decide how far the separation into Plateau and Central Types could be relied upon. The line sampled started at Port a' Ghlinne and led by way of Beinn na Sròine to near the Manse at the head of Loch Uisg. A few slides had to be discarded as of intrusive types inadvertently collected, but the remainder grouped themselves as follows :—(18901-6   18909-10)   Plateau ;   (18913) Central ;   (18915) Plateau ;   (18916-29,  18931-42) Central. Thus interstratification of the two types seems quite subordinate in this line of traverse. It may also be added that among sliced specimens the non-porphyritic Central Types are about twice as numerous as the porphyritic.                                          E.B.B., G.V.W.

Though subordinate, the interbedding of Plateau and Central Types is recognizable as a feature of the geology of the neighbourhood of Gortendoil in the north part of the peninsula. Lavas with small felspar-phenocrysts are here very clearly interbedded with others of non-porphyritic Plateau Type towards the top of the Plateau Group.                                                          E.B.B.

In the western part of the Croggan Peninsula, the Pale Group of Ben More begins to be recognizable, but it has not been possible to map a definite line for its base. Its presence is emphasized by a thin band of mugearite, which is to be seen at several isolated exposures on both sides of the Lochan an Daimh Syncline (*cf.* one-inch Map and Plates III. and V., pp. 91, 165). Northwards from here, the mugearite is repeated by folding, and its outcrop can be traced for a distance of over a mile along the northern flank of the Beinn a' Bhainne Anticline.

In this district, as is also the case in Area 7, the mugearite occurs at no great depth below the main mass of Central Types. It is likewise associated with an underlying big-felspar basalt, which occurs directly beneath it at an exposure about half a mile north-east of Lochan an Daimh, and again, a short distance below, near the lochan at the head of Allt a' Bhàird. G.V.W.

Of the lavas lying north of Loch Uisg, little need be said. A faulted outcrop of big-felspar basalt has been traced for about a mile on the hill-side near Sròn Gharbh. The band runs north-east with a well-marked dip towards the south-east, so that such lava-exposures as occur farther west along Loch Uisg presumably belong to lower positions in the Plateau Group. East of Sròn Gharbh, on the other hand, Central Types are soon met with, and very typical compact representatives are exposed west of the wall leading south-west from Kinlochspelve Farm. As neighbouring exposures at Kinlochspelve consist of Triassic sandstone, it appears almost certain that a great fault intervenes. The rocks of the district are very broken and disturbed, so that there seems no difficulty in admitting such a dislocation. E.B.B., G.V.W.

Although trap-featuring is never strongly marked, it is sufficiently evident to enable one to see in the landscape a well-marked development of arcuate folds (Plate V., p. 165). These folds are exposed in section, for instance, in the cliffs either side of Loch Buie.

G.V.W.

One of the most striking features of the district is the baking of the lavas by the underlying Loch Uisg Granophyre (p. 153). In the field it arrests attention owing to its influence upon the scenery. Looking at the peninsula from the road, one is immediately struck by bold crags, due to the baked lavas, above Loch Uisg, as contrasted with grassy slopes leading to Loch Spelve.

E.B.B., G.V.W.

AREA 7, SHEET 44 : LOCH BUIE TO LOCH SCRIDAIN AND GLEN MORE (INDEX MAP, p. 91).

The lavas of this area are characterized by the absence of fresh olivine with a consequent lack of rusty weathering on exposed surfaces. In the eastern part of the area, near Cruach nan Con and Beinn nam Feannag, trap-featuring is unrecognizable. Westwards towards Glen Leidle, however, as has been pointed out by Mr. Lightfoot, this type of scenery becomes increasingly evident, partly owing to the fact that we are approaching the Pneumatolysis Limit of Plate III. (p. 91), but mainly because we have passed out of the zone of cone-sheets (one-inch Map).

In the inner part of the area, the lavas are often strung through with veins of epidote, which, together with epidotized amygdales,

124 *Chapter VIII.—Lavas between Pneumatolysis Limit & Calderas.*

stand out as prominences on the weathered surfaces. Excellent examples can be seen near the bridge over the Coladoir River, at the head of Loch Scridain.

Cinnamon-coloured garnets occur in some of the amygdales, and good specimens can be obtained on the south coast about a quarter of a mile north-east of Rudha Dubh.     (C.T.C., B.L.), G.V.W.

Both Plateau and Central Types are well-represented in the area, and the latter are clearly seen to be the more recent. To test the degree of separation of these types, Mr. Tait collected two suites of specimens radiating out from Glenbyre. The first (18951-61) was along a line a mile and a quarter long taken through the summit of Beinn nan Gobhar, and the second (19086-19100) along the coast from a point 200 yds. south-west of Glenbyre to 1200 yds. north-east of the same.

Microscopic examination of the specimens demonstrates the presence of :—

    (1) Plateau Types in the neighbourhood of Glenbyre (18953, 4 ; 19086-9).

    (2) An important outcrop of mugearities on the coast 300-700 yds. north-east of Glenbyre (19090-19093). This outcrop is shown on Pl. III. (p. 91), but not on the one-inch Map.

    (3) An assemblage of Central Types in Beinn nan Gobhar, above the level of An Sithean (18955-61).

Central Types occur among the specimens collected on the coast north-east of, and almost directly above, the mugearite already referred to ; but it is not quite certain where the boundary of the main outcrop of Central Types should be drawn.

It was not found possible in Area 7 to map out a definite band to represent the Pale Group of Ben More (p. 125), though rocks of this group are manifestly present, including both the Ben More Mugearite and a big-felspar basalt at a slightly lower level. A few observations regarding these two interesting rocks will now be given.     G.V.W.

The continuation of the Ben More Mugearite from Area 8 was traced by Mr. Lightfoot for three miles south-wards from the shore of An Leth-'onn, near Kinloch Hotel. Later investigation has shown this rock to be a well-characterized lava. At the roadside-exposure in a small bay some 200 yds. east of the Hotel, the top of the mugearite is hidden, and the portion seen is intensely platy. In more complete sections about half a mile inland, the platy band is overlain by a thick, vesicular, brecciated, and reddened top. The dip is easterly, and some of the immediately overlying lavas, exposed between the mapped outcrop of mugearite and Rossal Farm, are themselves platy and of mugearitic aspect.     E.B.B.

An outcrop of the same, or approximately the same, mugearite-zone has been recognized in the Cruach Inagairt Syncline, and the Beinn na Croise Anticline, north-west of Loch Buie ; it has been traced by a number of isolated exposures from Cruach Inagairt, around the head of Glen Byre, to the coast north-east of Glenbyre Farm, where, as already stated, it is shown on Plate III. (p. 91)—though not on the one-inch Map.  Details are as follows :—

    At Cruach Inagairt, mugearite is exposed on both sides of the syncline and probably more than one flow occurs.

    Northwards from here, the outcrop on the east side can be traced across the

Leidle River to Leac an Staoin, where underlying big-felspar basalt is exceedingly well-exposed, and truly deserves its name, seeing that individual crystals of felspar range up to 10 inches in length. The dip here is to the west, and the mugearite is a well-marked platy rock.

Big-felspar basalt is again exposed a few yards south of the path and about 400 yds. east of Lochan Tana, whence the outcrop can be traced a distance of almost a quarter of a mile towards the head of Glen Byre. At this locality the dip is north-east, and the big-felspar basalt is apparently both overlain and underlain by pale rocks which on slicing would probably be found to be mugearites.

Platy rock of mugearitic aspect is seen about 50 yds. above the path some 760 yds. south-east of Priosan Dubh, and at no great distance below the base of the Central Types, which are admirably exposed on Beinn nan Gobhar.

A wide outcrop of mugearite is laid bare on the coast from 300-700 yds. north-east of Glenbyre Farm. In all probability more than one flow occurs, and mugearitic ash has also been noticed.

At a narrow road-cutting (The Split Rock) near Coill' a' Chaiginn, not far east of the mugearite exposures last mentioned, lavas of Central Types are well seen, the lowest of them with its base chilled. G.V.W.

A considerable exposure of mugearite, with associated tufaceous material of similar composition, occurs near Derrynaculen in Glen More. These rocks, quite clearly, emerge on the crest of an anticline from beneath the neighbouring basaltic flows, and in all probability are the continuation of the Ben More zone. The overlying lavas are mainly of porphyritic Central Types, but a specimen from a small outlier on top of the mugearite, half a mile west of Derrynaculen, is a typical Plateau Basalt (15596). (C.T.C.)

AREA 8, SHEET 44 : BEN MORE FROM GLEN MORE TO LOCH NA KEAL (INDEX MAP, p. 91).

In its southern part, the Ben More district continues in all its features the geology of Area 7 just described. Its westerly half, along the coast of Loch Scridain, is occupied by the Plateau Group of olivine-rich basalts ; at Ardvergnish is the mugearite ; east of this, up Glen More, one soon enters upon the region of Central Types, poor in olivine.

Though the occurrence of Central Types, both porphyritic and non-porphyritic, is well established within the area indicated in Plate III. (p. 91), the margin of the belt has not been fixed very precisely by detailed collection. It approximately follows a fault-line along most of its course. Sliced examples of the compact non-porphyritic type have been collected just east of the Glen More road, due west of the sharp bend of the river near Uluvalt (16587), and several of the flows of this neighbourhood are markedly porphyritic. E.M.A.

The main interest of the district is found in the lava-sequence of Ben More, where the Plateau Group reaches its highest development so far as is known. The vertical section of the Index (p. 91) is based upon the Ben More succession. The Plateau Group consists mainly of olivine-rich basalts, and is divisible into two portions, pale above and dark below, each of them roughly speaking about 1500 ft. in thickness. The Pale Group generally includes a thick zone of mugearite, and on the north face of the mountain its representative sections read from below upwards as follows :—pale basalts 400 ft. ; mugearite 300 ft. ; and pale basalts 800 ft.

The Pale Group of Ben More has been mentioned already as vaguely recognizable in Areas 6 and 7. In the Ben More district, it has been found possible to draw a fairly satisfactory line for its base which is shown both on Plate III. and also on the one-inch Map. The characteristic pale tint is sufficiently marked to be a scenic feature of the upper portion of the ridges north of Ben More, especially Beinn Fhada.

The Ben More Mugearite is, as already stated, an important element of the Pale Group. It weathers very pale indeed, and is generally easily distinguishable owing to its platy jointing determined by fluxion.

The basalts of the Pale Group are found on microscopic examination to be typical olivine-rich Plateau Basalts. So too are the underlying dark basalts. The cause of the distinction of tint only became apparent as a result of special investigation. With this end in view, Mr. Tait took a series of 72 specimens from the seashore at Dererach, on Loch Scridain, to the summit of Ben More. These may be classified as follows :—17751-17780 mainly dark basalts ; 17781-17795 mainly pale basalts ; 17796-17801 mugearite ; 17802-17822 mainly pale basalts. The distinction between dark and pale shows at once in the slides if they are arranged serially on a white background, and is due to a profusion or otherwise of magnetite—mostly derived from the decomposition of olivine. It appears then that the dark basalts probably contained, when fresh, a more ferriferous olivine. Other differences noted in this suite of slides were the frequent coarser texture of the lower basalts, and also a marked tendency of the ophitic augite of the Pale Group basalts to occur in distributed crystals giving rise to a mottled structure appreciable in transparent slices (p. 138). E.B.B.

In the Pale Group north of Ardvergnish, a highly porphyritic basalt can be followed above the mugearite for half a mile along the outcrop, and is shown on the one-inch Map as a big-felspar basalt. Other examples of porphyritic basalts have been noted occasionally, as for instance a little below the mugearite in Lag a' Bhàsdair, north-east of Ben More, but have not been separately mapped. The great majority of the Plateau Basalts are here, as elsewhere, not noticeably porphyritic.

Since the Ben More Mugearite is likely to attract particular interest, a few details may be given concerning its outcrop. Owing to its platy structure, the mugearite is easily distinguished near Ardvergnish (the farm itself stands on a cone-sheet of craignurite, Chapter XIX.), and can be traced for a mile and a half up Gleann Dubh where its thickness must sometimes exceed 100 ft. For about a quarter of a mile north from the sudden bend of the river above Ardvergnish, a band of breccia, mostly made of mugearite, interposes between the fluxional mugearite and the big-felspar basalt already noted. Unless this band belongs to some linear volcanic vent—which would be a curious coincidence—it strongly supports the view that the accompanying mugearite is a true lava.

The Ardvergnish mugearite has no present-day surface-connexion with the outcrop of Ben More. The latter starts somewhat abruptly in Coir' Odhar, whence it can be traced round the spur of Maol nan Damh cut by numerous basic sheets characteristic of the district (Chapter XXVII.).

From Maol nan Damh, the mugearite continues with sinuous outcrop right round the northern face of Ben More. It is about 300 ft. thick and forms crags with columnar jointing conspicuous at a distance. The mugearite has also been mapped round Beinn Fhada, except where its outcrop is interrupted by granophyre. Near the granophyre it is notably altered, so that it is difficult to identify; but its occurrence has been confirmed by microscopic examination (16636, 16651).

On the eastern side of A'Chioch, the mugearite has also been recognized and can again be shown to terminate suddenly along the outcrop, this time in a southerly direction.   E.M.A., J.E.R.

The local evidence that the Ben More Mugearite is a lava is as follows :—On the north face of Ben More, the mugearite is overlain by a foot or two of fissile black shale abounding in mugearite-fragments. A similar bed is seen at the top of the flow on the eastern declivity of A'Chioch (17149, 17150). Moreover what seems to be the top of the mugearite, north of the granophyre of Beinn Fhada, is thoroughly scoriaceous.   J.E.R.

Two other instances of sedimentary intercalations among the lavas of the Ben More district have already been referred to in Chapter III.   E.M.A.

It is interesting that red boles have only been noticed close to the line taken as the outer margin of the district—the Limit of Pneumatolysis as it is called on Plate III. (p. 91). An example of one of these exceptional red boles may be seen south-west of Rudha na Mòine on Loch na Keal. The general absence of such red boles is the more noteworthy, since the capital exposures of the precipitous western slopes of Ben More have been searched for them without result.

Present-day rusty weathering, such as is characteristic of many of the flows of the plateaux outside, is also wanting in Area 8. The lava-surfaces are of various shades of gray and brown. In keeping with this the microscope has failed to detect fresh olivine; although its pseudomorphs occur in 113 slides of the Survey collection.

As already pointed out, the decomposition of the olivine is interpreted as a result of pneumatolysis (Chapter V.). An accompanying characteristic, easily recognizable in the field, is the development of epidote. This mineral has been noted in amygdales as far west along the shore of Loch na Keal as the mouth of the river at Derryguaig, where it occurs, as so often, in association with chlorite. From here inwards it is found in increasing abundance in the amygdales of the lavas; and throughout much of the interior of the district it serves as a very conspicuous infilling to a series of reticulate cracks.

Thus the alteration of the lavas within the present area is very definitely proved. It is of importance, therefore, to note that trap-featuring though subdued is quite commonly apparent. The northern slopes of the Ben More range show a sufficiently obvious succession of lava scarps. On examination, it can be seen that these are due to the normal occurrence of slaggy top and solid interior such as is so often met with in the case of lava-flows. The featuring of the Pale Group is more closely spaced and weaker than in the case of the Dark Group below, and individual flows seem to be thinner on the average. J.E.R.

Reference has already been made to the wide-spread occurrence of epidote and chlorite in the amygdales of the district as a whole. The subject will be dealt with again in Chapter X. Albite is the most common associate of the two minerals just mentioned, but scolecite is sometimes very strongly developed. Many of the lavas occurring rather above the base of the Pale Group have vesicular cavities up to 6 inches or so in length, with their central portions filled with radiating fibrous masses of scolecite. Examples of this character have been noted, sometimes *in situ*, more often in scree, on Beinn Fhada, An Gearna, and southwards to Meall nan Damh. In the latter locality, very good material is furnished by a conspicuous scree, at about 1500 ft. above sea-level, on the west face of the hill. The base of the Pale Group in Maol nan Damh is shown on the map at about the 1800 ft. level, well above the scree; but it is an ill-defined line and re-examination might show that it has been drawn rather too high in this particular neighbourhood. An outlier of lavas, with amygdales of scolecite, albite, epidote, etc., occurs on the top of Coirc Bheinn in Area 2 (p. 104), not far outside the Pneumatolysis Limit as drawn on Plate III. (p. 91). The height of this outlier above sea-level is between 1650 and 1800 ft. It probably represents a western continuation of the Ben More suite —whether referable to the Pale Group or somewhat earlier it would be hazardous to say. E.M.A., J.E.R.

While it will probably be admitted by all that the general albite-epidote infilling of the amygdales of the lavas of Area 8 is part of the complex pneumatolytic phenomenon of central Mull, Dr. M'Lintock has argued for a particularly early date of pneumatolysis in the case of the albite-epidote-scolecite amygdales of Ben More. He suggests that they are a product of auto-pneumatolysis. The subject is discussed later (p. 141). The main difficulty is that Dr. M'Lintock's arguments, followed to what seems at present their logical conclusion, would appear to ascribe the constant alteration, and the peculiar amygdales, of the lavas encountered within the Pneumatolytic Limit to auto-pneumatolysis—but this one may say at once is incredible. It may further be added as a field-observation that a large proportion of the minor intrusions traversing the lavas of Area 8 are to all appearance altered in precisely the same manner as the lavas themselves. This remark applies to very numerous undulating basic sheets or sills in Ben More (Chapter XXVI.), and also to about a quarter of the North-West basaltic dykes of the district as exposed along the shores of Loch na Keal opposite Beinn a' Ghràig (Chapter XXXIV.). The amygdales of these dykes, for instance, contain epidote, albite, and, sometimes, garnet and hornblende. On the other hand, it must be remembered that the albite-epidote-scolecite amygdale-assemblage seems to extend, in the lavas of Coirc Bheinn, rather beyond the limit of general pneumatolysis as drawn on Plate III. E.B.B., J.E.R.

On approaching the granophyre-margin at any point, increasingly pronounced alteration becomes apparent. A convenient locality to observe this is above the road south-west of Knock. A series of specimens collected here (14816-9) clearly show that the lavas were first pneumatolysed and then baked with, for instance, the production of biotite. W.B.W.

AREA 9, SHEET 44 : SALEN TO SCALLASTLE BAY
(INDEX MAP, p. 91).

The district is so accessible that it is likely to be regarded by many as a type-area. As already remarked in describing Area 4a (Chapter VI.), comparison of the condition of the lavas within and without the Penumatolysis Limit can nowhere be undertaken to better purpose than on the two sides of Fishnish Bay. In Area 9, it has been found impossible to obtain fresh olivine ; and, in keeping with this, present-day rusty weathering is wanting. Tertiary red tops and boles are not, however, entirely absent, at any rate for some little distance in from the limit : a few inches of red bole are exposed on the shore about 100 yds. south of the abandoned pier at Salen ; and a good example of a reddened upper surface is seen near high water mark opposite Doire Dorch, Fishnish Bay.

On the hill-slopes facing the Sound west of Fishnish Bay, trap-featuring is strongly marked—an unusual occurrence within the Pneumatolysis Area. The phenomenon is interrupted south of Fishnish Bay, partly as a direct result of intrusion. Thus on looking from the Sound of Mull steamer up the course of Allt Mòr Coire nan Eunachair, it is easy to appreciate the struggle for scenic expression between lavas and cone-sheets where the latter begin to appear. W.B.W., E.B.B.

Even if there were no intrusions to complicate the issue, it is certain that trap-featuring would fail in much of the district. For instance, in the upper part of the drainage basin of Allt Mòr Coire nan Eunachair, and on the slopes of Beinn Chreagach Mhòr and Beinn Chreagach Bheag, there are a great number of breccia-filled vents, many of them too small to show on the map. The breccia is characterized by fragments of gneiss so that it must have originated through explosions—it cannot be interpreted as a crush-breccia. Round about the vents, the lavas show wide-spread shattering and veining with epidote, while the cone-sheets do not. Clearly, after suffering such treatment, the lavas could not be expected to show trap-features. They weather with a network of epidotic veins projecting from their exposed surfaces.

The district belongs predominantly to the Plateau Group. The main type encountered is non-porphyritic basalt. To test the validity of this conclusion based upon field-observation and scattered specimens, Mr. Tait collected a series between Fishnish Bay and the summit of Maol Buidhe (18066-18120). The suite of slices may be classed in succession from below as follows :— 27 specimens Plateau Basalt (18066-72, 18074-93) ; 1 Central Basalt (18094) ; 14 Plateau Basalt (18095-18108) ; 5 Mugearitic Lava with Tuff (18109-13) ; 2 Plateau Basalt (18114-5) ; 1 Central Basalt (18116) ; 4 Plateau Basalt (18817-18120). It is noteworthy that epidote was found quite near the shore (18067), and fresh olivine, as already stated, not at all.

In the list given above, the mugearite-specimens probably represent a continuation of the Ben More zone, but the outcrop from which they were collected has not been separated on the Map.

In Plate III. (p. 91) a considerable area of Central Lavas is shown in Area 9. It is greatly complicated by intrusions, and its boundary-line is a mere approximation, but typical compact non-porphyritic olivine-free basalt can be obtained south-east of the summit of Beinn Chreagach Mhòr.

The two outcrops of big-felspar basalt indicated in Plate III. are based upon mere scraps surviving among cone-sheets. One group of exposures is above the bend of Allt Coire Fraoich, where a steep southerly dip is clearly discernable, the other in Allt nan Clàr, 500 yds. up from a sheepfold.

Before passing on, one may note an interesting little point which a visitor finds ready to hand on the foreshore east of Salen Pier. A lava, which is for the most part comparatively non-vesicular, is seen here enclosing intensely vesicular globular masses a couple of feet in diameter. There are no chilled margins in relation to these porous globes, but at the same time the appearance is slightly reminiscent of pillow-structure.
E.B.B.

# CHAPTER IX.

## BASALT LAVAS OF CENTRAL CALDERAS.

### SHEET 44.

ALL the 74 lava-specimens collected for microscopic examination from the two calderas (Area 10, of Index Map, p. 91) are of Central Types. Five of these specimens come from the north-western caldera, the rest from the south-eastern. Both porphyritic and non-porphyritic types are well represented, the former ranging from doleritic to compact in their groundmass; the latter are always compact, and sometimes very markedly so. Among the porphyritic lavas, there is often a nearer approach to the big-felspar basalt type than is at all common in the assemblages of Central Types met with outside the calderas. Good examples of these more coarsely porphyritic basalts are afforded by Bith Bheinn (15566, 17129) in the north-west caldera, and Cruach Choireadail (17184) in the south-east.

It is hardly necessary to remark that fresh olivine, rusty weathered surfaces, and trap-featuring are everywhere conspicuously absent. In harmony with this, epidote is very commonly to be recognized in amygdales and cracks.

The extremely interesting tectonics of the district are dealt with more particularly in Chapters XIII and XXXII.

(C.T.C.), E.B.B.

### The North-West Caldera.

The various patches of lava encountered in the north-western caldera are rendered so discontinuous by intrusions that nothing significant can be said of the distribution of the various types. Porphyritic basalts are prominent in Bith Bheinn (15566, 17129), and are recognizable again in Beinn na Duatharach (14684). In both cases, they are accompanied by typical non-porphyritic highly compact varieties (14683), of which another good example forms the precipitous slope of Na Bachdanan at the head of a stream draining down to Loch Bà. Na Bachdanan lies within three miles of Salen and is easily located, so that it is likely to be visited by geologists wishing to familiarize themselves with the type. The crags yield thoroughly characteristic material, black on the broken face, and about as compact in texture as rhyolite. A specimen (14824) collected here has been analysed (p. 17).

No pillow-structure has been noted among the lavas of the north-western caldera; but the vent-agglomerates are suspiciously free from gneiss-fragments (Fig. 29, Chapter XVI.), so that perhaps the inbreak may have started at a comparatively early date. On the other hand, the most conspicuous feature of the caldera, as we know it to-day, is its girdle supplied by one of the latest intrusions of Mull, a ring-dyke of felsite which has taken its name from Loch Bà (Chapter XXXII.). W.B.W., J.E.R.

## The South-East Caldera.

The south-eastern caldera has been intensively studied since it has yielded so much of interest in the form of pillow-lavas. Its contents fall into three concentric zones, almost certainly of increasingly late date as one meets them successively (1-3) on traversing the caldera area towards its centre from either west or south :—

(1) In the outer zone of lava, the Central petrological characteristics are often ill-marked. Porphyritic felspars are to be found in many of the flows, but seldom in profusion. There is a distinct tendency in many cases towards Plateau Type; but definite examples of the latter have not been found. Good exposures are afforded by Beinn Bhearnach (north-west of Sgùrr Dearg), Beinn Fhada, and Creag na h-Iolaire (west of Loch Airdeglais).
(C.T.C.), E.B.B., G.V.W.

(2) The middle zone consists of highly porphyritic lavas crowded with felspars. Some of the examples are rather coarser, approaching more nearly to the Big-Felspar Type, than is common in the Central Outcrop outside the calderas. Exposures of lenticles preserved at intervals between cone-sheets are to be met with on Beinn Talaidh and Maol nam Fiadh. South of this, the belt is replaced by intrusions for a space, but can be picked up again at Ishriff, and continues into Cruach Choireadail where it is most conveniently studied. (C.T.C.), E.B.B.

(3) The interior zone consists entirely of very compact non-porphyritic basalts seen at many places. Of these one may cite :—the southern stream at Doir' a' Mhàim; the junction of the two streams flowing south-west from Monadh Beag; the stream-banks between Ishriff and the main road; the first bend above the road of the little stream draining south-west from Tom na Gualainne; and so on to a little north of the summit of Cruach Choireadail. The exposures mentioned are from the outer margin of the area of compact lavas, where they come near their porphyritic fellows of Zone (2). The course of the line separating Zones (2) and (3) is indicated for a mile on Fig. 52 (p. 308).
(C.T.C.), E.B.B., J.E.R.

Pillow-structure, noted at many places (Figs. 18 and 53, p. 312), frequently characterizes the lavas of Zones (1) and (2), but it is wanting in the interior Zone (3). (C.T.C.), E.B.B., G.V.W.

The petrological characteristics of the lavas of Zones (2) and (3) are strikingly Central, even in field-exposures. Those of Zone (1), as mentioned above, are less defined. Accordingly, Mr. Tait collected 25 specimens of lavas belonging to Zone (1), from east to west along a mile spanning the summit of Beinn Bhearnach, north of Sgùrr Dearg. On slicing (18876-18900), about half of these specimens (18880, 18889-94, 18896-18900) proved to be typical Central Types. The remainder, mainly from the western end of the traverse, are intermediate in character between Central and Plateau Types, and will be referred to again in this connection in the petrological account (Chapter X.). E.B.B.

With reference to pillow-structure, it may be stated, in general, that individual pillows are often about 2 ft. long, and are clearly marked out by chilled margins and also by concentric zones of amygdales. Figs. 18 and 53 indicate several mapped exposures, but there is great difficulty in directing an observer with any confidence

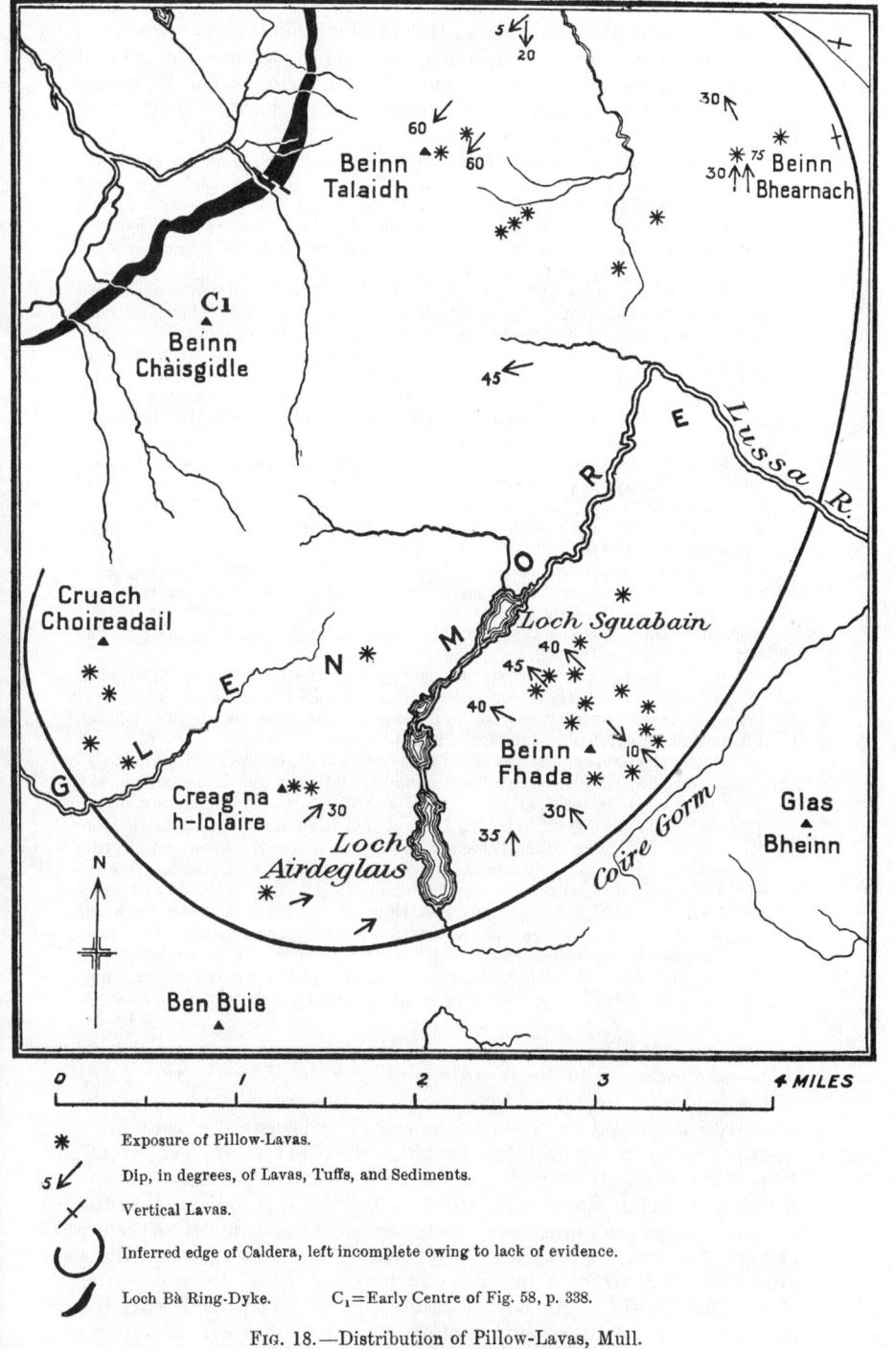

FIG. 18.—Distribution of Pillow-Lavas, Mull.
Quoted from 'Summary of Progress for 1914,' p. 40.

## Chapter IX.—Lavas of Central Calderas.

that he will find what he seeks in this matter ; landmarks are scarce, the phenomenon searched for is often inconspicuous, and the general geological complexity is apt to prove bewildering. The following examples are, however, specially selected with a view to their being discoverable if carefully looked for :—

(1) If one follows the path, shown on the one-inch Map, from Torness across the flat watershed into the Glen Forsa drainage-basin, one presently meets a small stream (coming in from Beinn Bhearnach), 1 mile north-north-west of Torness. One hundred yards short of the crossing, in the acute angle between path and stream, and about 20 yds. from either, is a little crag of non-porphyritic lava with very typical pillow-structure (Pl. IV A). E.B.B.

(2) A scenic feature of the Beinn Fhada ridge is a thick dolerite sill folded into a basin so that its outcrop surrounds the hollow of Coir' Odhar. At the north-east extremity of this outcrop (below the sill, but rather to the north-west) at the top of the slope from Coire Gorm, is seen a very thick coarsely porphyritic basalt with subordinate pillow-structure, and with pipe-amygdales in its base where it rests chilled upon a thin pumiceous tuff. This porphyritic flow has been mapped for a considerable distance and is about 120 ft. thick. Below it, in the section here considered, lies a non-porphyritic lava which affords a much more perfect example of pillow-structure. Its thickness is 35 ft., and, more particularly in its upper portion, it consists of pillows piled one on top of another with tuff between. From the base of each pillow spring pipe-amygdales, while small round amygdales in concentric zones occur at the sides and margins of the same (Pl. IV C).

The two lavas just described, the non-porphyritic again showing good pillow-structure, can be easily located once more about 300 yds. farther south-west, just inside the outcrop of the Coir' Odhar sill. The exposure is a little west of a small lochan, shown on the one-inch Map, as resting half on the lavas and half on the sill.

(3) The geologist is well advised to approach the exposures enumerated in (2) by way of the little streams that drain into Loch Scuabain ; for several examples of pillow-structure have been noted on the steep margin of the Beinn Fhada plateau overlooking the loch (Pl. IV B). G.V.W.

(4) On the southern slopes of Cruach Choireadail, between 300 yds. south-south-west and 500 yds. south of the summit, and again much nearer the road 1000 yds. south by west of the summit and 900 yds. east of Craig Cottage (that is about the 600 ft. level, in the angle of the outcrop of the Glen More Ring-Dyke, Fig. 54, p. 322), there are craggy exposures of coarsely porphyritic pillow-lava. Individual pillows measure 2-4 ft. across ; and, though rude in shape and devoid of zonally arranged vesicles, they are conspicuous and show very marked chilling at their margins (Fig. 21, p. 151). Between adjacent pillows, thin green or brown streaks with distinct fragmental structure have been frequently noticed together with harder patches of a pale-grey colour. Microscopic examination shows that these thin sedimentary and ashy infillings have been subjected to the same type of alteration as the associated lavas.

(C.T.C.)

References have been made in the preceding paragraphs to contemporaneous tuff and sediments associated with the lavas of the caldera. So far as one knows at present, the phenomenon is always developed on a very minor scale. The most considerable example as yet recognized has been described in Chapter III., p. 66. Two other very interesting cases may now be mentioned : one is situated on Beinn Bhearnach, where a dip of 75° is indicated on the one-inch Map, one mile west-north-west of the summit of Sgùrr Dearg ; the other is exposed in a stream where an arrow marks a dip of 45°, 600 yds. east and 200 yds. north of the south-west corner of Fig. 53 (p. 312). At both localities, tuffs interbedded with the basaltic lavas of the south-eastern caldera are seen containing fragments of fluxion-rhyolite. The Beinn Bhearnach example is

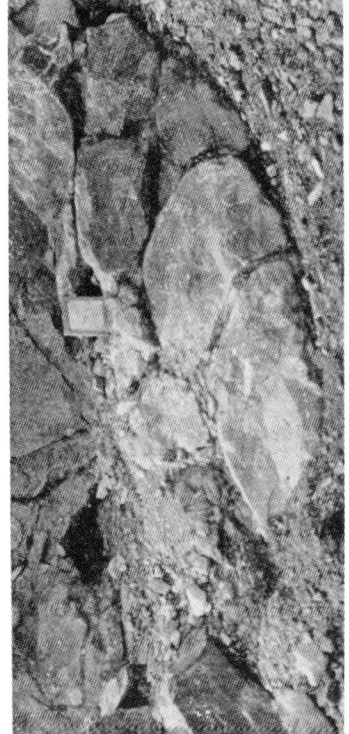

PLATE IV.—Pillow-Lavas of South-Eastern Caldera.

more readily found than might be expected, for it occurs at the foot of a little escarpment sufficiently marked to attract attention. Here one foot of fine sediment, along with the tuff containing rhyolite, separates two coarsely crystalline somewhat porphyritic pillow-lavas. The Doir' a' Mhàim tuff is easily recognized in stream-section where it passes beneath vesicular lavas of the compact non-porphyritic type. It is clear then that acid tuff was more than once available during the period of the basic lavas of Central Types found within the caldera. Very possibly, further research will show that in a few cases, where tuff has been assigned to necks on the one-inch Map, it is in reality interbedded among the lavas (p. 197).    E.B.B.

# CHAPTER X.

## PETROLOGY OF BASALT AND MUGEARITE LAVAS.

### INTRODUCTION.

A GENERAL introduction to the basalt and mugearite lavas of Mull is furnished in Chapter V. It is there pointed out that certain types of basalt-lava, Plateau Types, are particularly characteristic of the earlier part of the lava-sequence still preserved from erosion; that other types, Central Types, are equally characteristic of the later part of this sequence; that mugearite-lavas are interbedded among the Plateau Basalts, more especially on a comparatively high horizon, well-displayed in Ben More; and that early representatives of Central Types are also found associated with Plateau Types—these early Central Types include basalts of Staffa Type, occurring at the base of the Plateau Group in south-west Mull, and also big-felspar basalts, roughly synchronous with the Ben More Mugearites.

The chemistry of these different lavas is represented by eleven analyses, quoted and discussed in Chapter I. (pp. 15, 17, 24).

The differences, which in Ben More allow of a separation of the Plateau Basalts into a dark group below and a pale group above, have been dealt with in Chapter VIII.

The pneumatolytic changes and contact-alteration, by which both Plateau and Central Lavas have often been affected, have been discussed to some extent in Chapters V. and VIII., but are further considered in the following pages.

With the help of these preliminary remarks the reader will readily understand the grouping of the petrological material adopted in the sequel.

### PLATEAU TYPES OF BASALT.

(ANALS. I.-III.; TABLE I., p. 15.)

The lavas of the Plateau-region of Mull are, with relatively few exceptions, normal olivine-rich basalts (14913, 19071, 17781) or basaltic dolerites (15749, 16044, 14983, 18506). The latter lie in the debatable zone, where some authors employ the term basalt, others dolerite. As will be shown presently, the differences of texture do not necessarily correspond with the degree of development of ophitic structure. The majority of these rocks, whether basaltic or doleritic, are not conspicuously porphyritic. Abundant olivine is in many cases the only micro-porphyritic constituent. At the same time, there is a fairly well-marked tendency for both olivine and plagioclase-felspar to recur in two somewhat ill-defined generations (14976, 15777), as noted in Scottish Carboniferous basalts of Jedburgh Type. The analysed rock (19070, Anal. II. p. 15) is a basalt closely allied to the Jedburgh Type so far as structure is concerned, though in chemical composition it shows a characteristic relative deficency in alkalis. In some cases, there is a tendency towards a glomero-porphyritic grouping of the felspars either alone or with olivine (17779). The augite of the Plateau Basalts is generally of a purplish tint of varying intensity and sensibly pleochroic (15995, Anal. I., p. 15). The purple colour is apt to be particularly marked towards the margins of crystals and presumably bespeaks titanium. When not purple, the augite is pale-brown. In its structural relation to the felspar (labradorite), the augite varies from thoroughly ophitic (15686, 15995), to hypidiomorphic (18505). Where hypidiomorphic, the augite may occur either in long crystals,

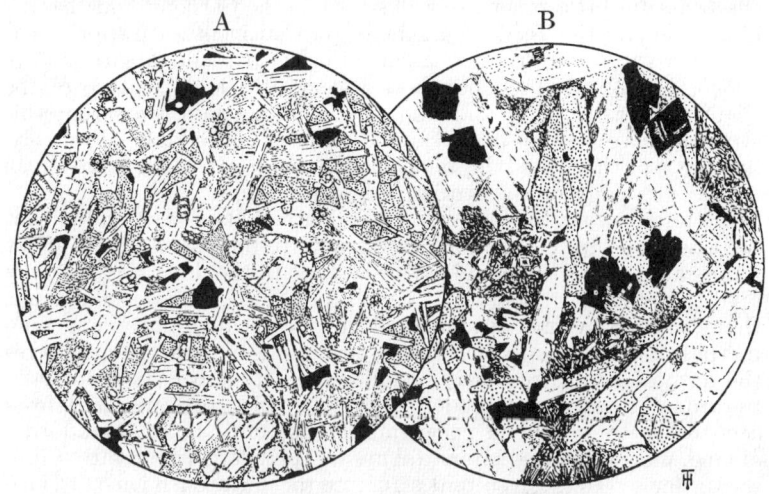

FIG. 19.—Lavas of Plateau Basalt Type.
A. [15686] × 17. Porphyritic olivine-basalt. The olivine is practically unaltered and occurs as porphyritic crystals (centre and bottom). The augite is titaniferous, having a lilac tinge, and is subophitic in its development with respect to the felspar. The felspar occurs as narrow elongated crystals of labradorite.
B. [20865] × 28. Segregation-vein in basalt-lava of Plateau Type. Titaniferous augite in large crystals that exhibit hour-glass structure and are zoned with aegerine-augite; partially analcitized labradorite; and conspicuous ilmenite. The residuum consists of microlithic alkali-felspar, aegerine-augite, and chlorite.

or equidimentional grains. In the latter case it determines a hypidiomorphic granular structure. In another type of granular structure, the augite granules are ophitic (15741). The analysed specimen (19071, Anal. III., p. 15) is an ophitic type where the augite-individuals approach dimensions characterizing granular types. Granular structure of either type is subordinate to ophitic structure among the Plateau Basalts as a group. The word granular is here substituted for Judd's term granulitic, as applied to the structure of basalt-lavas that have not suffered contact-alteration.

The tendency to hypidiomorphism of the augite is found both in the coarse (doleritic) rocks (14983, 18505, 18506) and in the definitely fine-grained basalts. Among the latter there are a few slides (17779, 17781, 18975) in which the hypidiomorphic granular

## Chapter X.—Petrology of Basalt and Mugearite Lavas.

augite occurs in minute and very abundant individuals that lack the purple tint of titaniferous augite. Such a granular base is highly characteristic of the Central Types, but is aberrant among the Plateau Lavas. Further points of similarity to the Central Types as exemplified by these slides are the tendency of the felspars to form definite phenocrysts, sharply separated by reason of their size from the rest of the felspars of the rocks, and also the occasional occurrence of small phenocrysts of augite. There is, it will be remembered, a conspicuous absence of augite-phenocrysts in the true Plateau Type of Mull lava (p. 16). Moreover the hypidiomorphism, noted above as a minor feature of the augite in the ground mass of the Plateau Types, never develops to such idiomorphism as is shown by the beautiful little prisms so well known in the Carboniferous basalts of Britain and the Tertiary basalts of the Continent. An approximation to a variolitic structure assumed by felspar and augite (15545, 15853, 18829) has been noted as a very occasional feature of the Plateau Lavas (p. 121), and is interesting as an indication that this structure may develop in rocks relatively rich in olivine. Probably, instances of this could be multiplied if more attention were given to the petrology of the scoriaceous tops of flows.

Before leaving the subject of the Plateau Basalts, a word must be said concerning the ophitic structure presented by a great number of the doleritic and basaltic lavas. In the doleritic rocks, the ophitic structure is moderately coarse and quite normal in character, though varied by an occasional tendency for the augite to assume idiomorphism. In the ophitic basalts, the augite most frequently takes the form of small irregularly bounded clots, that behave ophitically towards the narrow labradorite-crystals, and are separated from each other by a matrix consisting almost entirely of labradorite, olivine, magnetite, and interstital matter (now largely converted into serpentine). To the resultant structure the term o p h i m o t t l i n g may be applied. Generally, olivine is distributed more or less evenly without regard to the ophitic patches (16041, 16051, 18503); but in some instances there is a marked concentration of small crystals of the olivine in the interspaces beyond the limits of the ophitic augite (16042, 16049, 17789, 17813, 18151). This relative exclusion of olivine is due to an early crystallization of augite and felspar, followed without a break by felspar and olivine [1] (*see also* pp. 232, 240).

A certain amount of ground-mass, represented by zeolites and serpentine, is commonly present among the Plateau Basalts, and its failure to resist atmospheric erosion is probably responsible for the pimply-weathering of many of the basalt-flows outside the Limit of Pneumatolysis (Plate III., p. 91).

*Segregation-Veins.*—A few of the Plateau Basalts show a somewhat remarkable tendency to segregate contemporaneous veins consisting mainly of augite, felspar, and analcite, without olivine. These veins differ in texture and composition from the parent-lava, and afford interesting evidence as to the manner of differentiation

---

[1] This explanation of ophimottled structure was arrived at during a discussion of the evidence with Mr. A. F. Hallimond, and contributed largely in leading me to accept a crystallization-basis for the gravitational differentiation dealt with in Chapter XXX.                                                          E.B.B.

of the normal basalt-magma. They have an acicular crystallization, and are often associated with lines of amygdales, in which case there is generally an ingrowth of acicular crystals forming a first lining to the vesicles. The subject of their mode of occurrence has been dealt with more or less fully in Chapter VII., where it has been pointed out that the lavas affected by such veining are no less fresh than their unveined neighbours of similar composition and that they often retain their olivine in an undecomposed condition. The best examples of this phenomenon occur, beyond the limits of pneumatolytic action (Plate III., p. 91), in Sheet 43 of the one-inch Map ; and from them a small series of slides has been prepared.

A good instance is afforded by a lava (20782) in the cliff of Dùn Bhuirg on Loch Scridain. This lava is a normal ophitic olivine-basalt of Plateau Type with a violet titaniferous augite. Its olivine is mostly fresh, but there is some marginal serpentinization of the individual crystals. The felspars are short, twinned, crystals of labradorite. The rock, however, is traversed by minute veins of slightly turbid analcite, and this mineral, together with radiate zeolites, mostly lime- and soda-bearing varieties, occupies drusy cavities. The more definite segregation-veins (20783) traversing the lava are moderately coarse in texture. Their chief characters are a tendency towards panidiomorphism exhibited by their augite and felspar, and the presence of an analcitic and zeolitic matrix. The augite builds elongated crystals that are highly coloured and usually of a deeper tint towards their margins. The felspars are labradorite, and are often unduly elongated and zoned with varieties that rapidly increase in alkalinity. They are frequently analcitized either by strings or in patches. Both felspars and augite are sharply idiomorphic against the base, which consists of turbid analcite, chlorite, and a little alkali-felspar. Iron-ore, presumably ilmenite, is moderately abundant as thin plates and skeletal growths of triangular form. The alkalinity of the matrix is emphasized by the frequent presence of a green augite, either as a fringe to the large titaniferous augite-crystals, or as independent prisms in the matrix. Vesicular cavities are numerous, and these are filled by analcite, or by radiate growths of natrolite.

Veins from a similar lava (20865), from the shore of Ulva,[1] southeast of Ormaig Farm, are again of a relatively coarse rock. The augite is a deep-lilac coloured pleochroic variety with a zonal or hour-glass arrangement of the colouration (Fig. 19B, p. 137). It occurs in stout columns rudely ophitic towards the felspar, but elsewhere sharply idiomorphic with a good development of terminal faces and a border of pale green aegerine-augite. The felspars are narrow, zoned, laths of labradorite, strung and fringed with analcite. Ilmenite is abundant in large irregular crystals, more particularly in association with the augite, which latter is moulded upon it. There is a little interstitial base, now composed in large measure of chlorite and analcite. It contains microlites of alkali-felspar, a little aegerine-augite, and needles of apatite, and also small clear seemingly isotropic hexagonal crystals that sometimes appear to have hollow centres. These small six-sided crystals occur with others of similar dimension but rectangular outline, which in some

[1] Additional slides from Ulva are 21260-61.

instances show a low birefringence in their unaltered portions. Their form, refractive index, and manner of alteration, makes us inclined to regard them as either nepheline or pseudomorphs after that mineral; but it has not proved possible to apply such microchemical tests as would place the matter beyond doubt. The proportion of six-sided sections is certainly rather larger than would be expected in the case of nepheline, and it is possible that the crystals referred to may be in part analcite and in part alkali-felspar. At the same time, the probability of nepheline in rocks of this character must not be lost sight of; and undoubted occurrences may at any time be met with as further research on these segregation-veins is carried out. In the heart of the definitely igneous material of the rock described above, we encounter numerous amygdaloidal cavities lined with chlorite and filled either with analcite alone, or with chlorite, analcite, thomsonite, and natrolite.

Other examples may be drawn from the lavas of Tòn Dubh-sgairt (20784), just north of the Wilderness in the Gribun Peninsula, and from the shore east of the house on Little Colonsay (21258-9). Specimens from the last-mentioned locality are almost identical with those from Ulva, but are in a more intensely zeolitized condition.

It will be seen from the above description that these segregation-veins, which cannot be otherwise regarded than as normal differentiation-products of a basalt-magma, are in a general way mineralogically and structurally related to the lamprophyres. They distinctly recall the ocelli of the camptonites (Chapter XXXV.), a fact that must be reckoned with in discussing the age of the captonitic dykes of the Mull region (p. 380).

Another point of interest is the modification of the segregation-veins within the Limit of Pneumatolysis (Plate III., p. 91). This subject will be considered as soon as a brief statement has been given regarding the amygdales of the Plateau Region outside the same limit.

*Amygdales outside Limit of Pneumatolysis* (Anals. III.-X.; Table IX., p. 34).—Previous workers have collected considerable information concerning the amygdale-assemblages of the lavas (Plateau and Staffa Types) lying outside the Limit of Pneumatolysis of Plate III. (p. 91), It will suffice here to reproduce with verbal modification what Dr. Heddle says regarding the occurrence of a n a l c i t e in the Mull archipelago (*Mineralogy of Scotland*, vol. ii., p. 98).

Mull (Sheet 44): inland of the farmhouse of Carsaig (Compton); west side of Carsaig Arches; sea-cliffs below Beinn Chreagach, between Carsaig Arches and Carsaig Bay, associated with stilbite, mesolite, and scolecite (Goodchild); Loch Scridain, near Kilfinichen, on quartz (Mrs. Currie).

Mull (Sheet 43): Dearg Sgeir, with gyrolite, mesolite, and scolecite (Heddle and Goodchild).

Mull (Sheet 51): Calgary Bay (Currie).

Mull (Sheets 51 or 52): Northern division of island with stilbite, mesolite, and prehnite (Macculloch).

Staffa (Sheet 43): north end, on scolecite.

Ulva (Sheet 43): (Macculloch).

Treshnish Isles (Sheet 43): Bac Mòr, on stilbite and covered with natrolite; Lunga, with faröelite[1] and scolecite; Sgeir a' Chaisteil; Fladda, on scolecite and covered with natrolite.

[1] In British Museum Students' Index, faröelite is listed as a variety of thomsonite.

From its mode of occurrence in the segregation-veins just described, it is safe to regard much of the analcite of Mull as an auto-pneumatolytic product of the containing lavas.

*General Pneumatolysis* (Anals. XI. and XII. ; Table IX., p. 34). It has been pointed out in some detail in Chapter V. that within a line, termed the Limit of Pneumatolysis (Plate III., p. 91), all the lavas, and many of the intrusions, of Mull have undergone a characteristic type of alteration. The olivine of the lavas inside this pneumatolytic area has been entirely decomposed with a development of serpentine, chlorite, and magnetite. Analcite has also been completely replaced. Felspar has proved less susceptible, but all stages leading to total decomposition are common (17751, 18122, 18139). The slides just cited show a remarkable relative stability of augite. In the vicinity of amygdales, however, augite is replaced by chlorite. Occasionally augite is pseudomorphed by hornblende, or serves as a nucleus for horblendic outgrowths ; but in such cases, the change may perhaps more fitly be ascribed to low-grade contact-alteration (14814, 17365).

Albite and epidote have resulted from the decomposition of the original felspar, and with chlorite are the characteristic amygdale-minerals within the Limit of Pneumatolysis (2112, 17752, 17778-9). Generally speaking, they are disposed in circumferential zones ; but in two exceptional slides (17814 from Ben More, and 17829 from Loch Beg), it is clear that the bottom portion of vesicles had been filled with a fine deposit before the albite-epidote crystalline assemblage formed in the remaining spaces.[1]

Epidotic veins are highly characteristic of the more altered areas (114815, 18129, 18133).

It is particularly interesting to try and trace the fate of the segregation-veins, described above, within the Limit of Pneumatolysis. Dr. M'Lintock[2] has made a special study of certain amygdaloidal lavas of the Ben More Massif (Fig. 20, p. 142) ; and among them he has encountered segregation-veins, with associated outgrowths into vesicles (*cf.* 2109, 2629, 17753, 18122 from various localities), which so strongly recall the phenomena since investigated in the outer districts, that we venture to correlate them (*op. cit.*, pp. 17, 18, 21, and Plate I.). They have the same purple augite, and the same acicular structure ; but their felspar-laths, instead of being partially analcitized labradorite, are albite. Analcite is nowhere to be seen ; while chlorite has become an important associate. The interpretation we suggest is that the albite of the laths has replaced partially analcitized labradorite as a result of the general pneumatolysis of the district. Dr. M'Lintock, in his account, treats the original composition of the felspar-laths as an open question (*op. cit.*, p. 22) ; but he thought that any albitization that might have taken place was completed during the cooling of the containing lava.

We have stated above that analcite is altogether replaced within the Limit of Pneumatolysis, so far as the Mull lavas are concerned. The chief support for this claim is the complete absence of analcite

---

[1] *Compare*, M. F. Heddle, Mineralogy of Scotland, 1901, vol. i., p. 60.
[2] W. E. P. M'Lintoch, 'On the Zeolites and Associated Minerals from the Tertiary Lavas around Ben More, Mull,' *Trans. Roy. Soc. Edin.*, vol. li., 1915, p. 1.

## 142  Chapter X.—Petrology of Basalt and Mugearite Lavas.

inside the Limit, whether we look for it in segregation-veins, or in amygdales, or in the general base of the lavas affected. In addition, Dr. M'Lintock has in one case found albite pseudomorphing icositetrahedra of analcite [1] (*op. cit.*, pp. 10 and 19); but it must be admitted that he regarded the phenomenon as abnormal in the district he examined; and he did not think that analcite had ever been a common mineral in his lava-group. As already mentioned

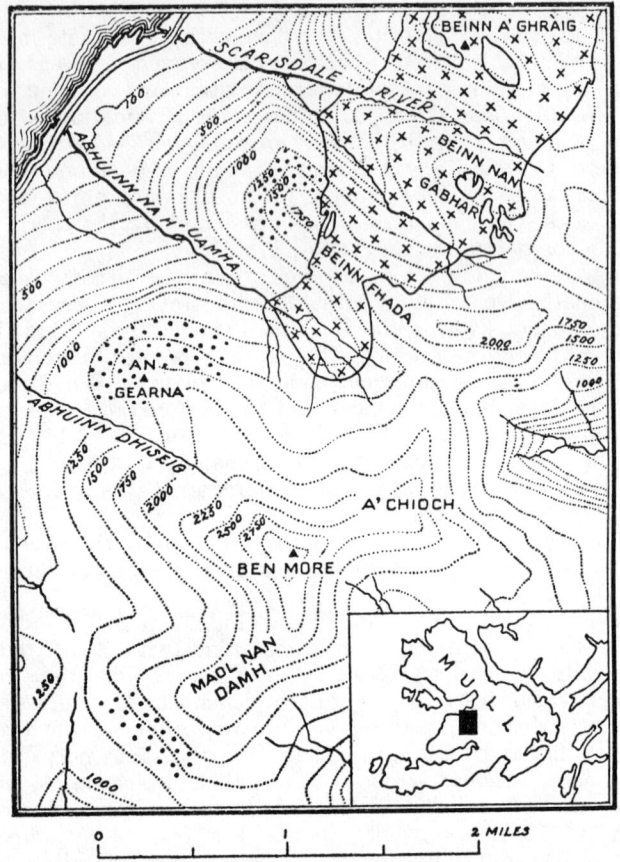

Fig. 20.—Map showing zeolite-localities (dotted) and granophyre (crossed). Quoted from *Trans. Roy. Soc. Edin.*, vol. li., 1915, p. 3.

(pp. 50, 128), Dr. M'Lintock assigns the alteration of the particularly vesicular zone of Ben More to auto-pneumatolysis; whereas we are disposed to follow Mr. Currie in referring it to the general pneumatolysis of the central region.

The outstanding feature of the vesicular zone of Ben More is its richness in scolecite, first described by Mr. Currie (p. 49). Some

[1] Dr. Heddle had previously recorded albite-pseudomorphs after heulandite, laumontite, and analcite in Carboniferous lavas of Central Scotland, *Mineralogy of Scotland*, 1901, vol. ii. p. 13.

of the fibrous agregates reach 10 cms. in length. Leaving out of consideration the segregation, or pegmatitic, minerals, Dr. M'Lintock arrives at the conclusion that the main sequence of amygdale-infilling has been : chlorite, followed in turn by albite, epidote, prehnite, and scolecite ; and he connects this time-scale with a gradual fall of temperature allowing of the deposition of increasingly hydrous minerals (except for the chlorite—*op. cit.*, p. 24).

Dr. M'Lintock's further work in elucidating the thermal metamorphism of the amygdale-assemblage in proximity with the granophyre of Fig. 20 is dealt with later (p. 152).

### MUGEARITES ASSOCIATED WITH PLATEAU BASALTS.

(ANAL. I. ; TABLE VII., p. 27).

Mugearites are represented amongst the lavas of Mull, more particularly in association with those of Plateau Type. Except for the main mugearitic horizon of Ben More and the plateau-region to the south and west, their occurrence is somewhat fitful. Petrographically, there are many examples which conform closely to the type as described from Skye ; but, as has been found in other districts where rocks of this nature are developed, there is a tendency for them to pass, by an increase in the basicity of the felspars, and by a greater proportion of ferromagnesian minerals, into rocks with more definitely basaltic characteristics, which must be described either as basaltic mugearites or mugearitic basalts. In the other direction, with an increase in alkalinity, and a falling off in the ferromagnesian constituents, the mugearites grade into oligoclase-trachytes.

That the relative alkalinity of the mugearites when compared with the basalts is an original specific character there can be no doubt ; but it is well to point out that rocks similar to mugearites in many respects can be, and are, produced by the more or less complete albitization of some of the finer textured basalts.

In their mode of weathering and their platy fracture, due in a large measure to the fluxional arrangement of the microscopic felspars, they are, as a rule, easily distinguishable from the basaltic rocks with which they are associated.

They are fine-grained holocrystalline dark-grey rocks, usually without phenocrysts. In some instances, however, small and infrequent phenocrysts of plagioclase may be detected with the unaided eye ; while in one case among the mugearites of Ben More distinctly porphyritic crystals have been observed (17797).

Microscopically, these rocks in their typical development consist of minute elongated crystals of an acid plagioclase, approximating to oligoclase in composition. The microlites are arranged with varying degrees of parallelism (trachytic structure) indicating that the rock retained considerable fluidity up to the final stages of its consolidation. Next in abundance is iron-ore, which is generally widely distributed throughout the rock as minute crystals of magnetite lying between the felspars or, less frequently, segregated into irregularly bounded patches. Olivine and, less abundantly, augite are the only original ferromagnesian minerals that occur in quantity; but in the case of the Mull examples olivine has not as yet been detected in an unaltered condition. It usually exists as rounded grains or minute elongated phenocrysts arranged parallel to the direction of flow as indicated by the felspar-microlites. In the majority of cases, it has been converted into highly birefringent chloritic pseudomorphs of deep-green colour (17796); but in other cases the pseudomorphs are of iddingsite (17840), or normal serpentine (14858). Chlorite and attendant finely divided iron-ore occur interstitially between the felspar-microlites and may conceivably represent in some instances a chloritized glassy residuum (16747). Small vesicular cavities are the rule, and these may be variously filled with chlorite (18111) or, less frequently, with secondary quartz. A specimen of the iron-stained vesicular top of the Ben More flow, as exposed two-thirds of a mile south of Kinlochscridain Hotel, shows fragments having varying grades of crystallization and containing a considerable proportion of what was once a glassy residue (20771).

Orthoclase has not been definitely recognized in the Mull rocks, nor have the pale-brown hornblende and biotite that occur in the Skye and Central Valley mugearites.[1]

Dr. Harker allows himself a certain amount of latitude in the interpretation of his type, and it appears quite clearly that slight departure of the dominant felspar from the composition of oligoclase, an increase or decrease in the percentages of olivine and augite, and the presence of micro-porphyritic crystals are, within limits, features of varietal significance only.

The Ben More mugearite at Kinloch (20582) is representative of the main mugearitic horizon of Mull, and is closely related to the Skye examples; but the analysis (p. 27) shows a decided difference in the somewhat higher percentages of silica and alkalis, and indicates an approximation to the composition of certain of the trachytes. The characters of the rock as a whole, however, link it with the mugearites, and Dr. Harker, who has seen the rock in question, is prepared to accept it as falling within the limits of his type. It is quite possible, as Dr. Harker points out, that many of the rocks which have been described as mugearites from other areas, particularly from the Scottish Carboniferous lavas, would, on analysis, show slightly greater alkalinity than the typical mugearite, although resembling it in general characters.

Certain specimens of basaltic mugearite from Ben More (17795, 17799) show their departure from the type mainly in the increased proportion of augite, which, when sufficiently abundant, assumes the micro-ophitic mottled texture of many of the finer-grained olivine-basalts. The felspars, too, although in most cases more acid than those of the basalts, are often more basic than oligoclase. Texturally, these basaltic mugearites are similar to the typical rocks and exhibit good fluxion-structure.

In the more acid direction again, a rock south-west of Cruach Inagairt (17328) is composed entirely of small narrow crystals of oligoclase, arranged fluxionally, with a green augite occuring between

[1] J. S. Flett, 'On the Mugearites,' *in* 'Summary of Progress for 1907,' *Mem. Geol. Survey*, 1908, p. 122.

the felspars in irregular granules and prisms. The amount of iron-ore is somewhat smaller than is typical of the mugearites. The colour of the augite and the general character of the rock recalls the soda-trachytes, and its analysis would probably indicate a higher percentage of alkalis than usual.

The mugearite of the mapped outcrop, north-west of the head of Loch na Keal, is almost typical (17991). It consists of the usual elongated oligoclase-microliths, moderately abundant small green pseudomorphs after olivine, a fair quantity of yellowish green augite which has suffered some chloritization, a moderate amount of magnetite in small crystals and patches, and a small proportion of chloritized material that probably represents residual glass. The rock has perfect fluxion-structure, beautifully shown by a parallel arrangement of the felspars and an orientation of the olivines and augites with their long axes in the direction of flow.

Another mugearite, not mapped but easily located, crops out from the raised-beach deposits at Ardchoirk (east of Lochdonhead). In some respects this rock approaches the soda-trachytes. The felspars are andesine zoned with oligoclase. The augite is pale green, slightly pleochroic, and has a tendency towards a micro-ophitic habit. Olivine is represented by minute, well-shaped, dark-green, highly birefringent, pseudomorphs; and magnetite is moderately abundant. The section (15555) is of special interest as it is crossed by a segregation, filling an elongated fissure or cavity, and composed of relatively large crystals of oligoclase, green augite, apatite, and an indeterminate zeolite. Probably this segregation is comparable as regards origin with those already described in connexion with the Plateau Basalts.

## Central Types of Basalt associated with the Plateau Basalts.

### (1). STAFFA TYPE.

(Anals. II. & III. ; Table II. p. 17).

The columnar lavas of the south-west of Mull are mainly of the Staffa Type, and are moderately compact fine-grained basalts, poor in olivine, that approach somewhat closely to the olivine-poor or olivine-free basalts of the central region. On the other hand, columnar structure reveals itself occasionally in olivine-rich rocks of normal Plateau Type which occur at a low geological position in the same south-western district.

As representative of the Staffa Type, we may cite :—the Fingal's Cave lava of Staffa (20874, 20875) ; the lava that surrounds Mac-culloch's tree (20581, Analysis, p. 17), and others in the same district (20774) ; the lavas from above and below the Ardtun Leaf-bed (20750, 20751) ; that exposed at the river-mouth at Tavool (20775) ; and the lowest lava seen at a point 4400 ft. west-north-west of Fionna Mhàm (20776).

All these rocks are distinguished in the field by fine-texture combined with columnar structure, generally of the double-tier type ; and their microscopic appearance is equally characteristic.

146   Chapter X.—Petrology of Basalt and Mugearite Lavas.

Their constituent minerals are augite, olivine, labradorite, and magnetite or titano-magnetite. The augite is typically non-titaniferous, and the felspars, for the most part, are microlithic in habit. The rocks consist of a felted mass of labradorite microlites, usually without good terminations, which include within their meshes small granules of augite, olivine, and magnetite. Augite is particularly abundant, but olivine is a minor constituent. There is usually no semblance of fluidal structure, but residual material in the form of a chloritized base representing glass is a constant feature. Infrequent microporphyritic crystals of olivine, augite, or felspar may be represented, and, in keeping with the position of these rocks beyond the limit of the central Mull pneumatolysis, olivine is in a more or less undecomposed condition.

Augite is the dominant ferromagnesian mineral, in fact, the dominant mineral. It has a granular habit, and shows a tendency to segregate into aggregates that may, or may not, behave ophitically towards the felspar. The structure is highly characteristic, and occasionally becomes almost intersertal (p. 280). The Staffa Type of basalt is essentially a fine-grained olivine-tholeiite, and, if it were more coarsely crystalline, it would agree closely with the Salen Type of tholeiite (p. 285). This relationship is clearly shown by the agreement of the analyses (p. 17). It is interesting to note that the Staffa Type includes the columnar basalt of the Giant's Causeway (I. 440).

The tendency to display columnar structure, generally in two tiers, is so marked a feature of the Staffa Type that it is necessary to emphasize the fact that all the conspicuous columnar basalts and dolerites of the Hebrides, or even of south-west Mull, do not belong to one petrological category.

The best known columnar sheet of Skye, that of Preshal More and Preshal Beg, approaches the Staffa Type in its preponderance of augite with which is associated felspar, olivine, and iron-ore; but the structure in the only Survey slide (9249) is of the ophi-mottled variety (p. 138). It will be remembered that Dr. Harker[1] considers this sheet a sill with, probably, a double-tier arrangement of jointing.

In four slices of columnar basalts from the south-western district of Mull, the rocks prove to be of Plateau Type (20746, 20748, 20749, 20777), with purple augite and with more abundant olivine than is characteristic of Staffa; but it is doubtful whether, in any of these four instances, double-tier jointing is developed. Three of them, it may be added, are certainly lavas, while the fourth is the columnar doleritic basalt of Réidh Eilean, outside Iona (Sheet 43), where exposures are too restricted by the sea to furnish decisive evidence of field-relations.

(2). BIG-FELSPAR BASALT TYPE.

Basalts with conspicuously porphyritic felspars are only occasionally met with amongst the Plateau Lavas. The most striking examples belong to the Big-Felspar Basalt Type and their

[1] A. Harker, 'The Geology of the Small Isles of Inverness-shire,' *Mem. Geol. Survey*, 1908, p. 119.

outcrops are indicated on the one-inch Map (Sheet 44) and Plate III. (p. 91). They occur towards the top of the Plateau Group, and may be regarded as the forerunners of the lavas of the Central Group. They are themselves definitely of Central Type, though their felspar-phenocrysts are distinctly larger than those of the porphyritic basalts of the central outcrops. The felspar-phenocrysts range up to 10 inches in length (p. 125), but this is very exceptional. An inch, or an inch and a half, is a more common measurement.

The chief porphyritic constituent is a basic labradorite approaching bytownite in composition, similar to, but showing more signs of zoning than, the smaller porphyritic felspars that characterize a large proportion of the later Central Lavas. These basic crystals are often edged with a more acid felspar (oligoclase), and in almost every case are albitized to a considerable extent.

Olivine in moderately large porphyritic crystals, often with good outlines, is a constant constituent, represented by frequent pseudomorphs in chlorite, calcite, and serpentine. It is always quite subordinate in amount to felspar.

These two porphyritic constituents make up by far the greater portion of the rock. In certain cases (19068), the olivine crystallized, in part, before the porphyritic felspars, but in the majority of rocks of this type (*e.g.* 18161) the whole of the olivine is of later separation.

In its content of a considerable proportion of olivine, the Big-Felspar Type approaches the Plateau Lavas, and comformably with this its augite assumes a definitely purple tint.

There is often no marked difference between the groundmass, in which the big felspar-phenocrysts are embedded, and that of an ordinary Plateau Basalt. Ophitic structure is frequently well developed (18839, 18840, 19068). On the other hand, there seems to be a marked tendency to variolitic structure, for, from a comparatively small collection, three slides exhibit it (14226, 18161, 18162).

In agreement with their position inside the Pneumatolysis Limit of Plate III., all specimens have their olivine completely decomposed. In one slide, chlorite, epidote, and calcite may be seen in crush-lines crossing porphyritic felspar and matrix alike.

### Central Types of Basalt occurring in the Central Group.

The Central Types of basalt are poorer in olivine than the Plateau Types; in fact, so far as one can judge from specimens, all of necessity collected within the Pneumatolysis Limit of Plate III., they are often destitute of that mineral. Chemically, as has been shown (p 25), this poverty in olivine corresponds with a poverty in magnesia. Two main types are recognizable, the one rich in porphyritic felspar and correspondingly rich in alumina and lime, the other essentially non-porphyritic.

An unusual proportion of the rocks show variolitic structure, a circumstance in keeping with the assumption that many of the Central Lavas were erupted into water. A description of these

148  Chapter X.—Petrology of Basalt and Mugearite Lavas.

interesting variolites will be furnished after the more normal porphyritic and non-porphyritic types have been dealt with.

All specimens show pneumatolytic affects such as have been described above (pp. 95, 141). These include a universal decomposition of olivine, a certain variable amount of albitization of basic felspars, and also the production of abundant epidote. It is interesting to note that quartz occurs more commonly in the veins and vesicles of Central Lavas (15598, 16721, 18656) than among the Plateau Types. In two slices, it appears as if the infilling of vesicles had been commenced by a sedimentary accumulation of chloritic waste at the bottom of the cavities. In one instance (16721), this early 'sediment' occupies the bottom of several adjacent vesicles and has been followed by crystallized quartz, epidote, etc. In (18960), bedded 'sediment' has choked the vesicles, and has developed a spotted structure probably due to subsequent thermal metamorphism.

(1). NON-VARIOLITIC PORPHYRITIC BASALTS OF CENTRAL TYPE.

(ANALS. III. & V.; TABLE VI., p. 24).

The chief characteristic of this type is an abundance of small phenocrysts of basic plagioclase, about 5 mm. across, and ranging in composition from basic labradorite to anorthite. A minor feature is the occasional occurrence of porphyritic augite (18049, 18889).

Where olivine pseudomorphs are recognizable, there is a tendency for the augite to have a somewhat purple tint and to occur with frankly ophitic structure in a ground of doleritic texture (16740, 17184, 18897, 18961), thus recalling the structure of the Plateau and Big-Felspar Basalt Types already described. In such cases, the relative scarcity of olivine (as compared with Plateau Types) depends upon the large part played by felspar-phenocrysts. The analysis of the Cruach Choireadail pillow-lava (Anal. 4; Table VI.) may be taken as representing this variety, for although the specimen analysed was a porphyritic variolite, others from the interior portions of the pillows are porphyritic dolerite (Fig. 21, p. 151). The analysis illustrates very clearly the large part that felspar-phenocrysts play in the constitution of the rock as a whole.

With further reduction in olivine, the augite of the ground-mass loses its purple tint, and becomes granular (18899), while the texture is moderately fine. Many of the lavas from Beinn Bhearnach, east of the head of Glen Forsa, are of this type, but have a very poor development of porphyritic felspar (18877-9). These lavas have been referred to in Chapter IX. as intermediate in character between the Plateau and Central Types.

With the disappearance of olivine, augite is almost universally granular and the texture of the base is very fine indeed. The analysed rocks from Derrynaculen (18469, p. 24), and from Cruach Doire nan Guilean (18471) are typical of this variety which is widespread (16487, 18050). They consist of abundant porphyritic and somewhat zoned crystals of basic labradorite, which show signs of albitization, in a fine-grained ground-mass that is made up of minute granules of augite, microlites of labradorite, and microscopic crystals of magnetite.

## (2). NON-VARIOLITIC NON-PORPHYRITIC BASALTS OF CENTRAL TYPES.

(ANALS. IV. & V.; TABLE II., p. 17).

The Non-Porphyritic Central Types, here considered, reproduce the structures and composition of those Porphyritic Types in which olivine is very scarce or absent. They are very closely allied to the Staffa Type (p. 145), but are generally poorer in olivine and finer in texture. An extremely compact type is prevalent (16441, 18921), of which the two analysed specimens (14824, and 18474, p. 17) are thoroughly typical. Small porphyritic crystals, a millimetre or so across, are occasionally met with. These are usually labradorite, but to a very minor extent augite (16441, 17242).

The matrix is microlithic, and consists of minute elongated microlites of labradorite, often altered, and microscopic granules of augite. Magnetite in a finely divided state is an abundant constituent and is largely responsible for the dark colour characteristic of these rocks in hand-specimens. Fluxion-structures are rare; more frequently the microlites are felted together without definite orientation.

## (3). VARIOLITES.

(PORPHYRITIC VARIOLITE: ANAL. IV.; TABLE VI., p. 24).

Variolites are common among such minor intrusions of Mull as correspond in composition with the Central Types of lavas (pp. 18, 24). In keeping with this, where the Central Type of magma has assumed the pillow-lava habit, that is, where it has been extruded as small masses into water, it has developed a distinctly variolitic facies—though not comparable in perfection with what is common among the minor intrusions. Instances of variolitic structure in Mull lavas which have not the pillow-habit have been already noted as rarities (pp. 121, 138). It has also been suggested that if scoriaceous tops were specially selected for examination further instances might be found in considerable number. The significant fact remains, however, that, among the Pillow-Lavas of Central Type, variolites are sufficiently abundant to require no special search to bring them to light.

Some of the variolotic pillow-lavas belong to the porphyritic class (18472, 18641, 18644, 18647, 18880); others are non-porphyritic (17859, 17900, 18032, 18039).

There are three main reasons why the Mull pillow-lavas are particularly interesting, as compared with other occurrences :—

(1) In their association of variolitic structure, on the microscopic scale, with pillow-structure on the large scale, they agree with what has been found in certain other parts of the world.[1]

(2) In their froth-invaded vesicles, to be described immediately, they reproduce another microscopic peculiarity which has already been found in pillow-lavas elsewhere.

---

[1] G. A. J. Cole and J. W. Gregory, 'The Variolitic Rocks of Mount Genèvre,' *Quart. Journ. Geol. Soc.*, vol. xlvi., 1890, p. 311. F. L. Ransome, 'The Eruptive Rocks of Point Bonita,' Univ. of California, *Bull. Dept. Geol.*, vol. i., 1896, p. 99. H. Dewey and J. S Flett, 'On Some British Pillow Lavas and the rocks associated with them,' *Geol. Mag.*, 1911, p. 203.

## Chapter X.—Petrology of Basalt and Mugearite Lavas.

(3) In their low soda-content (Anal. IV., p. 24), they contrast with what has been noted in other cases, more especially by Teall,[1] Dewey, and Flett.[2]

The phenomenon briefly styled 'froth-invaded vesicles' is quite a feature of many of the Mull pillow-lavas. It results from an invasion of vesicles by magma which after entry has frothed up *in situ* (Fig. 21D). The mere invasion of a vesicle by residual glassy magma is, of course, common, more especially among tholeiitic intrusions. What renders the phenomenon peculiar in the present instance is the subsequent frothing of the invading magma, which has led to a development of vesicles within vesicles (17859, 17897, 18032, 18039, 18646). Sir Jethro Teall[3] has figured vesicular complexes of this character from pillow-lavas of Arenig Age from the South of Scotland, so it appears likely that the phenomenon is of common occurrence. It is probably influenced by the chilled crust of the pillows resisting the external escape of vapours.

The variolitic structure of the Mull pillow-lavas is occasionally discernible under the microscope in a subradiate grouping of the felspar microliths (17900); but is only at all pronounced where augite takes a prominent share in the form of skeletal growths and bundles (17185, 18641, 18893). A particularly interesting suite was collected by Dr. Clough from the pillows of a porphyritic flow on the slopes of Cruach Choireadail (Fig. 21). The actual chilled edge of a pillow (17186; Fig. 21c) belonging to this suite is a dark grey, almost aphanitic, rock with small unzoned phenocrysts of basic plagioclase in a glassy base, which latter is crowded with irresolvable growths of augite accompanied locally by minute felspar-microlites sometimes radially arranged. Olivine is represented, if at all, by minute pseudomorphs. A neighbouring pillow has given a specimen of the marginal zone, just a little inside the extreme chilled edge. The augite and groundmass-felspar are much more clearly individualized. The former occurs for the most part in groups of narrow rods. The slide may be taken as typical of many of the rather ill-defined variolites found among Mull pillow-lavas (17185; Fig. 21B). The same pillow in its interior is porphyritic ophitic dolerite with augite in large crystals (17184; Fig. 21A). The felspar-phenocrysts of this interior dolerite are enlarged by marginal zoning, and pseudomorphs after olivine are well-developed.

A similar transition from variolite to dolerite is traceable in another porphyritic flow occurring on Sròn Dubh. Here the chilled base of a flow is found to be a variolite (17898), while three feet up the rock develops into a dolerite (17899, 17901). Once again, pseudomorphs after olivine are relatively very small in the variolitic portion. It may also be remarked, as a general feature of these rocks in Mull, that iron-ore is poorly represented.

The pneumatolytic changes in the pillow-lavas seem to be of the same kind and degree as are met with in other Central Types. As might be expected, there is generally a certain amount of albitization of the basic felspars.

[1] J. J. H. Teall in 'Silurian Rocks of Britain,' vol. i., Scotland, *Mem. Geol. Surv.*, 1899, p. 86.
[2] *Op. cit.*, p. 205.
[3] *Op. cit.*, Pl. XVIII.; instances of completely filled early vesicles are probably shown, Pl. XIX., Fig. 1.

## CONTACT-METAMORPHISM OF THE BASALT-LAVAS.

Lavas of both Plateau and Central Types have come within the influence of many later intrusions of considerable magnitude which

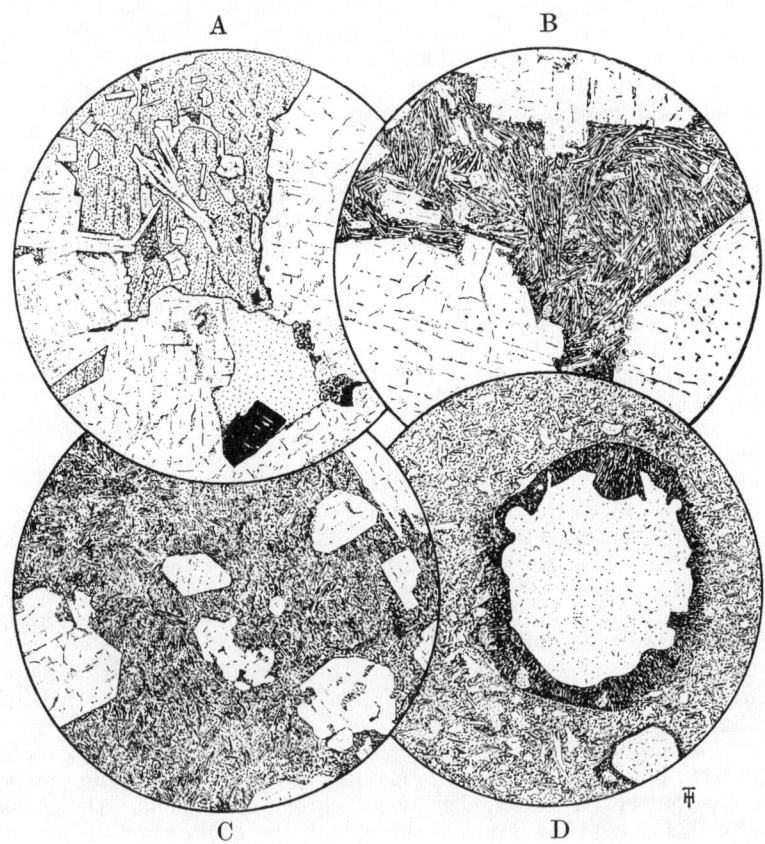

FIG. 21: A-C.—Pillow-Lava, Cruach Choireadail.  D.—Beinn Fhada.

A. [17184] × 17. Interior of Pillow. Moderately coarse doleritic rock with the augite and felspar in ophitic relationship.
B. [17185] × 17. Exterior of Pillow. The felspar occurs in two generations as porphyritic crystals of bytownite-anorthite, and as slender laths, which, with elongated crystals of augite, impart a variolitic structure to the matrix (compare with Fig. 23A, p. 163).
C. [17186] × 17. Chilled Margin of Pillow. Porphyritic basic plagioclase, near anorthite in composition, in a fine-textured matrix. The ground-mass is composed of small, elongated crystals of felspar, augite and iron-ore, with a chloritized residuum probably representing glass (compare with Fig 23B, p. 163).
D. [18039] × 17. Pillow-Lava, Beinn Fhada. Portion of the exterior of a pillow showing the characteristic invasion of vesicular cavities by mesostatic residual material which has subsequently frothed up *in situ*.

have been able to produce in them marked thermal effects. Sometimes the sphere of influence is but a narrow contact-zone, in other cases it embraces a wide and much less definite area.

As was found by Dr. Harker in Skye, and by Sir Archibald Geikie and other workers on the Tertiary volcanic rocks of Scotland in general, it is the low-temperature decomposition-products of the lavas themselves, and the minerals which fill amygdaloidal cavities, that first show the effects of contact-alteration. In extreme cases, however, every constituent of the rocks, whether original or secondary, may be recrystallized, and entirely new structures may be produced. In the field, an increasing toughness of the lavas is noticeable as an intrusive mass is approached, long before any actual signs of contact-metamorphism can be detected by the microscope.

In the majority of cases, the lavas that show the effects of contact-alteration most clearly had previously had impressed upon them low-temperature hydrothermal changes in common with the other lavas within the Central Area, and thus contact-alteration has given rise to obvious and interesting mineralogical features.

Contact-alteration of lavas of the Plateau Group can best be studied in relation to the Knock and Beinn a' Ghràig Granophyres; that of the lavas of the Central Types is especially well exhibited above the flat-topped Loch Uisg Granophyre, and in proximity to the gabbros of Corra-bheinn and Ben Buie.

*Plateau Basalts within the Contact-Zone of the Knock Granophyre.* The Knock Granophyre has produced interesting metamorphic effects on the Plateau basalt-lavas of its neighbourhood. In some cases (14816), augite has been replaced by pale-brown hornblende, and there has been a development of biotite from chlorite, and a crystallization of secondary felspar. In another instance (14817), the body of the rock has suffered partial granulitization with production of biotite in proximity to iron-ores; while vesicular cavities, presumably once filled with epidote and chlorite, have given rise to aggregates of biotite and fayalite in all respects similar to those which will presently be described in connection with the Loch Uisg Granophyre (p. 154).

Certain more coarsely crystalline rocks of doleritic characters (14818) have had their ophitic crystals of titaniferous augite partly converted into a brown, slightly pleochroic, hornblende. Occasionally this change has occurred throughout the augite-crystals, but at other times it is restricted to the external portions.

*Amygdales of Plateau Basalts within Contact-Zone of Beinn a' Ghràig Granophyre (Anals. XIII. and XIV.; Table IX., p. 34).* We have already drawn upon an account which Dr. M'Lintock[1] has given of the amygdales of a particular group of zeolite-bearing lavas exposed on the slopes of Maol nan Damh, An Gearna, and Beinn Fhada. The lavas concerned are olivine-basalts of Plateau Type. On the Map (p. 142), it is seen that one of the main localities studied, Maol nan Damh, is situated two miles from the Beinn a' Ghràig Granophyre, while An Gearna, and still more Beinn Fhada, are much closer. The thermal metamorphism of the amygdales corresponds with their nearness to the granophyre; and Dr. M'Lintock's main observations and inferences in respect to its distribution and character may be summarized as follows :—

Even in the field, obvious differences are apparent on comparison

[1] W. F. P. M'Lintock, 'On the Zeolites and Associated Minerals from the Tertiary Lavas around Ben More, Mull,' *Trans. Roy. Soc. Edin.*, vol. li., 1915, p. 1.

of the amygdales of Maol nan Damh with those of An Gearna. The dark marginal chloritic layers of the amygdales on Maol nan Damh lose much of their definiteness on An Gearna, and are largely replaced by confused zones of yellow epidote, sometimes with tufts of green hornblende. Moreover, the colour of epidote found in the two localities is generally different: deep bottle-green on Maol nan Damh, and pale yellow, brown, or pink on An Gearna. Finally, garnet is extremely rare on Maol nan Damh, and fairly frequent on An Gearna ; its tint varies, but is usually a pale wine-yellow.

On Beinn Fhada, the modification of the original amygdales is carried a step farther. Prehnite-tufts are in some cases found as pseudomorphs after scolecite. Beautiful specimens also occur where scolecite is sprinkled with groups of epidote and garnet, pale-pink, yellow, or even red, in colour. In other cases, nearer the granophyre, the amygdales merge at their margins into the containing rock : often prehnite is found veined and riddled with a pale yellow epidote and a garnet, pale-yellow to almost black in colour ; elsewhere amygdales are represented by pale-pink massive material consisting largely of garnet and epidote.

Microscopic investigation shows that the following minerals have arisen during the thermal metamorphism of amygdales :— prehnite, epidote, pyroxene, hornblende, garnet, sphene, albite.

The pyroxene and hornblende have been developed from the reaction of chlorite with scolecite or prehnite.

The sphene owes its origin to the titanium which in the original rock was contained in the augite and iron-oxides, and thence made its way into the epidote, and probably the chlorite, of the vesicles.

The prehnite, epidote, garnet, and albite have been derived from such minerals as scolecite and thomsonite. It is very interesting to note that their order of development, as revealed by the microscope, is the reverse of that which regulated the filling of the amygdales. Thus one finds : scolecite replaced by prehnite, followed in turn by epidote and garnet ; or thomsonite converted to prehnite and albite, followed as before by epidote and garnet.

The order of alteration just cited bespeaks, in Dr. M'Lintock's judgment, a rising temperature ; just as the order of their infilling corresponds with a falling temperature (p. 143).   H.H.T., E.B.B.

*Amygdales, Etc., of Central Basalts in the Roof of the Loch Uisg Granophyre.* The crags above the Loch Uisg Granophyre are particularly well-suited for the collection of Central Types of lavas in all stages of alteration. The simplest and most prevalent metamorphism is practically that of dehydration of those secondary hydrous minerals which existed in the volcanic rocks at the time of their metamorphism. As such it takes the form of a simple molecular rearrangement of a more or less isolated character and finds expression in the reconversion of soda-lime zeolites to soda-lime felspars, chlorite to biotite, etc. The anhydrous and pyrogenetic minerals such as the pyroxenes and felspars are as a rule unaffected.

In the body of the lavas, low-grade thermal metamorphism is made evident by the formation of minute crystals and patches of red-brown biotite, more particularly in the neighbourhood of particles of iron-ore.

## Chapter X.—Petrology of Basalt and Mugearite Lavas.

Many of the basalts, whether porphyritic or non-porphyritic, had already undergone some silicification, with a development of quartz in their vesicles and cracks. During metamorphism, the quartz behaved in a manner similar to quartz-xenocrysts caught up in a molten basalt. The interaction of quartz and basalt has, in such cases, given rise to secondary minerals and structures identical with those described in connexion with quartz-xenocrysts by Lacroix in his *Enclaves des roches volcaniques*. Partial assimilation of the secondary quartz of the vesicles has resulted in the residual quartz being surrounded by a reaction-border of augite, rhombic pyroxene, or both. Where assimilation has been complete, the place of the quartz is taken by clots and strings of pyroxene, which, but for the evidence of the less extremely altered rocks, might be taken for cognate pyroxenic nodules.

Partial assimilation of quartz and its envelopment by a reaction-border is exemplified by several sections (*e.g.* 18048, 18936, 18941). The reaction-border is of variable width and is composed of glass, usually turbid and coloured by iron, in which are abundant short well-formed prisms of augite and small patches of magnetite. The glass apparently had the composition of an acid plagioclase for such is the usual product of its devitrification. A more or less complete granulitization of the rock as a whole is a usual accompanying feature of the metamorphism. Elsewhere, enstatite-quartz areas represent amygdales and are contained in a granulitic matrix (18940).

Complete assimilation of quartz may be inferred where clots of granular augite occur in association with secondary felspar and chlorite (19085). In the larger patches, the augite has an acicular habit, radiating inwards from numerous points on the periphery of the vesicular cavity. A chlorite-nucleus is sometimes preserved. From the outline of such patches it would appear that they result from the metamorphism of amygdales composed of quartz and chlorite.

In other cases of complete assimilation of quartz, the original quartz-chlorite-filled vesicles and fissures are represented by zoned areas of biotite and rhombic pyroxene (18939). Generally, the outer portion of an amygdale is now composed of small prisms and slightly radiating crystals of rhombic pyroxene followed towards the interior by an aggregate of bright-brown biotite, pyroxene, and a little secondary felspar. Occasionally, secondary felspar alone occupies the central position. The containing fine-grained basaltic rock is more or less completely granulitized with a recrystallization of all its constituents other than the comparatively large porphyritic crystals of basic plagioclase.

The development of rhombic pyroxene in the two rocks quoted above (18939, 18940) presumably denotes a relatively intense phase of metamorphism.

Of all the metamorphosed amygdales, the most interesting occur in a compact porphyritic lava (18933). Here, circular or ellipsoidal areas that represent amygdales are occupied by a fine-grained aggregate of red-brown biotite into which elongated crystals of iron-olivine (fayalite) project from the periphery. The olivine is colourless in thin section but is usually partly decomposed with the separation of abundant magnetite. It is interesting to note that, contrary to custom, the crystals of fayalite are elongated parallel to the zone-axis (001) (100) and slightly flattened parallel to (100). Traces of the good cleavage parallel to (010) cross the crystals at right angles to their long axes. Usually in fayalite the elongation is parallel to the zone-axis (100) (010), but (100) generally shows a tabular development. It is suggested, though proof is lacking, that such a mineral-assemblage results from the metamorphism of vesicles filled with an iron-epidote and chlorite.

*Granulitized Basalts in the Contact-Zone of the Corra-bheinn Gabbro.* Many additional examples from other aureoles might be cited, but it is enough here to draw attention to three fine-grained granulites which represent lavas baked by the Corra-bheinn Gabbro. In such extreme cases, it is impossible to say whether the originals were of Plateau or Central Type. The specimens were collected in an area that is, for the most part, occupied by Central Types but it is quite possible that they really belong to sharply up-folded lavas of the Plateau Group.

One example (16494) consists of granular augite and hypersthene, recrystallized labradorite and magnetite, and, locally, brown hornblende. Patches of clear recrystallized labradorite enclosing biotite and strung with granules of pyroxene represent the partially chloritized original porphyritic felspars. The structure of the rock is typically granulitic. Another (16493) is similar but with the granulitic structure coarser and still more pronounced. The rock consists of a mozaic of hypersthene, augite, biotite, magnetite and plagioclase. Its vesicular cavities are filled with a crystalline mass of secondary felspar, near to bytownite in composition, and produced, in all probability, by the dehydration of some zeolitic mineral such as scolecite or laumontite. Such granulites are well-known products of the metamorphism of basaltic rocks and are identical with those described and figured by Dr. Harker[1] from amongst the Tertiary rocks of Skye.

The remaining specimen was presumably a lava of Plateau Type (16496) before its metamorphosis by the Corra-bheinn gabbro. It consists now of a granulitic mass of augite and labradorite veined with actinolitic hornblende. In addition, however, it has developed relatively large areas of clear secondary felspar and brown hornblende, the hornblende being in ophitic or poecilitic relationship to the felspar.  H.H.T.

---

[1] A. Harker, 'Tertiary Igneous Rocks of Skye,' *Mem. Geol. Surv.*, 1904, p. 52.

# CHAPTER XI.

## VARIOUS DOLERITES AND GABBROS.

### INTRODUCTION.

SUCH dolerite and gabbro-masses of Mull as are lettered eD on Sheets 43, 44, and 51 are covered by the present chapter, with the exception of the Loch Uisg Gabbro (Chapter XX.) and two small intrusions outside the Loch Bà Felsite at Coille na Sròine and Sròn nam Boc (Chapter XXXIII.). It will not be possible to refer to every occurrence, or to give a connected account of what is, after all, a heterogenous assemblage. The sequel is divided under two headings *Field-Relations* and *Petrology*, and, throughout, geographical classification is adopted for the sake of ready reference. All except two of the districts selected fall within Sheet 44.

There is good reason to believe that some of the intrusions considered here are of the same age as some of the lavas dealt with in preceding chapters. This is particularly true in regard to the small- and big-felspar dolerites shown in Fig. 22.  E.B.B.

### FIELD-RELATIONS.

*Dolerite-Plugs, 'S Airde Beinn and Loch Frisa (Sheet 51).* One of the most interesting features of the north end of Mull is the volcanic plug which forms the hill of 'S Airde Beinn, about 3 miles west-south-west of Tobermory. Good illustrated descriptions have already been given by Prof. Judd (p. 46) and Sir Archibald Geikie. From the hills south of Loch Meadhoin, the plug is seen as a large knob of rock protruding with more or less vertical walls out of the surrounding almost flat lava-flows. What immediately strikes the eye from this point of view is the fact that the knob contains a central depression, the site of a picturesque lochan. There is little wonder that this hollow is often wrongly interpreted as the crater of an extinct volcano. If one approaches the hill from the north-east, the intrusive nature of the mass is well demonstrated by the fact that the edges of several flows of basalt are seen to terminate abruptly against the vertical wall of the plug.

Let us now examine this intrusive mass in detail. It is found to run north-north-west, with a maximun length and breadth of 950 and 430 yds. respectively. Except for a short distance at the south-eastern corner, where the overflow-stream from the loch emerges, it has an almost vertical wall of complex nature. On examination, the material which forms the outer edge of this wall is seen to consist of a fine-grained hard splintery rock, which in places shows good amygdaloidal structure. Under the microscope, this outer portion proves to be the same type of basalt as that of the surrounding lava-flows, only in an altered condition due to its having been baked by the intrusion of the volcanic plug. No abrupt junction can be seen between this altered basalt and the dolerite of the plug ; in fact, the one passes almost insensibly into the other. Apparently the temperature of the molten dolerite in the plug was sufficient to melt up the edge of the adjoining basalt-lava, with a consequent local mingling of material from the two sources. The dolerite, followed in from its edge towards the loch, is seen to become gradually coarser in grain for a certain distance, and then to retain an almost uniform texture throughout.

The origin of the central loch now demands our attention. No rock except dolerite is exposed near its margin, and this dolerite is not different in composition from that 20 yds. or so away from the edge of the loch ; nor are any fragments of ashy material exposed along the shore. The only visible reason for the presence of the loch is the occurrence of a north-north-west line of fracture, which can be traced for miles across the country, and is well seen, in a cleft, on the northern face of the hill. At the south-east end, the stream from the loch emerges through an opening along the same line of fracture, so that in all probability the hollow of the loch has been eroded in a belt of shattered dolerite determined by faulting.

A mile and a half north-north-west from 'S Airde Beinn, and on the south side of the road leading from Tobermory to Glengorm Castle, a small plug forms a prominent crag known as Cnoc a' Chrocaire. It is only about 100 yds. in diameter, and is cut by a thin dyke and an irregular sheet of fine-grained dolerite.

In the other direction, a mile and a half south-south-east of 'S Airde Beinn, there are two more small plugs of dolerite outcropping above Loch Frisa. The more southerly is interesting as showing good almost horizontal columns. It is a medium to coarse-grained dolerite.                                                                G.V.W.

*Dolerite-Plugs, Dùn Mòr, Ulva Ferry (Sheet 43).* Two small dolerite-plugs occur a mile east of Ulva Ferry. The larger of them constitutes Dùn Mòr, a very conspicuous landmark contrasting strongly with the terraced lavas of the neighbourhood. The dolerite of Dùn Mòr is fairly coarse. That it is a plug is clear from its appearance as a whole, but no satisfactory junctions are exposed in confirmation of this interpretation. In fact, on its western side, a part of the dolerite can be seen at one point (under a little 2-ft. basalt-sill, which cuts it) resting with an unchilled base upon basalt-slag, just as if it were a lava. A combination of intrusive and superficial characteristics is, of course, a common feature in vent-intrusions.

E.B.B.

*Dolerites and Gabbros of Ben More (Sheet 44).* On Ben More, a few small masses of early dolerite, or gabbro, are lettered eD on the one-inch Map. They are generally of more or less sheet-like form, but are much thicker and more laccolithic than the later sheets of the neighbourhood lettered D on the Map and described later on in Chapter XXVI.                                             J.E.R.

*Gabbro of Beinn nan Lus (Sheet* 44). There is little to be said about the two small outcrops of gabbro mapped on Beinn nan Lus, beyond the fact that the rock is of early date as compared with the Beinn a' Ghràig Granophyre (Chapter XXVII.) with which it is in contact.

*Beinn na Duatharach and other Gabbros and Dolerites within North-West Caldera (Plate V. and Sheet* 44). There are several groups of gabbro and dolerite outcrops within the North-West Caldera outlined on Plate V. (p. 165). On the north-east side, the more important masses occur at A' Bhog-àiridh and Beinn na Duatharach; and on the south-east side, on Bìth-bheinn and west of Clachaig Cottage. All these masses have suffered conspicuous alteration, and in the majority of cases are cut at one point or another by the Glen Cannel Granophyre of Chapter XXXI.     W.B.W.

*Dolerites and Gabbros, Coire Mòr, Maol Uachdarach and Beinn Talaidh (Sheet* 44). A considerable mass of coarse dolerite or fine gabbro, cut to ribbons by Late Basic Cone-Sheets (Chapter XXVIII.) reaches in a south-east direction from Coire na Lice Duibhe along the top of Coire Mòr into Beinn a' Mheadhoin. It is of finer texture than the neighbouring Corra-bheinn Gabbro.

What may be part of the same mass of dolerite reappears along the summit of Maol Uachdarach where it is locally broken up to yield agglomerate. Its exposures show it cut by Late Basic Cone-Sheets running east and west, and by ring-dykes (Chapter XXIX.) running north and south. The dolerite gives rise to little crags and thus greatly assists in the mapping of the ring-dykes.     J.E.R.

On the one-inch Map, Sheet 44, a marginal line is inserted including within its scope the summit of Beinn Talaidh, and bounding, an area where lenticles of dolerite are found between the prevalent cone-sheets of the mountain. The fullest exposures are on Maol nam Fiadh, where the dolerite, though sometimes strikingly vesicular, shows a chilled contact against neighbouring porphyritic lavas (Fig. 53, p. 312).

An outcrop of gabbro, two-thirds of a mile north-west of Beinn Talaidh summit, is a black and white gabbro like much of the Beinn Bheag mass (Chapter XXII.), of which very likely it is an isolated part.

Separated from Beinn Talaidh by the pass connecting Glen Forsa with Glen More, one may note two strips of dolerite on the slopes of Beinn Bhearnach, one mile west-south-west of the summit of Sgùrr Dearg. The more north-easterly band is a dark rock that may fairly be classed as fine gabbro. It is interesting because in clear exposures it may be seen breaking up to yield neck-agglomerate.

*Small-Felspar Dolerites between Allt nan Clàr and Sgùrr Dearg (Sheet* 44). A group of dolerites characterized by small felspar-phenocrysts will now be considered (Fig. 22). They correspond in character, and probably in age, with neighbouring lavas of Central Type (Chapter IX.). Their field-relations show that they are sometimes sills, and sometimes plugs.

A large outcrop occurs in Allt nan Clàr. It is intensely cut up by cone-sheets as shown on Sheet 44, and it is of coarse crystallization so that it is not easy to separate it from the Beinn Bheag Gabbro (Fig. 37, p. 244).

More interesting are smaller outcrops on the north-east slope of Beinn Bheag. The small-felspar dolerite occurs here as massive sills intruded into fine Tertiary sediments (p. 66). The dolerite and sediment are broken up as they approach vent-agglomerate exposed on their north-east side.

A large mass of small-felspar dolerite runs roughly parallel with Abhuinn an t-Stratha Bhàin. It is chiefly noteworthy as having yielded blocks constituting a very prominent median moraine in Glen Forsa.

Several small occurrences may be noted in the gneiss north of

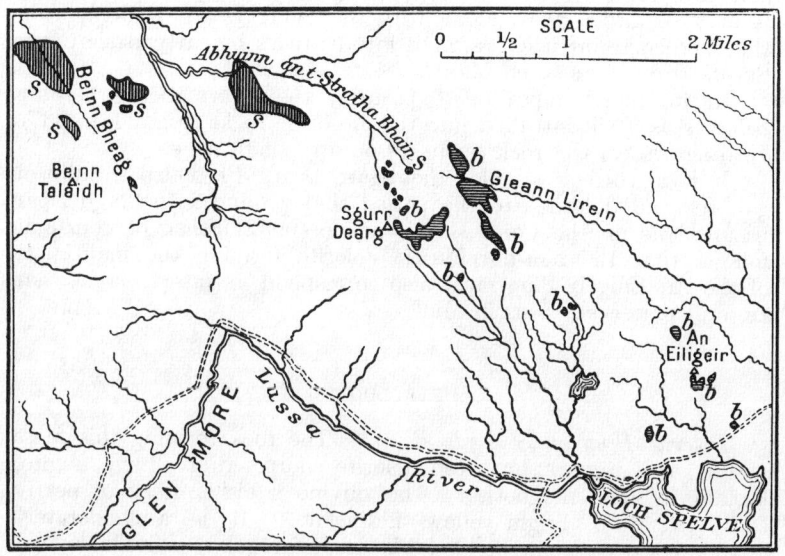

Fig. 22.—Map of Big-Felspar Dolerites *b*, and Small-Felspar Dolerites *s*, Sgùrr Dearg District.

Sgùrr Dearg. They are so baked and crushed that it is certain they are of an earlier date than the neck-agglomerate of the Sgùrr Dearg vents (Chapter XVI.).

*Big-Felspar Gabbros and Dolerites between Sgùrr Dearg and Loch Spelve (Sheet 44).* Continuing the same line as the small-felspar dolerites just described, comes an equally definite suite of big-felspar gabbro and dolerite intrusions (Fig. 22). They correspond in character, and probably in age, with the big-felspar basalt-flows found near the top of the lava-pile in the Coire Mòr Syncline close at hand (p. 121). They occur both as sills and plugs.

The more conspicuous masses of big-felspar gabbro are closely associated with the Moine gneiss of Sgùrr Dearg (Fig. 30, p. 204), and are themselves so terribly crushed, that it is necessary to state the evidence for the Tertiary age of the suite. It is as follows :—

Mesozoic sediments sometimes intervene between the gneiss and the gabbro even where these two occur close together; a tongue of the gabbro—or dolerite—clearly chills against Tertiary lavas on Beinn Bheag (Figs. 30 and 35, pp. 204, 237); minor masses occur well out in the lava-area—two of them at An Eiligeir intruded with chilled margins into a larger plug of non-porphyritic dolerite; and finally, as already stated, finer-grained, but otherwise identical, rocks are found as characteristic members of the Tertiary lava-sequence.

The evidence for a relatively early date of intrusion is afforded by the crushing and alteration shown by the outcrops at Sgùrr Dearg, and also by the fact that the rocks of these outcrops are clearly older than the vent-agglomerates of the neighbourhood. This last point will be dealt with in detail in Chapter XVI. Here it may be added that, along with the Sgùrr Dearg agglomerates, the big-felspar intrusions are cut into lenticles by early Basic Cone-Sheets (b I. of one-inch Map).

In its more important exposures, the Sgùrr Dearg Gabbro behaves as a sill intruded into the Mesozoic rocks (Fig. 35, p. 237). In other cases, the rock seems to occupy small necks.

If it is correct to refer the Sgùrr Dearg Big-Felspar Suite of gabbros and dolerites to the period that produced the big-felspar basalt-lavas of the Coire Mòr Syncline (p. 121), then of course it follows that the non-porphyritic dolerite forming the main part of the An Eiligeir Plug must also correspond in date to lavas that are still preserved to us in Mull.                                E.B.B.

### PETROLOGY.

*Dolerite Plug of 'S Airde Beinn.* The rock forming this mass (18064) is a remarkably fresh dolerite composed of olivine, augite, labradorite, and magnetite. The olivine is either fresh or partly converted into bright-yellow iddingsite. It is a moderately abundant constituent, builds fairly large irregular individuals, and appears to be in part of earlier and in part of later generation than the felspars. The felspar is a zoned basic labradorite that occurs in somewhat small stumpy crystals, two or three times as long as they are broad, and is ophitically enclosed by large crystals of slightly titaniferous augite. A second generation of augite of acicular habit, and in close association with minute crystals of magnetite, occurs in a chloritized residuum that fills all interspaces between the felspars not already occupied by the earlier ophitic augite.

This rock, in its felspathic nature and in the mode of occurrence of the augite, has points that connect it genetically with some of the pillow-lavas of the Central Region (Chapter IX.). Specimens illustrating its metamorphic action on the basalt-lavas with which it is in contact are easy to obtain (18065, 19268), and they show that the dolerite has produced more or less complete granulitization of the older rocks. In a compact basalt (18065), much granulitic augite has been formed, biotite has been produced around iron-ores, and there has been thorough recrystallization of the felspar-microlites.

In another specimen, amygdaloidal basalt (19268), the granulitization is equally well-marked, and lime-zeolites that occupied the vesicular cavities have, in all cases, been converted into felspar of corresponding composition.

*Dolerites and Gabbros of Ben More.* An interesting rock from a point 530 yds. north-west of Ben More summit is a coarse olivine-dolerite (17145) allied to the Big-Felspar Dolerites subsequently described (p. 164). It is noteworthy for the rich colour of its large ophitic titaniferous augite-crystals. The augite is a pale-plum colour and has crystallized from a magma rich in titanium in close association with large crystals and plates of titano-magnetite and ilmenite. It is usually fringed by a narrow zone of green hornblende, a mineral which also enters into the composition of the pseudomorphs after olivine. The felspar is labradorite and occurs as large, occasionally well-formed, crystals.

*Gabbro of Beinn nan Lus.* The more westerly mass (17983) is an olivine-gabbro. It is composed of yellowish brown augite in coarse ophitic relation to labradorite felspar that is zoned with oligoclase and strung with albite. The felspar appears to have separated in two stages, first as relatively small crystals of more basic composition often entirely surrounded by augite, and later as less basic crystals of larger size on which the augite, with its early felspar-inclusions, is only marginally moulded. The olivine is represented by somewhat ill-defined pseudormorphs in serpentine, the peripheral portions of which, and to a less extent the interior, have been converted into green fibrous hornblende. Occasionally better-formed pseudomorphs are entirely enclosed by augite.

The more easterly outcrop (14702-4) is, similarly, a coarsely ophitic olivine-dolerite or gabbro. Original augite has been largely converted into fibrous green hornblende and the labradorite albitized. This rock has come within the metamorphosing influence of the later Beinn a' Ghràig Granophyre. The augite where it shows most change (14704) has given rise to fibrous hornblende and magnetite. A fine-textured vein, probably a crush, traverses the slide, and shows the effects of metamorphism more markedly, for it has been transformed into a mass of small fibrous and ragged crystals of pale pyroxene. In addition, a calcareous patch, the original nature of which is obscure, has given rise to a fine-grained aggregate of pale brownish garnet and leek-green pyroxene, surrounded by an indefinite zone of diopside. The assemblage of minerals and their mode of occurrence is similar to that observed in certain metamorphosed amygdales encountered in the lavas of An Gearna and elsewhere (p. 152).

*Beinn na Duatharach Gabbro of the North-West Caldera (Anal. II.; Table VI., p. 24).* This mass is represented by the analysed rock (14846, p. 24) and other specimens. It is a moderately fine-grained gabbro that originally contained olivine and has a fairly coarse ophitic structure developed between diallagic or normal augite and large plate-like crystals of zoned labradorite. The felspars are strongly zoned and much twinned, pericline lamellation being a striking feature. [Ilmenite passing into leucoxene is the only note-worthy accessory.

The rock locally shows signs of crushing, disruption (14565-6), and widespread chloritization.

Contact-alteration, probably by the Glen Cannel granophyre, has affected the rock subsequently to its crushing to a variable degree. Serpentine, that formerly was present both as pseudomorphs after olivine, and as interstitial growths between the felspars, has, in the lower grade of metamorphism, yielded epidote, chlorite, and hornblende (14707). In the more intensely altered rocks, the olivine's decomposition-products have been transformed into a fine-grained aggregate along with red-brown biotite, green hornblende and recrystallized magnetite (14846); while crushed augite has been granulitized (14706).

The chemical relationship of the Gabbro to the Porphyritic Central Lavas of Mull and the Cuillin Gabbro of Skye is discussed in Chapter I. (p. 24).

*Bith-bheinn Dolerite within the North-West Caldera.* A mass on the northern face of Bith-bheinn (17131) differs from those described above in its obvious relationship to the quartz-dolerites and tholeiites of Talaidh Type (Chapter XXVIII.). It is a moderately fine-grained rock and consists of a hypidiomorphic to stoutly columnar augite in association with abundant magnetite and somewhat elongated crystals of labradorite. The felspar is turbid with decomposition-products and has suffered considerable albitization. In more acid fine-grained patches, all the recognizable felspar is albite, while augite is represented by secondary hornblende.

*Dolerites West of Clachaig within the North-West Caldera.* A small mass (15568) at the head of the stream to the south-west of Clachaig is of a somewhat unusual type in that it has a strongly developed microporphyritic character. Its larger constituents are augite and labradorite: the former is usually turbid with finely divided magnetite, and is ophitically intergrown with its companion which occurs as more or less isolated crystals strongly zoned with oligoclase and perthite. These larger constituents are enveloped by an acid matrix of fine texture, which has produced some disruption of the more coarsely crystalline portions of the rock, and consolidated as a mixture of perthite, micropegmatite and free quartz, the last-named in well-defined clear patches. Apatite is abundant throughout the rock as fairly large crystals but shows an abnormal concentration in the acid matrix. It would appear that we are here dealing with a gabbro that has been acidified by the incorporation of granophyric matter such as is discussed later in connexion with rocks of hybrid character (Chapter XXXIII.). The rock exhibits evidence of contact-alteration by the Glen-Cannel Granophyre, more particularly in the acid matrix, where finely divided iron-ore has recrystallized, while abundant minute scales of biotite and fibres of green hornblende have formed at the expense of chlorite.

A larger mass, west of Clachaig (17125), presents characters that ally it to the Talaidh Type of intrusion and is best described as a quartz-dolerite. It consists of a hypidiomorphic slightly titaniferous augite that occasionally shows traces of salitic lamellation; elongated, once-twinned, crystals of zoned labradorite; fairly abundant iron-ore; and a fine-grained chloritized matrix of more acid character. The matrix shows signs of contact-alteration in the development of a fibrous secondary hornblende.

*Small-Felspar Dolerites of Fig. 22.* This 'small-felspar' group of rocks is chiefly remarkable for reproducing, in some cases, the petrographical peculiarities of the porphyritic pillow-lavas of the central region (p. 150). As exemplified by the interior of an intrusion to the north-west of Sgùrr Dearg (16472), these rocks show a closely

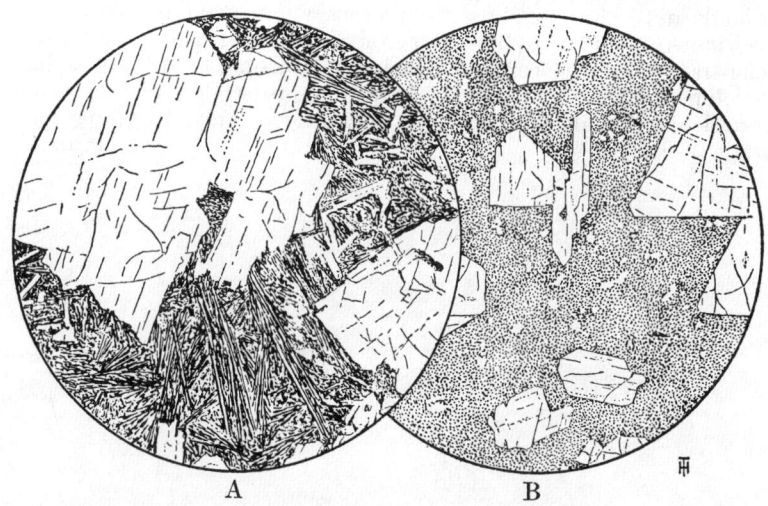

Fig. 23.—Basaltic facies of Small-Felspar Dolerite intrusions, N.W. of Sgùrr Dearg.

A. [16472] × 17. Interior of an intrusion, showing porphyritic crystals of albitized labradorite-bytownite in a variolitic matrix composed of narrow crystals of titaniferous augite, finely-divided iron-ore, a little plagioclase, and a chloritized residuum. (Compare with Fig 21B, p. 151.)

B. [18652] × 17. Chilled edge of a sheet, showing porphyritic felspars of identical character but having an aphanitic matrix in which all structure is suppressed and which presumably consolidated as glass. (Compare with Fig. 21 C, p. 151.)

packed assemblage of moderately well-formed crystals of zoned and partially albitized bytownite-labradorite, in a variolitic matrix composed of long narrow crystals of a titaniferous augite, finely divided iron-ore, a little plagioclase and a chloritized residuum (Fig. 23). In the chilled margins of these rocks (18652), the porphyritic crystals are of an identical character, but the matrix has solidified as an aphanatic mass, no doubt originally glass, in which all crystalline structure is suppressed.

Coarser representatives of the same group are genuine ophitic dolerites. An example from Allt nan Clàr (14356) is composed essentially of equal proportions of labradorite and augite in ophitic elationship. Occasionally the augite shows a distinct lilac tinge

164 *Chapter XI.—Various Dolerites and Gabbros.*

due to the presence of titanium. The rock contains a little enstatite associated with the augite, and also large plates of ilmenite (18654).

A small mass on Cruach Choireadail (17199), not included in Fig. 22, is mainly of the coarsely ophitic dolerite type described above; but, locally, it shows a tendency to develop a variolitic structure.

*Big-Felspar Dolerites and Gabbros of Fig.* 22. The distribution of these rocks is shown in Fig. 22. As represented by the intrusion one-third of a mile east of the summit of Sgùrr Dearg (16006), they are coarsely crystalline rocks that the microscope shows to be composed of large crystals of zoned and partially albitized labradorite, ophitic slightly titaniferous augite, and indefinite serpentinous pseudomorphs after olivine. The partly albitized felspars frequently contain secondary epidote; the augite has developed patches and fringes of uralitic hornblende; and a fibrous hornblende is often present in the serpentinous areas that represent original olivine. H.H.T.

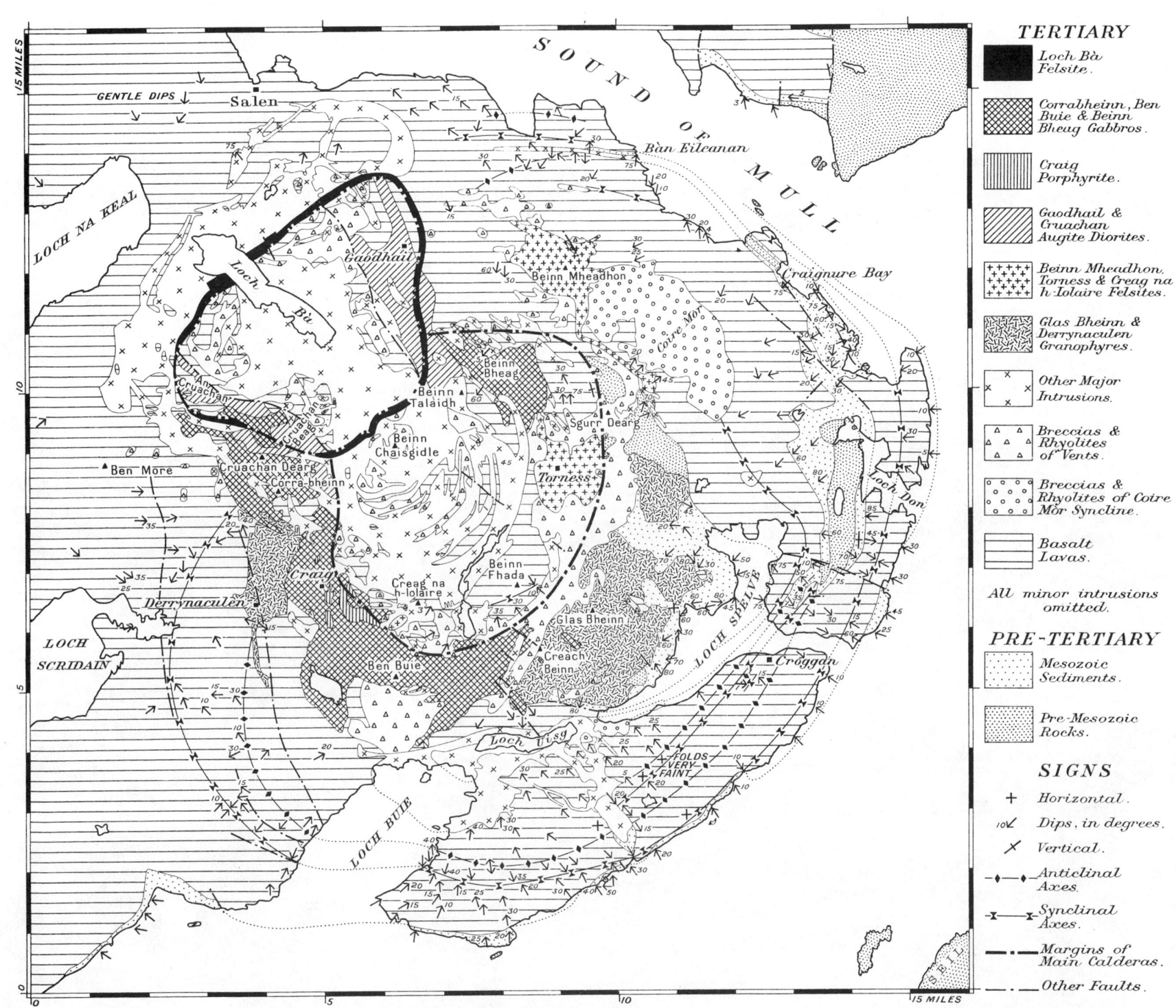

PLATE V. TECTONICS AND MAJOR INTRUSIONS ABOUT MULL CENTRES.

# CHAPTER XII.

## EARLY GRANOPHYRES.

### Introduction.

An account will now be given of three important granophyre-intrusions lettered oG on the one-inch Map, Sheet 44. They lie outside the margin of the south-eastern caldera of Central Mull (Plate V.). The two first occur close together, between the margin of the caldera and Loch Spelve, and may be grouped under one name—the Glas Bheinn Granophyre. The remaining mass occurs near the western end of Glen More and is known as the Derrynaculen Granophyre.

A fluxion-rhyolite, seen on the foreshore near Bàn Eileanan on the Sound of Mull, is also dealt with. Like many other felsites, it is lettered F on Sheet 44.

The subject-matter will be discussed under three headings :—

The field-relations of the intrusions.

Their possible connexion with the arcuate fold-system which surrounds the south-eastern caldera of Pl. V.; the folding itself is considered in detail in Chap. XIII.

Petrology.

### Field-Relations.

*Glas Bheinn Granophyre.* Plate V. and the one-inch Map, Sheet 44, show quite clearly the position of the two main outcrops of the Glas Bheinn Granophyre, separated, along the course of the Lussa River, by a tract of sandstone referable to the Trias.

Throughout the greater part of its extent the granophyre is so invaded by dolerite-sheets belonging to the Early Basic Cone-Sheet assemblage (Chapter XXI.), that its outcrop is reduced to a multitude of discontinuous strips and lenticles which are reproduced in somewhat diagramatic fashion on the one-inch Map. The only considerable exposures free from such interruptions occur near the Loch Spelve road in the neighbourhood of Maol Odhar and Cnoc na Faoilinn.

Most of the masses mapped as Glas Bheinn Granophyre are of one general type : a normal granophyre intensely crushed. In Maol Odhar, where the granophyre can be most conveniently examined, the crushing is less pronounced than usual.

In the Cnoc na Faoilinn neighbourhood, the Glas Bheinn Granophyre is seen to be a complex, though this is, so far as is known, of quite local significance. Fig. 24 shows that what is generalized as granophyre in Sheet 44 includes, at this place, a porphyritic felsitic-looking rock which the microscope proves to be inninmorite. West of the Cnoc na Faoilinn Inninmorite, the Glas Bheinn Grano-

166    *Chapter XII.—Early Granophyres.*

phyre is perfectly normal in composition; east of it, it is distinctly more basic than usual. It would seem probable that one is dealing here with a complex of three intrusions, all of them crushed and of manifestly early date.

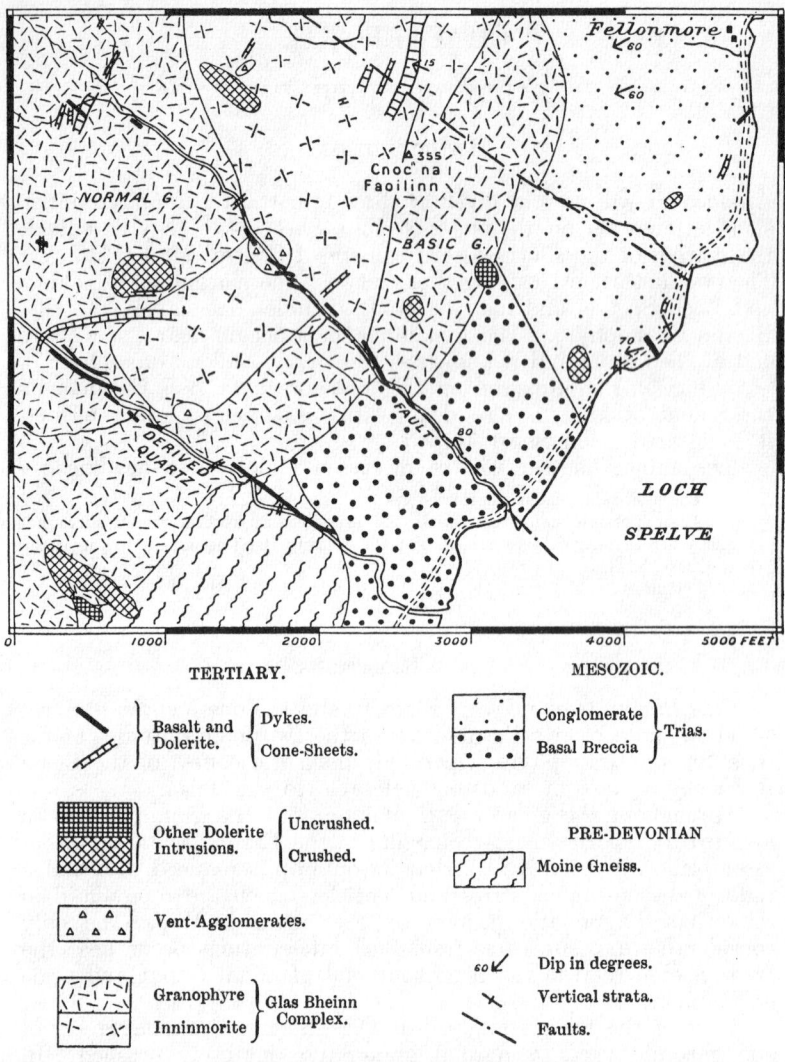

Fig. 24.—Map of Cnoc na Faoilinn, Loch Spelve.

The crushing which so constantly affects the Glas Bheinn Granophyre equally characterizes the associated Moine gneiss and Triassic sandstone. The granophyre does not show satisfactory contacts with Tertiary lavas; but the crushing does not betoken Pre-Tertiary age, for it extends very definitely into the lava-area, as in Killean on the east side of Loch Spelve.

The Tertiary date of the Glas Bheinn Granophyre is established on the strength of field-evidence supplied by exposures along a little stream, which drains past Seanvaile Cottage into Camas an t-Seilisdeir on Loch Spelve. About 300 yds. up from the road, there are excellent exposures of a permeation-zone connecting the Glas Bheinn Granophyre with the sandstone of the neighbourhood. The correlation of the Loch Spelve sandstones and conglomerates with the Trias is almost a certainty; and so too is the Tertiary age of any intrusion which cuts a Mesozoic sediment in the Hebrides. In agreement with this, one may add that the petrological type of the granophyre is distinctly Tertiary. Another stream-exposure of a somewhat similar nature to that of Seanvaile lies half a mile south-west of Cnoc na Faoilinn (Fig. 24). It is interesting as affording an additional example of granophyre full of foreign quartz, but it is only analogy with the Seanvaile exposure that suggests derivation of this quartz from Triassic sandstone.

The relatively early date of the Glas Bheinn Granophyre as compared with other Tertiary intrusions is shown by the following facts : the crushed granophyre passes insensibly in many exposures into mixed volcanic breccia occupying vents (p. 205); both granophyre and breccia are cut by innumerable uncrushed dolerite sheets of the Early Basic Cone-Sheet suite (p. 207); a large proportion of these early cone-sheets are in turn cut by the Ben Buie Gabbro (p. 245); the Ben Buie Gabbro is itself earlier than a host of other Mull intrusions.

The relations of the granophyre to the breccia or agglomerate of the many associated vents suggest that it owes its widespread crushing largely to explosions which opened up the vents. The agglomerate of the vents, it may be added, is, everywhere in the neighbourhood of the granophyre, very largely made up of granophyre-fragments.   E.B.B.

*Derrynaculen Granophyre.* The position of this intrusion, a couple of miles up Glen More from the head of Loch Scridain, is easily recognized on Plate V. (p. 165) and the one-inch Map, Sheet 44. Like the Glas Bheinn mass, the Derrynaculen Granophyre is freely cut by Early Basic Cone-Sheets, and many of these in their turn by Ben Buie Gabbro (p. 246). The main differences from the Glas Bheinn Granophyre are the general absence of crushing, and the fact that the intrusion invades the Plateau and Central Groups of the Tertiary-lava sequence of Chapter V.   (C.T.C.)

*Bàn Eileanan Rhyolite.* Very little need be said about this rhyolite. It is exposed on the shore and in the cliff of the low raised beach south of Bàn Eileanan, where the axis of the Craignure Anticline crosses the coast-line north of Scallastle Bay on the Sound of Mull. It is characterized by irregular and contorted fluxion with an apparent anticlinal disposition. The exposures suggest that one is dealing with the top of an acid intrusion that might be expected to have considerable lateral extent a little below the surface. On the ground of difference of character, it is improbable that there is any intimate connexion between this rhyolite and a granophyre-dyke which has been traced for a couple of miles inland along the same line of strike (*cf.* one-inch Map, Sheet 44).

The rhyolite may possibly be connected in origin with the Craignure Anticline ; but very little significance can be attached to this suggestion.

POSSIBLE CONNECTION OF EARLY GRANOPHYRES AND FOLDING.

Plate V. shows an arcuate system of folds which characterizes south-eastern Mull. Figs. 25 and 35, pp. 174, 237, bring home the degree of disturbance connected with the development of this fold-system. A remarkable feature of the fold-system as a whole is that it centres upon an area of profound subsidence, the more south-easterly of the two calderas of Plate V. It seems certain therefore that the fold-system as a whole cannot be accounted for by any definitely central intrusion.

If one looks outside the inferred margin of the south-eastern caldera, one finds two suites of major intrusions which are worth considering from the point of view of the origin of the folding : on the one hand, one notes the Glas Bheinn and Derrynaculen Granophyres ; and on the other hand, the Ben Buie Gabbro. The latter cannot be held to be intimately connected with the folding, for it freely cuts Early Acid and Basic Cone-Sheets, which are themselves certainly later than the main folding-epoch (pp. 222, 236). This limits the enquiry very considerably.

In considering the case of the Glas Bheinn and Derrynaculen Granophyres, it is well to remember that the Ben Buie Gabbro may in large measure mark the site of a former extension of the granophyre-complex. Vent-agglomerate of pre-Ben-Buie age is a conspicuous feature of the country between Loch Fuaran and Gleann a' Chaiginn Mhòir, and it consists very largely of granophyre-fragments.

That the Glas Bheinn Granophyre may be intimately connected with the folding-movement is definitely suggested by its position in the heart of a very pronounced broken anticline, the Loch Spelve Anticline of Chapter XIII. Its associates at the surface are almost always Pre-Tertiary rocks—Moine gneiss or Triassic sandstone. In the same way, the Derrynaculen Granophyre is found in the core of an anticline which locally brings to view a mugearite, associated with basalt of Plateau Type, from beneath a great covering of the Central Group of Lavas (Plate III., p. 91).

The main uncertainty is whether the granophyre-intrusions are not of too early a date to be correlated with the folding. The Glas Bheinn Granophyre is certainly earlier than a great series of volcanic vents which opened up in its neighbourhood more or less along the line of the caldera edge (Fig. 30, p. 204). Now these vents seem to have supplied the agglomerate which occupies the Coire Mòr Syncline (Chapter XV.)—itself richly charged with granophyre-fragments. Some may hold that the agglomerate-filling of the syncline antedates the folding that made the syncline, in which case, of course, the exposed portion of the Glas Bheinn Granophyre would seem to have consolidated before the folding commenced

(p. 196). On the other hand it is suggested in Chapter XV. that certain local circumstances are more in keeping with the view that the Coire Mòr agglomerate was showered down into a synclinal hollow already formed ; and this allows of the view that the granophyre as a whole and the syncline are contemporaneous.

If one adopts the conclusion that the granophyre is responsible for the folding, it is not difficult to picture the sequence of events. The granophyre may have risen from the depths in dyke-like form guided by the fracture bounding the caldera, until, on approaching the surface, it came into an environment where expansion was possible by virtue of the complex yielding of the masses constituting its walls and roofs.

## PETROLOGY.

*Glas Bheinn Granophyre.* The general character of the Glas Bheinn Granophyre is represented by eight slices (15053, 15556-8), taken from the outcrop north of the Lussa River, and (15063, 15868, 17416, 17418) from that lying south of the river. Both outcrops seem to show the same varieties. None of the material is fresh, a feature in keeping with the relatively early date of the granophyre combined with its position well inside the Pneumatolytic Limit of Plate III., p. 91. The most satisfactory specimens can be gathered on the hill-face south-west of the main road at the summit of the rise, one quarter of a mile south of Ardura Farm. A slice from this locality (15063), which may be considered typical, contains a considerable proportion of zoned labradorite-crystals fringed by micropegmatitic growths of quartz and orthoclase merging outwards into a general coarse granophyric matrix. The original ferromagnesian minerals are augite with basal striation (salitic), and greenish-brown hornblende. Both minerals occur as ill-formed crystals moulded, as a rule, on their felspathic companions. Neither of them are acicular in habit. As accessory minerals, one may note apatite in abundance, and iron-ore ; the latter includes both pyrites and magnetite, and is commonly associated with the augite. Most of the augite has been uralitized, both here and elsewhere ; in fact, only one other slice of the series (15556) shows this mineral in an unaltered condition. As regards the other slices, it may be noted that an acicular habit never becomes a strongly marked feature of the augite although it is by no means wholly wanting.

Two stream-sections, both of them close to the Loch Spelve road, have already been noted as affording exposures of marginal assimilation of foreign quartz by the Glas Bheinn Granophyre. The exposure one-sixth of a mile north-west of Seanvaile Cottage is represented by four slices (17411-4), and that west of the gneiss in Abhuinn Coire na Feòla by two (17420-1). Of these slices (17412) may be taken as a simple type. It shows a moderately fine-grained sandstone composed of subrounded grains which have, in many cases, been separated by growth of alkali-felspar. There is practically no micropegmatite. In (17414), the felspathic growths are on a larger scale, and the quartz is relegated to the ground-mass. Often the individual grains of quartz are separated, but equally often several remain in contact furnishing little clusters. The felspars are chiefly perthitic orthoclase with a certain amount of independent albite. In another slide (17421), perhaps more definitely igneous, there are phenocrysts that once consisted of basic plagioclase now represented by albite and epidote.

Another slice from the granophyre-margin near Seanvaile Cottage (17415), a third of a mile south-west of the cottage, strongly suggests a final stage of incorporation of the neighbouring sandstone. The rock is exceedingly rich in quartz, but most of this mineral is in definite, though simple, micrographic relationship with orthoclase, so that one cannot be certain of its clastic origin. In keeping with the richness in quartz, one notes that augite has been displaced by hornblende and biotite. The hornblende is strongly pleochroic brown and green, the biotite brown and yellow. Both minerals occur in irregular plates and are often intergrown. Early crystals of basic plagioclase also occur.

Attention has been directed to two minor subdivisions which are locally separable from the main mass of the Glas Bheinn Granophyre at Cnoc na Faoilinn, above Loch Spelve (Fig. 24). Each of these subdivisions is represented by three slices, the outer, lying to the west of the 355 ft. cairn, by (15064, 17061, 17063), the inner by (15869, 17062, 17417). The outer or more easterly Cnoc na Faolinn intrusion approaches closely to the normal Glas Bheinn Type but it appears to be richer in labradorite, and might perhaps be better styled a quartz-augite diorite, with hornblende quite subordinate. A fair proportion of the augite, now wholly uralitized, crystallized simultaneously with the basic plagioclase in graphic intergrowth.

The inner Cnoc na Faoilinn intrusion is a rather acid inninmorite (p. 282).

*Derrynaculen Granophyre.* The Derrynaculen Granophyre of the Glen More District is represented by seven slides (15612-7, 15649). Here again the material shows considerable alteration of the pneumatolytic type resulting in the uralitization or chloritization of the augite. Magnetite accompanies these secondary products and some of it is clearly secondary. In (15612), contact-alteration by the Ben Buie Gabbro has produced aggregates of biotite and iron-ore in place of the pneumatolytic products. The Derrynaculen Granophyre is evidently a fairly homogeneous mass. It differs from the Glas Bheinn Granophyre mainly in showing a greater tendency to acicular structure as regards its augite pseudomorphs and lath structure as regards its felspars. It is a granophyre of craignuritic affinities and in composition and in general character agrees well with the analysed specimens (15550, 16803) described on p. 226.

*Bàn Eileanan Rhyolite.* This little intrusion north of Scallastle Bay, on the Sound of Mull, is illustrated by the slide (16465), which shows clearly the fluxion-structures so characteristic of the intrusion in the field. The ground-mass of the rhyolite was evidently glassy at the time of its consolidation. It is now devitrified, yielding cryptocrystalline aggregates and fairly definite little crystal-patches consisting in the main of alkali-felspar. Small well-shaped pseudomorphs in chlorite and calcite after augite are common, and minute strings of iron-ore probably represent needles of the same material. An interesting feature is the occurrence of little xenoliths of soda-granophyre and, less frequently, of fine-grained basalt.     E.B.B.

# CHAPTER XIII.

## TERTIARY TECTONICS OF MULL AND LOCH ALINE.

### INTRODUCTION.

A SERIES of rather irregular subsidences, in part determined by faulting, has assisted in the preservation of Tertiary lavas and Mesozoic sediments in the west of Scotland and in the north-east corner of Ireland. Mull may be taken as representative of one of the most important of these subsidences, including in its scope not only Mull itself, but also :—on the north-east, neighbouring portions of Morven and Ardnamurchan ; on the west, Staffa and the Treshnish Isles ; and, on the north, Muck, Eig, and the Loch Slappin district of Skye (*cf.* Sir A. Geikie's Geological Map of Scotland).

The boundary of the Tertiary-Mesozoic outcrop of the Mull district is well-enough defined on the north-east, east, and south, by an emergence of Pre-Mesozoic rocks in mainland and island exposures extending interruptedly from Ardnamurchan (Sheet 52) to the western extremity of the Ross of Mull (Sheet 43).

From the last-named locality, a gentle anticline, running east-north-east, carries the outcrop of Pre-Mesozoic rocks to Inch Kenneth (Sheet 43). Erisgeir lies on the course of this anticline.

North-west of the Inch Kenneth Anticline, there follows an important extension of the Mull depression to which must be referred the lavas of Ulva, Staffa, and the Treshnish Isles, and also, if it be a lava, the columnar basalt of Réidh Eilean off Iona. The north-westward limit of the under-sea continuation of the Treshnish outcrop cannot be fixed precisely ; but it lies south-east of the island of Tiree, which just enters the corner of Sheet 43.

Tiree, as may be sufficiently realized on inspection of Sir Archibald Geikie's map of Scotland, is situated on a north-easterly trending anticline responsible for an emergence of Lewisian Gneiss in Tiree and Coll, and of Torridonian Sandstone in Rum, Soay, and, at the head of Loch Scavaig, in Skye. Eig and Muck lie on the Mull side of the Tiree Anticline, while Canna, on the other side, manifestly belongs to the main Skye depression.

Generally speaking, except near the plutonic centres of Mull, Ardnamurchan, and Skye, the Mesozoic sediments and Tertiary lavas of the Mull depression are gently and variously inclined. They are traversed by considerable faults, which, east of Loch Aline, run north and south, and, elsewhere, more or less north-west and south-east. Apart from this, and the marked subsidence of the region as a whole, the general tectonics of the district present few points of interest. Accordingly, the main part of this chapter will be devoted to the tectonics of the plutonic centre of Mull as illustrated in Plate V., p. 165. The outstanding feature, here, is the occurrence of two cauldron-subsidences, or caldera, the more southeasterly of them in large measure surrounded by concentric folds. After the more central tectonics have been disposed of, attention will be directed to a few other points of tectonic interest including the conspicuous Inninmore Fault (Fig. 26), and also the Wrench-Fault of Port Donain. Another matter of negative import touched upon is the passive rôle played by the ancient Great Glen Fault during the development of Tertiary disturbances in Mull and its neighbourhood.

### South-East Caldera and Concentric Folds.

In Plate V., two confluent tracts of Central Mull are outlined under the title of the Two Main Calderas. The more south-easterly is shown in large part surrounded by a series of concentric folds. The main elements of the tectonic scheme may be grouped as follows from without inwards :—

> Marginal Tilt, the outer limit of the disturbed country, traceable from near Craignure Bay to beyond Carsaig (M.T. of Fig. 25).
> Duart Bay Syncline, traceable from near Craignure Bay to near Corra-bheinn.
> Craignure Anticline, traceable by Bàn Eileanan and Craignure Bay to Eilean Trianach (C.A. of Fig. 25).
> Loch Don Anticline, continuing in relay the Craignure Anticline, and traceable from near Craignure Bay through the head of Loch Don (L.D.A. Fig. 25). Before reaching Loch Spelve, the Loch Don Anticline split into two, and one branch can be followed through the Croggan Peninsula and across Loch Buie to Derrynaculen.
> Coire Mòr Syncline, traceable from Coire Mòr to Loch Uisg past the west extremity of Loch Spelve.
> Loch Spelve Anticline, traceable through Sgùrr Dearg and Creach Beinn (L.S.A. of Fig. 25).
> The South-Eastern Caldera, as indicated on Pl. V.

These various elements will be considered in the order stated. It has already been argued (Chapter XII.) that the folds may have arisen as a sequel to the marginal intrusion of the Glas Bheinn and Derrynaculen Granophyres outside the already established South-Eastern Caldera. Their relatively early date in the history of the Mull complex is demonstrated by the manner in which they are traversed by Early Acid and Basic Cone-Sheets, not to mention the Loch Uisg Granophyre (Figs. 25, 34, and 35, pp. 174, 231, 237). This feature of the folding is discussed to some extent in the succeeding paragraphs, and is returned to later in connection with the various intrusions concerned (pp. 222, 231, 236).

Before splitting up the subject into its various elements, it is well to emphasize the opportunity offered by the peninsula south of Loch Don to any one wishing a field-introduction to the more peripheral parts of the fold-system. The two lower sections of Fig. 25 are drawn to true scale across this peninsula, and these combined with the one-inch Map (Sheet 44) will serve as efficient guides. Exposures are excellent, and erosion has developed the scenery in diagrammatic fashion, so that structure can be read to singular advantage. This is particularly the case about Loch a'

174　　　　　　　*Chapter XIII.—Tertiary Tectonics.*

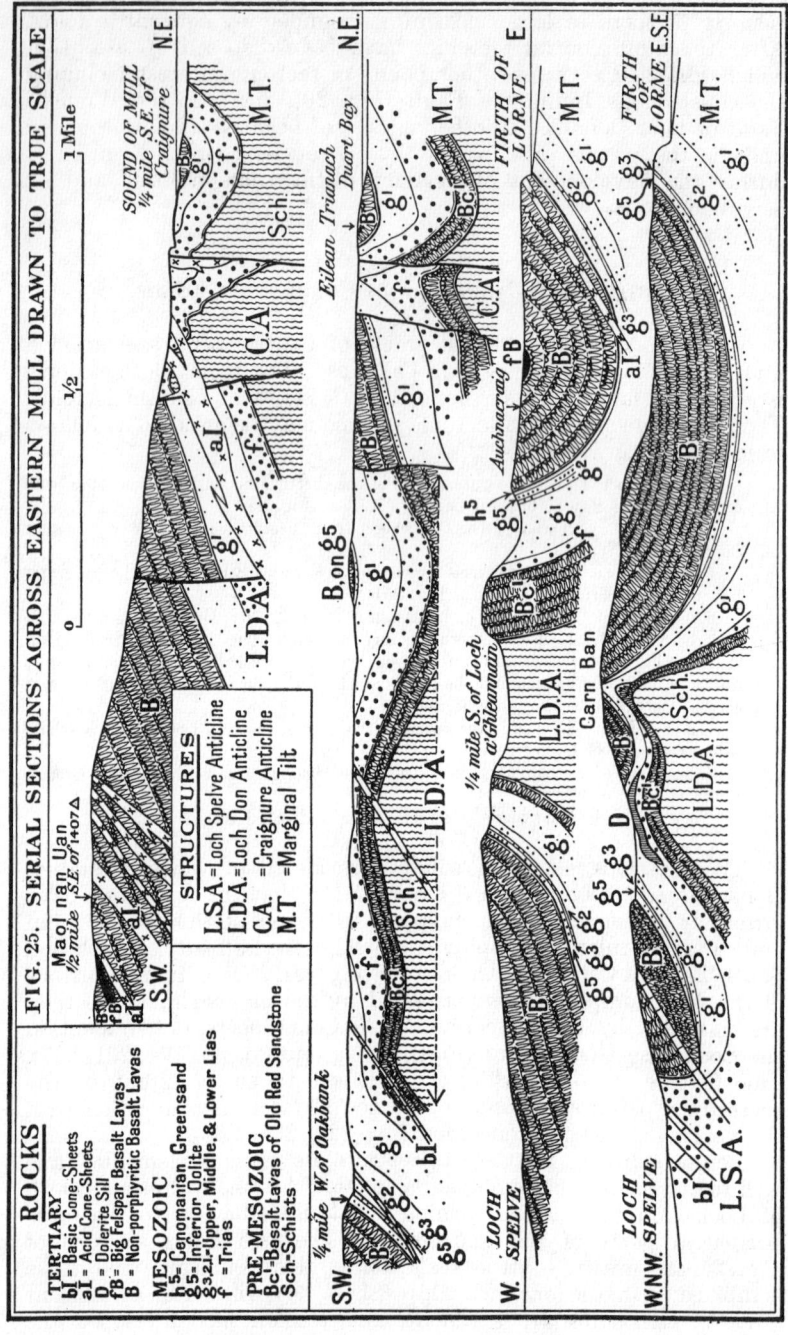

Fig. 25. Serial Sections across Eastern Mull drawn to true scale.

Ghleannain, which lies in a hollow excavated along soft 'schists' (slates and limestone), and is flanked by ridges of Old Red Sandstone lavas, in their turn separated by hollows, worn in Mesozoic sediments, from uplands of Tertiary lavas on either side. The shores of the Firth of Lorne and Loch Spelve help to complete the structural presentation.

The best district to study the relationship of the Loch Spelve Anticline to the South-East Caldera lies about Sgùrr Dearg, and is illustrated in Figs. 30, 35, pp. 204. 237,

In reading the following descriptions, it is important to bear in mind that, despite the time-intervals involved, there is no appreciable difference of dip between the Tertiary lavas of Mull and the underlying Mesozoic sediments. This fact often greatly simplifies the study of the Tertiary tectonic features of the district.

*Marginal Tilt.* The Marginal Tilt, as it is styled in Fig. 25, furnishes the outer limb of the Duart Bay Syncline, and must be held responsible for the preservation of basalt-lavas in the Java Point peninsula as compared with gneiss in Sgeir nan Gobhar, Glas Eileanan, and Eilean Rudha an Ridire, in the Sound of Mull.

From Duart Point southwards, the Marginal Tilt is clearly discernable in the lie of the lavas as shown in Plate V. The arcuate trend of the structure is also particularly well-shown in the course of the coast-line between Duart Point and the entrance to Loch Buie; for this gently curved coast marks approximately the line along which the lavas affected by the Marginal Tilt pass through sea-level.

West of Loch Buie, there is some difficulty in deciding what to refer to as Marginal Tilt. Near Carsaig Bay, the tilt clearly must form part of the eastern limb of a faint anticline which, assisted by faulting, leads to an important exposure of Mesozoic sediments; but, whereas this anticline is vaguely traceable east-north-eastwards (p. 181) into country unaffected by arcuate folding, the marginal tilt seems to continue as an independent and often well-defined flexure northwards across Loch Beg at the head of Loch Scridain.

*Duart Bay Syncline.* The continuity of the Duart Bay Syncline from Craignure Bay to near Corra-bheinn as shown on Plate V. is easily verified in the field, except where, for a space south of the entrance to Loch Spelve, the fold is very shallow. Sometimes the lavas are so clearly seen dipping in from either side towards the axis of the syncline that the structure becomes quite a scenic feature, as for instance north of Loch Spelve and in the coastal cliffs either side of Loch Buie.

Comparison of Plates III. and V. shows that an additional indication of the syncline is afforded by the preservation of relatively late rocks in its embrace; thus in the north-east part of its course, one finds Tertiary lavas flanked by Pre-Tertiary rocks, while north of the Port Donain Fault, and again, both east and west of Loch Buie, there are outcrops of big-felspar basalt, or mugearite, as the case may be, contrasting with the normal unbroken sequence of Plateau Basalts met with either side.

## Chapter XIII.—Tertiary Tectonics.

*Craignure Anticline.* From the steamers which ply the Sound of Mull, it is possible to detect the Craignure Anticline affecting the Tertiary lavas of the hill-face above Fishnish Bay, though appearances are considerably blurred by numerous cone-sheets. In the lower ground, intrusions are much less abundant, and on foot one can trace the anticline without much difficulty through Cnoc Damh, and so to the sea-shore at Bàn Eileanan ; in fact, for the last three quarters of a mile, the fold brings Mesozoic rocks to the surface. The inland exposures consist of portions of the Lower and Middle Lias dipping north at 30°. At the coast, only the Lower Lias is found, dipping at 75°.

An interesting feature, both inland and on the coast, is the asymmetrical development of the anticline ; the Liassic rocks are exposed only on the northern limb of the fold ; on the southern limb they are evidently cut out by faulting. Another feature is the complexity of the folding leading to a double outcrop of the Lias ; for a small exposure of Middle Lias Sandstone occurs isolated among basalt-lavas at the road-side 100 yds. north of the main outcrop.

The Craignure Anticline can be recognized again in the Java Peninsula, where its centre is defined by Mesozoic sediments and Moine gneiss exposed in the west shore of Craignure Bay, almost in the condition of a crush-breccia. The Mesozoic sediments of this side of the bay are not fully laid bare ; but they seem considerably reduced by crushing, and the only horizon recognized is Trias conglomerate. A little farther south, Lower Lias is seen cut through by an early acid cone-sheet (aI on one-inch Map, Sheet 44), which makes much of the south-west shore of Craignure Bay. The section at the base of the cone-sheet exposed on the shore (where the letters f and g' are printed on the Map) is particularly interesting for it shows the sheet cutting across obvious crushes affecting the underlying sediments.

Another noteworthy exposure on the south-west side of the Craignure Anticline is afforded by a stream 400 yds. south-south-west of Craignure Church, where basalt-lavas are seen faulted against Triassic sediments. The fault is steep, but with a well-defined reverse inclination towards the north-east. A little north-west of this exposure, it is nearly certain that a tongue from the acid cone-sheet has risen up along the fault as shown in the Craignure Section of Fig. 25.

What is very likely the same fault brings basalt-lava and Lower Lias together at the surface, in the south-west limb of the Craignure Anticline, a little south of Eilean Trianach (*see* one-inch Map). The fault is drawn in the Eilean Trianach section of Fig. 25 as of normal inclination, but the exposure does not show whether such is the case or not.

Similar evidence to that outlined above is met with again on the north-east side of the Craignure Anticline. Thus, the acid cone-sheet at the pier contrasts in its uncrushed condition with adjacent crushed Triassic conglomerate and associated basalt-intrusions. Moreover, a dyke-like branch of this acid sheet occupies a very well-defined fault exposed in the cliff of the raised beach 200 yds. north-east of the pier. This fault brings Lower Lias limestone against a repetition of Trias conglomerate on its north-east side (*cf.* Craignure Section of Fig. 25 and one-inch Map). Farther south-east, what is probably the same fault is responsible for bringing Trias conglomerate against the upper part of the Lower Lias (*cf.* Eilean Trianach Section of Fig. 25 and one-inch Map).

Generally speaking, all the exposures of Mesozoic rocks in the core of the Craignure Anticline, as exposed in Camas Mòr, are greatly shattered. It must not be thought, however, that this condition is universal; in fact, the sediments along the coast between Rudha na Sròine and Duart Bay are not particularly broken.

The Craignure Anticline must flatten out somewhere in the ill-exposed ground south-east of Camas Mòr.

*Loch Don Anticline.* The next pronounced element of the structural complex is the Loch Don Anticline which takes on in relay for the Craignure Anticline. Towards the north-east, the Loch Don Anticline passes into a faulted monocline illustrated in the Craignure Section of Fig. 25. The exposures are quite clear and need not detain us.

At Loch Don, and for some distance farther south, the anticline is both symmetrical and compressed, as may be judged from the one-inch Map, and the Loch a' Gleannain Section of Fig. 25. How far faulting accounts for the observed impersistence of some of the rock-groups is uncertain; but it does not seem to be altogether responsible for the inconstancy of the Trias; at any rate, what is taken for an unbroken contact between Lias limestone and Old Red Sandstone lavas is exposed 200 yds. south of Loch Don bridge.

Southwards, the Mesozoic rocks pitch beneath the lavas in a double crested fold illustrated in the Càrn Bàn Section of Fig. 25. The more easterly crest is very sharply compressed as is strikingly shown 300 yds. south-west of Càrn Bàn summit, where Mesozoic rocks are last exposed nipped up between the lavas.

Another interesting feature of the Loch Don Anticline is its bodily horizontal displacement by the Port Donain Fault (p. 183). Still another is the crushed condition of some of the rocks affected by the folding. The slates and schistose limestone are not noticeably shattered, nor are the Mesozoic sediments; but the lavas, both Old Red and Tertiary, and a basalt-sill intruded into the Mesozoic sediments near Càrn Bàn, are greatly brecciated. The Tertiary lavas only show this crushing in the southern part of the area, *e.g.* from the centre of the anticline westwards along the north shore of the entrance to Loch Spelve. The Old Red Sandstone lavas on the other hand are very generally crushed, in fact, to such an extent that their original nature is to a considerable extent masked. Quite possibly, in their case, the brecciation is partly of Pre-Mesozoic date and connected with movement along the Great Glen Fault.

The continuation of both branches of the Loch Don Anticline can be traced south-westwards half way through the Croggan Peninsula; but, as indicated in Plate V., the lava-dips are very gentle, and finally the more north-westerly branch of the fold fails altogether. E.B.B.

In Glen Libidil, the south-east branch of the Loch Don Anticline, co-operating with a fault, leads to an exposure of Mesozoic rocks (*cf.* Fig. 34, p. 231 and one-inch Map). Westwards, this branch crosses Loch Buie, where it is clearly recognizable in views of the coastal cliffs both sides of the bay. An interesting fault modifies its southern limb as exposed on the west shore. The locality is shown on the one-inch Map under the name An Coileim and occurs 100 yds. south of the axis of the anticline. A steep reversed fault is seen in section with vertical Lower Lias on its southern side. The fault has been traced inland and seems to partake in the curvature so characteristic of the plan of the fold-system.   G.V.W.

The Loch Don Anticline can also be followed inland to near Derrynaculen. Comparison of Plates III. and V. shows its effect upon the distribution of the lavas. This is more especially clear in the emergence of mugearite, and associated Plateau Type of basalt, near Derrynaculen.   (C.T.C.), G.V.W.

*Coire Mòr Syncline.* The Coire Mòr Syncline is clearly demonstrable in the Coire Mòr district by observed dips of lavas, and also by the mapping of the Big-Felspar Type of basalt (*cf.* Plates III. and V.).

North of the entrance to Loch Spelve, the syncline is obvious owing to the basalt-lavas along its course being flanked by outcrops of Mesozoic rocks referable to the Loch Don and Loch Spelve Anticlines respectively.

Along Loch Uisg, the syncline is visible in the dips of the lavas (Plate V.). Its northern limit seems affected by a powerful fault near Kinlochspelve Farm, where Triassic sandstone comes in contact with Central Types of lava as shown in Plate III.

There is good reason to believe that the agglomerate which occurs along the centre of the syncline at Coire Mòr, and again at the west end of Loch Spelve, rests upon the top of the adjacent lavas. It is shown (p. 196) that very likely it is of later date than the development of the syncline which it occupies.

*Loch Spelve Anticline.* Moine gneiss and Mesozoic sediments are exposed as the country-rock of a large triangular region of upheaval which has its apices near Kinlochspelve Farm, Rudha na Faing, and Sgùrr Dearg, situated roughly 5 miles from one another. The individual outcrops of these Pre-Tertiary rocks are quite small owing to an extraordinary profusion of Tertiary intrusions. Another feature of their occurrence is the excessive shattering they have sustained—how far their fracture is due to folding, and how far to explosion cannot be decided.

Triassic sandstone and conglomerate, broken by innumerable dislocations, predominates in the Pre-Tertiary assemblage bordering Loch Spelve; and Moine gneiss near Sgùrr Dearg. Only at three localities is any Lias found, and they are all interesting from the structural point of view:—

(1) On the shore of Loch Spelve, nearly a mile south of Rudha na Faing, fossiliferous Lias limestone occurs between Trias conglomerates and Tertiary lavas. The sediments and lavas are roughly vertical and greatly broken up (Càrn Bàn Section, Fig. 25). It is uncertain whether the absence of higher zones of the Jurasic should be referred to faulting. Uncrushed Early Basic Cone-Sheets cut the crushed sediments and lavas (p. 236).

(2) A little down the north-east slope of Beinn Bheag, three-quarters of a mile east of the summit of Sgùrr Dearg, steeply inclined baked sandy shale, and, at the northern end of the exposure, white sandstone and quartz-conglomerate, are exposed between Moine gneiss and basalt-lavas (Fig. 30, p. 204). There can be no hesitation in recognizing these rocks as Lower Lias and Trias respectively. The absence of the Lower Lias limestone strongly suggests faulting, as also does the failure to recognize any Mesozoic rocks at all between gneiss and lavas in the mile of reasonably good exposure farther to the south-east.

It is a feature of the Beinn Bheag district that its many Early Basic Cone-Sheets are not affected by the conspicuous folding of the Mesozoic sediments and Tertiary lavas which they traverse (Fig. 35, p. 237).

(3) After an interval of a mile, where exposures are mainly of agglomerate, Mesozoic rocks make their appearance in the bed of Abhuinn an t-Sratha Bhàin, between North Beinn Bhearnach and Dùn da Ghaoithe (Figs. 30 and 35). Trias and Lias are both seen. The former is restricted to patches of white sandstone entangled in brecciated gneiss. The latter is represented by baked fossiliferous concretionary limestones. At the most westerly exposure in the burn, a very impressive section is afforded of vertical limestones cut into thin slices by recurrent Early Basic-Cone Sheets inclined without reference to the folding (Fig. 35).

The structural complexity of the anticline in this district is brought home by the finding of Lias limestone in the heart of the gneissic area. The limestone is seen on end in the bottom of a gorge, the walls of which consists of gneiss; and, farther along, the same limestone occurs as minute lenticles preserved among cone-sheets. The limestone here is probably bounded by a fault on the south-east (Fig. 30).

The Loch Spelve Anticline is markedly asymmetrical, and its inner side is furnished by the great boundary-dislocation of the south-eastern caldera, which latter now calls for attention.

*South-east Caldera.* Intrusions are so abundant in Central Mull that the piecing together of the scattered tectonic data is a matter of difficulty. The evidence is perhaps most easily presented in a series of rather disconnected paragraphs. The district discussed at this juncture measures 5 or 6 miles across, and it can easily be recognized on Plates III. and V., where it is shown bounded by a broken line including both Beinn Talaidh and Beinn Fhada in its circuit.

(1) All the non-intrusive rocks of the area consist of Tertiary basalt-lavas of Central Types (Pl. III, and Chaps. V, IX, X). This in itself shows that the area belongs to a region of central subsidence; for, in the rest of Mull the prevalent non-intrusive rocks are Tertiary basalt-lavas of Plateau Type, and it has been shown in preceding chapters that the Central Types are, generally speaking, of later date than the Plateau Types. On the other hand, the occurrence of Central Types in the area does not suffice in itself to distinguish it from various synclines outside, where, as shown in Pl. III, such types are strongly represented. There is, however, satisfactory ground for distinction as will be shown later in paragraph (4).

(2) The importance and definiteness of the central subsidence is very clear along its south-east side. Here, outside the limit assigned to the caldera, one encounters the Moine gneiss and Mesozoic sediments of the Loch Spelve Anticline. It is true that only in the Sgùrr Dearg neighbourhood, illustrated in Fig. 30 (p. 204), is the approach of Central Type of lava to gneiss very close; but elsewhere (Pl. V), the two are only separated by Glas Bheinn Granophyre (Chap. XII) and Sgùrr Dearg Vent-Agglomerate (Chap. XVI). Of the former, it can be said that its associates are so commonly Pre-Tertiary that it is fair to regard it as an intrusion lying almost wholly beneath the level of even the Plateau Basalts; and of the latter, that it is naturally interpreted as the result of an explosion roughly guided by the marginal fault of the subsidence.

## Chapter XIII.—Tertiary Tectonics.

(3) The only exposure of the boundary of the caldera occurs north of Sgùrr Dearg, and is illustrated in Fig. 30. At three places, a little crushed gabbro intervenes along the line separating gneiss from lava. A scrap of Triassic sandstone is associated with the most southerly of these outcrops. Otherwise, there is no Mesozoic sediment represented on this side of the Loch Spelve Anticline; and the great group of Plateau Lavas is entirely absent. One of the first lavas met with west of the gneiss is a curious flow-banded basalt, either vertical or dipping south-westwards at 80°. Farther in, away from the gneiss, the dip moderates greatly (north section, Fig. 35, p. 237).

(4) When it was said in paragraph (1) that the supposed caldera is occupied by lavas of Central Types, no attention was paid to a feature of the lavas peculiar to the locality. A considerable proportion of the lavas show pillow-structure in the field (Fig. 18, p. 133); and this peculiarity is combined with a marked tendency towards variolitic crystallization (pp. 98, 149). It is clear, therefore, that something distinguishes the representatives of the Central Lavas found within the caldera from their neighbours in adjacent synclines. The explanation given in Chapter V is that the area was frequently the site of a crater-lake, dependent for its renewal upon repeated central subsidence, and that the pillow-lavas acquired their peculiar structure through flowing into the lake. According to this theory, the caldera was the earliest of the concentric structures, while the surrounding folds developed through subsequent peripheral uprising of the Glas Bheinn and Derrynaculen Granophyres as outlined in Chap. XII.

(5) Gneiss-fragments are abundant in vent-agglomerates, outside the supposed calderas, and almost completely absent in vent-agglomerates inside the same (Fig. 29, p. 201). As explained in Chapter XVI, this fits with the view that the calderas are regions of special subsidence, although alternative interpretations are of course available.

If it be admitted that the foregoing arguments prove the existence of a cauldron-subsidence, it will also be admitted that they serve to fix its approximate limits—only the south-eastern caldera is considered here, its north-western neighbour is very clearly defined as will appear presently. It is stated in paragraph (5) that the margin is actually exposed north of Sgùrr Dearg. Northwest of this, it must turn almost due west if it is to avoid some minute exposures of big-felspar basalt cut by cone-sheets in Allt nan Clàr (these exposures are not shown on the one-inch Map, but are inserted in Plate III.). Apart from this, there is no evidence for the exact line chosen for the caldera-margin in a westerly direction, but it probably lies south-west of certain gneiss-laden agglomerates indicated in Fig. 29 (p. 201).

In the other direction, south-westwards from Sgùrr Dearg, the limit of the caldera is pretty definitely fixed, for it must pass between the pillow-lavas of Beinn Fhada and the granophyre of Glas Bheinn. This places it in the belt of agglomerate which reaches past Glas Bheinn from Sgùrr Dearg.

In the Beinn Buie district, it is clear that the limit of the caldera separates the pillow-lavas on the north-east of the Ben Buie Gabbro from the terrestial lavas on the south-west. Accordingly, the caldera-edge must roughly coincide with the smooth concave north-east margin of the gabbro.

A further justification for drawing the caldera-edge where suggested by the evidence just cited is afforded by the concentric grouping of the line, thus drawn, and many much more easily traced geological features of the district.

The features which most readily occur to the mind in this connection are :—

The arcuate folds discussed above.

The Early Acid and Basic Cone-Sheets lettered aI and bI respectively on the one-Map, Sheet 44, and described presently in Chaps. XIX. and XXI.

The inner edge of the Ben Buie Gabbro (Pl. V. and Chap. XXII).

The series of ring-dykes centred on Beinn Chàisgidle (Pl. VI. and Chap. XXIX.). E.B.B.

## North-West Caldera.

The evidence of the north-west caldera is very much on a par with that advanced above in regard to its south-eastern neighbour. It may be summarized as follows, leaving discussion for Chapter XXXII., where the encircling Loch Bà Felsite is dealt with :—

(1) Loch Bà Felsite has risen along a line of fault-brecciation.

(2) All basalt-lavas sliced from within the circuit of the dyke are of Central Type (Chap. IX.), while those outside, from Ben More to Glen Forsa, belong to the Plateau Group.

(3) Apparent faulting of Early Basic Cone-Sheets near Gaodhail (Fig. 36, p. 238).

(4) Apparent faulting of Glen Cannel Granophyre near Beinn Chàisgidle (p. 341).

(5) Distribution of gneiss in vent-agglomerates (p. 201).

W.B.W., J.E.R., E.B.B.

There is practically nothing comparable to the ring-system of folding which surrounds the south-eastern caldera. In fact, only at one place has any marked disturbance been noted outside the north-west caldera, and this is at Glac a' Chlaonain south-south-west of Salen : here Liassic sediments emerge dipping at 75° away from neighbouring granophyre.

Inside the caldera, two interior lines of crush-rock have been traced running roughly north-west and south-east across Beinn na Duatharach and Cruachan Beag respectively. W.B.W.

### Structures unconnected with the Plutonic Centres.

Under this heading, a few remarks will be offered concerning folds and faults which have not been sufficiently covered in the introductory statement. As regards folds, it is worth noting that a very gentle syncline runs up Loch Scridain, roughly parallel with the Inch Kenneth Anticline of Sheet 43. This syncline is crossed by the vague Carsaig Anticline, which, starting from Carsaig Bay (Sheet 44), enters Sheet 43 near Tràigh nam Beach, then crosses Loch Scridain to pass a little north-east of Tavool House, and thus, by way of Creach Bheinn, reaches the cliffs between Aoineadh Thapuill and Tòn Dubh-sgairt. The Carsaig Anticline fails to bring Mesozoic rocks to the surface anywhere on the shores of Loch Scridain owing to the effect of the syncline which runs up that Loch. E.B.B.

Of faults, by far the most arresting is the north and south fault which brings Tertiary lavas, and underlying Mesozoic and Carboniferous sediments, against Moine gneiss at the angle of Inninmore Bay, almost at the entrance to the Sound of Mull. Mr. Manson's sketch (Fig. 26) of this well-known dislocation illus-

182　　　Chapter XIII.—Tertiary Tectonics.

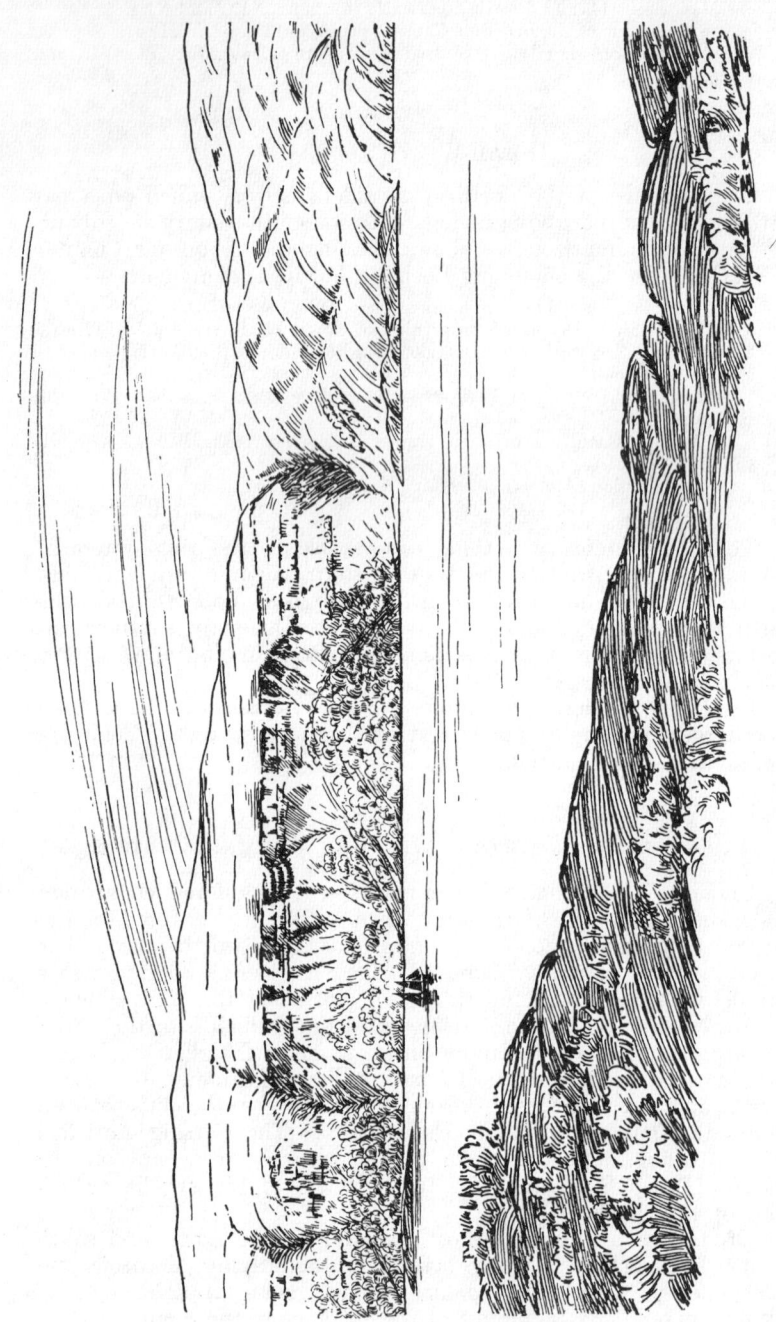

Fig. 26.—Inninmore Fault, seen from South across the Sound of Mull.

## Structures unconnected with the Plutonic Centres.

trates the marked contrast of scenery on the two sides : the trap-featuring of the lavas on the west, and the amorphous slopes of gneiss on the east. Although this fault completely alters the face of the country by bringing into conjunction highly contrasted rock types, it does not lead to any significant change of level : the downthrown lavas on the west rise to an elevation of 1573 ft. in Glais Bheinn ; while the gneisses, less than a mile and a half away, reach 1517 ft. in Màm a' Chullaich. It is impossible to gain an accurate knowledge of the throw of the fault, but the 1000 ft. contour in the gneiss-country just east of the dislocation approaches so closely the head of Inninmore Bay, where Carboniferous sandstone extends below sea-level, that 1000 ft. may be safely assumed as a minimum. Inland, there is less chance of judging, for the gneissic floor rises gently northwards under its sedimentary and lava cover, and emerges, at the base of the Glais Bheinn escarpment, at about 1000 ft. above sea-level. This does not necessarily mean that the fault diminishes in throw when followed towards the north ; but it certainly becomes a less conspicuous tectonic feature.   G.W.L.

On the other side of Mull at Bunessan (Sheet 43), a west-north-west fault serves as the boundary of the Mull lavas for a distance of 5 miles measured across the Ross. The Bunessan Fault is never exposed, but its course is easily followed, since it is marked by a continuous hollow, with Moine schists, or gneisses, on the south-west, and basalt-lavas on the north-east. These latter belong to the lower part of the Tertiary sequence, but not to its base ; and it is quite clear on the ground that their linear termination is determined by faulting.   (C.T.C.)

The course of this fault after reaching the bay at Bunessan is not very clear. Probably it continues up Loch Caol and across the Ross of Mull Granite to Dearg Phort, while at the same time it sends a branch up Loch na Làthaich, skirting the north-east shore of Eilean Bàn. It is certain, at any rate, that a fault intervenes between the gneisses of Eilean Bàn and the lavas of Ardtun ; whereas the gneisses of Na Liathanaich may very well form part of the floor upon which the Ardtun lavas rest.   E.B.B.

Before leaving this district, it is worth while noticing certain north-westerly faults conspicuous in the cliffs of the Gribun Peninsula (Sheet 43). The most obvious bounds Tòn Dubh-sgairt on the north-east, and has a throw of about 400 ft. An interesting feature of the north-westerly faults of both the Gribun and Ross Peninsulas (Sheets 43 and 44) is the clear manner in which they cut the sills and sheets of the Loch Scridain district described in Chapter XXIII.   E.B.B., E.M.A.

It is often impossible to decide how far there has been *lateral movement* along a steep fault. In the case of some of the north-west faults of Mull, there has certainly been a considerable amount. The best local instance of a wrench, or tear-fault, or flaw, as dislocations of this type are variously styled, is afforded by a well-marked west-north-westerly fault which runs from Loch Spelve across to Port Donain (Sheet 44). In its course, it cuts through the Loch Don Anticline, and displaces this structure bodily for more than 100 yds., measured horizontally. The Port Donain Fault is easily recognized westwards in a crush-line followed by the

Lussa River, while on the mainland it is possible continued by a fault passing Kilninver; but in neither of these directions is there a chance of seeing whether it maintains its wrench-character.

E.B.B.

## QUIESCENCE OF THE GREAT GLEN FAULT.

One of the most interesting features of the Tertiary Tectonics of Mull is of a negative character. The north-east fault known as the Great Glen, or Loch Linnhe Fault, is generally regarded as chief among Scotland's high-angled fractures. It passes through the foundations of Mull, where, as explained in the Memoir dealing with the Pre-Tertiary geology of the district, its course can be followed fairly accurately by paying attention to the extremely contrasted metamorphism of certain Pre-Devonian inliers showing among the Tertiary lavas: on the one side of the fault, are the gneisses of the west shore of Loch Spelve and Sgùrr Dearg, evidently belonging to the same gneissic complex as is exposed in Morven and the Ross of Mull; on the other side, are the roofing slates and limestones of the Loch Don Anticline, the obvious continuation of similar rocks occurring in Lismore and Lorne, and, like the latter, accompanied by Old Red Sandstone lavas. In striking contrast with these observations, there is clear evidence that no important displacement of Tertiary date has taken place along the north-east line. A trivial fault has indeed been traced from Lochan an Doire Dhairaich to Camas Mòr, but that is all. The arcuate folds of Plate V. have been developed boldly across the line of the Great Glen Fault. In fact, it is almost certain that this fault, instead of controlling the development of Tertiary structures, has been involved in them, and deflected outwards from the centre of the island; it is difficult otherwise to account for finding a gneiss outcrop so far south-east as Balure, on the shore of Loch Spelve.

The quiescence of the Great Glen Fault in the Mull district is of old standing, for the preservation of Jurassic strata does not seem in general affected by it. An interesting exception to this rule is discussed in the Memoir dealing with the Pre-Tertiary rocks, where it is suggested, on Dr Buckman's determination of fossils, that some group of strata (? Callovian), otherwise unknown in the district, has here been preserved from early Cretaceous erosion by slipping down into a fault-fissure along the line of the stream leading north-east towards Camas Mòr. This is the same line, which, in Tertiary times, showed a slight movement, so that it is quite reasonable to imagine that a gaping fissure may have locally opened along it in Cretaceous times (*see* also p. 395).

E.B.B.

## CHAPTER XIV.

### SYENITE, TRACHYTES, AND BOSTONITES.

#### Introduction.

A FEW alkaline intrusions are considered at this juncture. They are lettered as follows on the one-inch Map : syenite S ; trachyte O ; bostonite bO. There is no suggestion made that all these intrusions belong to a single epoch ; since there is little known in regard to their precise age-relationships. On the other hand, the evidence in every case, except that of the Gamhnach Mhòr Syenite, does conclusively demonstrate a Tertiary date, and often supplies further information as indicated presently under the heading of Field-Relations. The fullest record is furnished in regard to the bostonite-suite ; and there is some justification for referring these particular intrusions, in part at any rate, to a phase of the epoch of maximum explosive activity dealt with in Chapters XV. and XVI. It is this circumstance that has decided the position of the present chapter in the general scheme of the Memoir.

The Gamhnach Mhòr Syenite cannot be seen to cut Tertiary or Mesozoic rocks—in fact it is surrounded by the waters of Carsaig Bay—but there are two good reasons for considering it as of Tertiary date : in the first place, the syenite occurs in a region replete with Tertiary intrusions, some few of which are closely akin to it in composition ; in the second place, there is no intrusion of Pre-Tertiary date anywhere in the West Highlands with which this syenite can be equally well compared.

The syenite, trachytes, and bostonites of this chapter are grouped together for petrological convenience. It must not be thought, however, that they are petrologically isolated in the Mull assemblage. They merely mark an extreme of alkali-variation corresponding with a certain range of silica-percentage ; they have, on one side, a close connexion with the mugearite-lavas described in Chapter X. ; and, on the other side, with certain granophyres, for instance the Glen Cannel and Beinn a' Ghràig Granophyres (Chapters XXXI. and XXXII.). This aspect of the subject is developed further in the introductory chapter, Fig. 4, p. 26.

A brief statement of the field-relations of the various occurrences is given below, and is followed by a petrological account.

#### Field-Relations.

*Gamhnach Mhòr Syenite, Carsaig (Sheet* 44). This syenite forms a low island and reef at the entrance to Carsaig Bay on the south coast of Mull. It is probably intruded into Mesozoic or Pre-

## Chapter XIV.—Syenites, Trachytes, and Bostonites.

Mesozoic rocks, but the sea conceals its margin. It is traversed by three of the numerous basic sheets of its district and a couple of dykes. The sheets run roughly east and west, and one of them (14597) is described by Dr. Thomas as a tholeiite with cognate xenoliths, exactly recalling the basic marginal portions of the Rudh' a' Chromain sill (p. 286). It is thus sufficiently clear that the syenite is earlier than the Loch Scridain sills of Chapter XXIII.; but as yet the age of these latter is only fixed within somewhat wide limits (p. 279). (B.L.)

*Trachytic Vent and Sill of Bràigh a' Choire Mhòir (Sheet 44).* A mile west of Salen, rising above the lava-slopes which overlook the Sound of Mull, there stands a conspicuous little crag consisting of trachyte, surrounded on three sides by trachytic agglomerate. The trachyte is further marked out from the basalt-lavas by its felspathic appearance and its twisted platy structure. There can be no question that the associated agglomerate marks the site of an explosion-vent piercing the basalt-lavas, and measuring some 400 by 200 yds. The trachyte-crag is thus almost certainly part of a central plug belonging to the vent. W.B.W.

Although the trachyte of the plug is isolated as a scenic feature, it appears probable on close examination that it unites on its east side with a trachyte-sill which continues for about three quarters of a mile beyond the limit of the vent. J.E.R.

As regards age, it is obvious that the trachyte must be of later date than the lavas which it cuts, while it must be of earlier date than certain north-west basalt-dykes which have been traced across it. One of these dykes may be seen traversing the vent, and two others cut the attendant sill. The time-limits set by these observations are of course rather vague. W.B.W., J.E.R.

*Trachyte-Plug of Ardnacross (Sheet 52).* A considerable mass of intrusive trachyte is well-exposed about four miles north-north-west of Salen, close by the farm of Ardnacross. It extends for a mile along the coast of the Sound of Mull, and forms the promontories of Rudh' an t-Sean-Chaisteil and Rudh' a' Glaisìch. The country-rock of the district is furnished by basalt-lavas.

The trachyte is characterized by platy joints, which in most places vary rapidly in inclination. Within a distance of 80 yds. from the lavas, however, they become regular so as always to incline steeply from the line of junction. The actual contact of trachyte and basalt is not exposed, for at the only place where the edge of the trachyte is seen (in coast-section, on the south side of Rudh' a' Ghlaisìch) a narrow lenticular band of agglomerate separates it from the lavas. The usual platy jointing fails within a foot of this agglomerate, and the trachyte adopts a definitely chilled texture. Both exposed junctions, between trachyte and agglomerate, on the one side, and agglomerate and basalt-lavas, on the other, appear vertical. The trachyte-margin obviously maintains this verticality inland, since its course pays no regard to the shape of the ground.

The coastal strip of agglomerate contains subangular blocks of basalt up to four feet in length, and also a mass of red bole. It may be interpreted as broken-up basalt-lava, shattered by an explosion which preceded the intrusion of the trachyte-plug. There is thus a close connexion between the mode of occurrence of the Ardnacross Trachyte and that of Bràigh a' Choire Mhòir just described. The similarity is increased by the fact that the Ardnacross Plug is freely cut by north-west basalt-dykes. J.E.R.

*Lochan na Cille Trachyte* (*Sheet* 44). The only other considerable trachyte-intrusion to be mentioned is a mass of biotite-trachyte crossing the northern border of Sheet 44 near Lochan na Cille, east of Savary Glen. It builds conspicuous crags running north-north-east. Towards its margins, it develops a strong streaky structure with steep inward dip. Perhaps the mass is best described as an elongated plug. It has not been sliced, so that no mention of it will be found in the petrological section. (H.B.M.)

*Bostonites between Ben More and Carsaig* (*Sheet* 44). An interesting minor feature of the geology of the district reaching from Ben More southwards to the sea-coast near Carsaig is the occasional occurrence of bostonite as sheets and more irregular intrusions. In most cases, the bostonite is porphyritic. The most prominent example is a gently inclined sheet, 50 ft. or more in thickness, shown on the one-inch Map a little south of Rossal Farm, near the head of Loch Scridain. It has conspicuous phenocrysts of felspar in a very felspathic ground-mass, sometimes weathering a characteristic pink. Farther north, two similar occurrences are shown on the Map towards the head of Gleann Dubh, and another half a mile east of this glen terminates southwards a little north-east of Dùn Breac. As will appear presently, it is not quite certain that the last-named is not a lava. Another rather important outcrop, though not indicated on the one-inch Map, is exposed on the south shore, where it is cut across by the xenolithic composite sill of Rudh' a' Chromain (Fig. 45, p. 267). In this instance, the bostonite is non-porphyritic. E.M.A., G.V.W

A comparatively early date of intrusion for these bostonites is shown by the fact that the type-occurrence at Rossal is cut across by thin examples of the Early Basic Cone-Sheets of Chapter XXI., and also by a subacid cone-sheet probably belonging to the Early Acid Suite of Chapter XIX. This evidence is strengthened, and the age-limit carried one stage further back, by the finding of bostonite-fragments (16274-5) in a vent situated in the Sleibhte-coire, between Tòrr na h-Uamha and Guibean Uluvailt (p. 207). The Sleibhte-coire Vent is a good example belonging to the maximum period of explosive activity (Chapter XVI.); and its agglomerate, which contains many fragments of gneiss, is cut through alike by Early Basic Cone-Sheets and Corra-bheinn Gabbro, the latter described in Chapter XXII. (C.T.C.), G.V.W.

It will be shown almost immediately that there is a possibility of the bostonites being in part contemporaneous with the Central Basalt lavas of Chapter V. At the same time, there is considerable circumstantial evidence that some of them at any rate are of later date, and belong to the maximum explosive phase described in Chapters XV. and XVI. There is nothing contradictory in these two views, for, as pointed out at the close of Chapter IX., rhyolitic tuffs are occasionally found intercalated among Central Basalts, despite the fact that rhyolite figures much more prominently in the subsequent paroxysmal outbursts of Chapters XV. and XVI.

The evidence for associating bostonites with the paroxysmal phase in its more westerly manifestations is as follows :—

(1) If a bostonite-magma has not shared in the explosions which give rise to the Sleibte-coire Vent, it is a curious coincidence that Dr. Clough's specimens of tuff, chosen haphazard from this vent, should show an abundance of bostonite-fragments. The vent falls roughly within the area characterized by bostonite-outcrops ; but even in this area, bostonite bulks to an insignificant extent in the Mull Complex, and in the vicinity of the vent it does not show at the surface at all.

(2) The coincidence mentioned above is heightened by the fact that bostonite, along with gneiss, etc., figures in a small patch of tuff (15618-9) entangled among Early Basic Cone-Sheets on Tòrr a' Ghoai, between Glen More and Loch Fuaran. The tuff-patch is too small to show on the one-inch Map ; but in its relations to the Early Basic Cone-Sheets, and in its content of gneiss-fragments it exactly resembles the Sleibte-coire and other vents of the maximum period of explosive activity. Here again, the specimens of tuff were collected by Dr. Clough without any thought of their containing bostonite.     E.B.B.

(3) Near the type-exposure of bostonite south of Rossal Farm, a minute patch of breccia containing bostonite-fragments may be seen in contact with amygdaloidal basalt-lavas. Appearances suggest that the breccia rests upon the lavas ; but, taken in conjunction with somewhat similarly situated patches of breccia in other cases in Central Mull, it is fairly safe to interpret this particular occurrence as a product of explosion filling some small irregular cavity. The exposure lies half-a-mile east-south-east of Rossal Farm between the two branches of acid cone-sheet which are shown on the one-inch Map as uniting in the tributary stream south of Allt a' Mhàim. The evidence in this case suggests that where bostonite is known to occur it has a tendency to develop explosive phenomena, which supports the suggestion made to account for the phenomena detailed under headings (1) and (2).     G.V.W.

The evidence noted above, as suggesting a possible contemporaneity of bostonite with Central Basalts, is furnished in a stream east of Beinn nan Gobhar, a mile and a half north of the bridge at the head of Loch Scridain. On the one-inch Map, the locality is most easily recognized as the eastern margin of a bostonite-sheet, where this latter lies north of a branching east and west fault and east of a north and south fault. The bostonite is of the usual porphyritic type with anorthoclase phenocrysts, and was regarded at the time of mapping as an intrusion, largely on account of its petrological analogy with other bostonites of the district, themselves clearly-marked intrusions. Above it, at the side of the stream, 5 ft. of sediment is exposed, including a band of ashy grit. This gritty layer, on examination by Dr. Thomas, proved to be to all appearance a crystal-tuff of anorthoclase (16939). As the sediment seems interbedded in the Central Basalt sequence, its constitution affords strong presumptive evidence that bostonitic explosions sometimes intervened between outpourings of Central Basalt lavas. Also, its association with a bostonite-sheet raises the question whether bostonite in this case may not be a lava. Since the recognition of these possibilities, no opportunity has presented itself for a re-examination of the field-evidence.

It may be added that the same sheet of bostonite has been traced southwards to where it is shown on the one-inch Map as terminating north-east of Dùn Breac. At this point, it is cut across by a small outcrop of volcanic breccia (not shown on Map) containng fragments of bostonite visible in hand-specimens. The volcanic nature of the breccia is fairly certain, for its boundaries seem steep, and some of its fragments have vitreous margins. There is, however, nothing to show that its association with the bostonite is more than accidental. It is probably the contents of a small explosion-vent.

E.M.A.

## PETROLOGY.

*Syenite of Gamhnach Mhòr, Carsaig (Anal. II.; Table VII., p. 27).* The rock of Gamhnach Mhòr is definitely an alkaline syenite (14596-98, 15991). Its general colour is yellowish grey-brown, with dark ochreous spots. The texture is holocrystalline, though rather fine for a syenite; and phenocrysts of any size are wanting.

The fractured surface is rough, and glistens with small crystals of a somewhat greasy-looking felspar, which lie at all angles and are mostly from one to two millimetres in length.

In the hand-specimen, the syenite is quite unlike any other rock from Mull, and the microscope proves it to consist essentially of alkali-felspars with subordinate amounts of alkaline pyroxenes and hornblendes. Plagioclase is entirely wanting. The structure of the rock is hypidiomorphic-granular, and in some sections is reminiscent of the coarser-grained bostonites that are often devoid of conspicuous fluxion-structure. Felspar forms at least 75 per cent. of the rock, and the rest consists of basic silicates, accessory minerals such as magnetite and apatite, and limonitic decomposition products. Coarser-textured segregation-veins are present, but they show no special departure from the normal mineralogical assemblage (14598). A considerable variation in the relative proportions of the basic silicates is a noticeable feature of this rock, as it is indeed of many rocks of the class; in fact, one basic silicate may predominate locally to the almost complete exclusion of the others.

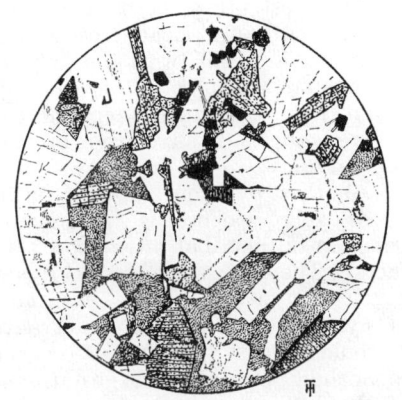

FIG. 27.—[15991] ×20. Alkali-Syenite of Gamnach Mhòr, Carsaig Bay: aegirine and aegirine-augite, alkali-hornblende (centre), magnetite, and colourless soda-orthoclase, with a yellow chloritized residuum.

The felspar is elongated in a positive direction parallel to the *a* axis, and is tabular parallel to the plane of symmetry; while carlsbad twinning is the rule. Unlike normal orthoclase, but in common with certain felspars of the alkali-syenites, the axial plane lies in the plane of symmetry and is thus parallel to the trace of the good

## Chapter XIV.—Syenite, Trachytes, and Bostonites.

cleavage (010). The mineral is optically negative with 2E between 60° and 70°.

The habit appears to be that of sanidine, but, as found by Williams [1] in the pulaskite of the Fourche Mountains, the chemical analysis of the rock indicates clearly that the felspar is not a normal orthoclase but a cryptoperthite or soda-orthoclase.

Calculations based upon the bulk analysis of the rock (p. 27), combined with the judged percentage of felspar present, would suggest that the mineral has a composition approximately as given in Column I. below :—

|  | I. | II. |
|---|---|---|
| $SiO_2$ | 67·4 | 66·08 |
| $Al_2O_3$ | 19·1 | 18·77 |
| $K_2O$ | 7·3 | 7·68 |
| $Na_2O$ | 6·8 | 6 54 |

This compositon compares closely with that usually assigned to cryptoperthite, and, for comparison, an analysis of cryptoperthite from Fredriksvarn, Norway, is given in Column II.[2]

The most prevalent basic silicate is a pale green aegerine-augite which occurs in elongated prisms that have a maximum extinction of from 35° to 40° with the long axis, and are usually edged and terminated with pleochroic grass-green aegerine extinguishing at about 7°. Occasionally aegerine constitutes a whole crystal. The aegerine-augite is enveloped by the felspar and is clearly of early separation.

Another pyroxene, also of early generation, is represented by small yellow ochreous pseudomorphs that have a somewhat irregular outline, but often show traces of the prism and pinacoidal faces. They are somewhat elongated and, at first sight, might be taken to represent olivine, were not the characteristic angles those of pyroxene. While the exact nature of this mineral is obscure, I would suggest that it has been either a rhombic pyroxene or a monoclinic pyroxene of enstatite-composition, the clino-enstatite of Wahl (*cf.* innimorite Chapter XXV.).

The other essential constituents of the rock are amphiboles, which are, for the most part, of later separation than the felspar, although they occasionally occur intergrown with the aegerine-augite. They are deeply coloured minerals, strongly pleochroic in various shades of blue-green, and brown, and occur as small somewhat irregular crystals between the felspars, on which they may be moulded, or in an ochreous residual material in which they assume a more idiomorphic habit.

The majority of the crystals are greenish blue, or greenish brown, with intense dispersion that renders it impossible to obtain a definite extinction in white light.

There is considerable variation in these amphiboles, both as regards colour and composition, often in one and the same crystal; and while the greater portion of them may be referred to the soda-amphibole arfvedsonite, those with indigo and yellowish tints partake

[1] J. F. Williams, 'The Igneous Rocks of Arkansas,' *Ann. Rep. Arkansas Geol. Surv. for* 1890 (1891), vol. ii., pp. 58-60.
[2] W. C. Brögger, 'Die Mineralien der Syenitepegmatitgänge der Südnorwegishen Augit-und-Nephelinsyenite,' *Groth's Zeit. fur Krystallographie und Mineralogie*, 1890, Bd. 60, pp. 527, 528.

more of the nature of riebeckite. A deep-brown hornblende with intense absorption occurs sparingly, and, as was suggested by Rosenbusch in the case of the Fourche Mountain 'granite' (pulaskite), is closely allied to Brögger's barkevikite.

The accessory minerals are magnetite and apatite. Both are of early separation and occur enclosed in the felspar, the magnetite as well-shaped regular crystals and the apatite as slender needles.

The rock in its alkaline character, and from the nature of its component minerals, is clearly allied to the nepheline-syenites, more particularly to the pulaskite of Williams.

Although the alumina-content is somewhat low, the presence of nepheline might be expected, but diligent search has failed to detect it, and there is none of the usual secondary minerals which

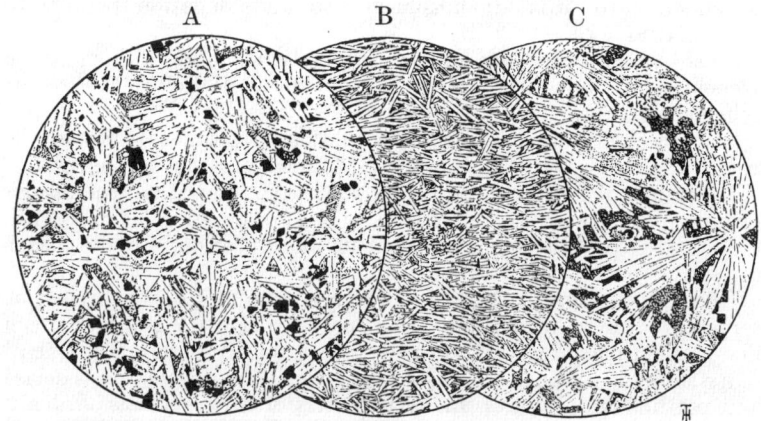

Fig. 28.

A [15756] × 15. Phonolitic Trachyte of Ardnacross (Rudh' an t-Sean-Chaisteil). Slightly elongated prisms of alkali-felspar, small plates of biotite, and a little magnetite, in a chloritized base. The rock contains a small amount of analcite-pseudomorphs suggestive of nepheline.

B [14335] × 15. Trachyte of Bràigh a' Choire Mhòir. Small prisms of alkali-felspar arranged with some parallelism in a dark base largely composed of a green alkali-pyroxene and chlorite.

C. [18477] × 15. Bostonite of Rudh' a' Chromain. Prisms of alkali-felspar, often arranged in sheaf-like or radiate aggregates, in a dark-green chloritic base that contains some calcite and quartz.

would indicate its original existence in the rock. If it occurs, it must be present merely as a sporadic accessory.

*Trachytic Vent and Sill of Bràigh a' Choire Mhòir (Anal. IV.; Table VII.; p. 27).* The agglomerate is represented by two slides (14334, 14436); the plug by three (14335, 14335a, 14821a); the sill by three (17101-3). The analysed specimen quoted on p. 27 belongs to the plug.

The plug is composed of a dark-grey, fine-grained, and somewhat splintery rock, weathering to a browner tint, and breaking with a rough surface, in one direction, and with a smooth satiny surface, in another. The smooth surface is due to the parallel arrangement of tabular felspars.

192  Chapter XIV.—Syenite, Trachytes, and Bostonites.

The microscope shows a plexus of orthoclase-crystals, tabular parallel to the plane of symmetry and usually twinned according to the carlsbad law. The spaces between the felspars are filled by a pale-green to yellow pyroxene (aegirine-augite), and soda-amphiboles that are strongly pleochroic in shades of indigo and yellow and are clearly allied to riebeckite. Pyroxenes and amphiboles are about equally represented.

The felspars are elongated in a positive direction, and are seldom more than 0·5 mm. in length. They are somewhat irregularly terminated, and their extinctions are often undulose. They are orientated in a manner indicative of flow.

The pyroxene occurs both within the felspar as granules and in the interstices as moulded grains; only occasionally, does it show a tendency towards idiomorphism and an elongation parallel to the vertical axis.

The amphiboles are always interstitial, and although the pleochroism is not so intense as in the typical riebeckite the colour-scheme is more or less the same.

Of accessory minerals, apatite in slender needles pervades the rock, while magnetite as small well-formed crystals is plentiful.

A transparent mineral of higher refractive index than the felspar occurs sparingly in the interstices. It appears to gelatinize with acids and is possibly nepheline, but its identification is by no means certain.

The sill-portion of the little Bràigh a' Choire Mhòir complex, as represented by a moderately fresh specimen (17101), is a grey-green fine-grained platy rock with brown-stained surface-films. The microscope shows it to consist of small infrequent phenocrysts of orthoclase and perthite in a typical trachytic matrix of little crystals of alkali-felspar arranged with good parallelism; between the felspars lie grains and prisms of grass-green aegerine-augite, occasional small pseudomorphs after olivine in iddingsite and chlorite, and scattered crystals of magnetite. In the decomposed portions, interstitial chlorite is more prevalent.

*Ardnacross Trachyte* (*Rudh' an t-Sean-Chaisteil.* Anal. III.; Table VII., p. 27). The Ardnacross Plug is a phonolitic trachyte represented by three slides (15753, 15756-7).

The rock is pale to dark yellowish grey colour and of medium grain. It is typically non-porphyritic, and consists of a plexus of positively elongated felspars that are mostly twinned on the carlsbad plan but occasionally show albite-lamellation. The felspar-laths are soda-orthoclase and albite-oligoclase, and are arranged with a certain amount of parallelism according to the direction of flow, so that the resulting structure lies midway between the types known as orthophyric and trachytic. In some cases, the rock is traversed by narrow veins, probably primary, that contain idiomorphic crystals of water-clear analcite; but for the most part this mineral has an allotriomorphic habit, and occurs interstitially.

Fresh nepheline has so far not been detected but sharply bounded rectangular patches of analcite are present in certain portions of the rock and are presumably pseudomorphs after this mineral.

Ferromagnesian minerals are quite subordinate to felspar in

quantity, and the only ones remaining in an unaltered condition are orange-brown biotite and to a less extent aegerine-augite. The biotite builds small somewhat ragged plates that occur interstitially and occasionally envelop the felspars in an ophitic manner. There are also small hexagonal pseudomorphs in chlorite after hornblende; the regular hexagonal form of such crystals as are cut perpendicular to the vertical axis is due to an equal development of prism and pinacoidal faces. Other similarly orientated small pseudomorphs have been noticed with the characteristic octagonal form of augite. The original nature of the hornblende cannot be determined, but it seems to have been a variety rich in iron. The augite, judged from a few grains that have escaped alteration, appears to thave been the alkaline aegerine-augite.

Of accesssory minerals, apatite and magnetite are both fairly prevalent. Apatite is especially abundant, and occasionally forms large crystals in the neighbourhood of the biotite. Generally speaking, the rock is fairly fresh, but here and there it has suffered a certain amount of decomposition resulting in the separation of interstitial calcite, chalybite, chlorite, and zeolitic minerals, and to a certain extent the analcitization and chloritization of the felspars.

The analysis given on p. 27, shows that the rock is more alkaline than the syenite of Gamhnach Mhòr which it resembles in some respects. Its high percentage of soda finds expression in the occurrence of a soda-felspar and analcite. It differs chiefly from the generality of trachytoid phonolites in the presence of biotite as the chief ferromagnesian mineral, a feature that allies it more closely to the trachytes.

*Bostonites betweeen Ben More and Carsaig.* The Rudh' a' Chromain bostonite exposed on the shore near Carsaig[1] (Fig. 45) is a fine-grained non-porphyritic grey rock similar to those from the type-locality of Marblehead, Massachusetts (18475-9, Fig. 28c).

The sheet at Rossal Farm differs in being conspicuously porphyritic (14898, 16734, 16692). It is dove-grey in colour, compact and fine-grained, but its chief character is an abundance of dirty-white felspar-phenocrysts that average about 4 or 5 mm. in length but occasionally reach a centimetre. Under the microscope, the porphyritic felspars prove to be cryptoperthite (anorthoclase) with a narrow external zone of orthoclase, and the matrix to consist of a plexus of somewhat indefinitely bounded small crystals of orthoclase arranged with rough parallelism.

Ferromagnesian minerals appear to be practically unrepresented, but the matrix contains a good deal of greenish epidote, a fair amount of magnetite, and abundant apatite in the form of slender needles.

It is interesting to note that practically identical rocks were met with and described by Dr. Harker[2] from among the Tertiary minor intrusions of Skye, more particularly from the south-east of Elgol (7485), and 350 yds. west-north-west of the north end of Loch na Creitheach (7486). These rocks were classed by him with the

[1] As explained under Field-Relations (p. 187), this sill is not shown on the one-inch Map.
[2] A. Harker, 'Tertiary Igneous Rocks of Skye,' *Mem. Geol. Surv.*, 1904, p. 289.

bostonites or bostonite-porphyries. The rock from Loch na Creitheach is identical in the hand-specimens with the Mull examples, but differs under the microscope in containing a small amount of fresh augite, a mineral so far undetected in the rocks under description.

An interesting rock (17160), strikingly similar to the bostonites of Rossal Type in the hand-specimen, occurs 725 yds. east 32° north of Ben More summit, but is not shown on the one-inch Map. It exhibits the same texture and colour as the Rossal sheet and contains similar porphyritic crystals and glomeroporphyritic groups of cryptoperthite. A difference, however, which, if real, would appear to connect it with the trachytes rather than with the bostonites lies in the occurrence of a strongly pleochroic blue-green amphibole that builds small ragged crystals between the orthoclase-felspars of the matrix. This hornblende occasionally has the pleochroism and other properties of riebeckite ; but, for the most part, the characteristic indigo-blue colour of this mineral is wanting and its place taken by shades of greenish-blue. There is no doubt that this rock belongs to the Rossal Type of intrusion, and it is not impossible that its amphibole may be a product of metamorphism.

H.H.T.

# CHAPTER XV.

## SURFACE-AGGLOMERATES LATER THAN THE BASALT-LAVAS.

### Introduction.

Volcanic breccia, or agglomerate, plays a very conspicuous part in the Geology of Central Mull. The various occurrences are of diverse dates, but they generally agree in being clearly later than the basalt-lavas described in Chapters V.-X. Careful mapping and observation have led to the conclusion that this agglomerate has two main modes of occurrence. It is found in many instances filling volcanic vents (Chapter XVI.), while in other cases, considered below, it is in all probability superimposed upon adjacent basalt-lavas. There are in addition several minor occurrences of ash and agglomerate interbedded with the basalt-lavas, but these have already been sufficiently dealt with in connexion with the associated lavas. All the Tertiary surface-agglomerates are lettered Z on the one-inch Map.

The surface-agglomerates dealt with in the present chapter appear to follow, in more or less conformable sequence, the basalt-lavas. They contain abundant fragments of these latter, along with debris derived from some of the early plutonic rocks of Chapters XI. and XII. On the other hand, they are freely cut, according to locality, by the Beinn Mheadhon Felsite of Chapter XVII., the Early Acid and Basic Cone-sheets of Chapters XIX. and XXI., and the Loch Uisg Granophyre of Chapter XX. Their age-relationship in regard to the arcuate folds of Chapter XIII. is more doubtful, and is discussed in the sequel.

There are two main outcrops of agglomerate to be considered in this chapter. Both of them are situated along the course of the Coire Mòr Syncline of Chapter XIII.: the one in the Coire Mòr district; the other at the west end of Loch Spelve. These two outcrops can easily be recognized on Sheet 44, and are also shown on Plates III. and V. (pp. 91, 165). In addition, a couple of small exposures of agglomerate in Glen More will be discussed, although the agglomerate in these two instances may be contemporaneous with, and not later than, the associated basalt-lavas.

The chapter will be closed with a short petrological description of certain rhyolitic rocks, which as tuffs and lavas (lettered R on the one-inch Map), form a characteristic part of the Coire Mòr accumulation. Mention will also be made of the effects of contact-alteration by the Loch Uisg Granophyre of some sandstones and mudstones which accompany the agglomerate at Loch Spelve.

## Chapter XV.—Surface-Agglomerates.

### FIELD-RELATIONS.

*Coire Mòr.* The agglomerate of the Coire Mòr outcrop is a coarse breccia consisting for the most part of blunted blocks and fragments of gneiss, granophyre, gabbro, basalt, and sandstone. The sandstone is no doubt Mesozoic; the basalt is of types occurring among the Tertiary lavas and includes typical representatives of big-felspar basalt; the gabbro, apart from texture, corresponds in type to big-felspar basalt, and both it and the granophyre can be matched in Tertiary intrusions cut through by volcanic vents in the neighbouring Sgùrr Dearg (Fig. 30, p. 204); the gneiss is of sedimentary Moine type, also visibly broken through by the Sgùrr Dearg vents. The only constituent not obviously attributable to the foundation upon which the agglomerate appears to rest is rhyolite. Certain parts of the breccia are made up almost exclusively of dark-grey rhyolite with small felspar-phenocrysts. The same type of rhyolite occurs in the solid form along with the tuffs, and is accordingly interpreted as lava contemporaneous with the agglomerate. Good exposures in Allt an Dubh-choire and Glen Lìrein are shown on Sheet 44, but relations in the latter case are greatly obscured by drift. Similar rhyolite occurs both as fragments and in solid form in many of the vents of Mull. It is suggested that the great agglomerates of Mull, however situated, have resulted for the most part from explosions emanating from acid magma.

While the contents of the Coire Mòr agglomerate vary from place to place, there is nothing that permits of the mass being separated into zones. Gneiss-fragments, for instance, are equally abundant in the marginal and central portions of the outcrop. In the main, the deposit is unbedded.

The relationships of the Coire Mòr agglomerate to the neighbouring lavas are much obscured by a profusion of cone-sheets, which cut both indiscriminately, and also by the unbedded character of the agglomerate. The view that the agglomerate accumulated on the surface, and not in a vent, is based mainly upon detailed mapping. In the first place, it is fairly easily recognized that the agglomerate occupies the centre of a syncline of lavas dipping away from Pre-Tertiary rocks exposed on either side (Plate V., p. 165). In the second place, when the lavas are zoned by mapping outcrops of big-felspar basalt, the general conformity of the agglomerates and the lavas becomed sufficiently obvious to make superposition of the agglomerate almost certain (Plate III. and Fig. 30, pp. 91, 204).

A point of great theoretical interest in building up a history of Mull (pp. 5, 168) is the age-relationship of the Coire Mòr agglomerate to the folding which has given rise to the Coire Mòr Syncline (Plates III. and V.). Some may hold that the comparatively close conformity of the agglomerate and the folded lavas is most easily interpreted as an indication that the agglomerate has shared in the folding of the lavas. On the other hand, it is suggested here that the agglomerate is rather later than the folding: the conformity is incomplete; the agglomerate approaches more closely to big-felspar basalt-lavas on the steep south-western side of the syncline than it does on the gentle north-eastern side—in fact it is sometimes seen in contact—and in keeping with this, it is, on the south-west side, particularly rich in big-felspar basalt debris, as if erosion had lent a hand in its production. These indications of a post-folding date for the agglomerate receive support in the Barachandroman outcrop which now falls to be described.

*Barachandroman, Loch Spelve.* The continuation of the Coire Mòr Syncline runs along part of Loch Spelve (Plate V., p. 165). At Barachandroman, there is an outcrop of breccia with a considerable thickness of quartzose sandstone and mudstone (both greatly baked) towards its base. The breccia here consists very largely of an unbedded assemblage of angular fragments of basalt and quartzite, which latter is often pebbly and is probably baked Triassic sandstone. Pieces of gneiss and other rocks occur less frequently.

On one side, the margin of this breccia is hidden by the waters of Loch Spelve, but on the other, a junction with olivine-free basalt-lava is well-exposed for a distance of about a mile. The lavas are dipping at 25° towards the breccia, and almost certainly continue under it. The district is one of fair relief, and includes but few intrusions, so that conditions are favourable for accurate observation. The main difficulty arises from the fact that the lavas merge into the breccia without a sharp plane of demarcation. This, combined with the fairly certain appearances of superposition of breccia on lava, suggests that the lavas were breaking up under subaerial decay at the time the breccia gathered upon them. Such an interpretation is in keeping with the view that the lavas were tilted before they were covered up, as this would tend to scree-formation. It is noteworthy, too, that the bedding of the sediments near the base of the breccia is intensely contorted as if the sand and mud had formed under water on a sloping surface and then slid down towards the bottom of the hollow.

*Tom na Gualainne, Glen More.* Agglomerate, or breccia, is seen in two small neighbouring exposures in Glen More resting upon the top of compact olivine-free basalt-lava. These two exposures are both shown on Fig. 52 (p. 308), where they can be located by reference to a small lochan on Tom na Gualainne, half a mile west of Loch Sguabain. The first is situated 200 yds. west-south-west of the lochan, and in it agglomerate lies with a smooth and fairly flat base on lava. The second occurs 300 yds. north-east of the lochan, but in this case, while the base of the agglomerate remains even, it has a steep north-westerly inclination. It has only been possible to show the agglomerate of the more northerly exposure on the one-inch Map where it is coloured as vent-agglomerate; this is the colour used, often in lack of evidence, for all the agglomerates and breccias of Central Mull, and it seemed undesirable to differentiate so small a strip.

In both exposures, the agglomerates referred to above consist largely of fragments of basalt and small-felspar rhyolite. They correspond in type with what is commonly encountered in Central Mull, so that their mode of occurrence suggests that a fair proportion of the agglomerates of this complex region may lie on top of basalt-lava instead of occupying vents. The further question arises whether, in such case, the agglomerates should be interpreted as contemporaneous with, or later than, the associated basalt-lavas. It has already been pointed out that, in two very minor instances, ash-beds containing ryholite-fragments can be seen interbedded with basalts of Central Type within the precincts of the south-eastern Caldera (p. 134). Beyond this, it is impossible at present to venture an opinion.

## PETROLOGY.

*Rhyolitic Associates of Agglomerates.* Two slides of rhyolite, one from Glen Lìrein (14427) and another from Allt an Dubh-choire (15537), were cut to represent the rocks interpreted as lavas associated with the Coire Mòr agglomerate. They carry fairly numerous, more or less square, phenocrysts of microperthite, measuring about 1-2 mm. across, in a very fine-textured groundmass. In (14227) the orthoclase is accompanied by a pale brownish-yellow augite, the two forming together small glomeroporphyritic aggregates. The base of (15537) is highly fluxional. In both cases, the rocks have suffered changes in accordance with their position well inside the pneumatolysis-limit.

A slide (14229) from near the mouth of the stream from Coire nan Dearc, east of Coire Mòr, affords a very convincing example of an acid tuff on account of its perfectly developed *aschen-struktur*. The rhyolite that yields all the fragments in this particular case is non-porphyritic.                                                                                        E.B.B.

### CONTACT-ALTERATION BY THE LOCH UISG GRANOPHYRE.

A series of slides (18946-18950) was made to illustrate the contact-alteration induced upon the sandstones and mudstones of Barachandroman, Loch Spelve, by the Loch Uisg Granophyre.

In the sandstones, the quartz occurs either as small grains with fritted margins, or as minute prisms and aggregates of prisms which experience of silica-bricks teaches us, must have developed as tridymite. In keeping with what is commonly observed in such cases, the several members of a prism-aggregate (now quartz) extinguish simultaneously irrespective of the directions of the long axes of the constituent prisms (18947).

In the mudstones, a perfect granulitic structure is a feature. Little augite granules and crystals of a spinellid mineral, probably magnetite, are sometimes very conspicuous (18949), and rhombic pyroxene is also well-represented as somewhat larger crystals that are moulded on neighbouring elements (18950). A very important constituent, apt to attract less attention, at first sight, is cordierite (18949, 18950). It builds small, somewhat indefinite, rectangular crystals, occasionally showing characteristic triple twinning, and also still smaller individuals that enter largely into the body of the rock. Recrystallized basic plagioclase felspar can be recognized here and there as a minor constituent. Altogether it is fairly clear that these baked mudstones were originally formed largely from weathered basaltic detritus.                                                                        H.H.T.

# CHAPTER XVI.

## AGGLOMERATES OR BRECCIAS OF VENTS.

### INTRODUCTION.

It has been pointed out in Chapter XV. that tumultuous agglomerate or breccia, plays a very important part in the geology of Central Mull ; and that,.while considerable masses of it seem to be remnants of surface-accumulations, others again certainly occupy eruptive vents. These latter are lettered V on the one-inch Map. Here and there, the difference of field-relations in the two cases has been obscured by later happenings, but Plate V. (p. 165), supplemented by Plate III. and Figs 30 and 35 (pp. 204, 237), emphasizes the frankly transgressive behaviour of the vent-breccias as compared with those that occupy the heart of the Coire Mòr Syncline. In considering Fig. 30, it should be realized that the Sgùrr Dearg Big-Felspar Gabbro, the Glas Bheinn Granophyre, and the Torness Felsite can all be clearly proved to be of earlier date than the adjoining breccia.

When the transgressive relationships of much of the breccia were first realized, two interpretations seemed possible : either the transgressive breccias might be a scree-formation blocking valleys of erosion, or they might be—as all now agree they are—a volcanic agglomerate choking funnels and fissures opened by explosion. The evidence in favour of the latter alternative may be illustrated under two heads :— E.B.B.

(a) If one supposes the breccia, wherever found, to be a superficial deposit, one should be able to reconstruct from its many isolated outcrops a fair picture of the pre-breccia land-surface. However, if one attempt any such reconstruction, it is only to find progress blocked by what appear to be insuperable obstacles. Thus, the hypothetical pre-breccia river-system south-east of Ben More would have to correspond altogether too closely with the modern river system to be at all probable *when considered in relation to the number of post-breccia intrusions* developed in the district. The reader will appreciate the force of this argument when he learns that several patches of breccia, too small to be indicated on the one-inch Map, intervene between the two outcrops shown at 1½ and 2 miles, respectively, east-south-east of the summit of Ben More. Of these two, the more easterly is exposed on the floor of Sleibhte-coire, and the more westerly lies 600 ft. higher up and just beyond the summit of a ridge. Exactly the same difficulty presents itself in other localities. For instance on the two slopes of Glen More, facing one another in Cruach Choireadail and Creag na h-Iolaire, patches of breccia are constantly found in association with older rocks. One cannot resist the feeling that these patches are not restricted to a surface, but extend through and through the framework of the country. (C.T.C.), E.M.A.

(b) A recurrence of breccia-conditions can sometimes be demonstrated under circumstances which seem to render an erosion-origin for the deposit quite impossible. The most striking instance is afforded in the Ben Buie district (Fig. 38, p. 247). On the southern slopes of the mountain, the Ben Buie Gabbro cuts across a great mass of breccia which had previously been traversed by a whole suite of cone-sheets. The relationship is everywhere clear, and is emphasized locally by the gabbro developing a compact edge against the breccia and sheet-complex. But on the northern slopes, west of Loch Airdeglais, the gabbro meets another though similar breccia, and the tables are turned ; here the gabbro, as it approaches the breccia, is shattered to pieces and yields abundant fragments to be incorporated in its neighbour. In considering the weight of this evidence, it is important to understand that the breccia of the south face of Ben Buie is full of granophyre-fragments derived, without a doubt, from some portion of the Glas Bheinn Granophyre-Complex (Chap. XII ). An erosion-theory is therefore faced with the following sequence of events :— (1) the baring of the Glas Bheinn Granophyre; (2) the accumulation of the southern breccia of Ben Buie ; (3) the further accumulation of cover for the Ben Buie Gabbro and the intrusion of this mass ; (4) the baring of the Ben Buie Gabbro and the accumulation of a northern breccia of the same unbedded, unassorted character as its southern predecessor. Surely, such a recurrence of unusual erosion and accumulation is incredible. Accordingly, one is driven to admit that the transgressive breccias of Central Mull occupy vents opened by explosion at more than one stage in the history of the igneous focus.   G.V.W.

A complete account, or even enumeration, of Mull explosion-vents is not attempted in this chapter. Certain examples have already been dealt with in connexion with the alkali-rocks considered in Chapter XIV. Many more must be passed over in silence, except in so far as they are covered by general statements in the sequel. The subject-matter will be divided under four main headings which give prominence to various aspects of the matter :—

(1) The Materials characteristic of Mull explosion-vents are discussed from the point of view of field-observation. Special attention is directed to the frequent occurrence of rhyolite or rhyolitic debris, and also to the local distribution of gneissic fragments (Fig. 29).

(2) Under the heading of Repeated Explosions, the whole question of age-relations is dealt with.

(3) The Sgùrr Dearg Vents, reaching from Beinn Chreagach Bheag through Sgùrr Dearg, Glas Bheinn, and Ben Buie, to Sleibhte-coire, are given detailed consideration, not only on account of their great intrinsic interest, but also because they may be regarded as good representatives of the explosion-vents of Central Mull taken as a whole.

(4) The Petrology of the vent-agglomerates is treated on the same lines as that of the surface-agglomerates of Chap. XV. Descriptions are almost confined to acid rocks associated with the agglomerates, either in solid or fragmental form. In regard to the non-fragmental rhyolites it may be said that they are lettered R on the one-inch Map, just the same as their fellows occurring among the surface-agglomerates. The petrological description is closed with an account of contact-alteration induced by subsequent intrusions.

In describing the surface-accumulations of agglomerate (Chapter XV.), it has been pointed out that, within the caldera area, some small proportion of the agglomerate coloured on the one-inch Map as vent-agglomerate really rests upon basalt-lavas. The geological complication of the district where this occurs is in fact too great to permit of satisfactory differentation of surface- and vent-agglomerate in every instance, but the uncertainties thus introduced are of very secondary importance.

### Materials as Observed in the Field.

The contents of the vent-aggglomerates vary from place to place. Tertiary basalt-lavas, Tertiary plutonic rocks, Mesozoic sediments, and Moine gneisses are locally abundant. As in the case of the surface-agglomerate (Chapter XV.), rhyolites occur both in mass and as fragments; in fact it is thought that most of the agglomerates of Mull have resulted from the breaking up of country-rock by explosions emanating from acid magma. Rhyolite is particularly abundant in the north-western caldera on both sides

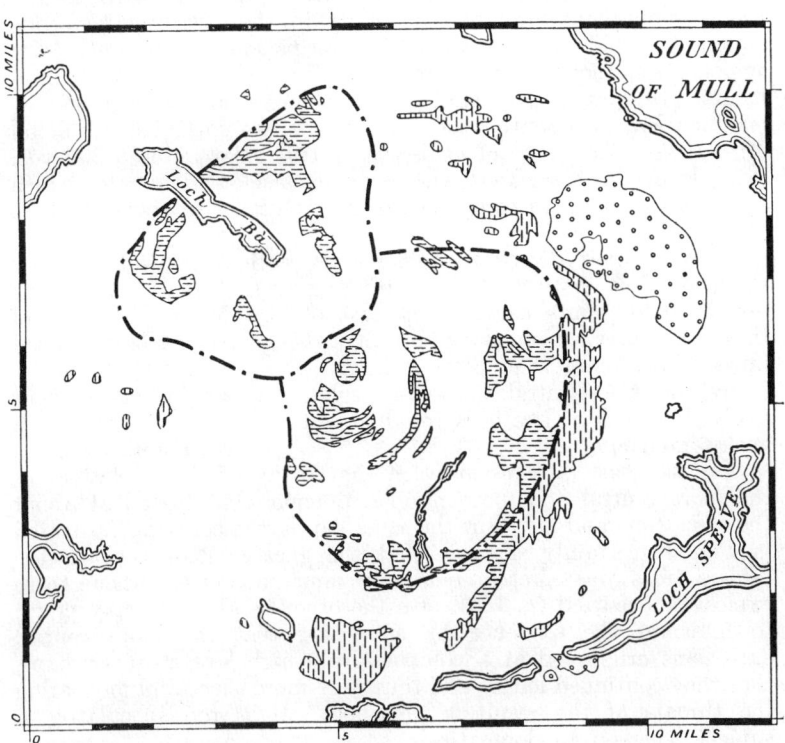

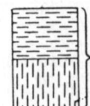 Agglomerates and associated Rhyolites in Vents. { Where agglomerate does not contain gneiss-fragments. Where agglomerate does contain gneiss-fragments.

 Agglomerates and associated Rhyolites in Coire Mòr Syncline; agglomerates contain gneiss-fragments.

—·— Margins of Main Calderas.

FIG. 29.—Distribution of Gneiss-Fragments in Mull Agglomerates.

of Loch Bà where large outcrops are shown on the one-inch Map.
W.B.W., E.B.B.

A feature of very general interest is the virtual limitation of gneiss-fragments to the more peripheral vents. This point is illustrated in Fig. 29. Many of these peripheral vents follow closely the region of upheaval surrounding the south-eastern caldera, and are often seen in contact with the old gneiss-floor, nowadays reexposed by erosion (Plate V. and Fig. 30, pp. 165, 204). This is the case in the district reaching from the west end of Loch Spelve northwards to Sgùrr Dearg. The abundance of gneiss-fragments in such a situation is of course exactly what might be expected; elsewhere it is more surprising. The data are as follows :—

For more than four miles north-westward from Sgùrr Dearg, where vents figure fairly prominently in the lava-country lying outside the calderas, gneiss is still a conspicuous component of the agglomerate, though for the most part represented by small fragments. Of quite exceptional size is a huge block shown on the one-inch Map just south-west of a fault on the north face of Beinn Mheadhon. This mass of gneiss has a triangular outcrop 200 yds. long. It occurs associated with ordinary neck-agglomerate, which seems at this point to be piercing lavas 1500 ft. or so above the base of the volcanic pile. E.B.B.

In the other direction, agglomerate with recognizable gneiss-fragments is a feature of the southern slopes of Ben Buie and of several small vents encountered east of Ben More. In some of these exposures, one must admit that the gneiss has been carried upwards for more than 3000 ft. E.M.A., G.W.L.

Within the Central Calderas, agglomerate or breccia is very freely developed; but it is to all intents and purposes free from gneiss-fragments. W.B.W., E.B.B., J.E.R.

On the reasonable assumption that many of the explosions of the more central and more peripheral regions originated at about the same time and at about the same depth, the contrast illustrated in Fig. 29 certainly suggests that the gneissic floor beneath the caldera-areas was already relatively much depressed when these explosions occurred (p. 180). On the other hand, some may prefer to interpret the evidence by supposing that the more central explosions originated at a comparatively high level; or, perhaps, that they continued longer and thus were more successful in clearing the throats of the resulting volcanoes. It is too speculative a subject whereon to dogmatize.

Although it does not affect the matter discussed above, it is proper, perhaps, to direct attention here to a little string of gneiss-bearing breccia seen traversing the basalt-lavas near the mouth of Loch Spelve at a point indicated by a note on the one-inch Map. The peculiarity of this occurrence is that the schist-floor at this point consists almost certainly of relatively lowly metamorphosed sediments unrepresented in the breccia-vein. Accordingly it seems probable that the vein was filled from above with material supplied by some explosion which occurred in the gneiss-floored district lying farther west. E.B.B.

## Repeated Explosions.

Before passing on to the consideration of local detail, let us return for a moment to the vitally important matter of the recurrence of explosive activity at widely different periods in the history of Central Mull.

The earliest [1] certain epoch of paroxysmal explosion is responsible, among other things, for the breccias that occupy, not only the Coire Mòr Syncline, but also the long row of gneiss-characterized vents lying more or less outside the caldera limits from Beinn Chreagach Bheag through Sgùrr Dearg and the southern face of Ben Buie to Sleibhte-coire (cf. one-inch Map and Plate V., p. 165). These vents are shown by their field-relations to be of later date than the Glas Bheinn Granophyre (Chapter XII.) and Torness Felsite (Chapter XVII.); but they are earlier than the Beinn Mheadhon Felsite (Chapter XVII.), the Early Acid and Early Basic Cone-Sheets (Chapters XIX. and XXI.), and the Ben Buie and Corra-bheinn Gabbros (Chapter XXII.). E.B.B., G.W.L.

The next great explosive epoch to be distinguished has already been singled out for special notice in the introductory remarks of this chapter. It is responsible for the agglomerate-vent, which has burst its way along the northern margin of the Ben Buie Gabbro west of Loch Airdeglais (Fig. 38, p. 247). G.V.W.

Perhaps to this second period should be attributed a strip of breccia, about a third of a mile west of Cruachan Dearg, and too small to show on the one-inch Map. The agglomerate is clearly later than the Corra-bheinn Gabbro, with which it is seen in contact. It is largely made up of gabbro-fragments, together with little pieces of an acid igneous rock. C.T.C.

Fig. 37 (p. 244) illustrates a clear case, where agglomerate in contact with an intrusion may at one point be earlier, and at another later than its associate. In this case, it is the Beinn Bheag Gabbro (Glen Forsa) that is involved. E.B.B.

The more central agglomerates or breccias are, in certain clear cases, later than neighbouring intrusions; but, at the same time, they are almost altogether earlier than the ring-dykes and other masses shown in Plate VI. (p. 307). Whether they are in the main earlier or later than the Ben Buie, Corra-bheinn, and Beinn Bheag Gabbros, is an open question.

It has just been stated that almost all the agglomerate of Central Mull is earlier than the ring-dykes of Plate VI. It is therefore particularly interesting to find what appears to be tuff made of rhyolitic and granophyric debris, associated with the Loch Bà Felsite (Chapter XXXII.), the very latest of the suite. Suggestive exposures in this connexion are afforded by the streams draining either side of Coill' an Aodainn, where the ring-dyke reaches the most southerly part of its course. There can be very little doubt that the Loch Bà Ring-Dyke was in communication with the surface, and that its vapours were thus enabled to escape with the violence required to produce volcanic ash (p. 345). W.B.W., E.B.B.

A few small agglomerate-vents have been found in significant association with some of the relatively narrow North-West Dykes of Mull. These occurrences are extremely important as evidence that at any rate a few of the North-West Dykes communicated with the surface. Their consideration is deferred till Chapter XXXIV. E.B.B., G.V.W.

[1] If the widespread basal mudstone is a weathered ash, as is probable (p. 54), it betokens considerable explosive activity at the very beginning of the volcanic history of the district.

204     Chapter XVI.—Vent-Agglomerates.

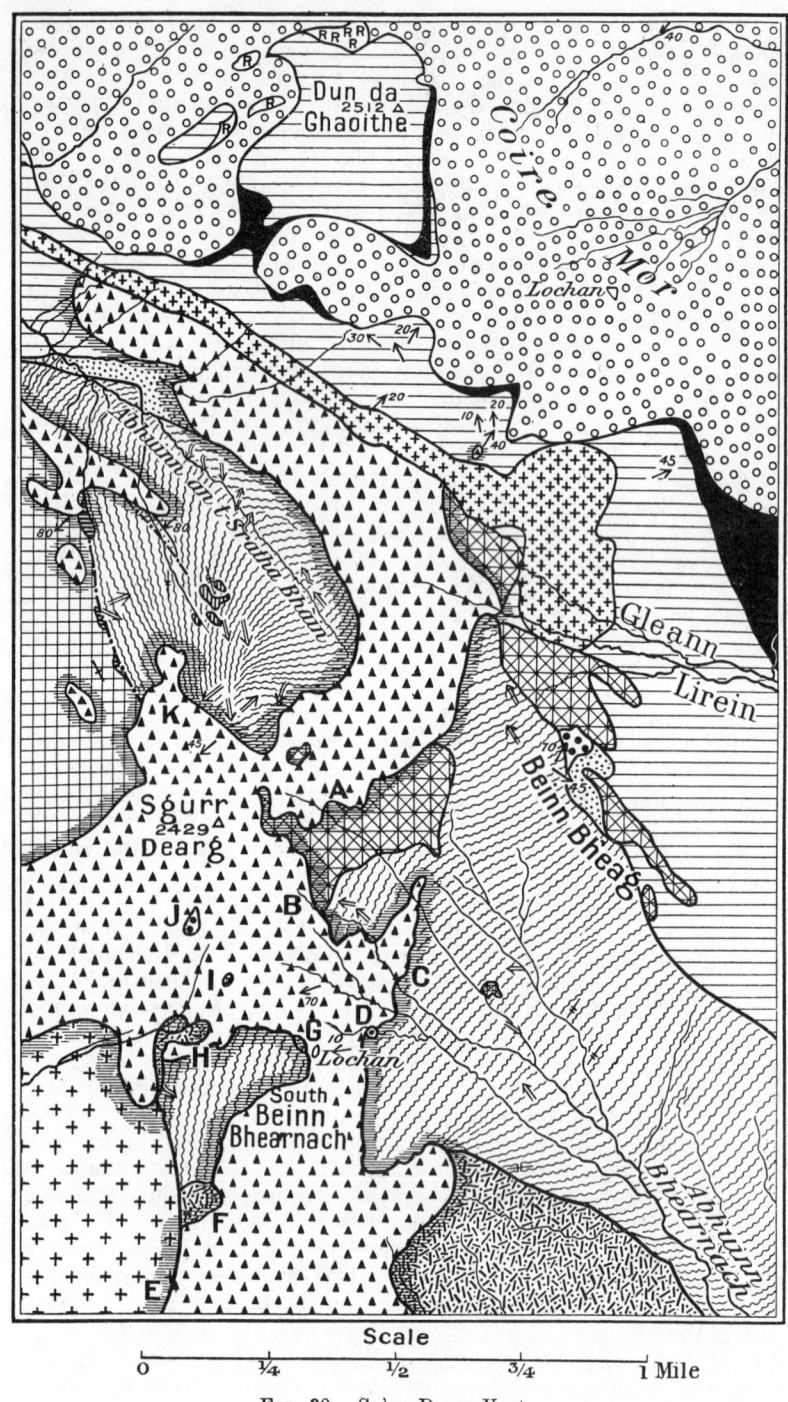

Fig. 30.—Sgùrr Dearg Vent.

## EXPLANATION OF FIG. 30.

For the sake of clearness, only one post-agglomerate intrusion—the Beinn Mheadhon Felsite—is shown. A few dykes and a large number of sheets (mostly Early Basic Cone-Sheets, *cf.* one-inch Map Sheet 44) are omitted. The following rocks are distinguished:—

TERTIARY IGNEOUS ROCKS.
*Post-Agglomerate Intrusion.*
Beinn Mheadhon Felsite.

*Rocks of Agglomerate Period.*
Surface-Agglomerate.
Vent-Agglomerate.
Rhyolite.

*Pre-Agglomerate Intrusions.*
Torness Felsite.
Glas Bheinn Granophyre.
Big-Felspar Dolerite and Gabbro.
Small-Felspar Dolerite.
Crushed Gabbro along Caldera Edge, the most southerly outcrop accompanied by a scrap of Trias Sandstone.

*Pre-Agglomerate Lavas.*
Ill-defined Central Types of Basalt.
Big-Felspar Basalt.
Plateau Types of Basalt.

MESOZOIC SEDIMENTS.
Lower Lias Shales and Limestone.
Trias Sandstone and Quartz-Conglomerate.

HIGHLAND SEDIMENTARY SCHISTS.
Moine Gneisses.

SIGNS.
Faults.
Vent-Margins.
Dips } Bedding.
Vertical.
Dips.
Undulating Dips. } Foliation.
Vertical.

### NOTES ALONG MARGIN OF SGÙRR DEARG VENT.

A. Big-Felspar Gabbro breaks down to coarse powder at contact with Agglomerate.
B. Scrap of Trias Sandstone at termination of Gabbro.
C. Scrap of Trias exposed 30 yards down-stream from Vent.
D. Small outcrop of Trias just outside Vent.
E. Torness Felsite breaks down to Agglomerate in stream just south of map.
F. Glas Bheinn Granophyre breaks down to Agglomerate; also small patch of Trias Conglomerate 30 yards within Vent.
G. Glas Bheinn Granophyre breaks down to Agglomerate; the dips show inclination of Shale bedded in Agglomerate.
H. Glas Bheinn Granophyre breaks down to Agglomerate or Breccia with associated Shales.
I. Small outcrop, or large boulder, of Big-Felspar Gabbro breaking down to Agglomerate.
J. Trias Conglomerate, perhaps a boulder in Agglomerate.
K. Bedded quartzose Breccia of Agglomerate period; dip steep and irregular.

## Chapter XVI.—Vent-Agglomerates.

### The Sgurr Dearg Vents—Detailed Account.

Under the heading, Repeated Explosions, it has been pointed out that a series of early vents can be traced from Beinn Chreagach Bheag through Sgùrr Dearg and Glas Bheinn to the south slopes of Ben Buie and northwards to Sleibte-coire. In most respects, apart from the matter of age already discussed, this series is thoroughly typical of the Mull vents in general—of which indeed they constitute a very large part.

*Beinn Chreagach Bheag to Beinn Mheadhon*—On the basaltic slopes above Fishnish Bay, there are numerous irregular outcrops of breccia and tuff, consisting largely of broken up basalt with a fair proportion of gneiss, rhyolite, and other foreign ingredients. Except for the immense block of gneiss, already described (p. 202) as occurring near the fault on Beinn nam Meann, the fragments throughout are of moderate to small size. In fact, in the biggest vent, as exposed in Allt Mòr Coire nan Eunachair, tuff frequently grades into well-bedded volcanic grit and mudstone.

In addition to rhyolite-ash, there are many outcrops of unbroken small-felspar rhyolite of the same type as occurs associated with the agglomerates of the Coire Mòr Syncline (Chap. XV.). Some of the more important of these outcrops are inserted on the one-inch Map.

The transgressive relationship of the ashes of this district is shown by their not mapping out as regularly interstratified beds would do. The only outcrop, which seems at first sight to be that of an intercalated ash, contours the hill-side for about half a mile, as indicated on the Map, two-thirds of a mile south of Bailemeonach Cottage. South of this outcrop, there rises a low lava-scarp as if belonging to a later overlying flow. However, the appearance can be shown to be misleading, for near the eastern extremity of the exposure the breccia is seen to cut up across the lava-scarp in a most striking fashion. It is interesting to note that the breccia here is richly charged with gneiss and granophyre, thus suggesting an underground extension of the Glas Bheinn Granophyre suite of intrusions (Chap. XII.).

One of the most striking features connected with these vents, taken as a whole, is the way the neighbouring basalt-lavas go to pieces in their vicinity. This is capitally seen round about the largest vent. The shattered basalt is veined with epidote, and it is significant that both shattering and veining antedate the Beinn Mheadon Felsite (Chap. XVII.), and also the Early Acid and Early Basic Cone-Sheets of the district (Chaps. XIX. and XXI.). It may be added that, at their numerous contacts with the agglomerate-filled vents, all these intrusions clearly cut through the agglomerate.

*Sgùrr Dearg*.—The agglomerate of Sgùrr Dearg consists for the most part of a coarse unbedded breccia made of blocks of gneiss, granophyre, and basalt, with subordinate big-felspar gabbro, Triassic sandstone, rhyolite, etc. At two points marked H and K in Fig. 30 it is associated with more or less finely bedded sediments such as might have been deposited in local pools of water.

In considering Fig. 30, it is essential to realize that all the intrusions shown, with the exception of the Beinn Mheadhon Felsite, are of Pre-Breccia age. This sufficiently emphasizes the transgressive relationship of the great breccia, or agglomerate, of the Sgùrr Dearg outcrop.

The brecciation of the gneiss and early intrusions everywhere within Fig. 30 is intense. As the mixed breccia is approached, it is often difficult to decide whether to call the rock one stands on breccia, or merely brecciated. It is common to find the mass shattered to bits yet with so little displacement of the parts that one can still trace geological boundaries, for instance, where broken sills or dykes of basalt run through gneiss.

Evidence that the vent-agglomerate is later than the Sgùrr Dearg Big-Felspar Gabbro of Chap. XI. is afforded by contacts lettered A and I on Fig. 30. In keeping with this it may be added that, though as a rule scarce in the vent-agglomerate, Big-Felspar Gabbro is a conspicuous feature of the neighbouring mass enclosed in the Coire Mòr Syncline (Chap. XV.).

Evidence that the vent-agglomerate is later than the Glas Bheinn Granophyre of

Chap. XII. is supplied by the contacts lettered F, G, and H (Fig. 30). This evidence is supplemented by the universal brecciation of the granophyre, coupled with the fact that, from Sgùrr Dearg to the southern slopes of Ben Buie, granophyre-fragments are one of the main constituents of the breccia. Two additional contacts will be mentioned presently from the Loch Spelve district.

Evidence that the vent-agglomerate is later than the Torness Felsite of Chap. XVII. is afforded in stream sections to which the reader is referred in the explanation of the letter E, Fig. 30. On the one-inch Map, the critical exposures can be located as they lie within a sharp westward inflection of the agglomerate-boundary, due east of Torness Cottage.

The vent-agglomerate is earlier than the Beinn Mheadhon Felsite of Chap. XVII., but the latter is itself so much cut up by later intrusions within Fig. 30 that this area is not as suitable as the district already described farther north-west for demonstrating the point. On the other hand, what is abundantly clear within the limits of Fig. 30 is that the breccia of Sgùrr Dearg, along with the brecciated gneiss and associated intrusions, is traversed by a vast number of dolerite cone-sheets belonging to the Early Basic Suite of Chap. XXI. These cone-sheets are not shown on Fig. 30, but many of them are indicated on the one-inch Map, Sheet 44. They have suffered enough change to have all their olivine decomposed, but most of them are unbrecciated and have distinctive chilled margins. On the south slopes of Sgùrr Dearg, however, many of these sheets enter an area within which they are broken, veined, and epidotized like the lavas neighbouring the vents of the Beinn Chreagach district. In fact, were it not that here they have as their country-rock the Torness Felsite, it would be hard to be sure whether some of them in this particular part of the district might not be metamorphosed lavas. Very probably, their local brecciation is a sign of a recurrence of explosive activity in connection with the relatively late intrusions which are known in plenty a little farther west.

*Glas Bheinn and Loch Spelve.*—There is such a mass of Early Basic Cone-Sheets in the Glas Bheinn country that little can be deciphered of the original field-relations of the continuation of the Sgùrr Dearg agglomerate-outcrop south-westwards past Glas Bheinn summit. In the more open country near Loch Spelve, there are a few easily accessible breccia-exposures to which attention may usefully be directed. In four of these, the agglomerate, or breccia, is seen in contact with, and in large measure derived from, intrusions of the Glas Bheinn Granophyre suite of Chap. XII. :—

(1) A small vent exposed in Abhuinn Coire na Feòla, two-thirds of a mile above road; it is shown on the one-inch Map just north-east of a north-west basalt-dyke.

(2 and 3) The southern margin of a vent exposed in two streams separated by Teanga Bhàn, respectively 200 yards north-west and 500 yards north-east of Kinlochspelve Farm.

(4) A minute, but rather conspicious, vent towards the head of Allt a' Ghoirstein Uaine, 1300 yards up from Creach Bheinn Lodge. The vent is not shown on the one-inch Map, but is easily located on the north-west bank of the stream, 100 yards up from the mouth of a little tributary that enters Allt a' Ghoirtein Uaine from the north.

The material of these Loch Spelve vents sometimes grades to fine sediments. Much of the deposit seen in the streams north-west of Loch Spelve is a bedded sandstone derived from the attrition of granophyre. The bedding is now generally at high angles. E.B.B.

*Ben Buie.*—Attention has already been directed to the vents opened successively on the south and north sides of Ben Buie (p. 199). The more southerly vent is an obvious continuation of the Sgùrr Dearg and Glas Bheinn series. Its material consists largely of granophyre-debris with many fragments of gneiss, and is cut across by Early Acid and Early Basic Cone-Sheets (Chaps. XIX. and XXI.) followed by Ben Buie Gabbro (Chap. XXII.). G.W.L., G.V.W.

*Sleibhte-coire.*—There are comparatively few peripheral vents north-west from Ben Buie, and most of them are of small size. The largest example is rather less than half a mile long, and is exposed in Sleibhte-coire between Tòrr na h-Uamha and Guibean Uluvailt. It is thoroughly typical of its kind, and its agglomerate, containing fragments of gneiss, is cut across by Early Basic Cone-Sheets (Chap. XXI and Corra-bheinn Gabbro (Chap. XXII.). (C.T.C.)

## Chapter XVI.—Vent-Agglomerates.

### PETROLOGY.

The most striking feature of the assemblage of rocks dealt with in the present chapter is the preponderance of what may be termed the foreign or accidental element in the constitution of the agglomerates. Naturally only the finer parts of the volcanic breccias can be examined microscopically, and on the whole there is little additional information to be gathered from the many slides that have been prepared to illustrate the phenomena recounted in the Field Descriptions. Moine gneisses and Tertiary granophyres, gabbros, and lavas are all abundantly represented as blunted rock-fragments, degenerating by further comminution into angular crystal-debris. Olivine is always decomposed, but felspar and augite, even where separately enclosed in the breccia-matrix, are sometimes unaltered.

Instead of elaborating this side of the subject, our attention will be directed more particularly to the acid rocks which, whether in mass or fragments, may be regarded as the probable representatives of the magma responsible for the explosions that produced the breccias and agglomerates.

*Associated Rhyolites.* The main development of rhyolite in association with vent-agglomerate was investigated by Mr. Wright, and occurs in the north-western caldera indicated in Plate V., p. 165. The outcrops figure quite prominently on the one-inch Map, Sheet 44.

A specimen collected west of Loch Bà, from the stream that flows out of the lochan one mile west-north-west of Clachaig Cottage, proves to be a partially devitrified obsidian (17126) containing small phenocrysts of albitized oligoclase. Quartz has formed around these phenocrysts, and also occurs elsewhere infilling cavities. There are, in addition, a few pseudomorphs that suggest the original presence of enstatite-augite. In the ground-mass, may be noted a conspicuous zonal arrangement of colouring matter connected with vaguely defined spherulitic growths that centre upon the quartz already mentioned.

Other specimens were collected on the east side of Loch Bà, and are located as follows :—(14847) from south of Na Bachdanan ; (17984-5) from the more southerly of the two outcrops east of A' Bhog-àiridh ; and (14687-8) from the north-west face of Beinn na Duatharach. Of these, one (14847) is a devitrified perlitic rhyolite with a few small phenocrysts of alkali-felspar ; others (17984-5) show lath-shaped phenocrysts of albite in a ground of microspherulitic structure consisting mainly of alkali-felspar—the whole approaching a bostonite in composition ; while those from Beinn na Duatharach (14687-8) are so contact-altered that they are difficult to interpret—possibly (14688) is an altered acid tuff.

In the south-eastern caldera, rhyolite has again been found associated with vent-agglomerate, at Beinn Chàisgidle, though not on a sufficiently large scale to show on the map. The rock (17931) is a small-felspar rhyolite, or felsite, similar to the Loch Bà Felsite described in Chapter XXXII.

Outside the two calderas, minor outcrops of acid rock have been encountered at several localities in connexion with vent-agglomerate, notably in the neighbourhood of Beinn Chreagach Mhòr. A specimen (16444) taken from a point 100 yds. north-west of this summit is of the small-felspar felsite, or rhyolite, type (*cf.* Loch Bà type, Chapter XXXII.), but is intensely altered by the gabbro cone-sheet of the mountain separately mapped in Fig. 36 (p. 238). This type of acid rock is characteristic of the district, but another specimen from the south-eastern side of the summit is non-porphyritic, and consists of minute oligoclase laths set in a felsitic matrix (16447).

Similar occurrences are met with in association with the vent-agglomerates on the south-west of the calderas, though they are too small to show on the one-inch Map. A specimen taken from the Sleibhte-coire Vent, midway between Tòrr na h'Uamha and Guibean Uluvailt, is again of the small-felspar type (16279) with phenocrysts of alkali-felspar and augite. It shows intense contact-alteration by the Corra-bheinn Gabbro with a development of microscopic biotite.

*Acid Fragments.* An examination of a large suite of slides shows that fragments of acid rocks of glassy or very finely crystalline texture are to be met with everywhere among the vent-agglomerates. Occasionally, they are the main constituent of a coarse volcanic grit (16476); more often, they provide isolated fragments set in the midst of other debris (16479).

More than one type is commonly met with in a single slice, so that, in general, specific localities are omitted from the following brief summary. The types represented include :—perlitic non-porphyritic rhyolites (14679, 17929); perlitic rhyolites with phenocrysts of andesine and augite, some showing conspicuous banding (17210); porphyritic rocks that have a patchy devitrified felsitic base (18051); rhyolites and felsites similar to that of Loch Bà, Chapter XXXII., with micro-porphyritic crystals of oligoclase and augite (17214); many fine-grained felsites (16480); and also a large number of somewhat trachytic aspect (14689, 18053), with minute felspar-laths developed in profusion, although without much parallelism. The Coire-sleibhte Vent, between Tòrr na h-Uamha and Guibean Uluvailt, is interesting on account of its containing fragments of a bostonite (16274-5), a matter already discussed (p. 188).

CONTACT-ALTERATION.

Several specimens in the collection illustrate the local baking of the agglomerates by later intrusions. A rock (14685), mapped with rhyolites that form the roof of the Glen Cannel Granophyre of Chapter XXXI. on the north-west face of Beinn na Duatharach, is a fine-bedded but completely reconstructed tuff, or possibly a fluxional rhyolite. It has been converted into a banded micro-granulite that consists of hypersthene, augite, biotite, alkali-felspar, quartz and magnetite. The hypersthene tends to cluster in the more quartzose areas, and the biotite, where of larger size than usual, builds poecilitic plates.

Another rock (16445) comes from near the Gabbro Cone-Sheet of Beinn Chreagach Mhòr. It was originally a mixed breccia of gabbro granophyre, basalt, and felsite. It shows the same mineral development as in the previous case. In some instances, an ophitic augite has been granulitised *in situ*, and the porphyritic plagioclase has been honeycombed by albite.

The vent-agglomerate of Sleibhte-coire situated between Tòrr na h' Uamha and Guibean Uluvailt is locally invaded by granophyre (16280-1). As will be explained in Chapter XXII., the granophyre concerned is intimately connected with the Corra-bheinn Gabbro, and veins in intricate fashion the country-rock immediately bordering the latter. In the present instance, the result has been the furnishing of an acid igneous matrix to an agglomerate that is in large measure composed of basic igneous fragments. Interesting hybridization has resulted, and is best illustrated in (16281), where pyrogenetic brown hornblende is an important constituent of the acid matrix. The changes involved are comparable with those dealt with in Chapter XXXIII.                                            E.B.B.

## CHAPTER XVII.

### BEINN MHEADHON, TORNESS, AND CREAG NA H'IOLAIRE FELSITES.

#### INTRODUCTION.

THREE early felsite-intrusions now fall to be considered. With many other felsites, they are lettered F on the one-inch Map, Sheet 44, but their positions are easily recognized on Plate V. (p. 165). The Beinn Mheadhon and Torness intrusions are of considerable bulk ; the Creag na h'Iolaire is of much smaller dimensions. All three are fluxional felsites or rhyolites. A comparatively early date is clearly indicated in the case of the Beinn Mheadhon and Torness Felsites by the fact that they are traversed by abundant representatives of the Early Basic Cone-Sheets described in Chapter XXI. The Beinn Mheadhon Felsite is also cut by the Early Acid Cone-Sheets of its neighbourhood (Chapter XIX.). The Creag na h'Iolaire Felsite does not come into contact with Early Basic or Early Acid Cone-Sheets, but it is cut to pieces by another set of basic sheets dealt with in Chapter XXVII. under the title of Creag na h'Iolaire Sheets. The Torness Felsite is probably rather earlier than that of Beinn Mheadhon. It breaks up and yields fragments to the Sgùrr Dearg vent-agglomerates (Chapter XVI.), whereas the Beinn Mheadhon Felsite is of later date than these latter. The only doubt attending this conclusion depends upon the possibility that the agglomerate in the vicinity of the Torness Felsite may be of somewhat later date than is usual in the Sgùrr Dearg neighbourhood.

The three intrusions are considered below under two headings : Field-Relations and Petrology.

#### FIELD-RELATIONS.

*Beinn Mheadhon Felsite.* The Beinn Mheadhon Felsite is sparingly porphyritic with small felspar phenocrysts, and much of it is highly fluxional and spherulitic. The fluxion-structure is often steep or contorted, but near the margins it helps to give an insight into the structure of the mass (Fig. 31). Careful examination shows that the Beinn Mheadhon portion of the outcrop is more or less flat-bottomed (with a protuberent tongue at its western end, north of Loch a' Mhàim, *see* one-inch Map) ; and that this flat-bottomed portion is connected south-eastwards with a dyke or inclined sheet which sinks eastwards and northwards under Dùn da Ghaoithe. The greater part of the flat-bottomed swelling lies in fairly open ground, geologically speaking, and, though there are several acid and basic cone-sheets cutting it, they do not obscure the original continuity of the mass. It is far otherwise, however, with the south-western margin of this swelling, and also with the narrow dyke-like outcrop running south-eastwards to Màm Lìrein. All along this line, nothing is seen of the felsite except isolated lenticles enclosed among a host of Early Basic Cone-Sheets. In Màm Lìrein, a minor swelling occurs giving rise to an easily recognizable outcrop.

Looking from the Sound of Mull, one can recognize the pale smooth country of the main mass of the Beinn Mheadhon Felsite, presenting a rather bedded appearance at a distance owing to the cone-sheets that traverse it. Some lava-crags east of Allt Achadh na Mòine are due to induration of the country-rock in the vicinity of the felsite, and serve to mark the top of the intrusion.

At the margin of the felsite, both top and bottom, a thin basaltic, or tholeiitic, layer is commonly met with (Fig. 31.). It is about 2 ft. in thickness; and, between it and the mass of the felsite, there intervenes a narrow belt of xenolithic hybrid rock resulting from partial digestion of the basalt by the felsite. The Beinn Mheadhon intrusion thus affords a typical example of a composite intrusion on a particularly large scale (pp. 8, 32).

The Beinn Mheadhon Felsite is later than all the breccia of its neighbourhood. This is very clearly seen at the head of Allt Achadh na Mòine where the characteristic basic margin of the intrusion is found in contact with agglomerate. It is equally well-shown by the unbrecciated appearance of the mass as a whole, which contrasts vividly with the condition of the neighbouring lavas.

*Torness Felsite.* The Torness Felsite is a white fluxional felsite represented by lenticles enclosed among Early Basic Cone-Sheets. Typical exposures are seen in the Lussa River, near Torness Cottage. Farther down-stream, half-way towards Arinasliseig Cottage, the felsite is crowded with small black xenoliths—but this is unusual.

The Torness Felsite is everywhere intensely shattered, and material of the same type is abundant as fragments in neighbouring outcrops of agglomerate, as, for instance, in the stream draining Coire nan Each, above Torness. A junction of the felsite and agglomerate is seen three-quarters of a mile east by south of Torness. The locality can be recognized on the one-inch Map, since the margin of the breccia takes a sudden bend westwards just at this point to pass through the critical outcrop. The felsite is here seen to break down into breccia, in which there is an admixture of foreign material.                                                                                       E.B.B.

*Creag na h'Iolaire Felsite.* The Creag na h'Iolaire Felsite is only visible as lenticles among basic sheets. Its original limits may be gathered from Plate V. (p. 165), and its present interrupted condition from a glance at the one-inch Map. It is a fluxional felsite, and the earliest rock of its immediate neighbourhood; but beyond this there is little to say.                           (C.T.C.)

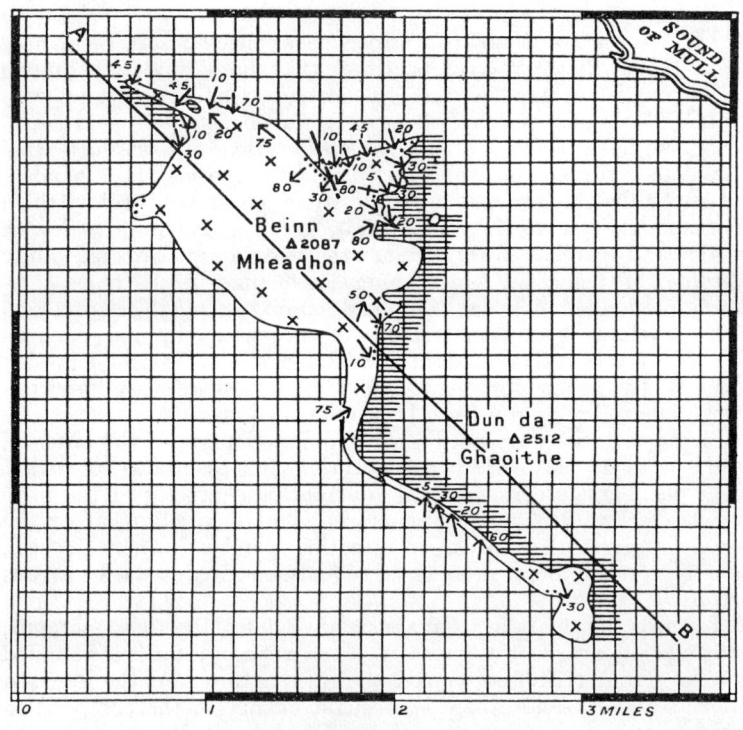

Country-Rock below Felsite.   Felsite   Country-Rock above Felsite.

..... Thin Basic Layer at margin of the Felsite.
Part of this layer is broken up by Felsite and converted into xenoliths.

Fault.   Dip in degrees. } Fluxion-Structures.
         Vertical.

SECTION TO TRUE SCALE ALONG A-B.

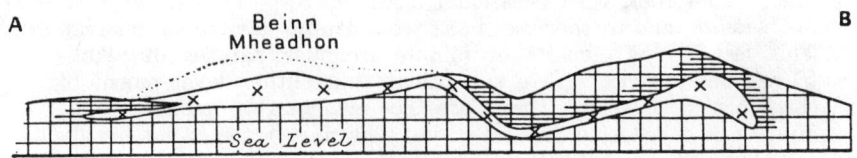

The left half of the intrusion as shown in this section is flat-bottomed; the right half continues downwards away from the observer.

FIG. 31.—Map and Section of Bheinn Mheadhon Felsite.

## PETROLOGY.

*The Beinn Mheadhon Felsite.* Specimens collected from Beinn nam Meann (15691-2) are of a compact rock that shows fluxional banding and a parallel arrangement of its occasional small felspar-phenocrysts. The micro-porphyritic crystals are oligoclase or acid labradorite, but have usually suffered some alteration with the formation of epidote and secondary quartz. Typically, the matrix is banded, and under the microscope shows a distinct separation of devitrified from more glassy material, the latter occurring as indefinite yellowish streaks and patches. Throughout the matrix are scattered pilitic microlites of oligoclase, which are enclosed alike in the more completely devitrified patches, composed mainly of alkali-felspar, and in the yellow glassy material. In the more vitreous portions magnetite is relatively abundant as minute crystals. Ferro-magnesian minerals are poorly represented, but a few small pseudomorphs after idiomorphic augite have been noted.

Slight variations are noticeable in different parts of the outcrop. At the head of Gleann Lìrean, the rock (15693) contains somewhat larger phenocrysts of felspar, the microlithic character of the little oligoclase-crystals is more pronounced, and devitrification of the base has given rise to larger felspathic patches. There is still, however, a residuum of more or less evenly distributed dark vitreous matter.

Towards its margin, south-east of Beinn Mheadhon summit, the composite nature of the mass becomes apparent, for it is bounded by a basic rock (16466) of typical tholeiitic nature. At the junction of felsite and tholeiite, there is a slight change in the composition of the former, and in addition the acid rock has taken up xenoliths of its more basic neighbour. There has also been some granulitization of the tholeiite at its contact with the felsite; but the line of demarcation is fairly sharp.

The tholeiite at its outer margin is chilled (16276-7), and takes the form of a dark finely vesicular chloritized glass, devoid of porphyritic constituents, but full of microlithic felspar. It has produced slight contact-alteration, for a millimetre or so, of the country-rock (agglomerate), bringing about the local transformation of chlorite into biotite.

*The Torness Felsite.* The Torness Felsite (15061, 18684) is a grey, fine-grained rock with occasional small micro-porphyritic crystals of perthite and orthoclase in a completely devitrified matrix. This latter now consists of minute irregular patches of alkali-felspar and quartz, with a subordinate microlithic development of early-formed felspar. There are no recognizable pseudomorphs after ferro-magnesian minerals, but chlorite is somewhat evenly distributed throughout the rock.

In the Lussa River, west of Arinasliseig Cottage, there occurs a local xenolithic facies of this intrusion containing abundant dark patches: some of these consist of albitized basic plagioclase in intimate association with serpentinous pseudomorphs after olivine (15060); others are composed of granular chlorite and albite with a little epidote (15059). We are evidently dealing here with fragments of basic rock picked up by the felsite and partly assimilated, but the source of the basic material is obscure.

*The Creag na h'Iolaire Felsite.* As represented by the specimen (17866) this felsite is a fine-grained rock that consists of microcrystalline quartz and alkali-felspar in a chloritic and epidotic base. The rock under the microscope has the appearance of having suffered much crushing and alteration with the separation of its crystalline components from each other. H.H.T., E.B.B.

## CHAPTER XVIII.

### GAODHAIL AND CRUACHAN AUGITE-DIORITES.

INTRODUCTION.

PLATE V. (p. 165) shows the position of two great plutonic intrusions known as the Gaodhail and Cruachan Augite-Diorites, and lettered H on the one-inch Map, Sheet 44. They lie in the course of a host of Late Basic Cone-Sheets (Chapter XXVIII.), and are only preserved as disconnected lenticles among these sheets. In spite of the complication thus introduced, it has been found possible to draw approximate boundaries for the diorite-masses; and it is very striking how nearly the two intrusions balance one another on either side of the axis of symmetry that runs north-west through Loch Bà. W.B.W.

The diorites are coarse-grained needle-rocks characterized by acicular augite. They vary considerably from point to point, and in one locality (pp. 217, 321) furnish what is regarded as an example of gravitational differentiation (Chapter XXX.). J.E.R.

Under the guidance of Dr. Harker, who visited the district with the writer during the early days of the enquiry, these rocks were at first regarded as hybrids, since they undoubtedly exhibit many of the phenomena of crystallization associated with hybridization. The phenomena observed have been subsequently proved, as will be shown in Chapter XXX., to be characteristic also of internal migrations accompanying gravitational differentiation. Accordingly, since Mr. Richey has shown that the rocks of the present chapter, in one locality, give indications of gravitational differentiation, it seems wise to regard each of the two great masses as the product of a single intrusion in which internal migration has occurred; though possibly their history may have been complicated by some absorption of xenolithic material. W.B.W.

The multitude of Late Basic Cone-Sheets cutting the augite-diorites is so great that it effectually obscures most of their other field-relations. In the Gaodhail River, however, just below Gaodhail, and again in Allt nan Clàr, at Tomsléibhe, the Gaodhail Diorite is seen to be cut by a massive craignurite cone-sheet with narrow basic margins. It is probable, though not certain, that this sheet belongs to a very early stage in the intermittent development of cone-sheets so characteristic of Mull (Chapter XIX.). In keeping with this, there is some evidence that the Gaodhail Diorite is earlier than the Early Basic Cone-Sheet that constitutes the gabbro of Beinn Chreagach Mhòr. Gabbro is found in the lower part of the Gaodhail River which seems to belong to the Beinn Creagach Mhòr Sheet (Fig. 36, p. 238), and a slice of the diorite taken from near the gabbro (the junction is not exposed) shows strong contact-alteration when examined under the microscope (14731).  E.B.B.

The remaining subject-matter of the chapter is discussed under two headings : Field-Relations and Petrology.

## FIELD-RELATIONS.

*Gaodhail Augite-Diorite.* Lenticles of the Gaodhail Augite-Diorite are well-seen in the following exposures :—

(1) For 100 yards above the upper path, west side of Glen Forsa, in the tributary stream adjoining the main river a little below the mouth of Allt an Eas Dhuibh.

(2) River Forsa, 200 yards below mouth of Gaodhail River, and just upstream from outcrop of a thick craignurite cone-sheet, lettered aI on the one-inch Map.

(3) From 600-1000 yards up the Gaodhail River.

(4) The two northern tributaries of the Gaodhail River. The country-rock downstream is mainly gabbro ascribed to an Early Basic Cone-Sheet (Fig. 36, Chap. XXI.).

(5) Slopes south and south-east of Lòn Bàn, Gaodhail River. The rock of the slopes south-east of Lòn Bàn is rather more acid than usual.

(6) The Tomsléibhe stream and the slopes of Beinn Talaidh. The largest of these exposures is where the path to Tomsléibhe Cottage crosses the stream. The diorite is unusually acid at this point.  W.B.W.

*Cruachan Augite-Diorite.* Stream-exposures of the Cruachan Augite-Diorite are afforded by the River Clachaig, Allt na h-Eiligeir, and in Coire Mòr. The Clachaig district is particularly interesting, for, on climbing from the bottom of the glen westwards on to An Cruachan, one finds that the lenticles of country-rock, showing between cone-sheets, become increasingly acid, so that at the hill-top one is dealing with granophyre and felsite in place of diorite—no attempt has been made to show these acid portions separately on the one-inch Map. Taken in conjunction with other examples of much the same kind this transition is regarded as due to gravitational differentiation, and is discussed under this heading in Chapter XXX.  J.E.R.

## PETROLOGY.

Under the title augite-diorite, for want of a better, we here group a number of somewhat coarsely crystalline rocks of intermediate to sub-basic composition. They are represented by the two above-described intrusive masses that are symmetrically disposed on either side of Loch Bà, and possess sufficient likeness to each other to render it certain that they have had a similar origin and solidified under similar conditions. They present in the size and character of their mineral constituents, and also in their type of crystallization, an exaggerated reproduction of rocks that approach the border-line between craignurite and the Talaidh Type of quartz-dolerite described in Chapters XIX. and XXVIII. There is so much evidence of suddenly disturbed equilibrium, accompanying pronounced changes in the composition of the magma during the period of its crystallization, that the normal character of the rocks

## Chapter XVIII.—Gaodhail & Cruachan Augite-Diorites.

might be doubted. It is felt, however, that a magma of basic craignurite-composition might furnish such rocks if allowed to consolidate under conditions of a more nearly plutonic nature; and that the disturbances in equilibrium might well result from the interaction of a partially solidified magma and its own acid differentiate. Owing to the evidence being complicated by the later intrusion of cone-sheets in great numbers, it is inadvisable to speculate too freely upon the true origin or nature of the rocks concerned; but it is worth recalling that according to Mr. Richey (p. 217) there is practically conclusive proof that the north-eastern mass becomes increasingly acid upwards in its north-eastern extremity.

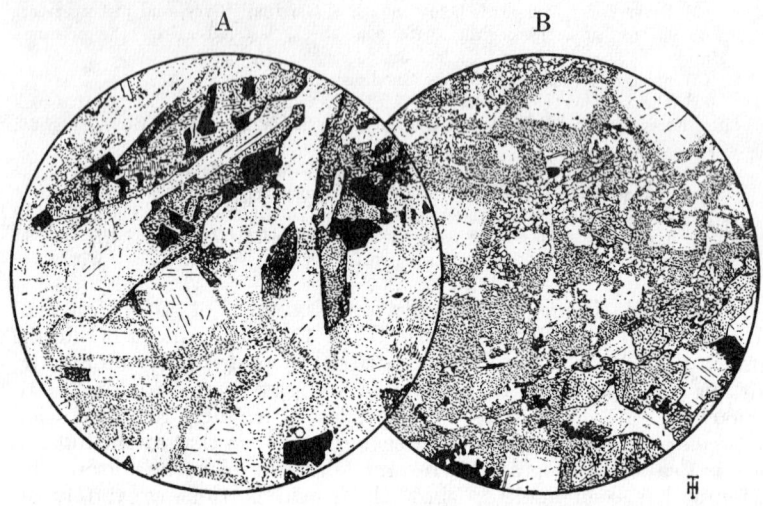

FIG. 32.

A [14740] × 17. Augite-diorite of Gaodhail mass (Tomsléibhe). Large columnar crystals of augite intimately associated with magnetite. Tabular crystals of oligoclase-andesine irregularly fringed with turbid alkali-felspar. Interstices frequently occupied by quartzo-felspathic matter in micrographic relationship.

B [14811] × 17. Augite-diorite of more acid character from the Gaodhill River, showing signs of interaction between its basic and acid components. The rock consists of partially resorbed ophitic augite and oligoclase in a micrographic matrix of quartz and turbid alkali-felspar. The oligoclase is fringed with perthitic orthoclase.

*Gaodhail Augite-Diorite.* A conspicuous feature of this intrusion (2117, 14740, 14742) is the occurrence of narrow columnar crystals of brownish augite (Fig. 32A), frequently greater than a quarter of an inch in length, intimately intergrown with magnetite and crossed by salitic striations in a manner identical with the columnar augites of the intrusions of Talaidh Type (p. 302). It must be remembered, however, that a similar augite often separates from the definitely hybrid zone that results from the interaction of granophyre and gabbro (p. 353). The felspar is most commonly in large plate-like or elongated tabular crystals, composed of andesine-oligoclase, and edged with an irregular zone of perthite with an edging of orthoclase.

Occasionally the felspar is perthite with an edging of orthoclase. The interspaces are filled with felspathic and quartzose matter, frequently in micrographic relationship. Large crystals of apatite are a feature, and are concentrated more particularly in the more acid portions. The columnar augites have in many instances been transformed into green fibrous hornblende (14746), and the plagioclase felspar increased in alkalinity by subsequent albitization.

These rocks sometimes contain fine-textured xenoliths that, apart from alteration, are of Talaidh Type, but field-evidence makes it quite clear that such xenoliths must be referred to an earlier source than the Late Basic Cone-Sheets. Under the microscope, they are seen to have suffered a considerable amount of granulitization and recrystallization of their augite, iron-ore, and felspar (14732). In other cases (14737), there has been much resorption of basic material with subsequent crystallization of a pyrogenetic hornblende within the limits of the xenolithic patches. Here, too, in addition to the usual columnar augite, we meet with that cervicorn variety so frequently developed in the Late Basic Cone-Sheets of Talaidh Type.

Turning now to a few examples that reproduce in clear fashion some of the features of interaction discussed in Chapter XXXIII., we note a rock (14357) in which original augite has been converted in whole or in part, with evident signs of corrosion, into a greenish-brown pyrogenetic hornblende. The bulk of the rock has a rather coarse granophyric matrix, including moderately large individuals of albite-perthite, with subordinate orthoclase. Apatite is extremely abundant in the alkaline plagioclase felspars and perthite. In another case (14811), a rock from the Gaodhail River composed mainly of fairly large crystals of oligoclase and ophitic augite, the oligoclase crystals have been attacked by an acid granophyric magma, that has added to them an irregular casing of perthite and filled the spaces between the crystals with micro-granophyric material (Fig. 32B). The augite likewise has in some parts of the rock been granulitized with the separation of magnetite, and in other parts, though resorption, has given rise to the usual greenish-brown pyrogenetic hornblende of the hybrid rocks.

Intense granulitization of augite and conversion to hornblende and magnetite may also be noticed in a granophyric rock (14731) that consists mainly of perthite and augite. This rock is clearly xenolithic and has basic clots of augite, magnetite and hornblende; but it is probable that in this instance contact-alteration may be responsible for some of the changes observed, since the exposure occurs close to a gabbro that is regarded as a continuation of the Beinn Creagach Mhòr Sheet (Fig. 36, p. 238).

*Cruachan Augite-Diorite.* The Cruachan mass as exposed to the north-north-west of Beinn a' Mheadhoin (19955-8) presents the same general characters as have been described above. It is equally coarse in its crystallization, and carries an identical assemblage of minerals. One specimen (17958) compares with the abnormal rock described above as coming from the Gaodhail River (14811). It shows well the attack of early formed oligoclase-andesine crystals by the granophyre that forms their matrix. The crystals have been irregularly resorbed and further modified by solid-transfusion of alkaline material, while new growths of alkaline character have been added. The original augites have beeen attacked in a remarkable manner, resorbed, and in some cases regrown with a separation of magnetite. In other cases the augite has passed to fibrous green hornblende. Pseudomorphs in magnetite, chlorite, and hornblende suggest the original presence of olivine in the basic portions of the rock, and it would seem in this instance that we are dealing with the results of interaction of an acid magma on an early consolidation of gabbroic composition. H.H.T., E.B.B.

# CHAPTER XIX.

## INTERMEDIATE AND ACID CONE-SHEETS.

### INTRODUCTION.

THE sheets considered here are lettered aI on the one-inch Map (Sheet 44), where they are described in the index as Acid Centrally Inclined Sheets or Cone-Sheets. The former title is unduly cumbrous, but it gives a good idea of a cone-sheet complex in which the characteristic feature is an assemblage of sheets inclined rather steeply towards a common centre. The term cone-sheet is employed because the sheets, viewed as members of a suite, suggest the partial infilling of a number of co-axial cone-shaped fractures with inverted apices united underground. In the case of the Mull intermediate and acid cone-sheets, the average inclination is about 30° or 40°. The centre about which most of them are grouped agrees approximately with $C_1$ of Figs. 18 and 58 (p. 133, 338).

The intermediate and acid cone-sheets are not so numerously developed as their basic analogues to be described later (Chapters XXI. and XXVIII.). Their two main localities are :—North-east of the Mull centre, on the lower slopes overlooking the Sound of Mull, especially above Scallastle Bay, and, south of the centre, on the southern face of Ben Buie.

The thickness of the individual sheets often exceeds 30 ft.

The subject-matter is treated below under three main headings:—

(1) Under Date of Intrusion, it is shown that many of the intrusions here considered are among the earliest cone-sheets of Mull ; at the same time, they are later than the arcuate folding of Chap XIII., and the early paroxysms of Chaps. XV. and XVI.

Where the term Early Acid Cone-Sheets is employed in the course of this Memoir, it implies intermediate and acid cone-sheets of at least as early a date as the Early Basic Cone-Sheets of Chap. XXI.

(2) Under Composite Intrusion, it is pointed out that, like other minor acid intrusions of Mull, their marginal layers, showing externally chilled edges, are often of relatively basic composition (pp. 8, 32).

(3) Under Petrology, it is explained that, though indexed on the one-inch Map as acid, the commonest type is a sub-acid or intermediate rock defined as craignurite.

### DATE OF INTRUSION.

The intermediate and acid cone-sheets of Mull frequently intersect and chill against one another. Close enquiry further shows that in some cases they are of widely different ages. At the same time, they all seem to be later than the great explosions that have left their agglomerates in the semi-circular tract leading from Beinn Chreagach Bheag, through Sgùrr Dearg, to the southern slopes of Ben Buie (Chapters XV. and XVI.). In fact five of their number have been traced through the Beinn Mheadhon Felsite of Chapter XVII., which is itself clearly later than the agglomerates of its neighbourhood.  E.B.B., G.W.L.

It will be remembered that it is open to discussion whether the agglomerates just mentioned are earlier or later than the arcuate folding illustrated in Plate V. (p. 165). The intermediate and acid cone-sheets are certainly of later date than this folding. An acid sheet-complex is well exposed on the Craignure shore (Fig. 25, p. 174), where its posteriority to the folding can be established on two grounds :—

> (1) The acid complex, apparently undisturbed, is intruded into sediments and intrusions which are tilted, smashed, and faulted. A very good section opposite the United Free Church Manse shows the base of an acid sheet cutting clean across obvious crush-lines (p. 176).
>
> (2) Dykes of acid material, apparently identical with that of the sheets, occur along both the faults shown on the map at the south end of Craignure Bay— the one inland behind the Inn, the other reaching the coast north-east of the Pier. (p. 176).

Similarly, in the Loch Don area, a sheet of craignurite, inclined at about 45° towards the south-west, can be traced from the Loch Don Bridge right across vertical Old Red Sandstone lavas situated on the east side of the Loch Don Anticline (*see* one-inch Map).

*Early Acid Cone-Sheets.* While later than the folding, many of these intermediate and acid cone-sheets seem to belong to a more or less initial stage of the long period during which the Early Basic Sheets of Chapter XXI. were intermittently injected. Evidence bearing upon this point comes from the district between the Sound of Mull and Glen Forsa, and also from Ben Buie. In the former, craignurite and granophyre cone-sheets lie for the most part outside the area of maximum development of the Early Basic Cone-Sheets ; but, at the same time, they are very commonly cut by thin dolerite and basalt sheets which probably belong to the Early Basic suite. A striking instance of intersection (shown on the one-inch Map) occurs in a stream three-quarters of a mile north-west of Lochan an Doire Dharaich, above Loch Don. In this case, the intersecting sheet is a fine gabbro, markedly vesicular. On the other hand, the largest and one of the earliest of the Early Basic Sheets of this district—the gabbro of Beinn Chreagach Mhòr (Fig. 36, p. 238)— is traversed by a couple of craignurite cone-sheets which chill against it.  E.B.B.

On the southern slopes of Ben Buie, the relatively early date of many intermediate and acid cone-sheets is clearly demonstrable owing to the gabbro of the mountain cutting right across the outcrop of a very representative suite of craignurite cone-sheets (one-inch Map and Fig. 38, p. 247). The Ben Buie Gabbro, as described in Chapter XXII., also cuts across very numerous representatives of the Early Basic Cone-Sheets, but at the same time it is itself cut by the later representatives of that assemblage.

G.W.L., G.V.W.

*Late Acid Cone-Sheets.* It must be admitted that a small and indeterminate proportion of intermediate and acid cone-sheets is of later date than the Ben Buie Gabbro. Thus, for instance, an acid, or intermediate, cone-sheet is one of the few to traverse the little isolated outcrop of Ben Buie Gabbro occurring in Coire na Feòla, between Glas Bheinn and Creach Beinn ; and another (not shown on the one-inch Map) cuts the Beinn Bheag Gabbro of Chapter XXII. in exposures reaching southwards from Coire Ghaibhre (Fig. 53, p. 312). E.B.B.

Some of the late acid cone-sheets are not of the ordinary craignurite-granophyre facies. Yellow felsite sheets on the slopes of Beinn Bhearnach, east of Torness (two are shown on the one-inch Map), cut the Early Basic Cone-Sheets with which they come in contact. Small-felspar felsite sheets are not uncommon in the south-east extremity of the Glen Cannel Granophyre of Chapter XXXI. One is shown on the map in the River Cannel, above the burial ground. W.B.W., E.B.B.

Small-felspar felsite sheets with basic margins occur commonly in the caps of the Beinn a' Ghràig Granophyre of Chapter XXXII. In this case, however, it is not clear whether one is dealing with true acid cone-sheets. The felsites seem to be in connexion with the underlying granophyre, and as the caps which they traverse are largely made of Late Basic Cone-Sheets, perhaps the felsites may be merely apophyses of the granophyre guided by joints up the middle of pre-existing cone-sheets. It would be extremely interesting if it could be established that the felsites here are genuine cone-sheets springing from the great Ring Dyke of Beinn a' Ghràig —for a connexion between cone-sheets and ring-dykes is not even hinted at in any other Mull exposure. J.E.R.

## Composite Intrusion.

A large proportion of the intermediate and acid cone-sheets are composite, in that their outer margins, extending inwards for some two or three feet, are more basic than their interiors. Often, the division between basic margin and more acid interior is quite conspicuous, but nothing that can be styled a chilled edge separates the two. Field-relations of a few typical cases will be considered there, while petrological details are taken later (p. 228).

Four representative examples may be cited from what may be styled in general terms the S o u n d  o f  M u l l  A r e a. The first reaches Scallastle Bay near Altcrich Cottage, where a note draws attention to it on the one-inch Map. The upper and lower margins of a thick felsite sheet are here seen constituted of 2 ft. of basalt (tholeiite). The latter is, as is the rule in such cases, chilled exteriorly, but not interiorly. Another interesting feature, observable on the shore at this point, is the veining of the upper basaltic layer by acid material from below. It may be added that the exposure is further noteworthy because of two large xenoliths situated just above the lower basic layer : one a slab, 6 ft. long, of *gryphea*-limestone ; the other a streaky rock which has resulted from a gneiss in which the micas have melted to glass (p. 229).

The second example is to be found on the shore of Craignure Bay, opposite the United Free Church Manse. Here the basal layer consists of 3 ft. of fine-grained basalt, and the acid layer immediately above is charged with fragments, up to 3 inches long, of a related intermediate rock.

The third example is chosen from a little beyond the Sound of Mull on the shore south of the entrance to Loch Don. The composite nature of this sheet is less conspicuous than in the two just described. It is possible that there has been a gradual merging from basic to acid due to digestion of the former by the latter.

The fourth example is taken from an easily reached and recognizable inland occurrence. A thick cone-sheet of intermediate composition, and rather coarse crystallization, has been traced intermittently for a couple of miles in Glen Forsa, where it passes under Gaodhail and Tomsléibhe Cottages. Its outer margins, top and bottom, consist of a foot or two of dolerite of tholeiitic type, well chilled against the augite-diorite of Chapter XVIII., which serves as country-rock. A feature of this Glen Forsa sheet is the fact that it is cut to pieces by the Late Basic Cone-Sheets of Chapter XXVIII. E.B.B.

The examples cited above come from the Sound of Mull Area, but the phenomenon of composite intrusion is widespread. A striking case is afforded by a thick craignurite cone-sheet which passes through the spur indicated on the one-inch Map by a southward deflection of the 1000-ft. contour on the southern slopes of Ben Buie. This great craignurite-sheet has two or three feet of intimately connected basalt of dolerite at both top and bottom, and as usual the basic margins are clearly chilled against the country-rock. G.W.L.

It is important to realize that the tendency to composite habit on the part of the intermediate and acid cone-sheets is not confined to any period of intrusion. The instances so far considered very probably all belong to the Early Acid Cone-Sheets. It has been pointed out that some of the acid cone-sheets are of later date, among them the yellow felsites east of Torness, and the small-felspar felsites cutting the Glen Cannel Granophyre. These later felsites also afford typical examples of composite sheets. E.B.B.

## Petrology.

The intermediate and acid rocks that occur as cone-sheets, although presenting numerous characters that indicate close genetic relationship, exhibit considerable variety in texture, and have a fairly wide range in chemical composition. Amongst their more acid members, they include felsites and fine-grained granophyres, but it is the intermediate and sub-acid members that call for special attention in so far as these rock-types are represented to an exceptional degree in Mull.

### CRAIGNURITE AND ITS ALLIES.

(ANALS. I. and VI., TABLE III., p. 19; II., TABLE IV., p. 20).

Of the intermediate cone-sheets the dominant type is that to which we apply the name craignurite because it is well represented in the neighbourhood of Craignure. Three rocks were chosen for chemical analysis. One of these is characteristic of the greater number of the sheets to which the name is applied, while the others mark approximately the respective limits of the type in an acid and basic direction. These three selected specimens (Fig. 33) will be described in some detail, and, then after a few supplementary remarks based upon other material, a precise definition of craignurite will be attempted.

The analysed typical craignurite (16802, p. 19) consists essentially of a highly characteristic network of narrow elongated crystals of augite, and skeleton-crystals of oligoclase and andesine set in a fine-textured acid divitrified base (Fig. 33A). There are a few

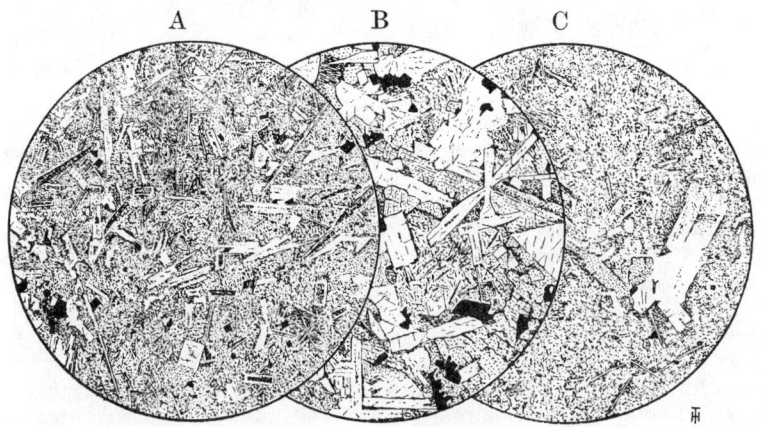

Fig. 33.—Craignurite and Allied Granophyre.

A. [16802] × 15. Normal Craignurite from Allt an Dubh-choire. The structure is highly characteristic. The rock consists of a network of narrow elongated crystals of augite and skeleton-crystals of oligoclase and andesine enclosing a fine-textured acid devitrified ground-mass.

B. [16800] × 15. Basic variety of Craignurite from Allt an Dubh-choire. This rock shows a coarser type of crystallization and differs from the normal type in the greater basicity of its felspar and in a diminution in the amount of acid matrix. The usual acicular type of crystallization is preserved.

C. [16803] × 15. Granophyre allied to Craignurite from Craignure shore. This rock contains the usual elongated crystals of augite and occasional small felspars of the craignurites. It differs, however, in its preponderance of acid matrix in the form of microperthitic and microgranophyric areas.

stumpy porphyritic crystals of labradorite and occasional large crystals of magnetite, but the rock cannot in any ordinary sense be termed porphyritic.

The long augite-prisms have been converted into green hornblende and serpentine, with an accompanying separation of granular sphene. The bulk of the felspar occurs as narrow zoned crystals of oligoclase and andesine, which give skeletal outlines in section, either box-like, hour-glass shaped, or forked, and indicate a rapid growth, more especially at edges and corners. The matrix is devitrified and of a brownish-grey colour. It is traversed in all directions by crystallites of oligoclase felspar and magnetite, and the ultimate products of crystallization are orthoclase, albite, and

P

quartz, in perthitic and micro-granophyric relationship. The grey colour of the ground-mass is due partly to turbidity of the alkali-felspar and partly to widely disseminated minute crystals of magnetite. Apatite is not an abundant accessory.

In the granophyre of craignurite affinity from Craignure shore (16803, p. 20), the ferro-magnesian constituents are less strongly represented, although of similar habit to those in the type-craignurite; and the felspars, though skeletal in form, are smaller in size and altogether less prominent. The bulk of the rock is made up of the material that constituted the ground-mass of the craignurite, and has been similarly devitrified into micro-perthitic and micro-granophyric areas, but with a somewhat greater separation of clear quartz.

The basic craignurite (16800, p. 19) owes its increased basicity to the stronger development of basic plagioclase, an increase in the amount of augite, and a diminution in the amount of acid matrix. The felspars, apart from those which belong more properly to the matrix, are elongated prisms of acid labradorite usually once twinned. They are associated with brownish-yellow hypidiomorphic prisms of augite and patches of iron-ore. The augite almost invariably has a fine lamellation parallel to the base (salite-structure). Twinning parallel to the orthopinacoid is a common feature. Intergrown with this augite, frequently occupying a central position, but occasionally in less definite relationship, is another pyroxene (see below) now almost always represented by serpentinous pseudomorphs. The matrix, as before, shows a characteristically immature type of crystallization: augite assumes a markedly acicular habit, and the felspars become more acid in composition and increasingly skeletal in form. Albite-oligoclase forms the larger individuals, but the ultimate product of crystallization is a perthitic and micro-granophyric residium that has crystallized in a patchy manner and has included slender needles of oligoclase, minute acicular crystals of augite, and finely divided magnetite. There is a tendency for the matrix to collect into areas, often of considerable size, that are more or less devoid of the larger crystalline individuals—a character often remarked upon in connexion with the acid matrix of quartz-dolerites, for instance, those of the Scottish Lowlands. It may be added that the Craignure Type at its basic end is linked closely in texture and composition with the quartz-dolerites of Talaidh Type described in Chapter XXVIII.

As has been stated, the greater number of Mull craignurites group themselves about the type-rock described above, but one point of considerable interest must be emphasized, namely, the almost constant occurrence, either within or in close association with normal aluminous augite, of another pyroxene that in most cases is represented only by pseudomorphs. In the elongated augites, it may be seen to form a narrow central zone of variable width, and, at other times, it may appear as a more complex intergrowth. In quartz-dolerites from other regions a similar relation is often noted between monoclinic and rhombic pyroxenes, and it is quite possible that a rhombic pyroxene has been present in some of the Mull craignurites, for it is often met with in the leidleites (Chap. XXV.) which differ from the craignurites in little but texture. At the same time, no fresh rhombic-pyroxene has been found in the craignurites, while, in more than one instance, it is possible to demonstrate from unaltered remnants (14879, 15539) that the pyroxene forming the core of the normal augite is also a monoclinic pyroxene in optical continuity with its envelope. In such cases, the earlier pyroxene undoubtedly belongs to the group of minerals known as enstatite-augite, and agrees with the pyroxene-phenocrysts of the inninmorites (p. 284).

Among the more acid craignurites, an approach to porphyrite is occasionally encountered (15544, 15542, 14924). Acicular and elongated augites are smaller than in the normal rock, and we note the incoming of glomero-porphyritic groups of labradorite felspar (usually albitized) along with rounded pseudomorphs after enstatite-augite, such as are characteristic of the inninmorite sills. The matrix, however, retains the character met with in the more basic rocks, but becomes greater in amount. There is a progressive suppression of its larger felspathic components, and a still more pronounced development of skeletal individuals (14222, 14223). Acid craignurites of this variety grade into closely allied felsites and granophyres. In these rocks, an acid character is demonstrated by an abundance of quartz, both free and in micro-granophyric growths; while craignuritic tendencies are shown by acicular augite, in the matrix, and also by glomero-porphyritic groups of albitized labradorite, enstatite-augite, and magnetite (16951, 16950, 16460).

Additional examples of the more basic varieties of craignurite, comparable with the analysed specimen, are (15554, 16593, 16804). Commonly, pseudomorphs after enstatite-augite, or possibly in some instances a rhombic pyroxene, form cores to the elongated crystals of aluminous augite, and less frequently occur as isolated crystals (16593). In exceptional cases, it may be the only ferro-magnesian constituent (14879). Olivine is sometimes represented in the more doleritic types as rare pseudomorphs (16364).

In conclusion the following definition of the type is offered:—

The craignurites range in silica-percentage from about 55-70, and their most characteristic representatives have about 65 per cent. $SiO_2$. They differ from the augite-andesites of Chapter XXV. in possessing a more mature type of crystallization, which at the same time has not developed so far as in the case of the augite-diorites of Chapter XVIII. In the field, they are grey moderately fine-grained rocks characterized by an acicular development of their chief constituents. Microscopically, they consist essentially of columnar and acicular aluminous augite and elongated skeletal crystals of a zoned andesine and oligoclase, in a matrix that appears to be the rapidly crystallized representative of a glassy base. The aluminous augite is generally associated with a non-aluminous variety which most frequently forms the central portion of the elongated crystals, or may exist as separate micro-porphyritic individuals. The matrix usually shows an intricate network of slender and skeletal acid plagioclase-crystals, set in the ultimate products of crystallization which are patches of microperthitic and microgranophyric material with minute areas of free quartz.

Among the other varieties of acid cone-sheets, we may mention a Small-Felspar Felsite Type, a description of which is given in the following section on Composite Cone-Sheets. An easily located sheet (13899), 700 yards south-south-west of Corrynachenchy House, south of Fishnish Bay, is a variant of the type, and has characters that ally it to the inninmorites. It consists of a small, but abundant, phenocrysts of albitized labradorite and pseudomorphed enstatite-augite in a felsitic matrix that locally assumes a sub-spherulitic structure.

A few cone-sheets crossing Allt an Dubh-choire above Scallastle Bay, Sound of Mull, may be regarded as variants of the Craignure Type (15533, 15534, 15538). Their matrix has a definitely microlithic character. Augite is moderately abundant, and retains the acicular habit as in the craignurites, but is more evenly distributed throughout, and usually of smaller dimensions than that met with in the typical rocks.

The acid cone-sheet that crosses the A'Chioch arête of Ben More (17159) is a compact fluxionally banded rock of trachytic character. It has practically no phenocrysts, except an occasional small albite, and is composed of microlithic felspar that has straight extinctions, with minute areas of devitrified felspathic matter and clear quartz and rusty iron ores.

Passing west of Maol na Samhna, south of the head of Loch Scridain, another sheet (14673, 14900, 16743) is a compact rock, more or less holocrystalline, with trachytic or orthophyric affinities. It consists mainly of stout prisms of albite-oligoclase and orthoclase, with interstitial quartz and chloritic pseudomorphous patches after augite. The augite mostly occurs as minute acicular prisms.

### COMPOSITE CONE-SHEETS.

Petrological data will now be furnished in regard to composite cone-sheets illustrated by slices in the Survey collection. Some of these cone-sheets have already been introduced to the reader's notice (p. 223).

In the cone-sheet at Altcrich on the shores of Scallastle Bay, the acid interior portion (16459) is a compact felsite showing, occasionally, a tendency towards a spherulitic structure. The more definite crystalline constituents are oligoclase in microlithic crystals and a subordinate decomposed acicular augite. The bulk of the rock, however, is made up of micro-perthitic and micro-granophyric patches resulting from devitrification. The margins (16460, 16457) are of a moderately basic tholeiitic rock that approximates to a fine-grained olivine-dolerite. The augite has a subophitic development, and remnants of olivine remain fresh.

The cone-sheet at Craignure (15550, 16803) has an interior that has already been described as a granophyre with marked craignuritic affinities (p. 226). Its margins (15552) are of moderately coarse doleritic olivine-tholeiite (Salen Type, p. 301) with a subradial grouping of augite and felspar.

The interior of the cone-sheet on the shore south of Loch Don (14222) is a typical craignurite of somewhat acid character, while the margins (14221) consist of a fairly coarse olivine-tholeiite of Salen Type, containing a small amount of alkaline residual matter.

The small-felspar felsites, mentioned on more than one occasion as late representatives of the acid cone-sheets, are frequently of a composite nature. Their interior portions consist of small phenocrysts of albite-perthite and albite, with less conspicuous pseudomorphs after enstatite-augite (14756, 14758, 14760), in a devitrified patchy granophyric and perthitic base. The matrix forms the bulk of the rock, and exhibits the same characters as the base of the acid craignurites, in that there is a finely acicular crystallization of augite. The margins vary in composition and texture from tholeiites (14757) to basic craignurites (14759, 14761).

A suite of specimens (14377-80), taken from a composite cone-sheet (not shown on the one-inch Map), 1000 yds. west-south-west of the summit of Beinn Talaidh, proves the central portion (14379) to be a pale soda-granophyre, with moderately large porphyritic albite and perthite crystals in a beautifully spherulitic base. The marginal portions approximate to the Talaidh Type of quartz-dolerite. The rock contains an abundance of hypidiomorphic to granular augite associated with somewhat elongated small crystals of once-twinned oligoclase, evenly distributed magnetite, and a little free quartz.     H.H.T., E.B.B.

### XENOLITHS.

Accidental xenoliths are a very rare feature of this suite of intrusions, but certain occurrences are worthy of mention. The gneiss xenolith (16451), that occurs immediately above the basic bottom layer of the Altcrich Sheet (p. 223), shows that the felspars have fluxed to a considerable extent, formed a copious melt with the magmatic material which has invaded the rock, and solidified as a clear glass. The remaining felspar exhibits evident signs of dissolution, and quartz has either dissolved with the formation of small crystals of rhombic pyroxene or, if incompletely dissolved, is fringed with this material. Biotite has been partly resorbed, and has given rise to an aggregate of small rectangular crystals of cordierite, which are crowded with minute dark-brown to black spinels. The cordierite is partly pinitized. The glassy base contains trichitic crystals of pyroxene, and has devitrified with a spherulitic tendency into feathery growths of alkali-felspar.

Another xenolith (16456) from a neighbouring sheet of craignurite, collected from the course of the main stream almost three-quarters of a mile above Corrynachenchy House, is of pegmatite with a glassy reaction-selvage. Glass has developed along the margins of the felspars, and between felspars and quartz. The felspar shows characteristic mottling due to reheating. The glass has devitrified giving rise to sub-spherulitic growths of alkali-felspar with, probably, some quartz.     H.H.T.

# CHAPTER XX.
## LOCH UISG GRANOPHYRE AND GABBRO.

### INTRODUCTION.

Two large intrusions, granophyre and gabbro (or dolerite) respectively, are considered together here, partly because they are associated at their outcrops, and partly because it has not been found possible to decide which of them is the older. The Loch Uisg Granophyre, like many later intrusions of similar type, is lettered G on the one-inch Map, Sheet 44. Its exposures are interrupted by the waters of Loch Uisg and Loch Buie, but there is no difficulty in realizing the unity of the intrusion : to assist the reader in this respect, notes have been inserted on the one-inch Map to indicate the base, margin, and top of the mass, as exposed on the shores of Loch Buie. The granophyre, throughout, is of medium-texture and rather needly crystallization. The unity of the Loch Uisg Gabbro, lettered eD the same as several intrusions described in Chapter XI., is perhaps more open to question : two quite isolated outcrops are met with namely, at Loch Uisg and Loch Buie, in the north, and in Glen Libidil, in the south. The southern outcrop is distinguished in the field by its conspicuous augite-clusters, but the microscope shows that this difference is of minor account. On other grounds, a correlation between the two exposures seems justified since :—

(1) The interval between the northern and southern exposure is only a mile.
(2) Both localities show extensive, and sufficiently similar, basic masses at the base of the Loch Uisg Granophyre.
(3) There are no exposures of the base of the Loch Uisg Granophyre where dolerite, or gabbro, does not occur in mass.

Further details of the two intrusions are given below under the headings Field-Relations, Age, and Petrology.

### FIELD-RELATIONS.

It has already been stated that the mutual age-relationships of the granophyre and gabbro have not been determined. Contacts have been examined in the cliff above Loch Buie, and also in a tributary of Glen Libidil descending Coill' a' Bhealaich Mhòir. There is no chilling of the one intrusion against the other, but rather, near Loch Buie, a suggestion of local admixture without the development of a xenolithic structure. Altogether, the appearances noted are too ambiguous to be interpreted with certainty. An indirect argument, however, suggests that the gabbro may be later than the granophyre, since no trace of gabbro is to be found attached to the roof of the granophyre.

## Field-Relations.

Thus, we are left in doubt as to whether the sheet-form of the Loch Uisg Granophyre, illustrated in Fig. 34, is original, or, so far as its bottom is concerned, determined by the position of a later intrusion. It is scarcely necessary to add that we have no knowledge of the course along which either granophyre or gabbro has found its way into its present position.

While the original base of the Loch Uisg Granophyre remains dubious owing to the uncertain age-relations of gabbro and granophyre, the same cannot be said of the top of the intrusion. Throughout most of its outcrop, the granophyre has a flat top revealed in spectacular fashion by erosion. A comparison of the one-inch Map with views obtainable from the main road along Loch Uisg brings home this circumstance more clearly than any written description.

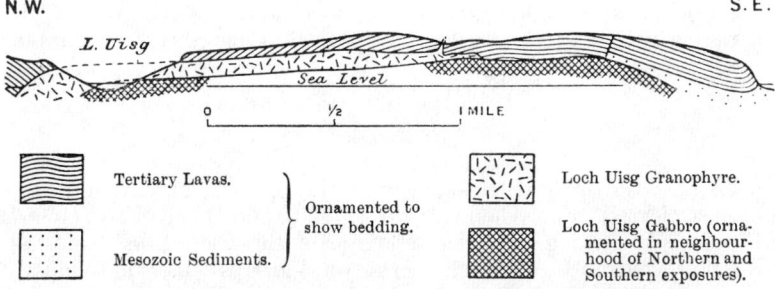

Fig. 34.—Section showing Loch Uisg Granophyre cutting folded Lavas.

The flatness of the granophyre top, and its close approximation with the present surface, enable us to recognize with ease two very interesting additional features :—

(1) The granophyre has everywhere profoundly baked the adjoining basalt-lavas; the microscopic affects have already been dealt with in Chap. X. (for alteration of overlying sediments, see Chap. XV.) What interests us here is the scenic contrast of the crags of contact-altered lava above Loch Uisg as compared with the grassy slopes of unaffected lava rising from Loch Spelve.

(2) The lavas, as can be realized by an inspection of scarp-features, have a very pronounced system of folding to which the flat top of the granophyre pays no attention (Fig. 34).

Near the west end of Loch Uisg, the top of the granophyre plunges steeply out of sight in a northerly direction, and in Cnoc a' Chrònain, on Loch Buie, it descends in like manner towards the south. Whether the marked lowering of the roof in these two localities indicates a dying out of the granophyre cannot be decided. In Glen Libidil, the roof and floor of the intrusion clearly come together as illustrated in Fig. 34—but of course it must be remembered that the age-relationship of the granophyre to its gabbro-floor is undetermined.

### Age.

The general position of the Loch Uisg granophyre in Mull chronology may be gathered from the following facts :—

# Chapter XX.—Loch Uisg Granophyre and Gabbro.

(1) The granophyre is intruded into the Plateau and Central Group of Mull lavas, and is clearly later than the arcuate folding which has affected these rocks (*cf.* Fig. 34 and Pl. V., p. 165).

(2) Near the eastern end of its outcrop, it cuts breccias referred to the superficial agglomerates described in Chap. XV. In its unbrecciated condition it contrasts very strongly with the Glas Bheinn Granophyre (Chap. XII.), which is broken to pieces, probably by the explosions responsible for the superficial agglomerates just referred to.  E.B.B., G.V.W.

(3) It shows a variable behaviour to cone-sheets that may reasonably be referred to the Early Basic Cone-Sheets: thus (*a*), at the west end of Loch Uisg and in Eilean Mòr, Loch Buie, it is freely cut by basic cone-sheets; (*b*), in its western apophysis, near Cameron Farm, it is cut by an acid cone sheet which is itself clearly cut by basic cone-sheets; but (*c*), in the same apophysis, it seems to cut some basic cone-sheets.

It is probable from the above that, along with many of the Intermediate and Acid Cone-Sheets described in Chapter XIX., the Loch Uisg Granophyre was intruded during some part of the period which gave rise to the Early Basic Cone-Sheets of Chapter XXI.  G.V.W.

## Petrology.

*Loch Uisg Gabbro.* Specimens of this gabbro, or dolerite, have been collected from Glen Libidil (17366) and Loch Uisg (15067, 17347). A striking field-characteristic of the Glen Libidil outcrop is the occurrence of augite in crystal-groups that individually measure about a quarter of an inch across. The central portion of these crystal-groups (17366) contains a few well-formed, but relatively small, crystals of labradorite; while the marginal part is crowded with similar, but rather larger, felspars. The material in which the augite-groups are disposed is formed of felspar, similar to that of the larger individuals mentioned above, magnetite, pseudomorphs after olivine, and a mesostasis of alkali-felspar, quartz, and epidote. The plagioclase outside the augite-groups is rather more zoned than that which occurs in intimate association with the augite, and has suffered a certain amount of albitization. Movement of the mesostasis with reference to the early formed augite- and felspar-groups is indicated by a small vein of residual material cutting through the augite. It is worthy of comment that this type of rock is easy to recognize in the hand-specimen, and that it presents, in a magnified degree, the ophimottled structure characteristic of many of the Mull lavas (p. 138). Where more completely and uniformly crystallized (17347), the rock has three main constituents, labradorite, olivine (as pseudomorphs), and augite. The felspar is generally similar to that described above. The olivine is more abundant, and is sometimes earlier than, and sometimes later than, adjacent felspar. The augite is considerably reduced in amount, and is definitely ophitic.

The gabbro, near its junction with the overlying Loch Uisg Granophyre, on the southern shore of Loch Uisg, shows three distinct types of crystallization within the limits of a single slide (15067):—

> (1) Labradorite is enclosed ophitically by an intergrowth of rhombic pyroxene and augite, the former now represented by pseudomorphs in fibrous hornblende. The ophitic structure is somewhat peculiar, in that the rhombic pyroxene, and

to a less extent the augite, has moulded the outer zones of the felspars, thus giving curious rounded contacts. The pyroxene, while showing a dominant tendency towards ophitic development, has, at the same time, an inclination towards hypidiomorphism. A few small crystals of olivine (pseudomorphed) are of earlier formation than the pyroxenes.

(2) Labradorite is enclosed in a subordinate acid mesostasis.

(3) A more or less patchy separation of early-formed constituents from mesostasis has occurred along the junction of the two types described above. An unusually large crystal of augite, moulded upon a pseudomorph after rhombic pyroxene is in contact with one of these mesostatic patches, and shows marginal corrosion, together with a border-precipitation of magnetite in the adjoining magma.

Microscopic study confirms the characters observed in the field, and shows that the Loch Uisg Gabbro is an olivine-gabbro, or perhaps more properly an olivine-dolerite, which, towards its contact with the granophyre, presents heterogeneous characters. The abundance of rhombic pyroxene in the marginal zone recalls a notable character of the hybrids described in Chapter XXXIII. If, eventually, it can be shown that the mixed zone at Loch Uisg has resulted from assimilation of granophyre by gabbro, its interest will be heightened; but at present it is uncertain as to whether the gabbro was acidified by later granophyre, or *vice versa*.

There is, it may be added, a general difference between the Loch Uisg and Ben Buie Gabbros, sufficient to support the belief that the Loch Uisg Gabbro is not a mere offshoot from the larger intrusion.

*Loch Uisg Granophyre.* A specimen, taken as representative of this intrusion, shows a microgranophyre having pronounced affinities with the craignurites, but in a poor state of preservation (18873). There is a moderate development of acicular augite, now represented by chlorite, and plagioclase zoned with orthoclase; the plagioclase of the elongated crystals is now albite. The ground-mass is micrographic with only a small proportion of readily distinguishable quartz. Although unrepresented in the Survey collection, patches with coarse graphic structure are a feature near Laggan, on Loch Buie.

Another specimen (18875), taken from the top of the mass, is of a similar variety, but bears evidence of more rapid cooling. A marked difference exists between the early acicular type of crystallization and that of the base. There is a strong development of felspar-needles (albite-oligoclase), and there are also a few long and sometimes very discontinuous pseudormorphs after augite. The ground-mass is felsitic rather than granophyric in nature, and is characterized by many curved skeletal growths which doubtless were originally formed of augite and magnetite.

This granophyre has produced striking metamorphic effects in its roof, as described already in Chapters X. and XV.

H.H.T., E.B.B.

# CHAPTER XXI.

## EARLY BASIC CONE-SHEETS.

### INTRODUCTION.

It has been explained in Chapter I. that there are innumerable basic cone-sheets in Mull, and that they are separable into two main groups. Members of the early group are lettered bI on the one-inch Map, Sheet 44, where they are for the most part treated as a complex, with seldom any attempt at showing intersections of one individual sheet by another. The Early Basic Cone-Sheets are mainly olivine-dolerites, and are inclined at an average angle of about 45° towards a centre (approximately $C_1$, Figs. 18 and 58, pp. 138, 338), that underlies the more south-easterly of the two calderas shown in Plate V. (p. 165). Individual cone-sheets are often 30 or 40 ft. in thickness, while, in the aggregate, the complex must, in Creach Beinn, exceed 3000 ft. The cone-sheets traverse a wonderful variety of country-rock, and, in addition, cut one another again and again ; but, in spite of this, their general regularity is very striking. Moreover, pre-cone-sheet boundaries can still be traced through the segmented country-rock into which the complex has been intruded, and these boundaries are marked by heavy lines on the one-inch Map. E.B.B.

Each sheet consolidated with chilled margins, no matter what the country-rock ; even where one sheet cuts another, a glance at the mutual junctions suffices to indicate which is the later of the two. Only where there is clear independent evidence that the sheets have been greatly altered since consolidation, are their chilled margins obscured or obliterated. In the vicinity of the Ben Buie and Corra-bheinn Gabbros, sometimes for a distance of 50 yds., chilled margins fail in the great majority of the local representatives of the early Basic Cone-Sheets ; contact-alteration has led to recrystallization, and this has masked the original fine texture of the margins of the sheets.

Further consideration of the Early Basic Cone-Sheets is divided in the sequel under the headings Distribution, Time-Relations, Convenient Exposures, and Petrology. (C.T.C.)

### DISTRIBUTION.

The outcrop of the sheet-assemblage is for the most part a conspicuous scenic feature of Mull geology, as well as a marked characteristic of the one-inch Map. It starts near Beinn Chreagach Mhòr, which rises above Glen Forsa in the north, passes through Creach Beinn, above Loch Spelve in the south-east, and returns northwards across Glen More, in the neighbourhood of Derrynaculen

Cottage. This course, viewed broadly, is of horse-shoe form, open to the north-west. Probably the north-west gap in the outcrop is original, though it may have been accentuated by subsidence of the north-western caldera of Plate V. (p. 165), and also by the intrusion of the Late Basic Cone-Sheets of Chapter XXVIII. and the Beinn a' Ghràig Granophyre of Chapter XXXII. The external limit of the assemblage-outcrop, though not abrupt, is very marked, and nowhere more so than on the slopes of Creach Beinn. The internal limit is also well-defined, but it can only locally be demonmonstrated as an original feature—where, for instance, there are comparatively simple tracts of lava in the northern Beinn Bhearnach, at the head of Glen Forsa, and again in the western part of Beinn Fhada, above Loch Sguabain in Glen More.

About Loch Spelve, one can recognize another interesting feature in the distribution of the Early Basic Cone-Sheets. There is a fairly consistent lack of parallelism between the margins of the assemblage-outcrop, on the one hand, and those of individual sheets on the other. The relationship has somewhat the geometry of false-bedding, and is quite clearly defined on the one-inch Map. The way in which individual sheets obliquely approach, and then die out along, the outer margin of the assemblage-outcrop is particularly obvious. That a similar relation holds also for the inner margin is best appreciated on comparing the sheet-complex exposed above Torness Cottage in Glen More with the lava-slopes of Beinn Bhearnach, north-west of Sgùrr Dearg—for though some of the detail here is difficult to unravel, the broad contrast is very marked indeed. E.B.B.

## Time-Relations.

Of superinduced irregularities in the assemblage-outcrop of the Early Basic Cone-Sheets, the most pronounced are the gaps due to the later intrusion of the Ben Buie and Corra-bheinn Gabbros. Details are reserved for Chapter XXII., but it may be stated here that a very large number of sheets, classed in the present chapter as Early Basic Cone-Sheets, are cut through by these two gabbros. It would afford a basis for a simple time-classification if the two gabbros concerned behaved precisely in the same manner in this matter. Almost certainly, however, the Corra-bheinn Gabbro is somewhat the later of the two, for it cuts off a series of porphyritic olivine-dolerite cone-sheets, many of which, on the other side of Allt Ghillecaluim, traverse the Ben Buie Gabbro (see one-inch Map, where a small selection of the evidence is presented). These particular porphyritic dolerite sheets are characterized by numerous small felspar-phenocrysts, a feature which is lacking in the great bulk of the Early Basic Cone-Sheets cut through by the Ben Buie Gabbro (and also in the Late Basic Cone-Sheets of Chapter XXVIII). An approximate time-scale may thus be adopted :—(1) Main Early Basic Cone-Sheets, non-porphyritic; (2) Ben Buie Gabbro; (3) Continuation of Early Basic Cone-Sheets, with small felspar-phenocrysts; (4) Corra-bheinn Gabbro (followed by Late Basic Cone-Sheets). (C.T.C.)

## Chapter XXI.—Early Basic Cone-Sheets.

Other time-relationships of the Early Basic Cone-Sheets are summarized in Fig. 35, which, among other things, shows the sheets traversing, undisturbed, the Loch Spelve Anticline and the agglomerate-masked margin of the early caldera (*cf.* one-inch Map and Plate V., p. 165). There is no more impressive exposure bearing upon this point than is afforded by the head-waters of Abhuinn an t-Stratha Bhàin, between north Beinn Bhearnach and Dùn da Ghaoithe, where lenticles of baked fossiliferous Lias limestone can be seen with vertical bedding cut abruptly across by much less highly inclined basalt and dolerite sheets. The same point is well-illustrated again in the more accessible sections of Creagan Mòr, east of Loch Spelve (Càrn Bàn Section, Fig. 25, p. 174). Here, two massive unbroken dolerite-sheets, dipping 45°-60° to west or south-west, cut through shattered Tertiary lavas and Mesozoic sediments, which, standing up on end, strike north and south. Moreover, between these two localities, close observation and detailed mapping reveal the continuous existence of the north-east limb of the Loch Spelve Anticline, with Tertiary and Mesozoic rocks steeply inclined towards the north-east; whereas the landscape of the district is determined by Early Basic Cone-Sheets steeply inclined towards the south-west.

Another feature of Fig. 35 which calls for comment is the clear fashion in which the Early Basic Cone-Sheets cut through the agglomerates of the Sgùrr Dearg district (Chapters XV. and XVI.). This relationship holds from Sgùrr Dearg to Beinn Chreagach Bheag in the north-west, and to Creach Beinn in the south-west. It is not only that the sheets again and again cut actual agglomerate in this region, but also they are often found in an unbroken condition separated by shattered bands and lenticles of such rocks as Moine gneiss, Trias sandstone, Tertiary lava, and Tertiary granophyre.

The immunity to shattering so commonly exhibited by the Early Basic Cone-Sheets occasionally fails. In the Lussa River and the neighbouring slopes of Beinn Bhearnach, south of Sgùrr Dearg, the sheets are somewhat broken, and are veined with epidote, as is so frequently the case with lavas in Central Mull. Indeed their sheet-nature would be difficult to realize, were it not that there are many exposures of lenticles of the Torness Felsite which here serves as country-rock. Actual breaking up of Early Basic Cone-Sheets to yield agglomerate has not been noted anywhere in Mull; but it is certain that the explosions, which followed the consolidation of the Ben Buie Gabbro west of Loch Ardeglais (Chapter XXII.), must be of later date than the main mass of sheets now under consideration.

Fig. 35 shows a few Acid Cone-Sheets cut by Early Basic Cone-Sheets. This is a relation very commonly met with, but, as pointed out in Chapter XIX., there is no evidence for a definite time-separation betwen the Early Acid and the Early Basic Cone-Sheets. It seems, rather, that most of the more prominent intermediate and acid cone-sheets of Mull were intruded during the earlier phases of the Early Basic Cone-Sheet period. This association of basic, intermediate, and acid in one long period, combined with the frequent combination of basic margins with acid or intermediate interiors, makes it fairly certain that some of the Early Basic Cone-Sheets

Time-Relations. 237

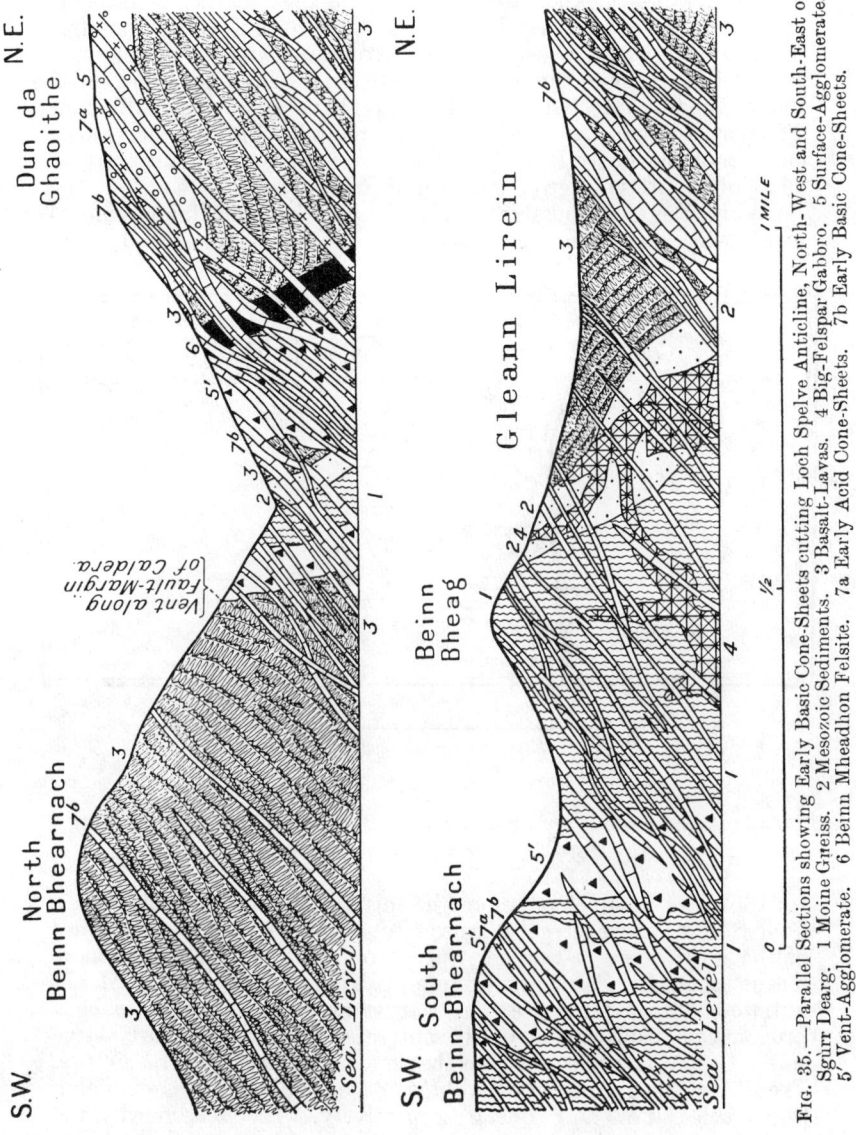

FIG. 35.—Parallel Sections showing Early Basic Cone-Sheets cutting Loch Spelve Anticline, North-West and South-East of Sgùrr Dearg. 1 Moine Gneiss. 2 Mesozoic Sediments. 3 Basalt-Lavas. 4 Big-Felspar Gabbro. 5 Surface-Agglomerate. 5' Vent-Agglomerate. 6 Beinn Mheadhon Felsite. 7a Early Acid Cone-Sheets. 7b Early Basic Cone-Sheets.

238   Chapter XXI.—Early Basic Cone-Sheets.

must be olivine-free basic or sub-basic rocks of the types which characterize the Late Basic Suite of Chapter XXVIII. Be this as it may, the field-observer soon realizes the validity of the general rule that Early Basic Cone-Sheets in Mull are olivine-dolerites, often of considerable thickness and exposed in craggy scarped faces, as in Creach Beinn; whereas, Late Basic Cone-Sheets are olivine-free basic to sub-basic rocks of finer texture, apt to be individually thin and to cover themselves beneath long slopes of scree, as in Beinn Talaidh. In bulk, the two suites have a somewhat different distribution—the Late Basic Cone-Sheets fall within the horse-shoe outcrop of the Early Basic Suite, but there is a sufficient overlap to allow of countless intersections of the Early by the Late.

While there is abundant evidence that the main folding-movement, which centred about the south-eastern caldera of Plate V

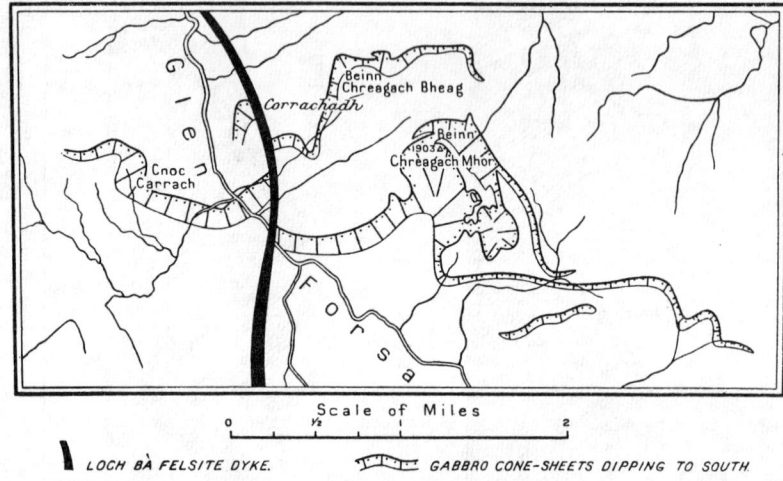

FIG. 36.
Early Basic Cone-Sheets of Beinn Chreagach Bheag and Beinn Chreagach Mhòr probably displaced at Loch Bà Felsite.

(p. 165), was completed before the introduction of the Early Basic Cone-Sheets, there is good reason to believe that the subsidence of the north-western caldera, ringed round by the obviously more recent Loch Bà Felsite of Chapter XXII., was developed, if not initiated, at a later stage. Fig. 36 shows what appears to be a pronounced displacement of the outcrops of two particularly large Early Basic Cone-Sheets where they encounter the Loch Bà Felsite. These two sheets are the gabbros, or coarse dolerites, of Beinn Chreagach Mhòr and Bheag respectively. The reason why the evidence is somewhat inconclusive is that exposures of the gabbros in the critical region are much obscured by later sheets and morainic drift. The details are as follows:—

(1) At the summit of the hill from which it takes its name, the Beinn Chreagach Mhòr sheet is about 200 ft. thick, and is a coarse black-and-white gabbro. Elsewhere, it is generally not so thick, and is a fine-grained dark gabbro. Its

outcrop is easily traced down-hill into Glen Forsa, along the line shown in Fig. 36, until the alluvium (see one-inch Map) is reached. Then, inside the Loch Bà Felsite dyke, there occurs a lenticle of the same type of dark gabbro between minor cone-sheets in a small tributary burn that drains from the east into Glen Forsa; another lenticle is met with in the main river; while others again (coarse-grained and black-and-white) appear in the Gaodhail River that enters the glen from the west. Of these exposures, only the small lenticle in the eastern tributary is important from the point of view of deciding whether or no a shift of outcrop occurs at the Loch Bà Felsite.

(2) The coarse dolerite of Beinn Chreagach Bheag runs, as a line of crag, down to the bottom of Glen Forsa to end abruptly in a little cliff just where neighbouring exposures show that the Loch Bà Felsite lies concealed beneath moraine. There is nothing to suggest a direct westward continuation of the dolerite once the dyke is crossed; whereas, about 300 yds. down the glen, near the southern end of a ruined village named Còrrachadh, bare rock of the Beinn Creagach Bheag character can be followed as a low ridge, 400 yards long, leading away from the Loch Bà Felsite until lost sight of beneath alluvium.

## Convenient Exposures.

All that remains, before turning to the petrology of the group, is to indicate a few sections of interest which do not happen to have been mentioned in preceding paragraphs:—

(1) Coire Bearnach, approached from the road that skirts the northern termination of Loch Spelve, supplies very good exposures of cone-sheets in gneiss.

(2) A section showing typical cone-sheets with well-exposed chilled margins is afforded near the road just mentioned, westwards for some little distance from Rudh' Àird a' Chaoil. E.B.B.

(3) An excellent coast and road section is met with east of Glenbyre Farm, on Loch Buie. In the same district, an unusually thick and gently inclined dolerite sheet (Beinn Chàrsaig) is regarded as belonging to the suite. G.V.W.

(4) Stream-sections show dark dolerite-sheets cutting pale Derrynaculen Granophyre on the north face of Glen More, and are particularly diagramatic on account of the colour-contrast. (C.T.C.)

## Petrology.

The Early Basic Conic-Sheets are represented by about 80 slides in the collection. They are on the whole a very uniform assemblage of non-porphyritic olivine-dolerites with ophitic lilac-coloured augite.

Only two Early Basic Cone-Sheets have furnished examples of fresh olivine. These are the most massive sheets of the group, the Beinn Chreagach Mhòr Sheet (16448), on the north side of the Mull centre, and the Beinn Chàrsaig Sheet (14916), on the south. Elsewhere, olivine has been converted, probably in large measure by posterior pneumatolysis, into chlorite and serpentine.

The felspar is zoned labradorite. The augite is of the purplish titaniferous variety, but is generally paler than that of the lavas (Chapter X.); in fact, strongly tinted augite is exceptional (18975).

In several of the examples, there are interspaces filled by chloritic and serpentinuous products. In others, a distinct acid mesostasis is discernible (18962, 18964), and the containing rock becomes practically a quartz-olivine-dolerite, allied in structure and composition to the more basic products of the differentiation-series discussed in Chapter XXX. More acid types are represented in the collection, but before considering them it will be well to say a few words concerning the structure of the predominant and more thoroughly basic varieties.

## MAIN TYPE.

As a rule, the structure displayed by these rocks is a variant of the ophimottled type already referred to as characteristic of a great proportion of the Plateau Basalts (Chapter X.). Of the three chief minerals, augite, felspar, and olivine, the felspar has had the most extended time-range of crystallization : for, starting in company with augite, it finished in company with olivine. The result is a separation of augite-felspar aggregates from surrounding felspar-olivine aggregates (15559, 16448, 18965, 18968).

The relatively early date of the augite is obscured by its almost complete refusal to exhibit crystal-boundaries. Zonal growth is shown by a marked deepening of the purple tint of the augite towards its irregular margins. Commonly the early formed augite-felspar crystallizations occur in clusters, and in such cases the zonal coloration of the augite is in relation to the margin of the cluster as a whole.

The felspar often changes its type of developement from the central portions of the augite-felspar aggregates towards the margin of the aggregates, or towards the felspar-olivine areas beyond. The central felspar-growths tend to be skeletal and sometimes rudely radiate (15637, 18965). They are, in some measure, a converse of the cervicorn growth of augite in felspar-crystals, characteristically developed in many of the sub-basic rocks of Mull (Chapters XXVIII. and XXX.). Wherever a felspar can be traced outwards towards, or across, the margin of an augite-aggregate, it not infrequently shows a marked increase of width, while the felspars of the felspar-olivine areas occur for the most part in well-formed lath-shaped crystals.

The olivine varies considerably in form and in its relation to associated felspar. It may be hypidiomorphic, but not infrequently definitely ophitic.

The only important accessory is titaniferous magnetite, which occurs as skeletal growths, sometimes earlier than the augite, but perhaps more often of later formation.

The above remarks cover most of the examples from a typical roadside collection made between Ardachoil and Ardura Farms on Loch Spelve (18962-79). Two specimens, however, call for attention. One (18974) introduces to our notice a subordinate porphyritic type ; the other (18963) a sub-basic type. There is no conclusive evidence to show that these two particular rock-specimens belong to the Early Basic Cone-Sheets, although their position, far from any recognizable development of Late Basic Cone-Sheets, makes it extremely probable. In the following descriptions, closely allied petrographical material has been employed, collected from other localities where there is little or no uncertainty regarding the age-relationships of the sheets concerned.

## PORPHYRITIC TYPE.

Dr. Clough has shown that there is a development of porphyritic Early Basic Cone-Sheets in the time-interval that separates the intrusion of the Ben Buie and Corra-bheinn Gabbros. These porphyritic rocks are typically represented by sheets (16381, 16392), which cut the main outcrop of Ben Buie Gabbro, and by others (17407-9), which traverse the satellitic mass in Coire na Feòla. It is unnecessary to enter into detail, since these porphyritic types are petrologically identical with the porphyritic Central Mull lavas described in Chapter X., except that their texture is distinctly coarser than is usual in the case of extrusive rocks. Sections show abundant small phenocrysts of very basic plagioclase felspar, marginally zoned, in a coarse granular base of olivine (as pseudomorphs), hypidiomorhpic augite, and felspar. Olivine and, less frequently, augite are sometimes associated with the felspar in glomero-porphyritic aggregates.

## SUB-BASIC TYPES.

The roadside exposure of Early Basic Cone-Sheets, between Ardachoil and Ardura farmsteads, shows how intimate is the connexion between the normal olivine-rich type and the sub-basic type that has quartz-dolerite affinities. One example (18976) showing augite-felspar aggregates, followed by felspar-olivine additions, has, locally, a mesostasis of the type common in quartz-dolerites, though in this case the texture is too fine for detailed interpretation. Two other examples from the same series (18962, 18964) show a more pronounced separation of acid mesostasis that is evidently, in large part, composed of quartz and alkali-felspar. These examples of Early Basic Cone-Sheets are less rich in olivine than usual, and their augite is not so highly coloured. The order of crystallization of their constituents, also, departs slightly from that normally observed in these rocks, for the augite and basic felspar continued to grow to a considerable extent after the formation of olivine had ceased, and hence there is no tendency for olivine to concentrate near the interstices occupied by the acid mesostatic material.

Three specimens (16361, 16365, 16366) from the foot of Ben Buie, almost certainly belonging to the pre-gabbro suite of cone-sheets, afford examples of transition between quartz-dolerite and craignurite. It is particularly interesting to note in them an important development of columnar augite in connexion with their mesostasis. As noted in the craignurites (Chapter XIX.), many of these columnar augites have a core, now preferentially replaced by hornblende, which, no doubt, originally consisted of either enstatite-augite or a rhombic pyroxene.

An example (18963), already mentioned, from the Ardachoil-Ardura road-section, is similar to some of the Late Basic Cone-Sheets of Chapter XXVIII., and will be referred to later in connexion with them as affording a connecting link between the ordinary Talaidh Type of quartz-dolerite and the variolitic type of Cruachan Dearg (p. 304).     E.B.B.

# CHAPTER XXII.

## BEN BUIE, CORRA-BHEINN, AND BEINN BHEAG GABBROS.

### Introduction.

The three olivine-gabbros dealt with in the present chapter are lettered E on the one-inch Map, Sheet 44. The Ben Buie and Corra-bheinn Gabbros lie outside the earlier of the two central calderas of Plate V. (p. 165), and the Beinn Bheag Gabbro inside, but otherwise the Corra-bheinn and Beinn Bheag Gabbros roughly balance one another on the two sides of the north-west axis of symmetry which runs through Beinn Chàisgidle. Certain small masses of allivalite, lettered U on the one-inch Map, and of porphyrite, lettered P, are also treated in the sequel—in connexion with the internal relations of the Ben Buie Gabbro.

The Ben Buie Gabbro is responsible for the finest rock-scenery of the island, with characteristic dark ice-moulded surfaces of particularly massive appearance. The Corra-bheinn Gabbro behaves similarly, though largely masked by later intrusions. The Beinn Bheag Gabbro gives rise to lines of crag in Beinn Bheag itself, where it is greatly cut up by the Late Basic Cone-Sheets of Chapter XXVIII.; but in Glen Forsa it is scenically quite inconspicuous.

A discussion of the many points of interest presented by these three gabbros is grouped below under the following headings :—Intrusion-Form, Internal Relations, Relations with Cone-Sheets, Relations with Vent-Agglomerates, and Petrology.

(C.T.C.), E.B.B., G.V.W.

### Intrusion-Form.

The well-marked agreement of the inner boundary of the Ben Buie and Corra-bheinn Gabbros with the general arcuate distribution of Mull geology must not be lost sight of. In Plate V. (p. 165), the inner boundary of the Ben Buie Gabbro is interpreted as approximately guided by the fault-margin of the early south-eastern caldera. Such a connexion, if firmly established, would involve a recognition of the Ben Buie Gabbro as an irregular example of a ring-dyke. The evidence, however, is not so precise that it can be relied on in detail. All that one knows for certain may be recapitulated under three headings :—

(1) Inside the gabbro, lavas with pillow-structure are easily found, but none has been met with outside (Fig. 18, p. 133).

(2) Outside the gabbro lies the Derrynaculen Granophyre, which seems to correspond naturally with the Glas Bheinn Granophyre, and this latter skirts the outside of the caldera (Pl. V., Chap. XIII).

(3) Outside the gabbro, the vent-agglomerates of the southern slopes of Ben Buie, with their granophyre- and gneiss-fragments, agree in type with the vent-agglomerate found farther east outside the caldera (Fig. 29, p. 201).

Another type of evidence suggests that the Ben Buie and Corra-bheinn Gabbros (but not that of Beinn Bheag) may have a closer analogy with cone-sheets than ring-dykes. Both masses occasionally show flow-banding, marked by a parallel arrangement of somewhat wavy layers of darker and lighter minerals, and this flow-banding is always inclined at high or moderate angles in *rough* agreement with the cone-sheets of the neighbourhood. Examples from the Ben Buie Gabbro may be cited as follows:—

(1) The banding of the complex margin, to be described presently, of Creag na Còmhla near Loch Fuaran, is inclined north-east. G.V.W.

(2) Good banding is exposed south of Uillt Gharbha between ¾ and 1 mile west-north-west of Ben Buie summit. The inclination is north-east at 30° or 40°.

(3) Steep banding striking north-west, or west, is shown by exposures within the angle of the path south-west from Craig Cottage across the river-junction.

(4) Banding inclined east, or north-east, sometimes at about 70°, is seen 300 yds. north of the Glen More Road, 150 yds. east of the gabbro-margin.

Banding in the Corra-bheinn Gabbro has been noted somewhat often in the more westerly stream below the Coir' an t-Sailein Quartz-Gabbro, and also on the hill-slopes farther east. The dip of the banding is persistently north-east, and the inclination varies from 25°-80°. (C.T.C.)

The Beinn Bheag Gabbro is locally well-banded, where it overlooks Coire Gaibhre from the north, but in this case the inclination of the banding contrasts with that of adjacent cone-sheets. In fact, the banding inclines north-east and the cone-sheets south-west. E.B.B.

Unfortunately, the available mountains are not sufficiently large to reveal whether any of the three gabbros have an inclination in bulk corresponding with the suggestions afforded by the local flow-banding. What they do show is that these great masses are of irregular outline and steeply bounded, and that they undoubtedly continue in depth for some considerable distance without serious diminution. (C.T.C.), E.B.B., G.V.W.

The difficulties with which the geologist is faced in accounting for the disappearance of pre-existent country-rock in the case of these great gabbro masses are epitomized between Creach Beinn Bheag and the south-east slopes of Meall nan Capull, where a series of miniature outcrops of fine-grained gabbro appear on the one-inch Map as outposts beyond the main Ben Buie front. Their individual form contrasts markedly with that of surrounding cone-sheets through which they abruptly rise. One feels that there must have been a considerable displacement of country-rock, either bodily or piecemeal, either up or down, to clear pipe-like cavities for rising gabbro-magma. Possibly explosion to some extent prepared the way.

## Chapter XXII.—Three Main Olivine-Gabbros.

### Internal Relations.

*Beinn Bheag Gabbro.* Fig. 37 summarizes the distribution of rock-types. There is no junction exposed between the coarse and fine varieties of black-and-white gabbros. Contacts of black-

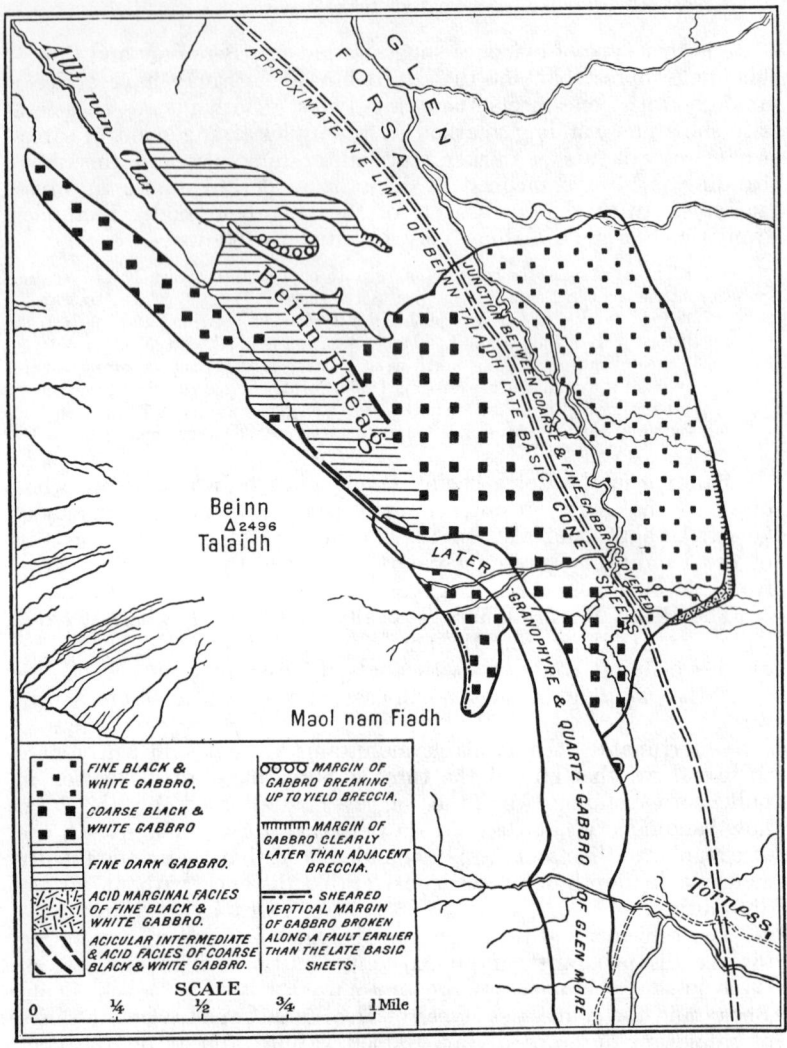

Fig. 37.—Map of Beinn Bheag Gabbro.

and-white with dark gabbro suggest, if anything, that the former is the younger of the two. The belts with acicular crystallization (recalling exactly that of the big augite-diorites of Chapter XVIII.) merge insensibly into adjacent coarse black-and-white gabbro.

E.B.B.

*Ben Buie Gabbro.* Three features of special interest may be noted in connexion with the Ben Buie Gabbro :—

(1) On the south of Loch Fuaran, and particularly along the ridge of rock which runs north-west from Creag na Còmhla, the margin of the gabbro shows a well-marked banding. Some of the bands are rich in magnetite, which is magnetized, and occurs in strings up to 4 inches wide, but of no great lateral extent. Other bands exhibit a pock-marked appearance due to the presence of olivine-rich portions. No single origin can be ascribed to the bands ; in some instances they are, no doubt, included, and highly altered portions of cone-sheets ; in others appearances indicate that they are magmatic segregations or the gabbro-complex. G.V.W.

(2) Several small *allivalite* masses (U one-inch Map) occur enveloped in the Ben Buie Gabbro of the Glen More district. The largest and most conveniently placed outcrop is exposed on either side of the road, about a third of a mile west of Craig Cottage. It measures 700 by 200 yds. Wherever seen, the gabbro-allivalite junctions are not quite sharp. In one occurence on Sròn Dubh, too small to show on the one-inch Map, gabbro clearly veins allivalite (17868). Elsewhere age-relationships are not obvious, but it is possible that the allivalite-masses are, in all cases, of somewhat earlier date than the surrounding gabbro.

(3) Another type of intrusion, associated with the Ben Buie Gabbro is the *Craig Porphyrite* (P on one-inch Map), named after the road-side cottage already mentioned. The porphyrite occurs in several detached outcrops shown on Pl. V. (p. 165). It is characterized by a comparatively fine texture, abundant small felspar-phenocrysts, and, also, by a profusion of xenoliths derived from the surrounding gabbro. It does not chill against the gabbro, though manifestly of later date, so that it looks as if the time-interval between the two was small enough to leave the gabbro warm when the porphyrite appeared. The porphyrite carries enstatite in place of olivine, and thus differs somewhat markedly in composition from the gabbro. (C.T.C.)

## Relations with Cone-Sheets.

As is clear on the one-inch Map, all three gabbros have an outer part comparatively free from intrusions, and an inner part cut to pieces by the Late Basic Cone-Sheets of Chapter XXVIII. (the words outer and inner are used with reference to the centre about which the three gabbros are grouped). In the Ben Buie Gabbro, the relatively simple outer part is far more extensive than the inner more complicated part ; whereas, in the other two gabbros, the division is fairly equal (Fig. 37). The simplicity which characterizes the outer portions of the outcrops of the Ben Buie and Coire-bheinn Gabbros has been attained at the expense of a pre-existing intrusive complex including Derrynaculen Granophyre, vent-breccia, and a host of Early Cone-Sheets. On the other hand, the simplicity of the outer portion of the Beinn Bheag Gabbro is much less informative : the gabbro in this case, where its out-crop is simple, is bounded by lavas which are equally free from other intrusions. (C.T.C.), E.B B,, G.V.W

It will be understood from the above that a fairly accurate age-comparison is possible between the Ben Buie and Corra-bheinn Gabbros. Detailed evidence will now be given to illustrate certain features of the time-scale already enunciated in Chapter XXI., namely :—(1) Main Early Basic Cone-Sheets—non-porphyritic ; (2) Ben Buie Gabbro ; (3) Early Basic Cone-Sheets, with small felspar-phenocrysts ; (4) Corra-bheinn Gabbro. It will be convenient to consider this time-scale backwards rather than forwards, and to deal first with the marginal relations of the Corra-bheinn Gabbro :—

*Corra-bheinn Gabbro.* Most of the hill-side south-west of the gabbro-margin consists either of basalt-lavas or of Derrynaculen Granophyre, in both cases cut by numerous olivine-dolerite cone-sheets dipping steeply north-east; non-porphyritic and porphyritic dolerites occur, the latter distinguished by crowded small felspar-phenocrysts. Quite commonly, these dolerite-sheets (Early Basic Cone-Sheets of Chap. XXI), make up about half the rock just outside the gabbro. Individual sheets may be 20 or 30 ft. thick, and they strike at a considerable angle against the gabbro-margin, there abruptly to terminate. The relationship is very obvious, since the gabbro, for 200 or 300 yds. in from the margin, is almost entirely free from intrusion by cone-sheets of any kind whatever. Moreover, examination of the marginal zone shows that about a mile south-west by south of Cruachan Dearg, and in various other localities, the cone-sheets just outside the gabbro are veined by numberless irregular granophyre-strings, giving rise to an injection-breccia ; and these strings are also cut off by the gabbro. In harmony with this, the gabbro, for 20 yds. or so in from its edge, encloses and veins numerous shapeless masses of earlier basic igneous rock. Contact-alteration has masked the original character of these xenoliths ; but it is natural to conclude that many of them have been derived from neighbouring cone-sheets.

There is little or no evidence of a recrudescence of Early Basic Cone-Sheets after the Corra-bheinn Gabbro. All, or almost all, the numberless basic cone-sheets, shown on the one-inch Map as a complex forming the summit region of Corra-bheinn and Cruachan Dearg, are of types referable to the Late Basic Cone-Sheets of Chap. XXVIII.

*Ben Buie Gabbro.* Anyone wishing to satisfy himself that the Ben Buie Gabbro is cut by dolerite-sheets with small felspar-phenocrysts, and that these sheets form part of the suite cut off by the Corra-bheinn Gabbro, should commence by ascending Uisgeacha Geala from the hill-track, one-third of a mile north-west of Craig Cottage. Several typical exposures of porphyritic dolerite sheets piercing gabbro are seen, both in the stream, and in the neighbouring slopes on the east. The same relation is again exposed near the gabbro-margin above and below the path about half a mile farther north-west. It is impossible to follow individual sheets with certainty from exposure to exposure, but the sheets, just mentioned, are identical in type with some of the suite, which, another half-mile farther on along the same line of strike, is cut across *in toto* by the Corra-bheinn Gabbro. This particular locality has been chosen merely because of its situation relative to the Corra-bheinn Gabbro. The mere fact that a certain number of olivine-dolerite sheets carrying small felspar-phenocrysts cut the Ben Buie Gabbro can be illustrated almost anywhere in the outcrop of the latter.

To realize the relationship that, in bulk, the Early Basic Cone-Sheets are cut off by the Ben Buie Gabbro, one has to leave the north side of Glen More and go, for instance, to the gabbro-edge on Tòrr a' Ghoai, about a mile south-east of Derrynaculen Cottage. A great mass of doleritic sheets occurs here just outside the gabbro; as a rule, sheet cuts sheet, but occasionally a lenticle of a pre-sheet granophyre is preserved in an interval. Against these lenticles, and against one another, the sheets, at a hundred yards and more from the gabbro, are for ever showing chilled margins. Sixty or seventy yards from the gabbro, such chilling begins to be lost sight of, and, within fifty yards from the same, it cannot be recognized. The last twenty yards consists of rather fine-grained basic igneous rock—no doubt contact-altered cone-sheets with blurred individuality—intricately veined and mixed with gabbro, and, to a less extent, with granophyre. A visible field-indication of contact-alteration in these dolerite cone-sheets is a development of reticulate cracks. The porphyritic type of dolerite-sheet is also represented by a few examples, which retain their chilled edges in proximity to the gabbro, and continuing across its margin actually chill against it. These manifestly later dolerites cut the granophyre-veins and do not show the characteristic crack-system just alluded to.   (C.T.C.)

The same general phenomena are met with, wherever, between Tòrr a' Ghoai and Creach Beinn, a contact of Ben Buie Gabbro with Early Basic Cone-Sheets is exposed. The evidence afforded by the south face of Ben Buie and Beinn Bheag is summarized in Fig. 38, where gabbro is shown cutting out a great development of Early Acid, Intermediate, and Basic Cone-Sheets (Chap. XIX. and XXI.) The few basic cone-sheets, which in Fig. 38 traverse the Ben Buie Gabbro, are in part referable to the Late Basic Cone Sheets of Chap. XXVIII., and in part to the porphyritic representative of the Early Basic Suite.  G.W.L., G.V.W.

The minor isolated outcrops of fine-grained gabbro, which are shown on the one-inch Map (three on the south face of Meall nan Capull, two on the south face of Creach Beinn, and one on Creach Bheinn Bheag) furnish additional illustrations; and these all the more striking because in them one can easily see the Early Basic Cone-Sheet Complex, interrupted on the one side of each little gabbro mass, and reappearing on the other. The largest of the Meall nan Capull gabbro out-crops is readily found at the head of Coire na Feòla. It is cut by about ten cone-sheets (one of them acid), but these ten are as nothing to the basic sheets that are obviously intersected.

*Beinn Bheag Gabbro.* The geographical position of this mass precludes a direct age-comparison between it and any appreciable proportion of the Early Basic Cone-Sheets. It is, however, clearly of earlier date than all the very numerous Late Basic Cone-Sheets of its locality.

## RELATIONS WITH VENT-AGGLOMERATES.

The age-relationships of the Ben Buie, Corra-bheinn, and Bheinn

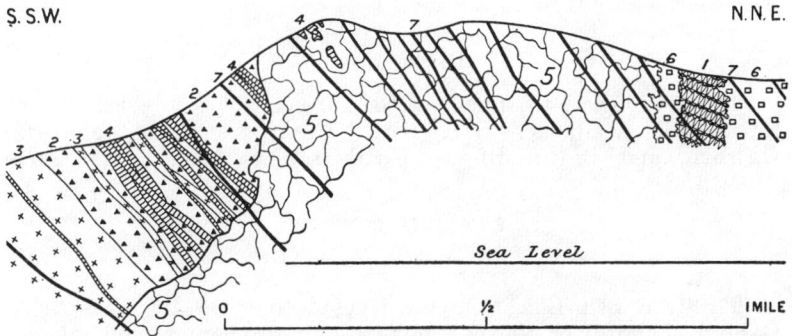

FIG. 38.—Section across Ben Buie.
1. Basalt-Lavas with pillow-structure.  2. Vent-Agglomerate of Early Paroxism.
3. Early Acid Cone-Sheets.  4. Pre-Ben-Buie Early Basic Cone-Sheets.
5. Ben Buie Gabbro.  6. Vent-Agglomerate of Post-Ben-Buie Date.
7. Post-Ben-Buie Early Basic Cone-Sheets and Late Basic Cone-Sheets.

Bheag Gabbros to the vent-agglomerates of their neighbourhood may be summarized thus :—(1) **main agglomerates**; (2) **gabbros**; (3) **subordinate agglomerates**. Direct field-evidence will be presented to prove that some agglomerates are older and others younger than the gabbros. That the former are the more important follows at once when it is remembered that most of the agglomerates in the region with which we are at present concerned are clearly earlier than the Early Basic Cone-Sheets of Chapter XXI., and, therefore, *a fortiori*, earlier than the Ben Buie and Corra-bheinn Gabbros.

*Ben Buie Gabbro.* A comparison of the north and south slopes of Ben Buie brings to light a most interesting dual relationship between agglomerate and gabbro (Fig. 38). On the south side, the Ben Buie Gabbro cuts through a great mass of volcanic breccia largely made up of granophyre-fragments. This circumstance is rendered particularly obvious, because the breccia is abundantly penetrated by massive pre-gabbro cone-sheets of various compositions; each sheet has a chilled top and bottom, except in the immediate vicinity of the gabbro where the latter cuts across the complex in clear exposures. Moreover, the gabbro-margin is quite unbroken at its contact with the agglomerate; in fact, where seen about the 600 ft. contour on the west slope of Gleann a' Chaiginn Mhòir, the gabbro develops a fine-grained crystallization against the agglomerate. On the north side of Ben Buie, all this is reversed, for here a later agglomerate is found at the edge of the gabbro. Not only is this agglomerate largely made up of gabbro blocks and fragments, but also there is often a marginal passage from mixed agglomerate into shattered gabbro, so that nothing could be clearer than that the Ben Buie Gabbro has been locally broken up by subsequent explosions. G.V.W.

*Corra-bheinn Gabbro.* The Corra-bheinn Gabbro is manifestly later than the vent-agglomerate that is shown on the one-inch Map traversed by Early Basic Cone-Sheets in Sleibhte-coire (pp. 187, 207). On the other hand, it is earlier than a narrow strip of agglomerate (too small to show) about a third of a mile west-south-west of Cruachan Dearg. This little agglomerate-patch is in contact with the Corra-bheinn Gabbro, from which it has derived most of its material. In addition to pieces of gabbro, it carries a small proportion of some non-local acid igneous rock. (C.T.C.)

*Beinn Bheag Gabbro.* Fig. 37 (p. 244) indicates where the fine-grained dark gabbro of the Beinn Bheag complex locally chills against agglomerate; and, on the other hand, where it locally breaks up into agglomerate. The fine-grained black-and-white gabbro of the same complex is seen, east of Glen Forsa, to be later than neighbouring agglomerate. E.B.B.

## PETROLOGY.

The petrological treatment of the subject will be divided under three main headings :—Ben Buie Complex, Corra-bheinn Gabbro, and Beinn Bheag Gabbro.

### BEN BUIE COMPLEX.

(ANAL. I. ; TABLE V., p. 23)

The title Ben Buie Complex is used to cover the Ben Buie Gabbro (Eucrite) of the earlier part of this chapter, and, along with this great mass, the attendant small intrusions of allivalite and porphyrite, and also the banded granulitic marginal assemblage developed near Loch Fuaran. The word gabbro as applied to the bulk of the Ben Buie Complex is generally replaced in the sequel by the narrower term eucrite, since the Ben Buie Gabbro, unlike the Corra-bheinn and Beinn Bheag Gabbros, is predominantly of eucritic character. The allivalite-masses, it will be remembered, are supposed to be of older date than the eucrite that envelops them ; in this case, the same general relations hold at Ben Buie as Dr. Harker found in Skye and Rum, where ultrabasic plutonic rocks have been succeeded by others of less basic, but allied character. The Craig Porphyrite is clearly later than the eucrite ; but, as pointed out already, it should be regarded, according to its field-relations, as part of the same complex.

The chemistry of the allivalite-eucrite assemblage has already been discussed (p. 21) in relation to analyses from Ben Buie and Rum.

The Ben Buie Complex has escaped pneumatolysis to such an extent that it retains a large proportion of its olivine in a fresh state. This is in keeping with the observations that the two most massive Early Basic Cone-Sheets also contain fresh olivine (p. 239).

*Allivalite.* Several small masses (p. 245), by reason of the dominating presence of olivine and anorthite (Fig. 39A), may best be designated anorthite-peridotites or allivalites (15630, 15624, 18452). As was noted by Dr. Harker,[1] the rocks of this group show considerable variation in the relative proportions of their essential constituents, but olivine usually makes up at least half of their bulk. In some cases, the almost complete suppression of felspar (15629), and the concentration of the usual chromiferous spinellid, produce a rock that approximates to dunite; while, in the other direction, an increase in the proportion, and a slight drop in the basicity, of the felspar (16499), taken in conjunction with the presence of diallagic augite and a rhombic pyroxene (2645, 15625), indicate a leaning towards the eucrites of which the Ben Buie complex is mainly composed.

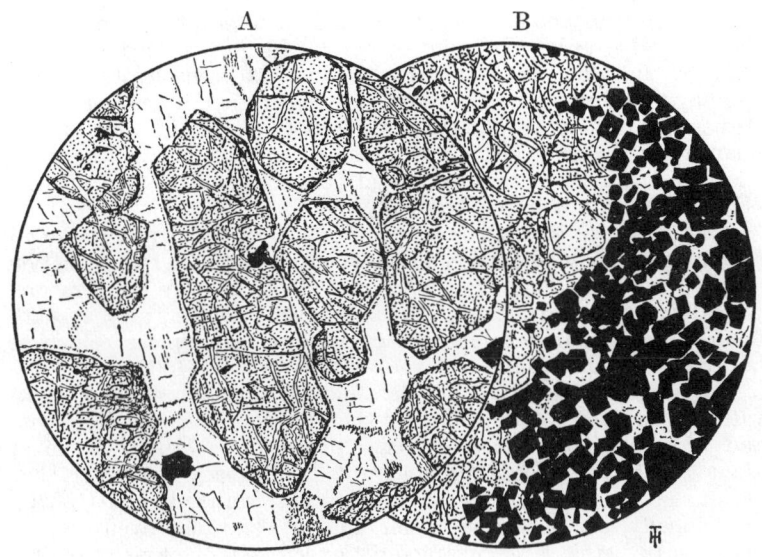

Fig. 39.

A. [18452]×20. Allivalite of the Ben Buie Complex. Hypidiomorphic olivine in a matrix of basic plagioclase felspar that approximates to anorthite in composition.

B. [16531]×20. Band of picotite in an olivine-rich allivalite of the Ben Buie Complex.

Olivine in these rocks is usually in an unaltered condition, and occurs chiefly as rounded to sub-idiomorphic crystals that often attain large dimensions. The crystals are characteristically colourless in thin section, but are locally rendered almost opaque by inclusion of dust-like particles and dendritic growths of magnetite. In its structural relations, the olivine commonly occurs in poecilitic fashion imbedded in large plate-like masses of felspar. As was pointed out by Dr. Harker, in those members of the allivalite group,

---

[1] A. Harker, 'Geology of the Small Isles of Inverness-shire,' *Mem. Geol. Surv.*, 1908, p. 89.

which are richest in olivine, this mineral is the first to separate from the magma, but as the proportion of olivine decreases we find it moulded upon, and often enclosing, an earlier-formed felspar (2645).

The felspar of these ultra-basic rocks is optically negative and has extinction angles that prove it to be near anorthite in composition. It is always twinned according to both carlsbad and albite laws, but pericline lamellation is not such a common feature as in the more normal gabbroic rocks subsequently described. Most frequently, it is unzoned, but occasionally, more especially when augite is present in appreciable quantity, it is zoned with an optically positive felspar of basic labradorite composition.

The constant presence of a chromiferous spinellid as an accessory is an essential character of this group of rocks. Generally, it builds well-shaped crystals of fair size which appear coffee-brown by transmitted light, and are referable to picotite (15631, 15624). But in some instances, it is opaque, and then we may assume that the mineral is either chromite or chrome-magnetite (16499).

This spinellid is an early product of consolidation, but, seemingly, did not always precede the separation of the other constituents. It may be observed enclosed in olivine in very limited quantity, but shows a distinct tendency to segregate in the more felspathic portions of the rock; and veins of anorthite and picotite (Fig. 39B) are not infrequent (15631).

The rocks as a whole are remarkable for their freshness, and seldom show any alteration other than a partial serpentinization of their olivine along cracks; some slight rise in temperature, probably on the intrusion of the subsequent eucrite, may account for the occasional development of fibrous and almost colourless hornblende at the margin of the olivine-crystals (15624).

As has been stated above, the genetic connexion of these rocks with the eucrites is established by the occasional presence of augite and of a plagioclase of less basic composition. Augite is never abundant in the rocks of undoubted ultrabasic character, but occurs in small quantity in association with olivine (15629). Occasionally, it appears in larger amount, and may even enclose, poecilitically, the large characteristic crystals of olivine (16499). A rhombic pyroxene occurs most sparingly, and is, as far as can be judged, an enstatite or non-pleochroic variety (15625).

*Eucrite* (Anal. I.; Table V., p. 23). Passing now to the more prevalent, and presumably later, type of intrusion, we find that, although it presents much variety in texture, and in the relative abundance of constituents, there is a fairly constant mineral-assemblage. It is characterized, usually, by an abundance of olivine, and by the presence of an ophitic diallagic or normal augite, and of a felspar near bytownite in composition (Fig. 40). The consistent basicity of the dominant felspar, as well as other characters described below, enables us to place the greater part of the Ben Buie complex with the eucrites rather than with the normal gabbros.

The olivine presents the same general characters as in the peridotitic rocks described above. It is usually an abundant constituent, and many of the eucritic rocks approximate to the allivalites by reason of the prevalence of this mineral and the basicity of their felspar (18451, 18453). With an increase in the amount of augite (16711), there is a corresponding decrease in the amount of olivine, and in some varieties of the gabbro it may fail completely (16720, 17903, 18454). It is usually an early product of crystallization, and occurs poecilitically enclosed by either felspar or augite (16715, 17386). Rarely, it shows schiller structure (17386). Sometimes a narrow reaction-border separates olivine from the enveloping felspar (17388), and, as is usual in such cases, consists of an aggregate of enstatite and colourless augite with a little actinolitic hornblende. Some of the olivine is distinctly yellow in thin sections, and entirely free from all striations other than those due the traces of the prismatic cleavages (16710).

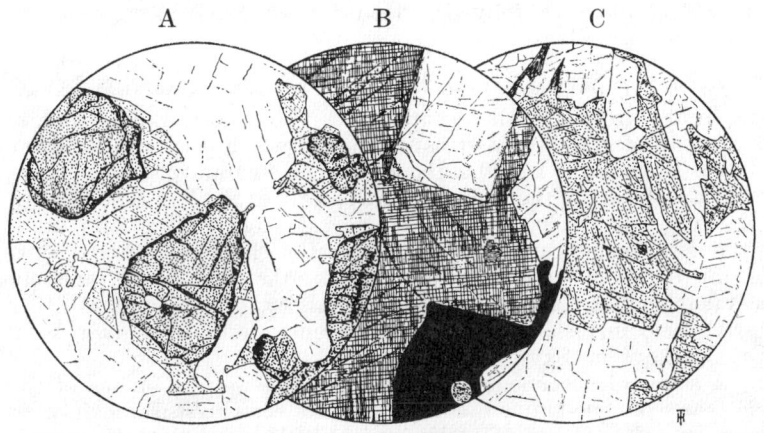

FIG. 40.—The Ben Buie Eucrite and its Varieties.

A. [16711]×15. Basic variety showing large crystals of olivine in association with ophitic augite. The colourless component is a zoned felspar with the average composition of bytownite.

B [17903]×15. Gabbro. Coarse olivine-free rock composed of diallagic augite, basic labradorite and iron-ore.

C. [16720]. Olivine-free coarse ophitic dolerite. The augite in this case is devoid of schiller-structures.

A rhombic pyroxene occurs but rarely; in some examples, it appears to be a colourless enstatite, but in others it has the pleochroism of hypersthene (16713). It is occasionally associated with a little fox-red biotite. Uusally, it is either intergrown with augite or occurs in close association with olivine.

The felspar is always more basic than normal labradorite, and most frequently ranges in composition from basic labradorite to near anorthite. It is generally of later consolidation than olivine, which it poecilitically encloses, but occasionally part is of an early generation and may occur within the olivine or with olivine moulded upon it (17386). It is usually twinned according to both carlsbad and albite laws, and quite often pericline twinning is a striking feature (17316). Zoning is much more frequent than in the felspars of the allivalitic rocks, and it is common to find an optically negative felspar (bytownite-anorthite) surrounded, with perfect transition, by a zone of optically positive material (labradorite-bytownite) (16710, 16711).

Accessory minerals are comparatively rare in these rocks, apart from scattered patches and crystals of magnetite, and, in the more basic rocks, a chromiferous spinellid. Apatite is of infrequent occurrence, but when present builds moderately large crystals (15623).

Before leaving the subject of the normal Ben Buie Gabbro, a word may be said concerning the banding which locally is a feature of the mass. This banding is due in part to textural, and in part to compositional, variation (15626, 17205). The observed structures are in every way comparable to those presented by the banded gabbros of Skye, so fully described by Sir A. Geikie, Sir Jethro Teall, and Dr. Harker.[1]

*Banded Granulites, Loch Fuaran.* Sharply separated petrographically from all other components of the Ben Buie complex, is a group of granulitic and banded rocks which occurs to the west of Ben Buie, in the neighbourhood of Loch Fuaran (p. 245). These rocks are shown on the one-inch Map as part of the Ben Buie Gabbro (Eucrite), but their appearance suggests recrystallization; and, although in some cases they reproduce the characters of that type of micro-gabbro known as beerbachite, they are most probably metamorphosed masses of an earlier gabbro, or remnants of other early basic intrusions. Similar occurrences were encountered by Dr. Harker[2] in Skye.

A series of specimens collected from the western shores of the loch are of fine-grained compact rocks that consist mainly of granules of strongly pleochroic hypersthene, granulitic augite, and small elongated crystals of labradorite, is a granulitic matrix of clear labradorite and magnetite (16864, 16736). In some bands the ferro-magnesian minerals may fail altogether, and magnetite become extremely abundant in a granulitic matrix of recrystallized labradorite (16799). The pyroxene, either hypersthene or augite, may combine, on occasion, an ophitic with a granulitic structure (16729).

As in Skye, the true nature of such rocks is a matter of some uncertainty; but, from a variety of characters that they present, it is clear that they have resulted from the recrystallization of earlier basic rocks. A somewhat coarser type of granulitic rock that may be termed a granulitic hypersthene-gabbro (16722), consists mainly of hypersthene in ragged to ophitic crystals, small fresh olivines, and labradorite. Evidence of partial fusion of an earlier rock is furnished by minute inclusions of hypersthene in the felspar arranged in straight and curving lines indicative of flow.

Again, an augite-granulite (16716), composed for the most part of equigranular augite, labradorite, and magnetite, has clear areas of recrystallized labradorite that probably represent the larger

---

[1] A. Geikie and J. J. H. Teall, 'On the Banded Structure of some Tertiary Gabbros in the Isle of Skye,' *Quart. Journ. Geol. Soc.*, vol. l., 1894, pp. 650. 654; A. Harker, 'Tertiary Igneous Rocks of Skye,' *Mem. Geol. Surv.*, 1904, pp. 117-120.
[2] A. Harker, *op. cit.*, pp. 115, 116.

felspathic constituents of the original rock. Also, a band of more augitic character running through the mass suggests that the rock was in all probability a banded gabbro.

From Garbh Shlios, north of Loch Fuaran, a rather coarser rock has been collected from just outside the margin assigned on the one-inch Map to the Ben Buie Gabbro. The specimen may be styled a granulitic gabbro, and is mainly composed of augite, hypersthene, and recrystallized labradorite (16521). A similar rock has been collected, away from the Loch Fuaran district, from just within the south-eastern margin of the Ben Buie Gabbro, north-west of Loch Uisg (17387). In both cases, fresh olivine is enclosed in a thoroughly granulitic matrix.

*Granulite-Inclusions.* Turning to granulitic rocks that can definitely be proved from their field-relations to be inclusions of earlier rocks, we find that a dolerite-sheet to the west-south-west of Sròn Dubh (17902) has been completely granulitized to a mass of augite, rhombic pyroxene, and olivine (decomposed), in a recrystallized mosaic of labradorite. A similar rock, in its more coarsely ophitic parts (16520), has given rise to recrystallized labradorite, ophitic hypersthene and augite, and a little biotite, more particularly in association with iron-ore. In its finer portions (16519), it has furnished what may be termed a granulitic gabbro composed of the same minerals. The rhombic pyroxene is occasionally pseudomorphed by hornblende. The type of granulitization and the mineral assemblage are similar to what is met with in basaltic lavas metamorphosed by the Corra-bheinn Gabbro (p. 155).

*Segregation-Veins.* The Ben Buie Gabbro, in common with most gabbros of similar type, is sometimes traversed by narrow anastamosing segregation-veins of lighter colour. South-west of Craig, such veins are well-developed, and are seen to be granophyre, sometimes almost pegmatitic in character (15628) and of extremely acid nature. They consist of plates of perthitic orthoclase, enclosing irregular patches of quartz in optical continuity with each other, and also of smaller rectangular perthitic crystals surrounded by a deep and coarsening fringe of granophyric material. In other cases, pale veins (15627), that behave similarly, appear, when sectioned, to retain the normal doleritic or gabbroic character of the surrounding rock, and to differ only in an intense albitization of the labradorite-felspar.

*Craig Porphyrite.* This type of rock, (p. 245), which clearly must be regarded as part of the Ben Buie complex on account of the field-relations, is distinctly abnormal in character, and presents peculiarities that have resulted from the absorption of gabbro-material by an invading magma of less basic composition. Many features of this rock are paralleled in the hybrid zones that result from the interaction of granophyres and gabbros (Chapter XXXIII.). For the most part, as developed on Sròn Dubh, it is a fine-grained rock of dark colour, characterized by a micro-porphyritic development of its ferromagnesian and more basic felspathic constituents (Fig. 41). It has a variable texture, and exhibits considerable differences in both mineral and chemical composition. In its most persistent development (16525), the porphyritic constituents are mainly augite, a rhombic pyroxene, and basic plagioclase, with less frequent olivine. The augite is a pale variety, and occurs in small sharply idiomorphic crystals; while the rhombic pyroxene and olivine (17880) are always in the form of well-shaped pseudormorphs. Labradorite builds individuals of all sizes, sometimes larger than, but usually of similar dimensions to, those of the other porphyritic constituents. Frequently, the felspars have the appearance of being more or less foreign to the matrix in which they lie (17881). Magnetite is fairly abundant as large irregular crystals and grains, and is scattered

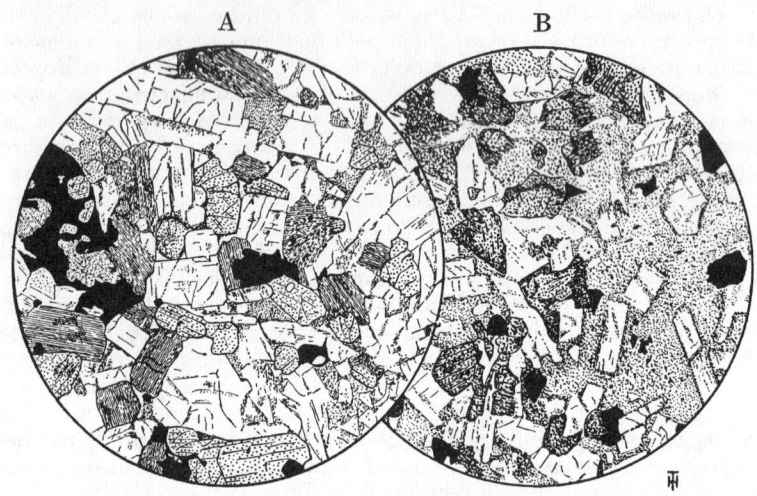

FIG. 41.—Craig Porphyrite.

A. [16525] × 17. Porphyrite, showing the normal development of porphyritic crystals of rhombic pyroxene, augite, and labradorite, with accessory magnetite, in a subordinate fine grained felspathic matrix.

B. [16523] × 17. Porphyrite of variable type, showing clots or relatively basic material similar to that figured in A, with a greatly increased amount of fine-grained acid matrix that has the characters of a soda-granophyre.

indiscriminately throughout the rock, but more particularly through the matrix. Occasionally, it occurs in intimate association with augite and olivine. The rhombic pyroxene is usually pseudomorphed by fibrous serpentine and hornblende, and the olivine by serpentine. The internal structure of the respective pseudomorphs is quite distinct. The plagioclase phenocrysts are commonly albitized; strings of albite traverse them in all directions, epidote has separated out, and chlorite occupies veins and patches within them.

The matrix is a turbid mass of felspathic matter into the composition of which alkali-felspar enters to a considerate extent. Generally speaking, the amount of matrix is not large (16525), but it varies in quantity (17881), and, in some instances, (16523), the rock becomes patchy in appearance, due to the separation of more basic clots by acid material that has the characters of soda-granophyre.

There can be little doubt that we are dealing here with a rock that has formed from a magma charged with more or less completely digested basic material. Most of its constituents, with the exception of certain of its felspars, have separated from the melt, and give the rock a somewhat delusive normal appearance.

### CORRA-BHEINN GABBRO.

The olivine-gabbro of Corra-bheinn, as developed to the west of Cruachan Dearg and in Glen Clachaig, is a moderately coarse ophitic rock, less basic in general than the eucrites of Ben Buie. It is composed of augite, olivine, and labradorite, with subordinate iron-ore. Olivine is usually present, but most often as pseudomorphs of irregular form composed of serpentine, fibrous hornblende, or both (14307). The pyroxene is a brownish-yellowish augite without schiller structures, and is usually in coarse ophitic relationship to the felspar (16501, 17961, 18450). Occasionally, it may assume a hypidomorphic habit, and is then frequently clustered with crystals of labradorite in glomero-porphyritic groups (14307). The felspar is always strongly zoned labradorite, and usually forms moderately large mutually interfering crystals. The optical sign is positive, and other characters indicate that the labradorite is of more or less normal composition, and seldom, if ever, becomes as basic as bytownite. Iron-ore occurs with fair regularity as patches and crystals of magnetite (14307, 18450); and apatite, when present, occurs as sparsely distributed large crystals (16501).

Here, again, a more basic magma seems to have preceded the introduction of the main mass, but the phenomenon is on a smaller scale, or at any rate is less evident, than in the case of the Ben Buie Complex. We find in the Corra-bheinn Gabbro small basic patches (16500) similar to the eucrite of Ben Buie. These patches contain large crystals of fresh olivine poecilitically enclosed in crystals of ophitic augite. The felspar, in its larger individuals, is anorthite zoned with bytownite, but there are smaller unzoned crystals, frequently enveloped by olivine, which are presumably anorthite. A little biotite occurs in association with the olivine.

About a quarter of a mile to the north-east of Corra-bheinn, what appears to be the Corra-bheinn Gabbro, sliced by Cone-Sheets, has the character of a quartz-gabbro (17961), and is indistinguishable from the later gabbros that form the more basic portions of the Glen More and similar Ring-Dykes (Chapter XXIX).

Felspathic veins (16541), traversing the gabbro and presumably segregated from it, are composed essentially of interlocking rectangular crystals of turbid alkali-felspar with areas of secondary iron-epidote and quartz. The felspars are mostly microperthitic or crypto-perthitic in character, but others are distinctly microgranophyric. Microgranophyric matter often fringes the perthitic crystals, and fills interspaces between the larger individuals. Dr. Clough has also recorded acid veins of this type (16528) immediately beyond the limits of the gabbro and cut off by the latter.

The older dolerites, where in contact with, and invaded by, the gabbro, have been granulitized in a manner similar to those involved in the Ben Buie Gabbro. They have developed a marked granulitic structure with the production of rhombic pyroxene, and frequently contain large areas of recrystallized felspar (16528).

### BEINN BHEAG GABBRO.

The mass, which occupies the slopes of Beinn Bheag to the east of Beinn Talaidh and stretches down into Glen Forsa (Fig. 37, p. 234), is extremely variable in character. It may be separated, but with indefinite division, into a coarse black-and-white gabbro, a fine-textured rock of similar composition, and a dark-grey fine-grained gabbro.

The typical coarse rock (18657), that occurs south-east of the summit of Beinn Bheag and along the western flanks of Allt nan Clàr, is a coarsely ophitic olivine-gabbro composed of occasional large serpentinized olivines, a yellowish brown augite without schiller structures, zoned basic labradorite, and large patches of magnetite—partly primary, partly secondary, in origin.

The finer black-and-white gabbro (18661-3), to the east of Glen Forsa, is also olivine-bearing, and petrographically resembles the coarser rock described above. The augite, however, while retaining its coarsely ophitic structure in a modified degree, has a tendency to assume idiomorphic outlines. In general character, this gabbro is comparable to the normal olivine-gabbro of Corra-bheinn.

We will now consider the series of more acid rocks that occurs intimately associated with the coarse gabbro on either side of the summit of Beinn Bheag (Fig. 37). The most striking feature of these rocks, when microscopically examined, is their granophyric (18660) or felsitic matrix, containing abundant free quartz, and strongly charged with apatite (18659). Among the crystalline constituents enveloped by this matrix, are large elongated crystals of augite (18660), associated with magnetite, and exhibiting salitic striation (18659). They are usually replaced, either in whole or in part, by green hornblende. Ophitic masses of augite, in association with intensely albitized labradorite, are of local occurrence. In these masses, the labradorite is frequently zoned with oligoclase (18659), but there is the clearest evidence that the basic felspars were unstable in the acid matrix, and were being attacked and eaten up. Further, it is evident that the matrix was to a considerable extent responsible for the intense albitization of the basic plagioclase, for where this mineral has been protected by augite it is unchanged, but where it lies in contact with the matrix considerable albitization has been effected. From appearances in the field, it is probable that these rocks have resulted from the interaction of the coarse black-and-white gabbro, with an acid differentiate provided by some hidden portion of the intrusion.                        H.H.T.

# CHAPTER XXIII.

## SHEETS EXCLUSIVE OF CONE-SHEETS: SOUTH-WEST MULL.

## INTRODUCTION AND LOCH SCRIDAIN.

### INTRODUCTION.

THERE is in south-western Mull (Fig. 42) a great field of sheet-intrusion, where individual sheets show no marked tendency to dip inwards towards the plutonic centre of the island, but often rather the reverse. The inclination of many of the sheets is markedly capricious, but in the majority of cases it is at quite moderate angles; and, since the lavas of the district are also gently inclined, it is natural to class the sheets as sills. It should be realized, however, that frankly transgressive relationships are everywhere the rule, and that several of the steeper sheets are only arbitrarily distinguished from dykes. Under these circumstances, the lowly-inclined habit, commonly adopted by the sheets, does not have any very close connexion with the bedding of the Tertiary lavas or Mesozoic sediments; and indeed, several lowly inclined sheets are met with traversing highly inclined gneiss in the Gribun and Ross of Mull peninsulas.     E.M.A., J.E R., (B.L.).

On the one-inch Map, the sills considered in this chapter are lettered D, where basic or intermediate, and F where acid. There are certain early sill-like masses of olivine-dolerite, on Ben More, characterized by an irregular laccolithic tendency, which, combined with their relatively early date, separates them conveniently from the normal sheets of their immediate neighbourhood. They have been lettered eD on the one-inch Map, and are treated in Chapter XI. A few bostonite-sills (lettered bO) also occur in south-western Mull, but they are considered along with allied alkali-types in Chapter XIV.

In Fig. 42, it has been found possible to show many sill-occurrences which are omitted from the one-inch Map. It will be understood, however, that, except along streams and the sea-shore, no attempt has been made to record the evidence completely, even on the six-inch field-maps. The boundaries of the south-western field of intrusion are best characterized on the north across Loch na Keal, and on the west in the Ross of Mull. Eastwards, the failing of the sheets is somewhat blurred by an abundance of other intrusions; and, southwards, the sea intervenes.

The petrology and field-relations of the sheets of the whole field are sufficiently harmonious to support the view that, with subordinate exceptions, the intrusions belong to a single complex—if this expression be employed in the same broad sense as when

# Chapter XXIII.—Sills of South-West Mull.[1]

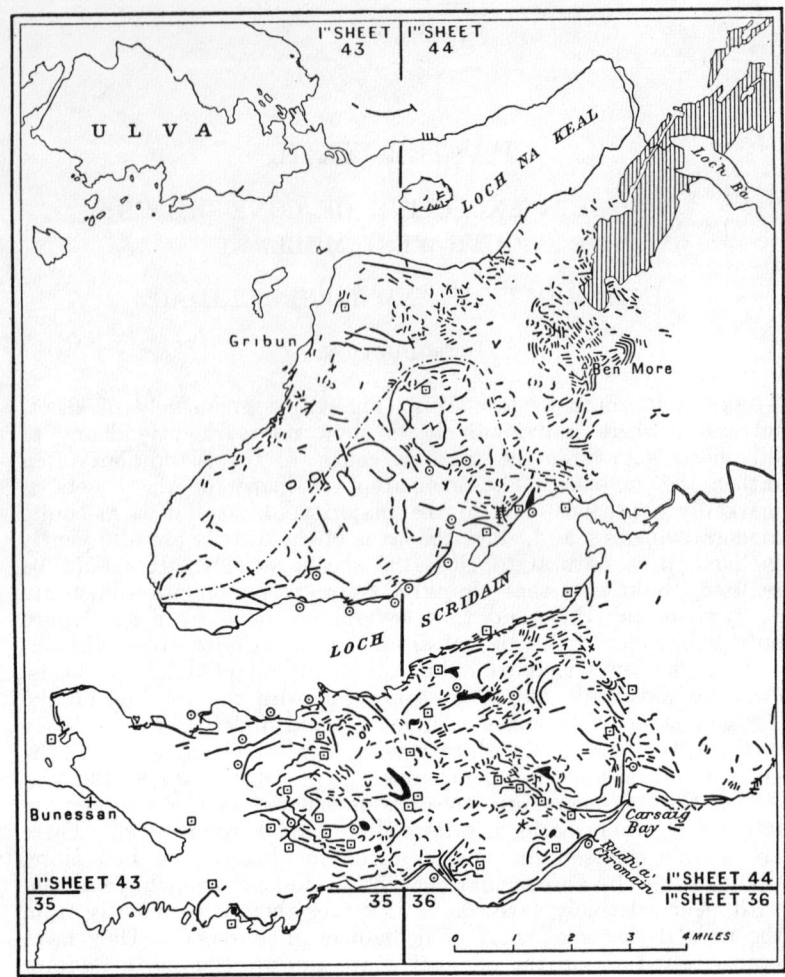

▦ Knock and Beinn a' Ghràig Granophyres.

∼ Sills and Sheets (other than Cone-Sheets).

–·–·– Northern boundary of Loch Scridain district, characterized by pitch-stones and xenoliths.

○ 'Accidental' xenoliths containing sapphire.

▫ Other 'accidental' xenoliths.

Fig. 42.

Map of South-West Mull, showing distribution of Sills and Sheets other than Cone-Sheets.

Introduction.     259

we speak of the Central Lavas, or the Late Basic Cone-Sheets of Mull. There are sufficient local differences, however, to render a subdivision of the field desirable. In Fig. 42, a line is drawn marking the northern border of the Loch Scridain district characterized by an abundant development of pitchstone, and also by an unusual profusion of accidental xenoliths. Probably, within this line, a bulk-analysis of the sills would show an andesitic composition, with rather more than 55% silica. Outside this line, there are two districts, the one named after Gribun, the other after Ben More. The dividing-line between these two coincides roughly with the junction of Sheet 43 and 44 of the one-inch Map, except that the Gribun district extends east into Sheet 44 to include Brimishgan. A bulk-analysis in both the northern districts would probably show a tholeiitic composition, with less than 55% silica. At the same time, the Gribun district is much more obviously connected with the Loch Scridain district than either of them is with Ben More.

Such little evidence as is available suggests that the sills of south-western Mull fall within the long period of activity characterized by the intermittent injection of cone-sheets. It is argued later (pp. 279, 290) that perhaps the most probable date is in the latter part of the cone-sheet cycle, when tholeiitic magma was more particularly available and was very prone to yield acid differentiates.

The remainder of the present chapter will be devoted to the field-relations of the sills of the Loch Scridain district, with particular attention paid to the prevalence of pitchstone. In Chapter XXIV., the highly characteristic xenolith-assemblage of the Loch Scridain sills is considered in some detail. In Chapter XXV., a full account of the igneous petrology of the sills is given. Chapter XXVI. passes on to the remainder of the south-western field included in the districts of Gribun and Ben More.

E.B.B.

## Sills and Sheets of Loch Scridain District (Sheets 35, 43, 44); Field-Relations.

The sheets of the Loch Scridain district south of the broken line Fig. 42 are most of them gently inclined, with a tendency to dip southward or south-eastward. This dip, however, is far from constant in direction; and many of the sheets are steep.

As will be explained in Chapter XXV., there is a wide petrological range among the sheets of the district; they include olivine-basalts and dolerites of various types, and also a suite of tholeiites, andesites (leidleite and inninmorite), and felsites. Most of the sheets are either tholeiite or andesite. The two can frequently be distinguished from one another in the field by a tendency of the tholeiites to be of greater thickness and coarser crystallization: they are often from 15 to 30 ft. thick; whereas the andesites are mostly thinner than 15 ft. Among the andesite-sheets between 5 and 6 ft. thick, many have glassy centres (pitchstone).[1]

[1] For Mr Bailey's view that, in a large number of cases, tholeiite and andesite are combined in composite-sheets, see pp. 264, 286.

260  Chapter XXIII.—Field-Relations of Loch Scridain Sills.

No wholly tholeiitic sheet is known to have a glassy centre. On the other hand it is possible, though it is not quite demonstrated, that such exceptional sheets of the district as have glassy margins may be tholeiite throughout. So far as observation goes, glassy centres and margins appear to be mutually exclusive phenomena; and they will be considered separately in the sequel under the two headings Pitchstone and Tachylyte Margins.   E.M.A.

On very rare occasions, some of the sheets fail to chill at their margins; and, in one case, conspicuous melting of marginal country-rock has been recorded. Both these phenomena are dealt with, presently, under the heading Chilled Margins—Sometimes Absent.   E.B.B.

A very general feature of the tholeiite-andesite assemblage of Loch Scridain is a tendency to carry 'cognate' and 'accidental' xenoliths (Chapter XXIV.). A few of the assemblage are obviously composite, with markedly acid interiors. A good example of a sheet which is both xenolithic and composite is described at the close of the present chapter under the title of the Rudh' a' Chromain Sill.

PITCHSTONE.

An account has been given in Chapter II. of the development of our knowledge regarding the occurrence of pitchstone among the Loch Scridain sheets. Figs. 43 and 44 have been borrowed from the *Quarterly Journal* of the Geological Society of London, Vol. LXXI., to illustrate the account given below. Since Fig. 43 was drawn, several additional occurrences of pitchstone have been mapped, including convenient exposures along Loch Scridain west of Glen Seilisdeir; but these additions are sufficiently indicated by the letters pD on the one-inch Map, Sheets 43 and 44, while the numbers on Fig. 43 serve as a useful basis for the notes which follow and cover all analysed material.

Most of the pitchstones are non-porphyritic andesites of the type defined as glassy leidleite (p. 281); a few are porphyritic andesites distinguished in the field by small phenocrysts of felspar and augite, and are called glassy inninmorites (p. 282); a few, again, are of rhyolitic composition.

In any particular sill, pitchstone is generally associated with more crystalline rocks, to which the descriptive term stony is applicable. The two types are often interbanded in layers parallel to the margins of the intrusion; and, within individual layers, there is frequently an intermingling according to the sheath-and-core arrangement described by Sir Archibald Geikie and others from occurrences elsewhere in Scotland. Of these two relationships, the interbanding will be considered first.

Although the transition from a glassy to a stony band in these Loch Scridain sills is, characteristically, abrupt, there is never any resemblance to a chilled contact. A typical non-porphyritic sill may contain a central 5 ft. of pitchstone, with 3 ft. above, and an equal thickness below, of finely crystallized rock, in some cases andesite, in others tholeiite. The stony margins may be thinner in proportion, and in one instance were observed to fail. In most cases, the pitchstone forms a single fairly regular layer in the centre of the sheet, but sometimes it is split by stony partings. Thus, in an exposure slightly north of east of Tiroran (Sheet 44), there are at least two bands of pitchstone, and farther west, on the east promontory of Slochd Bay (Sheet 43), parallel interbanding of glassy and stony layers is again well seen.

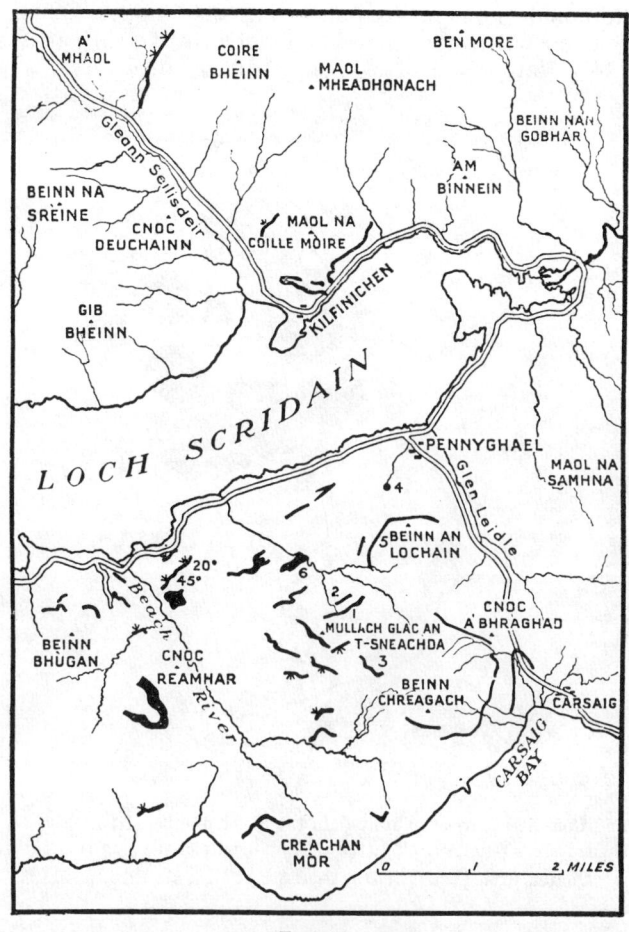

Fig. 43.
Map of some of the occurrences of Pitchstone in Loch Scridain district. Quoted from *Quart. Journ. Geol. Soc.*, vol. lxxi., 1916, p. 206.

Among the inninmorite-pitchstones, stony partings seem to be more common than is the case with their non-porphyritic fellows, and Nos. 5 and 6 (Fig. 43) show rapid alternations of glass and stone. No. 4, on the other hand, is inninmorite-pitchstone throughout.

Sheath-and-core structure is very frequently met with among the pitchstone-sills of Loch Scridain, whether leidleite or inninmorite. Fig. 44 illustrates a typical example, and may be briefly

## Chapter XXIII.—Field-Relations of Loch Scridain Sills.

described as follows :—The stony base and top of the intrusion appear to throw off arms, or, more accurately, narrow sheets of a similar material, which traverse the glassy portion in a branching and sinuous manner, completely dividing it into irregular rounded cores, which are not in visible connexion. In many cases the sheaths of crystalline matter, varying from a quarter of an inch to 3 inches in thickness, show a median joint or suture-line, along which terminate numerous minute transverse joints.

A point of particular interest was noted in the sill partly shown in Fig. 44. Here the median sutures can, in some cases, be traced

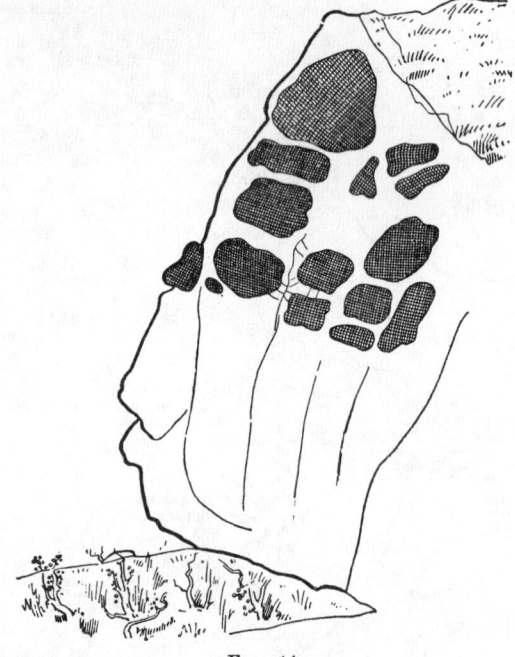

Fig. 44.

Sheath-and-Core Structure as exhibited in the Sheet numbered 1 in Fig. 43, p. 261. The shaded areas represent pitchstone. The height of the crag is about 5 feet.

Quoted from *Quart. Journ. Geol. Soc.*, vol. lxxi, 1916, p. 211.

downwards, and form a continuation of main joint-planes, which cut at right angles across the stony base of the sheet. The impression produced on an observer is that the transition from glassy to stony material has been brought about by the escape by some volatile material—in the first place from the stony margins, and, secondarily, along joint-planes developed in an already consolidated, though still in large measure glassy, magma. There is general agreement that this first impression is entirely correct in regard to the stony crystallization of the sheaths. Accordingly, we shall consider this side of the subject first.

There are probably only two alternative interpretations of

sheath-and-core structure available :—(1) the sheaths might be regarded as later intrusions, or segregations, surrounding and separating earlier-formed glass, or (2) the sheaths are derived from the glass by modification *in situ*. We think that the field-appearance decides definitely in favour of modification *in situ* ;[1] not infrequently, indeed, one can detect a ghostly reminder of sheath-and-core structure in a sill that has become wholly stony. Moreover, where the sheath is narrow, it is possible to slice across it (15998) and see with a pocket lens or a microscope that the limits of devitrification are not rock-boundaries, but are traversed in every direction by individual crystals.

TABLE XI.—WATER OF AUGITE-ANDESITES.

|  | Ia | Ib | IIa | IIb | IIIa | IIIb | IV | V |  |
| --- | --- | --- | --- | --- | --- | --- | --- | --- | --- |
| $SiO_2$ | 61·69 | 59·21 |  |  |  |  | 62·37 | 64·13 | $SiO_2$ |
| $H_2O+105°$ | 2·36 | 1·54 | 2·38 | 1·56 | 2·44 | 0·93 | 5·54 | 2·71 | $H_2O+105°$ |
| $H_2O$ at 105° | 0·25 | 2·05 | 0·45 | 1·34 | 0·38 | 1·64 | 0·44 | 0·36 | $H_2O$ at 105° |
| Cl | 0·02 | nt. fd. |  |  |  |  |  |  | Cl |
| Spec. grav. | 2·64 | 2·61 | 2·82 | 2·77 | 2·89 | 2·71 | 2·50 | 2·57 |  |

Analyses by E. G. Radley, quoted from E. M. Anderson and E. G. Radley, *Quart. Journ. Geol. Soc.*, vol. lxxi., p. 212.

Ia, Ib... (15996 and 15997; Lab. Nos. 384, 385). Glassy (*a*) and stony (*b*) portions of LEIDLEITE sill, ¼ mile N.N.E. of cairn on Mullach Glac an t'Sneachda. (Locality 1, Fig. 43; complete Anals. III. and II., Table III., p. 19).

IIa, IIb.. (17243 and 17244; Lab. Nos. 417, 418). Glassy (*a*) and stony (*b*) portions of LEIDLEITE sill, 730 yds. N.N.E. of cairn on Mullach Glac an t'Sneachda. (Locality 2, Fig. 43.)

IIIa, IIIb. (17245 and 17246; Lab. Nos. 419, 420). Glassy (*a*) and stony (*b*) portions of LEIDLEITE sill, 1000 yds. S.E. of cairn on Mullach Glac an t'Sneachda. (Locality 3, Fig. 43.)

IV. .. (15989; Lab. No. 286). Extremely glassy INNINMORITE sheet, near head of stream from Tòm a' Choilich. (Locality 4, Fig. 43; complete Anal. IV., Table III. p. 19.)

V. .... (15990; Lab. No. 387). Fairly crystalline INNINMORITE sheet with glassy base, 3/16 mile S.W. of Trig. Station on Beinn an Lochain. (Locality 5, Fig. 43; complete Anal. V. Table III., p. 19.)

The next point of importance, to which attention may be directed, is that the analyses of pitchstones by Mr. Radley, quoted in Table XI., show a notable content of water retained by the rock when heated up to 105°C. Probably this water is in a condition of molecular diffusion, or solution, in the glass. Its amount is particularly striking in IV. (15989), an inninmorite-pitchstone in which crystallization has proceeded to a much smaller extent than in the case of any of the other rocks analysed

[1] The same conclusion is expressed by Sir Archibald Geikie in regard to the Eskdale Dyke of Southern Scotland ; 'Ancient Volcanoes of Great Britain,' 1897, Vol. II., p. 134.

During crystallization of the magma, this dissolved water must of necessity be set at liberty. It may, in part, be expelled bodily from the rock, or it may be segregated into cavities or cracks. In either case, it is easy to see that jointing of a hot, though solid, glass might facilitate the removal of dissolved water, and thus open up the way to devitrification. One is, of course, faced with the difficulty of deciding whether, or no, the devitrification of the sheaths actually occurred while the glass was cooling (primary devitrification), or whether it took place long afterwards (secondary devitrification).[1] No quite positive answer can be given, but it is very doubtful whether any one familiar with sheath-and-core structure in the Loch Scridain field-exposures would hesitate in regarding it as primary; moreover, the microscopic appearances of the devitrified base of stony leidleites seems to indicate a natural step towards the indisputably igneous crystallization of the craignurites (described in Chapter XIX.).

If now we accept the hypothesis, that sheath-and-core structure is determined by devitrification of a hot glass sufficiently solid to admit of the formation of joints, the question naturally arises whether the stony marginal layers of so many of the sills of the district are not, in a sense, sheaths which have succeeded in ridding themselves of water owing to their external position. It is probable that this has been an important factor in many cases; and, in the opinion of the present writer, the facts, in so far as they have been investigated, do not warrant any further deduction. Mr. Bailey, however, believes that another factor can be recognized, and his argument is as follows. E.M.A.

It is common experience that andesite and rhyolitic magmas, under superficial, or comparable, conditions of cooling, are much more prone, than basalt, to glassy consolidation. In the petrological sequel, it is pointed out that, in several cases, the marginal portions of the Loch Scridain sills are more basic than their interiors, and this in itself would favour the occurrence of a central glassy layer flanked by stone.[2]

It would, however, be easy to exaggerate the importance of the basicity-factor. Not only does sheath-and-core structure demonstrate the possibility of one and the same magma presenting two different consolidation-facies side by side, but also it is established that some of the thoroughly glassy sills are andesite, while others that are stony are felsite. The analysed inninmorite from Tòm a' Choilich (15989), with its 62 per cent. $SiO_2$ (Anal. IV., Table XI.), is almost a pure glass with a very small proportion of crystals, mostly phenocrysts; whereas, the felsite (18464) has almost 71 per cent. $SiO_2$ (Anal. I., Table IV., p. 20), and yet is devitrified to stone. The difference would be less, it is true, if the two rocks were compared after dehydration, but it would still remain worthy of attention—at any rate, it would remain greater than the difference between the stony marginal and glassy central portions of the sill, Loc. 1 (Anals. Ia. and Ib., Table XI.). E.B.B.

[1] Cf. T. G. Bonney, Presidential Address, *Quart. Journ. Geol. Soc.*, vol. xli., 1885, p. 37.
[2] Mr. Anderson does not regard the composite character of the generality of the sills with stony margins and glassy interiors as established (p. 287).

## TACHYLYTE-MARGINS.

Glass-selvages in Mull have been, in the main, recorded from the North-West basaltic dykes, and will be further referred to in Chapter XXXIV. dealing with these. Of cases which fall under the present heading, the best known is perhaps that, recorded by the late Duke of Argyll, and fully described by Professor Cole (p. 47), bordering a roughly horizontal sill intruded into the assemblage which contains the leaf-beds at Ardtun. The tachylyte here forms an upper and a lower border seldom exceeding an inch in thickness, and, as in the other two cases to be recorded, is markedly sperulitic. From Professor Cole's analysis of the tachylyte, which contains 53 per cent. of silica, and the microscopic description of the remainder of the rock, there can be no doubt that the sill is tholeiitic. E.M.A.

A very interesting case has been described by Professor Heddle (p. 49) from the margins of an inclined sheet, which reaches the coast of the Gribun peninsula near Dearg Sgeir, as indicated by a note on the one-inch Map (Sheet 43). Inland, this sheet generally dips south at from 45° to 60°, but on the coast, where Professor Heddle examined it, it assumes an irregular habit, and figures as a little complex of connected sheets and dykes. At several points in the somewhat extensive coastal exposure, there is a conspicuous selvage of glass, which varies somewhat rapidly and capriciously in thickness. The maximum seen is about 2 ft., where glass constitutes the whole of a vein. Professor Heddle, in his text and illustration, shows that locally " the tachylyte has manifestly been dragged forward, *in a condition of plasticity, if not of fluidity*, while the flow of the more central portions was still continuing." E.B.B.

A hitherto unrecorded example occurs about 500 yds. to the west of Tiroran, where the lower border of a sill, dipping gently to the north, can be seen to be formed by about an inch of tachylyte, (18527). E.M.A.

## CHILLED MARGINS—SOMETIMES ABSENT.

When it is stated that the great majority of the sills of the Loch Scridain district have stony marginal portions, even where their interiors, as often happens, are glassy, it must not be thought that their actual margins are unchilled. The almost invariable rule is that, towards the contact with country-rock, the grain of crystallization of the sill is very markedly reduced. The result is an extremely compact layer, though seldom with any appreciable vitreous lustre (*see* previous section).

While conspicuous marginal chilling is almost invariably shown by the Loch Scridain sills, there are a few noteworthy exceptions. An easily located sill reaches westwards from Scobull Point on the north shore of Loch Scridain to the 104 ft. cairn at Slochd (Sheet 43). It is a composite sill, with tholeiite-margins and a leidleite-interior (20799), which latter often retains glassy cores. Like so many of the sills of the district it is highly xenolithic (20800), with ' accidental ' xenoliths ranging up to 6 ft. in diameter. It is intruded into vesicular basalts of Plateau Type. Both the top and the bottom of the sill show variable relations to the adjacent lava, sometimes chilling, and sometimes continuing their normal grade of crystallization (finely-doleritic) without modification right up to the point of contact. A series of micro-slides were taken to investigate the matter: (20798) shows the Plateau Basalt lava above the sill; (20798a and b) show unchilled tholeiite in contact with Plateau Basalt, and invading and apparently carrrying off metamorphosed amygdales, so that there can be little doubt that an appreciable solution of the lava has occurred; (20801) shows Plateau Basalt lava below the sill; and (20801b) shows merging of this lava into unchilled tholeiite of the sill.

What is probably another example of the same kind is exhibited farther west along the north shore of Loch Scridain at Port na Croise. In this case, the unchilled upper surface of an inninmorite-porphyrite (20796) is overlain by a rock which is almost certainly referable to the Staffa Type of columnar basalt-lavas (20797).

Dr. Clough seems to have found the same occasional failure to chill in examples examined by him on the south side of Loch Scridain (Sheet 43). He has noted on his field-map that a xenolithic sill with sheath-and-core structure, exposed along the shore of Port Mòr, is "sometimes distinctly chilled at the base, but apparently not always." Again he remarks, in regard to the base of a xenolithic sill north-east of Eilean Bàn, at the mouth of Loch na Làthaich, that the "junction is sharp but chilling doubtful."

The local absence of marginal chilling may be connected with high temperature of intrusion, or with some compositional peculiarity which tended to delay consolidation. It is worthy of note that a sill belonging to the Loch Scridain assemblage, and exposed at Tràigh Bhàn na Sgurra (spelt Traigh Bhan Sgoir in the Memoir on Sheet 35), has melted the pelitic gneiss, into which it has been intruded. A careful description has already been published by Messrs. Clough and Cunningham Craig in the Geological Survey Memoir on Sheet 35, and their account is completed by petrological notes by Dr. Flett. It is only necessary to state that the melting of the pelitic gneiss has proceeded in one extreme case to a thickness of 4 or 5 ft.; and that the banding of the softened pelitic gneiss, in another instance, has been sharply deflected in sympathy with the flow-movement of the sill. Intrusion of melted gneiss (buchite) is also recorded as a minor phenomenon. The molten gneiss, Dr. Flett has shown, is cordierite-sillimanite-buchite [1] with green spinel. The sill is one of the common tholeiitic assemblage, and is characterized by many xenoliths both 'cognate' and 'accidental.'   E.B.B.

### RUDH' A' CHROMAIN SILL.

Reference has been made, on more than one occasion, to the possibly frequent, though generally inconspicuous, composite character of the sheets of the Loch Scridain district. Very often, these sheets are also xenolithic. A list of composite sheets represented in the Survey collection is given, p. 286, and of xenolithic sheets, p. 272. The Rudh' a' Chromain Sill is taken at this

[1] *See* footnote, p. 268.

juncture as a very well-defined, clearly exposed, and easily located composite sill with strongly developed xenolithic character. It must be understoood that, while the composite character of this selected example is easily recognizable in the field, this is not true of the majority of the sheets of the district.

Rudh' a' Chromain is situated on the south coast of Mull (Sheet 44) just below the Nuns' Pass, west of Carsaig Bay, Fig. 42. Fig. 45 illustrates the shore-section as drawn by Mr. D. Tait. From above downwards, the section is roughly as follows :—

Upper tholeiite-band with chilled top. Below the compact surface-layer, come 4-6 ft. with abundant 'cognate' gabroid xenoliths; and, below this again, 2½-5 ft. densely crowded with aluminous xenoliths of all shapes, ranging from an an inch to 4 ft. in length.

Central felsite 20-30 ft., pale grey, slightly porphyritic, and locally glassy, with sheath-and-core structure. There is no sign of chilling on approach to the tholeiite-bands above and below : but, though there is a slight increase of basicity for a foot or two, shown by a darkening of colour, the contacts are well-defined. A few xenoliths occur, mainly, perhaps, in the less acid layers where they consist

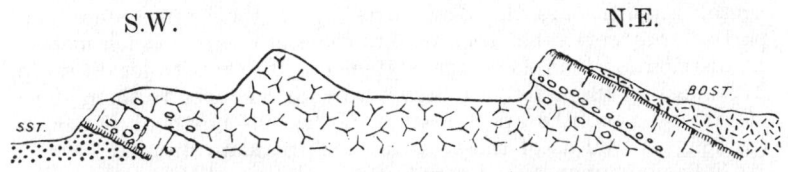

FIG. 45.

Section at Rudh' a' Chromain across xenolithic composite sheet, showing external chilled margins against sandstone (SST.) and bostonite (BOST.).

Quoted with minor alterations from *Quart. Journ. Geol. Soc.*, vol. lxxviii, 1922, p. 234.

of sandstone, and range up to 6 inches in length. The more central xenoliths include shale as well as sandstone, and are larger, sometimes attaining several feet in diameter and averaging 2 ft.

Lower tholeiite-band 2-5 ft., characterized particularly by the size and abundance of its 'cognate' xenoliths and the relative rarity of 'accidental' xenoliths. It shows a well marked chilled base.

The petrology of these layers is referred to later (p. 286). Here it may be pointed out that the marginal tholeiite is represented by Analysis IX., Table II., p. 17, while the acid interior is identical in type with the analysed specimen (I, Table IV., p. 20) from another sill of the district.

Although the Rudh' a' Chromain sill shows compact chilled margins, it has thermally altered the Carsaig Sandstone for a few millimetres from the contact. A melt has been developed between the sand-grains, and, on subsequent cooling, tridymite has deposited as fringes about the undissolved quartz-remnants. The tridymite has now reverted to quartz in optical continuity with the quartz of the original grains, but it still retains its characteristic crystal-form.

H H.T.

# CHAPTER XXIV.

## SHEETS EXCLUSIVE OF CONE-SHEETS: SOUTH-WEST MULL.

### LOCH SCRIDAIN XENOLITHS.[1]

#### INTRODUCTION.

The detailed account of the composite and xenolithic sill of Rudh' a' Chromain, given at the end of the preceding chapter, serves as a convenient introduction to the subject of the xenoliths of the Loch Scridain district in general. The xenoliths are referable to two distinct classes, 'cognate' and 'accidental', and these will be dealt with separately in the sequel. The accidental xenoliths are the more arresting in their appearance, and particular attention has been paid to them in regard to such matters as distribution; in fact, the statements of xenolith-localities (p. 272, and Fig. 42, p. 258) are drawn up with reference to them alone. 'Cognate' xenoliths are a feature of the sills north of Ben More (p. 289), as well as of those of the Loch Scridain district.

Following upon the descriptions of the xenoliths, a discussion is offered as to the c o n d i t i o n, p o s i t i o n, a n d d a t e o f t h e L o c h S c r i d a i n M a g m a - R e s e r v o i r.

'COGNATE' XENOLITHS (ENCLAVES HOMOGÈNES OF LACROIX).

The 'cognate' xenoliths, referred to above as occurring in the tholeiite bands of the Rudh' a' Chromain sill, and more particularly in the lower band, are dark coarsely crystalline glomeroporphyritic patches, clearly marked off from the fine-textured tholeiite that envelops them. They consist commonly of bytownite and hypersthene (16598), and more rarely of bytownite and green augite (16612a, 17173); in either case, the pyroxene tends to be idomorphic towards the felspar. Olivine seems to be represented by occasional pseudomorphs (17174).

Similar 'cognate' xenoliths have been recognized not infrequently in other sills of the district; and, in fact, were first noticed by Mr. Cunningham Craig, in the Tràigh Bhàn na Sgurra sill mentioned on p. 266. They are fairly abundant in the lower part of this sill, where they reach a foot or more in diameter. Dr. Flett, in the Memoir on Sheet 35, says that they consist of bytownite associated, sometimes with green augite, sometimes with enstatite or bronzite.

A few other instances of tholeiitic and andesitic sills with similar 'cognate' xenoliths are as follows (all in Sheet 44):—

[1] The 'sillimanite' of the Mull xenoliths has been investigated by N. L. Bowen, J. W. Greig, and E. G. Zies: 'Mullite, a silicate of alumina,' *Journ. Wash. Acad. Sci.*, vol. xiv., 1924, p. 183. They find that it agrees physically and chemically with the 'sillimanite' of artificial products. Its formula is $3Al_2O_3 2SiO_2$, whereas that of true sillimanite is $Al_2O_3 SiO_2$. They have named it mullite, with Mull as type-locality. Their paper was received after this memoir had gone to press.

A sill cutting the Gamhnach Mhòr Syenite in the Rudh' a' Chromain neighbourhood (14597).

A xenolithic rock half way between Rudh' a' Chromain and Loch Scridain, near the mouth of a tributary to Abhuinn nan Tòrr from Mullach Glac an t-Sneachda. The matrix of this rock (of which the margins are not exposed) largely consists of fused argillaceous material (buchite); the 'cognate' xenoliths are isolated, and often exist as broken, crystals of hypersthene and augite (16072).

A sill with sheath-and-core structure, and abundant 'accidental' xenoliths, on the north shore of Loch Scridain, south of Seabank Villa. Some of the augite-crystals are almost two inches long (18530).

A sill with sheath-and-core structure in its central part, practically at the northern limit of the Loch Scridain district (Fig. 42, p. 258), north of Coir' a' Charrain (17121).

## 'ACCIDENTAL' XENOLITHS (ENCLAVES ÉNALLOGÈNES OF LACROIX.)

A full account of this interesting subject is not attempted here, since it would entail too lengthy an elaboration on the mineralogical and physical side, coupled with constant reference to literature which has only an indirect bearing upon Mull geology. At the same time, a statement is offered of the main theoretical conclusions, the grounds for which are given in greater detail elsewhere.[1] Special attention is paid to such aspects of the problem as may be of interest to those who wish to investigate the field-evidence.

### MATRIX.

The sills which carry the 'accidental' xenoliths belong to the tholeiite-andesite (and occasional felsite) suite of Chapter XXV., and in most cases the andesitic rock is leidleite. A large proportion of them retain cores of glass in their more central parts, and a few of them, like the Rudh' a' Chromain sill, are strikingly composite. As stated above (p. 267), the 'accidental' xenoliths of the Rudh' a' Chromain sill are concentrated for the most part in the tholeiitic margins; though some are also found in the acid interior. Experience of other xenolithic sills, where a composite character is recognized or suspected, supports the view that this relative concentration of xenoliths into the more basic, upper and lower, parts (not including the actual chilled selvages) is a rule that is often observed; but further field-enquiry is desirable.

The matrix, in which the xenoliths are imbedded, is generally normal igneous rock with very little modification. A striking exception is afforded by an exposure already alluded to on account of its 'cognate' xenoliths (Locality 29, p. 273), an exposure which may also be remembered as having supplied Mr. Anderson (p. 52) with the first sapphire-bearing material sent in for determination (16072-3). The matrix here is a dark grey amygdaloidal rock of fine texture, and of igneous aspect, as seen in the hand, except that it is studded with small lustrous plates of sapphire. Under the microscope, its igneous appearance all but vanishes, for it is found to be a sillimanite-buchite with little referable to an igneous

[1] H. H. Thomas, 'Certain Xenolithic Tertiary Minor Intrusions in the Island of Mull (Argyllshire),' *Quart. Journ. Geol. Soc.*, vol. lxxviii., 1922, p. 229.

source, except xenocrysts of hypersthene and augite. Its apparent porphyritic felspars are broken bytownite and anorthite, derived, as explained in the sequel, from interaction between aluminous xenolithic material and tholeiitic magma. That the buchite has acted as an intrusion is indicated by the fact that it is packed with aluminous and siliceous xenoliths of all sizes up to 6 ft. long. Its vesicles, now filled with zeolites and other low-temperature minerals, are another indication of its fluidity. Similar amygdaloidal developments are common in buchite-xenoliths throughout the Loch Scridain district; and, also, in the marginal buchite at Tràigh Bhàn na Sgurra, where Messrs. Clough and Cunningham Craig recognized intrusive contacts on a small scale (p. 266).

### FIELD-CHARACTERS.

The main field-characteristics of the 'accidental' xenoliths are :—

(1) Frequency and size; xenoliths occasionally attain to a length of six feet or more.

(2) Diverse nature.

(3) In many cases, intense alteration; marginal interaction between aluminous xenoliths and igneous magma has tended to develop conspicuous crystal-growths until checked by the rapid cooling brought about by sill-injection.

H.H.T.

The xenoliths which have been identified may be grouped as follows :—

(a) Micaceous gneiss or granulite, and probably quartzite, belonging to the regionally metamorphosed pre-Devonian floor.

(b) Granite and pegmatite. These are possibly fragments of some granitic intrusion of Old Red Sandstone age. The nearest visible rocks of this type are in the Ross of Mull granite-area, but the origin of these fragments has not been specially studied.

(c) Sedimentary rocks later than the period of regional metamorphism.

(d) Basaltic lava.

Of these four classes, the types included under (c) are the most numerous, and have received most attention. They may be subdivided into original sandstones, shales (in a broad sense), and carbonaceous rocks. The last may have been, in many cases, bituminous shales, and, in some cases, coals.[1] The sandstone-xenoliths are usually the largest, often ranging up to several feet in diameter, while the intrusion, numbered (41) in the list given below (p. 273), contains a mass interpreted as altered sandstone, many yards in length.

As to the character of the xenoliths carried by the various sills, it is noteworthy that the degree of metamorphism is not always the same. Sandstone-xenoliths from Locality 43 appear but little altered, while cases of complete fusion (17997) have been encountered at Locality 5. Again, in different sills, although all aluminous xenoliths are probably in the form of buchite, there is variation in the amount and coarseness of the crystalline material (corundum-

[1] Dr. Heddle, *Mineralogy of Scotland*, vol. i., p. 1, gives an analysis of graphite found by Earl Compton (? Locality 27, p. 273) with 83·56 per cent. carbon and 14·93 per cent. ash.

anorthite-spinel) that we know to have been formed by the interaction of aluminous sediment and magma. For instance, at Locality 13, there is much less reaction-zone material, in conjunction with buchite, than in the case of xenoliths from Rudh' a' Chromain. Further, sapphires, which are large and abundant in the latter locality, are, in the former, small and relatively rare. A condition of relatively slight metamorphism is also characteristic of xenoliths from Localities 28 and 39. It is thus seen that the degree of alteration of the contained xenoliths is to some extent peculiar to the sill in which they occur.   E.M.A.

Reference has been made to the reaction-zone which often characterizes the outer portions of the aluminous xenoliths. This zone is composed of anorthite, sapphire, and spinel, with sillimanite enclosed in the anorthite. It is easily recognized in field-exposures, where the felspar is particularly conspicuous owing to its bulk, and the sapphire in small crystals attracts attention on account of its blue colour. The sillimanite cannot be seen individually, but often tints the enclosing felspar a pink or rosy hue—for instance at Rudh' a' Chromain, and, in Sheet 43, north-east of Eilean Bàn and west of Port Mòr. This phenomenon is all the more interesting as pink sillimanite is a great rarity in other parts of the world.

## Distinctive Aluminous Xenoliths.

The most typical occurrences of the argillaceous rocks, referred to above as shale in a broad sense, have had their origin in some sediment with the composition of a fireclay, except that, if Anal. XVI., Table IX. (p. 34), is representative, the clay would appear to have been abnormally rich in alkali, with soda predominating over potash. That they have come from one common source is indicated by their uniform alteration to sillimanite-buchite, with a conspicuous development of sapphire, especially in their marginal crystal-zones where modified by igneous permeation. They differ in their metamorphic facies from the buchites and hornfelses developed from the common pelitic schists and gneisses of the district: these, owing to their greater content of $SiO_2$ and $MgO$, generally yield abundant cordierite when heated by granite, basalt, or camptonite, or indeed when melted artificially. Dr. Flett[1] has described a very beautiful cordierite-buchite developed marginally to a camptonite-dyke in Ardmucknish (Sheet 45), and Dr. Pollard's analysis of the phyllite, which by melting has furnished this buchite, shows 2·04 per cent. MgO. Probably, the only near approach to the composition of the aluminous xenoliths, afforded by any argillaceous rock of Mull known *in situ*, is furnished by the thin Tertiary basal mudstone (Chapter III.). Very occasionally (p. 59), this mudstone is of a pale colour showing a poverty in iron. The analysed mudstone (Anal. XV., Table IX., p. 34) is of a dark tint corresponding with a high iron-content; and this, it must be admitted, is very much more typical of the deposit as a whole.   H.H.T.

---

[1] J. S. Flett, *in* 'The Geology of the Country near Oban and Dalmally,' *Mem. Geol. Surv.*, 1908, pp. 129-132.

# Chapter XXIV.—Xenoliths of Loch Scridain Sills.

## LOCALITIES.

The following is a list of localities for sills with 'accidental' xenoliths within the Loch Scridain district. The data have been supplied mainly by Mr. Anderson and Dr. Clough, as indicated in Chapter II., where the general progress of the discovery is sketched. Occurrences of sapphire and graphite are specially noted in this list, since these minerals are of great interest in themselves, and also because their wide distribution will be used presently in discussing the history of the magma that has yielded the Loch Scridain sills. It should be understood that most of the exposures detailed below belong to distinct individual sills (Fig. 42, p. 158).

### Inland: North of Loch Scridain (Sheet 43).

(1) North-South sill, Culliemore (? sapphire).

### Inland: North of Loch Scridain (Sheet 44).

(2) Coir' a' Charrain (graphitic shale).
(3) 2/3 mile N.N.W. of Allt a' Mhuchaidh Bridge and a little below sill shown on 1-inch Map (sapphire).
(4) Abhuinn Bail' a' Mhuilinn, 100-200 yds. above bridge (sapphire, 17993-5 20280).
(5) Allt a' Mhuchaidh, 1000 yds. above bridge (sapphire, 17996-9, 18529 20281).
(6) 100 yds. N.E. of Killiemore House.
(7) 200 yds. E. of cairn on Maol na Coille Mòire (this sill is inninmorite).
(8) Sill parallel with coast, ½ m. S.E. of same cairn.

### North Coast of Loch Scridain (Sheet 43).

(9) ¼ m. E.S.E. of Tavool House (sapphire).
(10) Slochd eastwards to Scobull Point (sapphire 20798-20801).

### North Coast of Loch Scridain (Sheet 44).

(11) S.W. of Tiroran House (sapphire, 18531-2).
(12) S. of Seabank Villa (sapphire, 18528, 18530).
(13) W. of Allt na Coille Mòire (sapphire, graphite, 17266-8, 17402-5).
(14) E. of Allt nam Fiadh (sill not shown on 1-inch Map, 17279).

### South Coast of Loch Scridain (Sheet 43).

(15) N.E. of Eilean Bàn (18007, 18024, 18029).
(16) Tòrr Mòr (sapphire, 18168-9).
(17) W. of Port Mòr (sapphire, 18001-5, 20276-9).

### South Coast of Loch Scridain (Sheet 44).

(18) ¼ m. E.S.E. of Eilean an Fheòir (sill not shown on 1-inch Map).
(19) N. of Kinloch House (sill not shown on 1-inch Map).

### Inland: S. of L. Scridain Northern Half of Ross of Mull (Sheet 43).

(20) Sill turning S. at Capull Corrach, 800 yds. W.N.W. from Loch a' Charraigein (sapphire, graphite, 18025).
(21) W.S.W. sill, running from Coillenangabhar to Capull Corrach (sapphire, graphite, 18006, 18027).
(22) N. of coast-road, W. of Allt Chaomhain (sapphire).
(23) 900 yds. W.S.W. of cairn on Beinn Bhùgan (sill not shown on 1-inch Map).
(24) Allt Chaomhain W. of Beinn Bhùgan (20763).

# Field-Characters of 'Accidental' Xenoliths. 273

*Inland: S. of L. Scridain, Northern Half of Ross of Mull (Sheet 44).*

(25) Near mapped sill in Beach River, N. of Cnoc Reamhar (16075).
(26) ½ m. S.E. of Goirtein Driseaeh (16067).
(27) Waterfall S.E. of Torrans (? graphite).
(28) 1 mile N.W. of Mullach Glac an t-Sneachda (16065, 16071).
(29) Near mouth of tributary to Abhuinn nan Tòrr from Mullach Glac an t-Sneachda (sapphire; this exposure shows a buchitic matrix, 16072-3, 20272-5).
(30) Junction of Uisge Fealasgaig with Leidle River (14595).

*Inland: S. of L. Scridain, Southern Half of Ross of Mull (Sheet 43).*

(31) Scoor House, S. of old chapel, on borders of Sheets 35 and 43.
(32) Beside road N. of east corner of Loch Assapol.
(33) Stream junction, 1300 yds. W. of cairn on Cruachan Mìn, and (perhaps same sill) 1500 yds. S.S.W. of same cairn.
(34) S.W. sill, 800 yds. S.S.W. of cairn on Cruachan Mìn (sapphire 20762, 20755), and (perhaps same sill) 700 yds. S. of same cairn (20761).
(35) In stream, 700 yds. W.S.W. of cairn on Cruachan Mìn.
(36) E. and W. sill, ½ mile S.S.E. of cairn on Cruachan Mìn.
(37) 500 yds. S.W. of cairn (1119 ft.) on Beinn Chreagach.
(38) 100 yds. N.E. of same cairn (sapphire).
(39) Big sill, at 500-600 ft. level, on margin of Sheets 43 and 44 (13845).

*Inland: S. of L. Scridain, Southern Half of Ross of Mull (Sheet 44).*

(40) E.W. sill, crossing path 800 yds. N.N.W. Àiridh Mhic Cribhain.
(41) Intrusion, crossing stream 500 yds. N.N.E. of Binnein Ghorrie.
(42) Two sills, E. of Airidh Fraoch, one traced ¾ mile E.S.E. to stream W. of Beinn Chreagach.
(43) In stream 1000 yds. W. by N. of cairn (1235 ft.) on Beinn Chreagach (16070).
(44) Above 600 ft. contour, ¼ mile, S.W. of same cairn (sill not shown on 1-inch Map).
(45) Above stream-head between Cnocan Buidhe and Dùnan na Marcachd.
(46) Streamlet, N. of Plantation, W. of River, 200 yds. N.W. of Feorlin Cottage, Carsaig (sapphire, 18492-5). Also old road 100 yds. N.W. of Feorlin Cottage (20283-9). Neither exposure shown on 1-inch Map.
(47) 700 yds. E. of Loch na Géige (lettered F on 1-inch Map).

*South Coast of Mull (Sheet 35).*

(48) Tràigh Bhàn na Sgurra (13839-41, 14388-90, 14,444 ; for full description see Memoir on Sheet 35).

*South Coast of Mull (Sheet 44).*

(49) East shore of Tràigh Cadh' an Easa' (this sill is an inninmorite, not shown on 1-inch Map, sapphire 16595).
(50) Rudh' a' Chromain and Nuns' Pass (sapphire, 14893-6, 16596-16612, 17170-17178, 18480-18491, 20271).

A visitor wishing to familiarize himself with the field-evidence is recommended to the coastal exposures on the north shore of Loch Scridain (Sheets 43 and 44), the south shore of Loch Scridain (Sheet 43), Tràigh Bhàn na Sgurra (Sheet 35), and Rudh' a' Chromain (Sheet 44). These are thoroughly representative, and easy, both to find, and to approach.   E.B.B.

## PETROLOGY.

The microscopic petrology of the 'accidental' xenoliths is summarized in the following pages. As already stated, the reader can find a fuller treatment in *Quart. Journ. Geol. Soc.*, vol. lxxvii., where the subject-matter is illustrated by five plates from microphotographs.

### Siliceous Xenoliths.

The siliceous xenoliths, as might be expected, show relative fusion of felspathic constituents, involving the marginal solution of quartz-grains and the production of glass (20763). On cooling, there has been an almost universal formation of tridymite around the relict grains of undissolved quartz (20283). The tridymite-fringes are recognized by their identity of form with what is commonly met with in furnace-linings; the tridymite itself no longer exists as such, but has reverted to quartz in optical continuity with the quartz-grains which have served as nuclei for its growth. Fringes of augite, or rhombic pyroxene, occur round the quartz-grains, where the interstitial melt has been of somewhat basic composition to begin with, or has been modified in this direction by magmatic transfusion. Cordierite has sometimes been developed to a small extent in contaminated interstitial melts (16067).

### Aluminous Xenoliths.[1]

(ANALS XVI. AND XVII.; TABLE IX., p. 34).

It has been pointed out that the characteristic *aluminous xenoliths* of the district started as shale, or mudstone, poor in lime, magnesia, and iron. The first step in their metamorphism has been the production of sillimanite-buchite (18005), a compact grey-blue or lilac-coloured rock consisting of glass crowded with minute felted or parallel needles of sillimanite from 1 to 2 mm. long. The sillimanite is peculiar on account of its pink pleochroism. It exists, partly as a residuum left during the vitrification of the clay, and partly as a precipitation during the cooling of the aluminous melt in which it lies embedded. Sapphire (blue corundum) is a common associate of the sillimanite of the buchites, and occurs as small isolated tabular crystals of deep-blue colour and brilliant lustre. These represent the excess of alumina over that required for the formation of sillimanite and melt, and are to be regarded as the earliest solid phase to separate out from the aluminous melt. They rarely form 1 per cent. of the buchite, and usually less than 0.5.

The general fate of the sillimanite of the buchite has been to serve at later stages of its history as a source of alumina for the production of other aluminous minerals—anorthite, spinel, sapphire, and cordierite. Very occasionally, it appears to have concentrated its resources, and to have developed relatively large rose-pink crystals, presumably as a result of long-continued heating (18001a). The sapphires, as a rule, during subsequent changes, seem to have added to their number at the expense of the sillimanite.

In general, all the Mull sapphires are of the same blue colour and tabular habit; and in exceptional cases they range up to $1\frac{1}{2}$ cm. across.

[1] See footnote, p. 268.

Anal. XVI., Table IX. (p. 34), may be taken as representing such a sillimanite-buchite as has been described above.

The further history of the buchite is very interesting. Obviously, its contact with the tholeiite-magma was marked by instability. So long as the sedimentary and igneous melts lay side by side, they must have tended to mix by diffusion; and diffusion would continue (given the requisite temperature conditions) until uniformity of composition was attained—except in so far as some mineral might be permanently precipitated, insoluble even in the final combined melt. In the case under investigation, cooling has

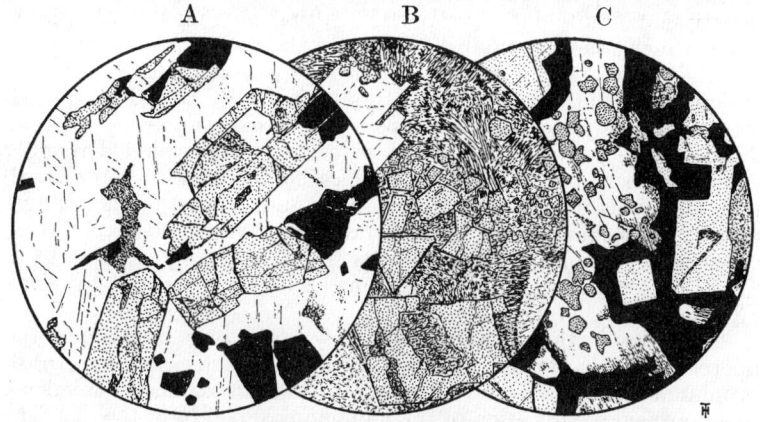

FIG. 46.—Aluminous Xenoliths.

A. [16612] × 15. Large idiomorphic crystals of sapphire with associated greenish-brown spinel, enclosed by coarsely crystalline anorthite, with a little residual brown glass.

B. [18493] × 15. Idiomorphic crystals of rose-pink spinel, and a large crystal of cordierite in a matrix of oligoclase felspar, an opaque spinellid and residual glass.

C. [18001A] × 15. The section shows a large irregular crystal of cordierite (centre) studded with brown spinel, also idiomorphic highly pleochroic rose-pink sillimanite (right and left) enclosed in a semi-opaque glassy matrix.

checked the process of diffusion long before uniformity was attained; and one can thus read a most interesting story of selective diffusion and temporary precipitation.

In the first place, as regards selective diffusion, it is common to find the sillimanite-buchite xenoliths extremely modified in their external parts through the selective immigration of lime, and, to a less extent, of magnesia and iron, from the adjoining tholeiite-magma. The outer portions of the xenoliths have thus developed crystalline aggregations composed for the most part of anorthite, accompanied by sapphire and spinel: the former more abundant towards the buchite, and the latter towards the tholeiite. These crystalline aggregates represent modified sediment, rather than modified magma, since their anorthite is crowded with sillimanite needles (16603, 17177-8, 17405, 20271)—there has indeed been a partial resorption of sillimanite accompanying the marginal modification of the buchite, but much has still not been incorporated.

The phenomenon of selective diffusion from magma to xenolith is evident, under the microscope, from the relative concentration of anorthite in these modified borders. Naturally, such a phenomenon is extremely difficult to follow up by chemical analysis ; for each minute layer, while liquid, had a composition peculiar to itself, and was undergoing continual modification by interchange with its neighbour-layers on either side. Moreover, once crystallization set in, it would lead to a local concentration of certain constituents, quite out of proportion to that obtaining in the liquid from which their crystallization commenced.

The only point, which emerges from comparison of the chemically modified xenolith (Anal. XVII., Table IX., p. 34) with the merely thermally modified part of another xenolith (Anal. XVI.), is that, while the alumina percentage has remained roughly constant, there has been a fall in silica combined with a rise in all non-aluminous bases, other than soda ; and, further, such changes cannot be accounted for by any process of unselective admixture of original xenolithic material (as represented by Anal. XVI., Table IX.) with original tholeiitic magma (as represented by Anal. IX., Table II., p. 17).

A little further detail may now be given regarding the microscopic characters of the highly characteristic anorthite-sapphire-spinel aggregates. This assemblage is extremely prevalent, as a marginal modification of the aluminous xenoliths of the district ; and it is evidently the normal result of modification of the xenolithic material by the magma at a certain stage of its career. Sometimes, it occurs completely enclosing a kernel of buchite to a depth of several inches. Wherever found in connection with other mineral developments, it retains its proper position next to the buchite. It need not, however, occur on all sides of a xenolith. Cases are common where the anorthite-sapphire-spinel development is met with on one side, while a different later development is found on the other. In such cases, one is probably often dealing with xenoliths that became detached from the lining of the magma-reservoir at a comparatively late stage, or else with early xenoliths subsequently broken.

The anorthite-sapphire-spinel aggregates are almost holocrystalline. The anorthite occurs in large crystals reaching a length of several inches ; they mutually interfere with one another's growth, except towards the buchite, where they are separated to some extent by glass and are more noticeably idiomorphic, or even skeletal. As already stated, the anorthite encloses much vestigial sillimanite. The sapphire becomes more abundant, as the buchite, the source of alumina, is approached. Its association with the anorthite is of a type that points conclusively to crystallization from a common solution of these two substances, in some cases under eutectic conditions. The spinel is of a deep-green variety (hercynite-pleonaste, Anal. XVIII., Table IX., p. 34). It occurs more especially towards the tholeiite-margin of the aggregates, and this is in keeping with its iron-magnesia content. Its manner of association with anorthite indicates contemporaneous crystal-growth with a gradual convergence towards a eutectic composition of melt.

Altogether, the modification of the external parts of the xenoliths points to a preferential immigration of lime, and, less rapidly, of iron and magnesia, from the adjoining tholeiite-magma. The resultant melt dissolved some portion of the sillimanite-felt, and, with very gradual fall of temperature, precipitated anorthite, sapphire, and spinel, more or less simultaneously.

For a long time, transfusion from either side of the zone of crystallization seems to have maintained the requisite composition for an approximate sapphire-anorthite-spinel eutectic condition. It must be remembered, however, that this crystallization was proceeding in one particular liquid of a whole series of liquids, all in unstable contact one with another. Thus, from the point of view of the complex as a whole, the crystallization of the border-zone of the xenoliths was premature.

Outside these xenoliths, one must suppose the tholeiite-magma itself in a modified condition, presumably enriched in almost every constituent, including magnesia, as compared with lime. Many micro-sections illustrate the interaction of this modified tholeiite, both with the modified xenoliths—the anorthite-sapphire-spinel association—, and also with the sillimanite-buchite. There has often been disruption of the coarsely crystalline border-zone, and resorption of the anorthite and its contained sillimanite. This has been followed by an additional crystallization, on the remaining anorthite, of an increasingly acid plagioclase that ranges through labradorite to oligoclase. Such later-formed felspar is free from included sillimanite, and the excess of alumina thus furnished has separated either as sapphire or spinel (16601, 17998a, 18492). The latest crystallization of felspar is usually oligoclase, either as an outer zone to pre-existing more basic felspars, or as skeletal growths (16601, 17177-8, 18493, 20271). The accompanying spinel may be black, brown, or plum-coloured.

Where the modified magnesia-rich tholeiite has come in contact with the sillimanite-buchite, it has, to some extent, permeated it and, to some extent, dissolved it. In the former case, large crystals of cordierite have grown into the buchite, keeping pace with the diffusion of magnesia. Much of the sillimanite of the buchite is resorbed, but a considerable proportion generally remains undissolved in the usual form of slender needles; these may often be seen passing across the boundary of a cordierite-crystal into the buchite beyond (18532).

Where the magma has dissolved, rather than permeated, the sillimanite-buchite, a special type of cordierite-buchite (17997), or cordierite-sillimanite-buchite (18529), results. The cordierite is colourless and free from pleochroism, and builds single rectangular crystals or small simple and complex twins.

The formation of cordierite in, or about, the xenoliths appears in all cases to belong to a comparatively late date. The proportion of cordierite to spinel is determined by the relative amounts of magnesia and silica available; absence of spinel indicates an excess of silica.

Where sapphire, spinel, and cordierite have crystallized from a contaminated igneous melt, the sapphire and spinel are sometimes enclosed by clear anorthite (17999), and the spinel by cordierite (17405).

This is as far as the story of the modification of the xenoliths takes us. Apparently, sill-intrusion was now initiated, and the rapid cooling that ensued brought the uncompleted assimilation to a close. To the intrusion-period we may reasonably refer the deformation, which characterizes many of the zenoliths, and also the frequent development of vesicles in the buchitic glass.

## Condition, Position, and Date of Loch Scridain Magma-Reservoir.

In a physical discussion of the evidence outlined above, the clusion is reached (*Quart. Journ. Geol. Soc.*, vol. lxxviii., pp. 250-4) that the metamorphism of the xenoliths was initiated at a temperature of about 1400° C., and was continued through a prolonged period of slow-cooling until about 1250° C.; and that, after this, rapid cooling intervened as a result of sill-intrusion of the magma. It is to the later part of the interval of slow cooling that we may reasonably assign the development of the 'cognate' xenoliths (p. 268). H.H.T.

The field-distribution of the 'accidental' xenoliths is certainly in keeping with the idea developed by Dr. Thomas that these xenoliths were mostly derived from the lining of a single magma-reservoir. The great majority of the sills are met with cutting lavas; and yet their common conspicuous xenoliths are of sedimentary, gneissic, or granitic origin: they are decisively far-travelled erratics. Moreover, the peculiarity of composition of the aluminous xenoliths, first recognized by Dr. Thomas, makes it very difficult to match them with any rock known *in situ* in the West Highlands (p. 271). It is easier to derive such peculiar material from the lining of some single reservoir, than to picture it caught up indiscriminately by individual sills. The same remark applies, though with less force, to the graphitic xenoliths: in this case, a few Tertiary coals and carbonaceous shales, with which the xenoliths may well be compared, are known in such widely separated districts as South-West Mull (Sheets 43 and 44) and Morven (Sheet 52). E.B.B.

Bearing in mind the non-metamorphic character of the sediments which furnished many of the characteristic xenoliths, we are justified in placing some boundary of the magmatic reservoir in the space-interval between the Tertiary lavas and the Pre-Mesozoic floor of gneiss and granite. It is suggested that the roof of the reservoir may have roughly coincided with the base of the lavas, in part of the subsidence-area of central Mull, east of the head of Loch Scridain. It seems that the reservoir was of considerable depth, since it was able, in many instances, to supply material of widely different composition to individual composite sills. It is pointed out (p. 33) that this probably indicates a marked difference of temperature between the upper and lower parts of the reservoir, so that crystals of felspar and augite, precipitated in the relatively cool upper regions, remelted as they fell into the relatively hot lower depths. A concentration of xenoliths, both 'cognate' and 'accidental' in the more basic portions of the magma is exhibited in the Rudh' a' Chromain sill and certain other good examples, and is indeed regarded as a feature of the district. Such a concentration is in keeping with the idea of gravitational differentiation within a magma-basin of the type here advocated.

An unsolved problem of considerable local interest is the date of the Loch Scridain Magma-Reservoir. The petrology of the typical sills, as described in Chapter XXV., virtually negatives any reference of them to the Plateau Lava period ; though one can never reach certainty on such evidence, since, as pointed out already, the Staffa Type of lavas are of tholeiitic composition, in spite of their occurrence (where present) at, or near, the base of the Plateau Group. If one accepts the argument, developed in Chapter XIV., that the bostonites as a group belong to the first paroxismal phase of the Mull centre, it follows that the Loch Scridain sills are, at any rate, no earlier than this epoch of explosions, for the Rudh' a' Chromain sill cuts a bostonite (Fig. 45, p. 267; the bostonite is omitted from the one-inch Map). Not improbably, they belong, like their fellows of Ben More (p. 290), to some stage of the long period of activity marked by intermittent intrusion of cone-sheets. The petrological assemblage of the Loch Scridain sills can be matched, either towards the beginning of this period, when composite sheets with craignurite-centres, were a feature (Chapter XIX.), or during the latter half, which was dominated by the intrusion of a tholeiitic magma with many acid accompaniments (Chapters XXVIII.-XXXII.). The xenolithic facies of the Loch Scridain magma has not been recognized among cone-sheets. The nearest approach to it is in the case of two presumably early cone-sheets of the Scallastle-Fishnish district above the Sound of Mull (p. 229). The xenoliths in these cone-sheets are of a far-travelled 'accidental' type, much altered ; but, as they are not strikingly numerous, and have not been noticed to include the typical aluminous variety, their agreement with those of Loch Scridain is not particularly close. On the whole, it is thought probable that the Loch Scridain Reservoir dates from the latter half of the cone-sheet cycle, and that it did not itself supply material to form cone-sheets at levels accessible to present-day observation. H.H.T., E.B.B.

The only other general point deserving attention is that the Loch Scridain sills seem uniformly earlier than the north-westerly faults which affect their district.

Such little evidence as is available shows that intrusions of inninmorite are cut by leidleite or tholeiite. This is the case in the bed of Abhuinn nan Tòrr (Locality 6, Fig. 43), and also at Culliemore and Allt na Crìche, on the two sides of Tavool House (Sheet 43). It would be rash, however, to generalize upon three examples. What appears certain is that the inninmorites are a very subordinate, though well-characterized, product of the Loch Scridain Reservoir. It is significant, in this connexion, that they sometimes occur with tholeiitic borders, and also occasionally carry typical 'accidental' xenoliths, with, or without, sapphire (Localities 7 and 49, p. 272). E.B.B., E.M.A.

# CHAPTER XXV.

## SHEETS EXCLUSIVE OF CONE-SHEETS:
## SOUTH-WEST MULL.

## LOCH SCRIDAIN PETROLOGY.

### INTRODUCTION.

BEFORE proceeding to detail in the matter of classification, it is necessary to state clearly how the types tholeiite and augite-andesite are distinguished in this memoir. In both cases, the essential minerals are plagioclase-felspar and augite, and there is, in addition, a residuum of glass, or such finely crystalline material as is commonly spoken of as a devitrification-product. Where the glass, or its devitrification-product, is seen as a fairly continuous matrix, it constitutes in Rosenbusch's vocabulary a g r o u n d - m a s s ; where, on the contrary, it is relegated to more or less angular interstices between the crystalline elements it serves as a m e s o s t a s i s ; and mesostasis characterizes what is known as i n t e r s e r t a l  s t r u c t u r e. The distinction between ground-mass and mesostasis is of necessity arbitrary, and we find it convenient to add that, in typical intersertal structure, the mesostasis has a patchy distribution. Olivine-free and olivine-poor plagioclase-augite rocks with intersertal structure are classed by Rosenbusch as t h o l e i i t e s ;[1] plagioclase - augite rocks with a definite groundmass are, generally speaking, a u g i t e - a n d e s i t e. One can readily see that, in general, such a difference in structure as is here relied upon for distinguishing tholeiite from augite-andesite has a chemical foundation, for the glassy residuum of any particular rock is relatively acid as compared with the early crystalline elements. We have found by experience in Mull that the term intersertal is of great convenience in describing the structure of rocks with somewhat less than 55 per cent. $SiO_2$, while it ceases to be applicable to rocks with more than that amount. The analyses, upon which this statement is based, are given in Tables II. and III. (pp. 17, 19). Moreover, our local experience in this matter seems to agree with that of other workers elsewhere, for Messrs. Eyles and Simpson[2] have found that there is a consensus of opinion that the names dolerite and andesite meet in their application to rocks having 55 per cent. $SiO_2$.

---

[1] Rosenbusch admits both non-porphyritic and porphyritic tholeiites. In the course of the present memoir the name tholeiite is employed in the restricted sense of non-porphyritic tholeiite; while some such description as porphyritic basalt with tholeiitic ground-mass is applied to porphyritic tholeiites. This practice is adopted for its local convenience, and does not clash with Rosenbusch's classification.

[2] V. A. Eyles and J. B. Simpson, 'Silica Percentages of Igneous Rocks,' *Geol. Mag.*, 1921, p. 437.

It may be asked, how does increased crystallization affect the structural distinction between tholeiite and augite-andesite ? The answer, so far as Mull petrology is concerned, is that coarse-grained tholeiite passes over to quartz-dolerite (Chapters XXVIII. and XXIX.), whereas coarse-grained augite-andesite, if non-porphyritic, or only slightly porphyritic, becomes craignurite (Chapter XIX.), and, if markedly porphyritic, it becomes porphyrite.

While 55 per cent. $SiO_2$ is taken, in this memoir, as the basic limit of augite-andesite, 70 per cent $SiO_2$, is taken as the acid limit, thus including rocks which, as far as composition is concerned, have their analogues among the dacites. True dacites, if porphyritic quartz is regarded as an essential feature, are absent from Mull.

The subject-matter is grouped in the sequel under four main headings :—Augite-Andesite, Basalt and Dolerite —including Tholeiite, Felsite, and Composite Sills.

## AUGITE-ANDESITE.

(ANALS II.-V., TABLE III., p 19; AND TABLE XI., p. 263)

The augite-andesites of Mull are for the most part non-porphyritic rocks belonging to a type to which the name leidleite has been given, after Glen Leidle in the Loch Scridain area. A relatively small number are porphyritic, and belong to another type which has been called inninmorite (p. 282), after Inninmore Bay near Loch Aline.

### LEIDLEITE.

Rocks of this type are dark-grey andesites, varying in texture from finely crystalline (14605, 14625, 14632, 14639, 15997, 17246) to glassy (15996, 17243, 17245). The silica-percentage ranges from 55-70. The constituent minerals are plagioclase, augite, hypersthene, and magnetite, to which glass, or its devitrification products, serves as a ground-mass. In the more stony types, the ground-mass is largely composed of an ill-defined crystallization of alkali-felspar and quartz ; whereas, in the glassy varieties (pitchstone), it is in large measure glass. In some of the stony leidleites, closely allied to the tholeiites, the stony texture is due to relative abundance of early crystal-elements, rather than to a devitrification of the matrix (17244).

The felspars are generally acid labradorite or andesine. They occur as narrow laths with somewhat ill-defined edges and ends, and are simply twinned and somewhat zoned. Skeletal growths are prevalent in the more glassy varieties.

The augite is a pale-yellowish green non-pleochroic variety, with an extinction of about 45°. It occurs in narrow blades and laths elongated parallel to the vertical axis, and in habit and dimensions agrees closely with the felspar. While often the augite-crystals are independent of each other, they are commonly grouped in a roughly stellate fashion, or form sheaf-like growths co-operating with the felspar to give rise to a subvariolitic structure (15996).

## 282  Chapter XXV.—Petrology of Loch Scridain Sills.

Skeletal growths, of the type so often described from variolites, are sometimes well-developed (14540, 17243).

Hypersthene accompanies the augite, at any rate in many of the slides showing fresh material (15996-7, 17243, 17245). In habit and appearance, it agrees so exactly with the augite that it is difficult to obtain an idea of the relative proportions of the two ; and we do not regard its occurrence as an essential feature of the type. It is distinguished only by its straight extinction and its definite, though weak, pleochroism.

The magnetite occurs, either as strings embedded in, or fringing, the pyroxene-crystals, or else, in the matrix, as rods made up of a series of adherent octahedra.

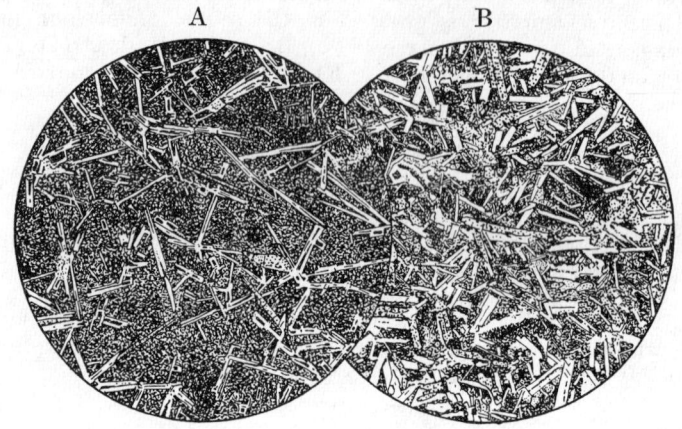

Fig. 47.—Leidleites. A from glassy centre and B from stony margin of sheet numbered 2 in Fig. 43, p. 261.

A. [17243] × 20. Narrow laths and skeletal growths of plagioclase, and blades of augite (and some hypersthene) in a matrix of brown glass.

B. [17244] × 20. Laths of plagioclase, and elongated crystals of augite (and some hypersthene), in a matrix of felspar-microliths, augite-granules, and interstitial glass. There is an approach to the intersertal structure of the tholeiites.

Quoted from *Quart. Journ. Geol. Soc.*, vol. lxxi., 1916, p. 208.

In typical leidleites, fluidal and perlitic structures on a microscopic scale are wanting, and so also is recognizable intergrowth of quartz and felspar.

The minerals of the stony leidleites are relatively decomposed compared with those of the glassy varieties. Another point of interest is the merging of stony leidleite, with increasing crystallization, into craignurite, a type also exemplified in the sills of the district (14600, 14627, 14878). One of the craignurites is worth special mention for the beautiful examples it affords of vesicles occupied, without loss of form, by residual magma (14600).

### INNINMORITE.

Rocks of this type when fresh are generally dark-grey or brown, and range in texture from finely crystalline to glassy. An essential feature is the occurrence of few, or many, minute porphyritic crystals of basic plagioclase and uniaxial augite in a ground-mass that closely reproduces the characters of the leidleites as defined above. It is naturally difficult to compare in detail two types, which themselves show a considerable range of variation, but it seems that the ground-mass of the inninmorites tends to differ, from the bulk-crystallization of comparable leidleites, in a much more restricted development of moderately basic plagioclase and, also, in a practical want of hypersthene. The result is that augite, often in skeletal curving growths, is the most characteristic mineral-development in the ground-mass of inninmorite (14591,

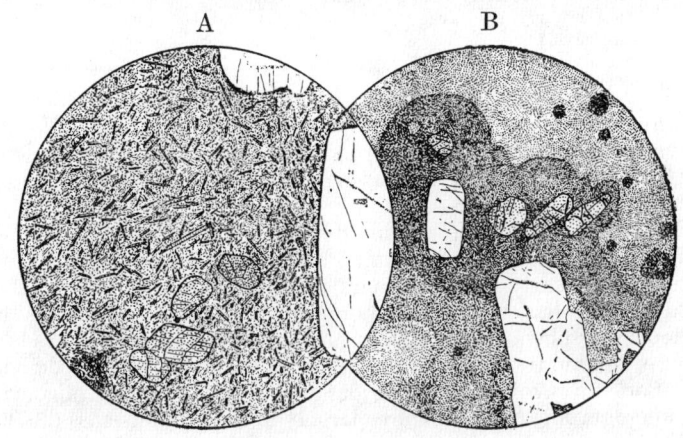

Fig. 48.—Inninmorites.

A and B from localities 5 and 4 respectively in Fig. 43, p. 261.

A. [15990] × 20. Small Phenocrysts of basic plagioclase and rounded crystals of uniaxial augite in a ground-mass of augite- and felspar-microliths, with interstitial glass.

B. [15989] × 20. Small phenocrysts of basic plagioclase and uniaxial augite in a matrix of brown glass. The glass is variable in colour, and locally almost opaque.

Quoted from *Quart. Journ. Geol. Soc.*, vol. lxxi., 1916, p. 208.

14593, 14603) ; in keeping with this, a completely glassy ground-mass is frequently met with (14590, 14662, 14671, 15989), whereas completely glassy leidleites have not been definitely recognized.

The range of silica-percentage for the type is taken as from 55-70.

The felspar-phenocrysts rarely exceed 3 or 4 mm. in length. The characteristic habit is a stumpy prism with somewhat rounded angles. In composition, the felspar approximates to anorthite. Zoning is uncommon. Twinning is invariable, and usually follows the carlsbad and pericline laws ; albite-twinning, though sometimes conspicuous, is perhaps less universal.

The augite-phenocrysts are rounded crystals generally less than 1 mm. in diameter, and are nearly always untwinned. The mineral is very susceptible to alteration, and yields serpentinuous pseudomorphs resembling those after hypersthene or olivine. In the field, little cavities are frequently found in place of the pseudomorphs. The uniaxial nature, and the mode of alteration of the augite, rendered a closer study desirable. This was undertaken by Mr. Hallimond.[1] The following is a summary of his results, based upon material separated from an inninmorite (15900, Anal. V., Table III., p. 19) :—

> Crystals rounded, unzoned, and having uniform extinction.
> Optically uniaxial, or nearly so.
> Refractive indices for sodium light: $\omega$ 1·714; $\epsilon$ 1·744.
> Birefringence: 0·03.
> Pleochroism: $\omega$ smoky brown; $\epsilon$ pale yellow.
> Extinction on the cleavage (110) $30\frac{1}{2}°$, and on the plane of symmetry approximately 41°.
> Chemically a ferromagnesian metasilicate (Anal. I, Table IX., p. 34).
> Specific gravity 3·44.

Such uniaxial augite agrees with the enstatite-augite of Wahl.

Very occasionally (14664), hypersthene is found among the phenocrysts of the inninmorites, playing exactly the same role as the uniaxial augite.

The devitrification of the inninmorites leads to a development of ill-defined quartz and alkali-felspar, with concurrent decomposition of the early-formed minerals (14593, 14599, 14603). Where crystallization has proceeded more slowly, and a porphyrite has resulted, the phenocrysts, even enstatite-augite, may, in large measure, remain fresh. Such a porphyrite is illustrated by specimens from Port na Croise, Loch Scridain (Sheet 43), where oligoclase, with micrographic fringes, figures largely in the ground-mass (20795). All the most maturely crystallized inninmorites of Mull are not classed as porphyrites, since their porphyritic structure may tend to be obscured, and, as pointed out in Chapter XIX., it is often convenient to group them as craignurites.

### BASALT AND DOLERITE, INCLUDING THOLEIITE.

(ANALS IX., TABLE II., p. 17; AND I., TABLE VI. p. 24).

#### THOLEIITE.

It has already been explained that tholeiite, as used in this memoir, denotes a non-porphyritic olivine-poor or olivine-free augite-felspar rock with intersertal structure and a silica-percentage less than 55. The tholeiites of the Loch Scridain district continue, in the basic direction, the less variolitic development of the leidleites (and to a very minor extent of the inninmorites too). Corresponding variolitic rocks may be called variolites of tholeiitic affiinity, and are well-represented.

In Mull, three main types of tholeiite are distinguished :—

The T a l a i d h T y p e (Anal. IX., Table II.) approaches leidleite in the absence, or extreme rarity, of olivine, and in the frequent tendency to elongation of the augite. The Talaidh Type

---

[1] A. F. Hallimond, 'Optically Uniaxial Augite from Mull,' *Min. Mag.*, vol. xvii., 1914, p. 97.

of tholeiite (and quartz-dolerite) is particularly strongly developed among the Late Basic Cone-Sheets, and is accordingly described more fully in Chapter XXVIII. The type is represented at Loch Scridain by a few slides (14663, 15993, 17170), and its close approach to the basic leidleites of the district is illustrated by comparison with (14914, 15993, 17244, 17246).

The Brunton Type also lacks olivine, but the tendency to elongation on the part of the augite is lost ; instead, this mineral segregates along with the felspar into crystalline groups, partly in contact with one another, and partly separated by mesostasis. The Brunton Type will be returned to in Chapter XXXIV., devoted to the dyke-rocks of Mull. Here, one may enumerate a few slides of Brunton Type among the Loch Scridain Sills, namely (14532, 14633, 14637, 16055, 16056, 17119, 18518).

The Salen Type is characterized by olivine as an essential, though minor, constituent, while the amount of mesostasis is often greatly reduced. The Salen Type of tholeiite is particularly well-developed among Mull dykes, and will be dealt with again in Chapter XXXIV. Only occasional examples have been met with among the Loch Scridain sills (14541).

In the account of the late Basic Sheets of Talaidh Type (Chapter XXVIII.), it is pointed out that variolitic developments are often encountered. In the same way, the Loch Scridain Sills, not infrequently, include variolites, in which the crystallization of plagioclase and augite so nearly occupies the whole slide, that it is safe to presume that the composition of the rocks is tholeiitic rather than andesitic (14543, 14545, 14581, 14656, 14852, 16061, 17276, 17279). There is often an obvious resemblance between these variolites and those described in Chapter X. as occurring among the Central Lavas.

A few rocks in the tholeiite-assemblage are perhaps best described as spherulitic tachylytes. They have consolidated with a well-marked spherulitic structure, in which clearly defined crystals take no part. A slice is divided up into rounded, or polygonal, areas, within which there are radiating crystalline fibres (probably associated augite and felspar) with an aggregate refractive index well-above that of balsam (16066, 17277, 18515, 18527). Similar rocks are described as constitutiing the Cruachan Dearg Type of Late Basic Cone-Sheet in Chapter XXVIII., and, in two analysed specimens, have been found to contain less than 55 per cent. of $SiO_2$ (pp. 304, 305).

#### OTHER BASALTS AND DOLERITES.

A small proportion of the Loch Scridain Sills are olivine-basalts, or dolerites, that are not conveniently grouped with the tholeiites. In several cases, they are of olivine-poor varieties characterized by porphyritic felspar (14622, 16645, 14669, 16737), and indistinguishable from the the porphyritic Central Types of lava described in Chapter X., except that, occasionally, the olivine is fresh (14611). In a few of the porphyritic sills, crystallization has proceeded further than is common in the case of the lavas (14880, 15994). In the analysed specimen (15994, Anal. I., Table VI., p. 24), slightly zoned basic labradorite, or bytownite, occurs in two generations, distinguished only by a little difference of size, and the matrix is furnished by augite and olivine (often fresh) crystallized together. Another specimen from the same sill (14666) shows a little residuum with skeletal crystallization. Altogether this rock greatly resembles the small-felspar basalt of Fig. 23 (p. 163), but is fresh.

Non-porphyritic olivine-basalts, or fine dolerites, are scarcely represented in the Survey collection of sills from the Loch Scridain area. There is one slice (14621) which approaches olivine-tholeiite in character, and another (14865) indistinguishable from olivine-rich Plateau Types of basalt-lava (Chapter X.). Both of these slices contain fresh olivine.

## FELSITE.

(ANAL. I.; TABLE IV., p. 20).

At the acid end of their range of variation, the leidleites and inninmorites join hands with closely related felsites and rhyolites —or acid pitchstones. The arbitrary limit of silica-content, which has been chosen to bound the application of the names leidleite and inninmorite, in this direction, is 70 per cent. The analysed rock (18464, 14661) is just on the acid side of this limit, and may be described as a felsite of inninmorite-affinity. It has occasional little phenocrysts of felspar (albitized), and of uniaxial augite (represented as characteristic pseudomorphs), set in a felsitic ground, in which skeletal growths of augite (also pseudomorphed) are less abundant than in typical stony inninmorites. The interior of the Rudh' a' Chromain Sill (14894-5) is of similar character, except that its felspar-phenocrysts are sometimes fresh labradorite, and sometimes replaced by calcite. In the cores of this sill, rhyolite (acid pitchstone) of analogous type is also found, and is extremely beautiful under the microscope on account of its skeletal growths and local devitrification, combined with perfect freshness (18486). The phenocrysts of the felsites, allied to the inninmorites, seem distinctly less prominent than in the true inninmorites. Where porphyritic structure fails altogether, felsitic equivalents of the leidleites are found (14604, 14618, 14631, 14658). Their main distinguishing mark, as compared with many other felsites, is a tendency to carry fairly numerous pseudomorphs after acicular, or skeletal, augite.

## COMPOSITE SILLS.

The close genetic relationship between the tholeiite-andesite-felsite assemblage of the Loch Scridain district is emphasized by a proneness of these rocks to occur in composite sills in which the rule is that the marginal portions are more basic than the interiors. The field-occurrence of the Rudha' a' Chromain sill has already been described in detail as a striking and easily located example (p. 266). In this case, Talaidh Type of tholeiite (17170, Anal. IX., Table II., p. 17) has a felsitic and rhyolitic interior of inninmorite affinity (14894, 14895, 18486), agreeing closely in composition with an analysed specimen from south of Coire Buidhe (18464, Anal. I., Table IV.,.p 20).

Another good example is exposed at the bend of the Carsaig road, east-north-east of Cnoc a' Bhràghad. Here, Brunton Type of tholeiite (14633) borders felsite of leidleite affinity (14632, 14631).

An 8-ft. sill on the same road, 50 yds. north-west of the wood above Feorlin Cottage, has variolitic margins allied to Talaidh Type of tholeiite (14581, 14584), and a glassy or stony interior of leidleitic and felsitic affinities (14582-3).

Coire Buidhe, in this same neighbourhood, at the 900 ft. contour, affords another good case, though somewhat less marked. Talaidh Type of tholeiite (14663, 15993) here borders glassy inninmorite (14664, 15992). The latter has a few phenocrysts of hypersthene.

Similarly, towards the northern limit of the area, 1000 yds. north-north-west of the cairn on Coirc Bheinn, tholeiite allied to the Salen Type (17119) occurs above and below a glassy andesite in which hypersthene again figures (17120). An extremely good example of hypersthene-bytownite cognate xenoliths comes from this sill (17121).

The Scobull Point sheet, already described (p. 265) on account of its local lack of marginal chilling, shows basic tholeiite (20798a, 20801b) with tholeiitic leidleite (20799) between.

In the three pairs of leidleites quoted, Table XI., p. 263 (15996-7, 17243-4, 17245-6), it seems to us that in each case the stony outer members of the sills are more basic than the glassy interiors. The microscopic contrast is least in regard to the first of these three pairs, and it is for this pair that full analyses are available. In the other two cases, the stony marginal part of the sills closely approaches Talaidh Type of tholeiite (Fig. 47B). It will be understood that comparison without analyses leaves room for uncertainty, especially in the case of rocks which are partially glassy, and Mr. Anderson is doubtful whether (1) we are correct in believing that most of the sheets with stony margins and glassy interiors are more basic in their marginal portions, and (2) even, if this be the case, whether the difference of composition may not be a segregation-phenomenon. He is therefore inclined to question the grouping of the generality of such sheets along with such admittedly composite sheets as those of Rudh' a' Chromain, etc.     H.H.T., E.B.B.

# CHAPTER XXVI.

## SHEETS EXCLUSIVE OF CONE-SHEETS: SOUTH-WEST MULL.

## GRIBUN AND BEN MORE.

### Gribun District (Sheet 43).

The sills of the Gribun District, as defined (Chapter XXIII.), are only separable from those of Loch Scridain by the relative infrequency of pitchstone and of 'accidental' xenoliths. Even this distinction is not absolute: a conspicuous sheet, descending the lava-escarpment above the 11th milestone from Salen, shows, in its interrupted northward continuation, cores of pitchstone, and also large quartzite-xenoliths; while another xenolithic sill occurs in the higher of the two escarpments at the north end of Fionn Aoineadh, but the xenoliths in this case look as though they may be 'cognate.'

As in the Loch Scridain country, glassy margins are a rarity; but an east and west sheet crossing Beinn an Lochain has two-inch vitreous selvages.

The Gribun Sheets are often about 5 ft. thick, and are sharply distinguished in appearance from the lavas which they cut, for their joint-system commonly divides them into upright slabs. Their inclination is varied, but is generally directed towards the south. Chilled edges are conspicuous.

Such sheets have been noted cutting gneiss, Mesozoic sediments, and Tertiary lavas, in the Gribun Peninsula, and also the mugearite-plug of Na Torranan and the dolerite-plug of Dùn Mòr, on the north side of Loch na Keal.

Only one sheet has been sliced from the district, as field-determination showed clearly that the prevalent type is non-porphyritic tholeiite, or allied stony leidleite. The selected specimen is from the summit of Creag Mhòr, overlooking the Loch na Keal road (17111). It is a tholeiite with a few porphyritic felspars, and, what is more interesting, very occasional crystals of fresh hypersthene.

### Ben More District (Sheet 44).

The Ben More district, as defined in Chapter XXIII., furnishes several points of interest which will now be discussed under the headings Field-Relations, Age, and Petrology.

## FIELD-RELATIONS.

The Ben More district lies for the most part within the Pneumatolysis Limit of Plate III, p. 91. In a general way, the sheets have suffered the same type of alteration, within this limit, as the lavas among which they occur; in fact, only one exception to this rule has been noted.

A few of these sheets are coarser and more basic than their fellows. Some of them perhaps belong to the same suite as the small laccolithic masses lettered eD on Ben More (Chapter XI.), which are freely cut by several of the normal sheets of their neighbourhood. Good examples of fine-grained olivine-dolerite sills are afforded by a group mapped on the southern slopes of Maol nan Damh (17281)—one of them, in particular, of massive dimensions. Another good instance is shown half a mile farther west, on the slopes leading down to Allt na Coille Mòire (18519). Other sheets of similar type are illustrated by (17262, 18520, 18521). The petrology of these sills reproduces, in abundant olivine and purple augite, the essential features of the Plateau Lavas of Chapter X. and Early Basic Cone-Sheets of Chapter XXI.

The most characteristic sheets of the Ben More district are an assemblage of thin sills, seldom exceeding 5 or 6 ft. in thickness, abundantly developed within a triangle, having Maol nan Damh at its apex and extending northwards to include the Scarisdale River and Eorsa. In the southerly part of their distribution, in Ben More, A' Chioch, and Maol nan Damh, these sheets are linked together by a tendency to dip west or north-west at gentle angles. They have been grouped on this basis under the title A' Chioch Sheets.

Even in their type-development, the A' Chioch Sheets are very irregular, and often follow the bedding of the lavas instead of maintaining an independent inclination. On the steep south-western slopes of Ben More, the conditions are particularly complex, as there are two groups of sheets which are locally separable from one another—although, for all we know, both groups may be included in the A' Chioch assemblage. One of these local groups consists of horizontal sills, some of them composite, showing an association of olivine- and olivine-free-tholeiite (18523-4); the other, and apparently later group, is distinguished by a northerly dip of its constituent sheets.

The A' Chioch assemblage dies out along its strike southwards from the Ben More massif. E.M.A.

Westwards and north-westwards from Ben More, the inclination of the A' Chioch Sheets becomes too irregular to serve as a ground for comparison with the type-occurrences. However, on the north faces of Ben More and A' Chioch, before the general west or north-westward inclination is lost sight of, another distinctive feature of the assemblage becomes prominent: this is the common occurrence of crystal-concretions, or 'cognate' xenoliths, towards the centre of the sills. Such concretionary sheets are a characteristic of the whole country lying between Ben More, Dishig, and the Scarisdale River. Good roadside exposures may be mentioned a mile north-west of Dishig between two streams flanking the 7th milestone from Salen. J.E.R.

## AGE.

An inferior limit to the age of the A' Chioch Sheets is provided by several of these sheets cutting an acid cone-sheet (Chapter XIX.), which crosses the coll between Ben More and A' Chioch. This relationship is illustrated by a single example on the one-inch Map. It is important as showing that the A' Chioch Sheets are almost certainly later than any of the lavas that are still preserved in Mull.                                                                E.M.A., J.E.R.

An upper limit to the age of these sheets is afforded by several intersections of the sheets by Late Basic Cone-Sheets (Chapter XXVIII.) at the east end of Beinn Fhada. In keeping with this, the sheets are absent from the Beinn a' Ghràig Granophyre (Chapter XXXII.), and are baked in its neighbourhood (16617). In Chapter XXVIII., it will be shown that the Beinn Fhada cone-sheets are probably among the latest of their kind, whereas the acid cone-sheet crossing the Beinn More arête may be one of the earliest; so all that seems established from their field-relations is that the A' Chioch Sheets belong to some part of the time characterized by the cone-sheets as a whole.

When one turns to petrology, there is some reason for supposing that these sheets belong to a phase of the Late Basic Cone-Sheet period. The magma is of tholeiitic type, much more comparable with that of the Late Basic, than the Early Basic, Cone-Sheets. So far as petrology is concerned, the sheets may equally well be compared with the Central Lavas, but, as pointed out, the field-evidence is strongly opposed to any suggestion of contemporaneity in this connexion.

## PETROLOGY.

The somewhat decomposed condition of most of the sheets renders detailed petrological description out of place. It is only necessary to say that the prevalent types (16626, 16635, 16638, 16640, 16652, 17141, 17162, 17163, 17261, 18522) are reminiscent of non-porphyritic Central Lavas, only, often, with a more advanced tholeiitic crystallization. A slice from Eorsa (17612) is a Salen Type of tholeiite (p. 285) with fresh olivine and purple augite. The fresh olivine, in this case, is in keeping with the position of Eorsa outside the Pneumatolysis Limit of Plate III. (p. 91). Occasionally, the slices show a porphyritic tendency (16620, 16968, 17263), and one of them (16637) is indistinguishable from porphyritic basalt-lava of Central Type. Only one good example (17158) of the 'cognate' xenoliths, so characteristic of many of the field-exposures, has been cut. It shows glomero-porphyritic aggregates of basic plagioclase and olivine (as pseudomorphs) in a variolitic base; and could be matched precisely among the pillow-lavas of Central Type (Chapter X.). In what is probably an A' Chioch sheet, at the western limit of the Ben More district, and well outside the Pneumatolysis Limit, similar glomeroporphyritic aggregates of felspar and olivine occur, and the olivine is fresh.

A few rather acid sheets are of acicular crystallization (16941,

16969, 17280), and are best classed as craignurite, defined in Chapter XIX. Two examples of composite sheets, centred by rhyolite or felsite, with small alkali-felspar phenocrysts, are represented in the collection. One of these (17275) is lettered F on Am Binnein; the other (17164), not shown on the one-inch Map, occurs 300 yds. north-west of the summit of Ben More.     E.B.B.

# CHAPTER XXVII.

## SHEETS EXCLUSIVE OF CONE-SHEETS: ELSEWHERE IN MULL AND NEIGHBOURHOOD.

### INTRODUCTION.

THE previous four chapters have been devoted to the sills of South-Western Mull. Attention will now be given to certain sills and sheets occurring in other parts of the island and its neighbourhood. A minor field of basic and intermediate sill-intrusion is encountered on the two sides of Loch Aline in Morven (Sheet 44). The sills here are lettered D on the one-inch Map, except that two intermediate examples are lettered F. Tertiary tholeiite-sheets (D) are occasionally met with in Lismore, and in Lorne, south of Oban (Sheet 44).

A quite distinct field of the same type includes the north coast of Mull between Caliach Point and Ardmore Point (Sheet 51). In this case, the sills (D) belong, on account of their position, rather to the Ardnamurchan than the Mull centre.

It is thought that the sheets, or sills, mentioned above, may be ascribed to intervals in Late Basic Cone-Sheet time (Chapters XXVIII.-XXXII.). Of earlier date, probably, than of some part of Early Basic Cone-Sheet time (Chapters XXI.-XXII.), is a sheet-complex of rather exceptional character. It constitutes much of the more westerly Creag na h-Iolaire, north of Ben Buie, and its component rocks are lettered cD on the one-inch Map. They are not cone-sheets, and their field-relations, and early date, separate them from most of the other sills or sheets of Mull. They are, however, dealt with in this chapter because they do not furnish material enough for separate treatment.

A final section is devoted to a couple of thoroughly early dolerite-sills (D)—earlier than the Early Basic Cone-Sheets. They are considered last, in spite of their age, because the Loch Aline sills have much in common with most of the sills of South-West Mull, described in the immediately preceding chapters.

A few trachyte-sills, lettered O, have already been discussed along with bostonites and syenite in Chapter XIV.   E.B.B.

### LOCH ALINE DISTRICT (SHEET 44).

A few sills have been shown on the one-inch Map on both sides of Loch Aline, and some of very minor importance have been omitted. West of Loch Aline, the bay east of Eilean na Beitheiche supplies an interesting section of a tuff-filled vent, cut through and veined by a basalt-sill. Two other sills, 10 and 2 ft. thick respectively, descend to the shore a little east of Savary Glen. All these three are seen cut by basalt-dykes. On approaching Loch Aline village, the reverse relation of a sill cutting a dyke is encountered, as shown on the one-inch Map.

On the east side of Loch Aline, the most noteworthy sills are those near Glais Bheinn summit, and in the valley of Allt na Samnachain. Here, the rock-types (14961, 14964, 14954) are such as can be matched among the porphyritic Central Lavas. Another sill (15808-9), lettered D on the one-inch-Map, west of Loch Teàrnail, is a rather basic leidleite (p. 281). Leidleite is so met with in the cliff above Rudha an t-Sasunnaich, but in this case is lettered F (14967). Another sill (or dyke) of leidleite (15794), lettered F, stretches for a couple of miles east from the angle of Allt Dubh Dhoire Theàrnait. Near the place where glacial striæ are shown on the one-inch Map, this leidleite cuts an unmapped north-north-west tholeiite-dyke of Salen Type (15783).

Inninmorite (p. 282) is found on the shores of Inninmore Bay beneath the keeper's cottage (14948). The rock occurs here as a sheet dipping north, and is lettered D on the one-inch Map. From this point, inninmorite has been traced northwards and north-westwards, interruptedly, as one or more thin dykes, to the head of Coire Slabhaig (14947, 14965, 14960). A little unmapped sill in the Trias, east of the fault which descends the cliff into this corrie, is interesting for having hypersthene as well as uniaxial augite among its phenocrysts (14959). On passing west, one finds a southwardly inclined sheet or dyke (lettered M) passing up the cliff westwards on its way from Coire Slabhaig (14958). For a mile, it continues in this direction, and then turns abruptly (the exact turn is concealed) into a north and south dyke which runs out to sea at Rudha an t-Sasunnaich (14955, 14966). It seems that the quadrilateral, outlined in the above statement, is bounded by sheet and dyke manifestations of a single intrusion. Some of the material is very suitable for slicing, for instance from the cliff above Rudha an t-Sasunnaich, or at the 1135 ft. level shown on the map west of Glais Bheinn. The district is of course the nameplace of the type (p. 281).

Inninmorite occurs elsewhere in the district as a sill, marked D, above Doire Daraich on the south-east coast (15805); in another sill (15820), north-east of the Table of Lorne; and in a west-north-west dyke (not mapped, but occurring at the head of the stream draining into Loch Aline, south of Allt Leacach). E.B.B., G.W.L.

### LISMORE AND LORNE (SHEET 44).

A Tertiary sheet allied to leidleite (13751) crosses Lismore at the south end of Loch Baile a' Ghobhainn with a southerly inclination. Near the loch, it cuts a basic dyke which runs slightly north of west. Contacts with other dykes are not exposed.

Two more sheets, with southerly inclinations, occur, one at the south end of the island, the other on Eilean Musdile. Their relations to dykes are not exposed. (H.B.M.)

In the Oban district of Lorne, thin Teritary tholeiite sills are met with in two main localities where they cut Old Red Sandstone sediments and lavas. The more northerly is about Gleann Sheileach, where good exposures are afforded on Druim Mòr, and also near the base of the Old Red Sandstone, along the south side of the glen. The sills are about 5 ft. thick, and of the Salen Type of tholeiite (p. 285), though with rather less olivine than is often the case (19042, 19046). The more southerly sills crop out on the high ground north and east of Lerags House, towards Loch Feochan. Wherever they occur, they are conspicuously cut by the north-westerly faults of the district, but their behaviour with reference to the north-east faults is less certain. E.B.B.

### NORTHERN MULL (SHEET 51).

A minor field of thin tholeiitic sheets extends inland for a couple of miles between Caliach Point and Ardmore Point. The sheets may be described as irregular sills with a marked transgressive tendency. They dip, throughout most of the area, towards the south-west; but, at Caliach Point, they incline in the opposite direction, towards the east or north-east. Sir Archibald Geikie, in his *Ancient Volcanoes* (Vol. II., p. 158), has figured an example, from Caliach Point, cut through by a north-west dyke.

Only one specimen has been sliced. It comes from the west side of Ardmore Bay, and is thoroughly fine-grained, suggesting a rapidly cooled rock of tholeiitic composition. G.V.W.

### CREAG NA H' IOLAIRE SHEETS (SHEET 44).

The southern slopes of the more westerly Creag na h' Iolaire, north of Ben Buie, consist largely of a complex of dolerite and basalt sheets, which are too irregularly disposed to be referable to any of the cone-sheet assemblages. Some of the sheets are porphyritic, others non-porphyritic. A sliced specimen (17194) proved to be an olivine-dolerite, of a type familiar among the Early Basic Cone-Sheets of Chapter XXI.

So far as can be judged, the intrusions, referred to this complex, are later than the intrusive felsite of Chapter XVII., as well as the neighbouring lavas. Their relationship to the many small outcrops of breccia, or agglomerate, that occur along with them, varies greatly. Sometimes sheets cut agglomerate, sometimes agglomerate cuts sheets. The Late Basic Cone-Sheets of Talaidh Type (Chapter XXVIII.) clearly traverse the Creag na h' Iolaire Complex. (C.T.C.)

### TWO EARLY SILLS, MULL (SHEET 44).

Probably the biggest basic sill in Mull is a coarse dolerite, with chilled margins, which lies somewhat transgressively among the lavas of Beinn Fhada, a couple of miles north of Loch Uisg. This dolerite gives rise to conspicuous crags, and it is easy to recognize that it has been folded, along with the lavas, into a syncline centring about Coir' Odhar (*cf.* one-inch Map and Plate V., p. 165). It is natural to regard the syncline as part of the early arcuate folding south-east of Mull : but more direct evidence of the early date of the Coir' Odhar sill is shown by the manner in which the intrusion is freely traversed by Early Basic Cone-Sheets of Chapter XXI. The sill itself, under the microscope, proves to be an olivine-dolerite (17099, 17323), with somewhat purple augite as its main constituent, and a fairly well-defined ophimottled structure (p. 138). Its resemblance to some of the Plateau Basalts is an example of reversion ; since the lavas, among which it is intruded, are definitely, of Central Types. E.B.B., G.V.W.

The olivine of the Coir' Odhar sill is entirely decomposed, as might be expected from the intrusion's position well within the Pneumatolysis Limit of Plate V.

The other sill, selected for notice in this paragraph, occurs intruded into the Lias and Trias of the anticlinal core, exposed north and west of Càrn Bàn (*cf.* one-inch Map and Fig. 25, p. 174). It has shared in the intense folding of the sediments with which it is associated, and has been correspondingly crushed and altered. Owing to its condition, it would never have attracted attention as an intrusion, if its outcrop had been situated among the neighbouring basalt-lavas, instead of among Pre-Tertiary sediments ; so that it is safe to assume that the apparent freedon from sill-intrusion of the Mull lavas, in many parts of Central Mull, is in some small measure deceptive. It would be unwise, however, to push this conclusion very far, since sills do not figure prominently in the Mesozoic sediments of the region. The Càrn Bàn sill is, like the Coir' Odhar sill, a very early intrusion, for the folding, which affects it, is definitely of earlier date than any cone-sheets still recognizable as such (*cf.* Chapter XIII.). E.B.B.

# CHAPTER XXVIII.

## LATE BASIC CONE-SHEETS.

### Introduction.

The Late Basic Cone-Sheets are lettered tI on the one-inch Map, Sheet 44 : t because Beinn Talaidh is virtually built of them ; and I because they are steeply inclined towards a centre, or series of centres, on the line $C_1$ $C_2$, Fig. 58 (p. 338). Their distinguishing petrological character, as compared with the Early Basic Cone-Sheets of Chapter XXI., is an absence of olivine. The average thickness of individual sheets is perhaps not more than 10 or 15 ft., and is certainly less than in the case of the Early Basic suite. Their texture too is finer ; in fact, in Cruachan Dearg, it is often vitreous, but elsewhere this is uncommon. The most general type is a fine-grained quartz-dolerite or tholeiite (always albitized), which shows a great tendency to break up into brown-weathering slabs. Instead of giving prominent scarps like the Early Basic Cone-Sheets, these Late Sheets are apt to cover themselves with featureless scree. Where, as so often, they cross scarps due to the early series, their habit is to weather back in nicks ; and the same is characteristic of their outcrops among the gabbro crags of Ben Buie, Corra-bheinn, and Beinn Bheag. As in the case of other cone-sheets, the inclination is often about 45°; but, perhaps, the average is rather steeper than this. Nearly vertical examples are by no means rare, but they are definitely non-typical. Every sheet has consolidated with a chilled top and bottom. (C.T.C.), W.B.W

No better introduction to the Late Basic Cone-Sheets could be desired than is afforded in the Gaodhail River, above the track which leads up Glen Forsa from the Sound of Mull road, east of Salen. An accurately measured section is illustrated in Fig. 49. It runs at right angles to the strike of the sheets for about 600 yds. Of this distance, 128 yds. are obscured by gravel, and, in the remaining 472 yds., there occur 124 Late Basic Cone-Sheets with 73 interspaces of coarse intrusions, classed in Fig. 49 as country-rock. The total thickness of Late Basic Cone-Sheets actually exposed is 660 ft., and that of country-rock 342 ft.—a ratio of about 2 : 1. In one part of the section, for a distance of about 400 ft. this proportion stands at 12 : 1.

The measured section of Fig. 49 includes less than half of the belt of Late Basic Cone-Sheets shown on the one-inch Map as crossing the River Gaodhail ; and it is certain that 2000 ft. is not an excessive aggregate thickness to assign to the sheets of this neighbourhood. Their great aggregate bulk is the more readily appreciated on turning from the River Gaodhail to Beinn Talaidh, where one finds a mountain 2496 ft. high, for the most part built of thin Late Basic Cone-Sheets. W.B.W.

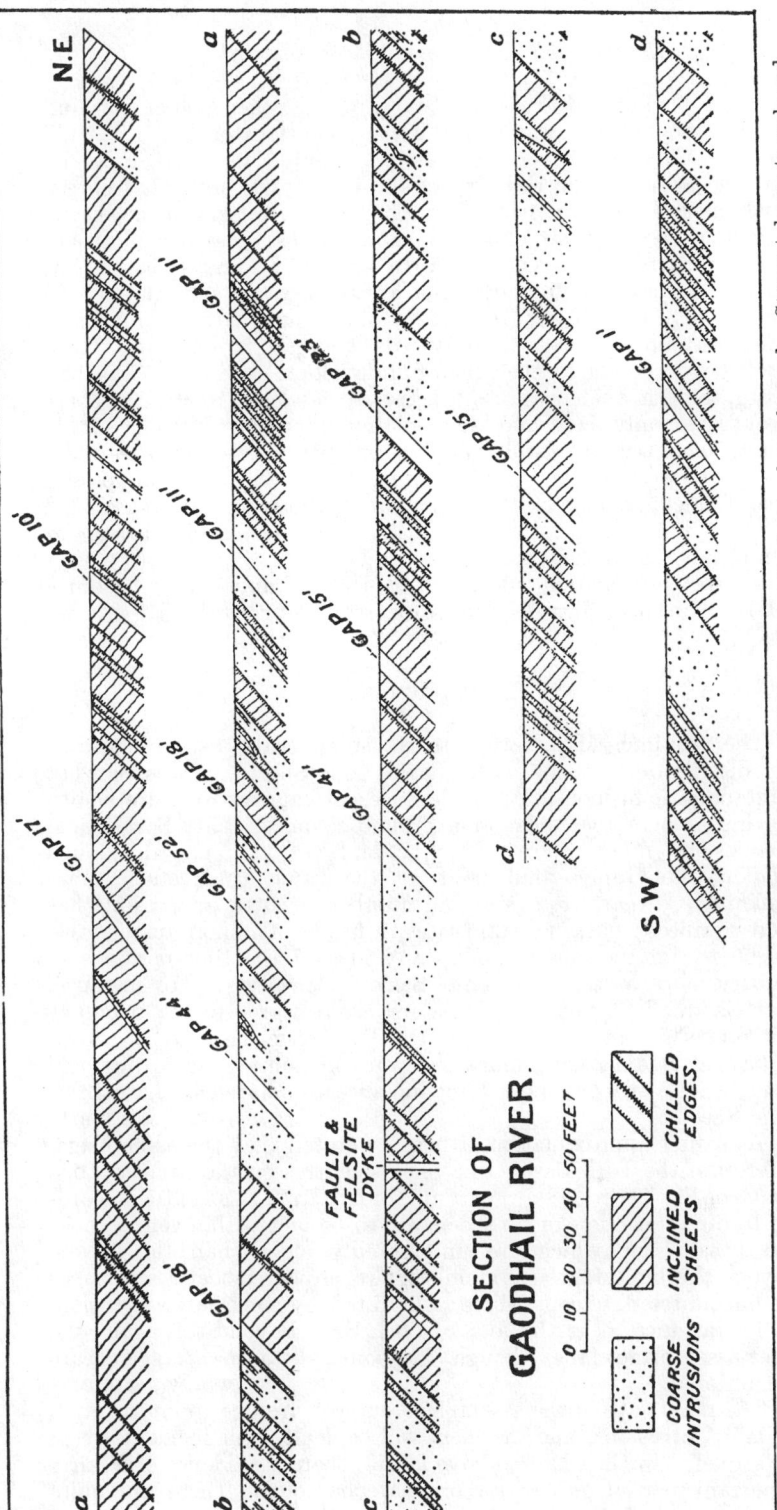

Fig. 49.—Late Basic Cone-Sheets (Inclined Sheets) exposed in Gaodhail River. Chilled edges only shown where observed. Gaps not drawn to scale and figures given refer to horizontal distances measured in feet.

Quoted with slight alteration from *Summary of Progress for 1910*, p. 36.

Corra-bheinn and Cruachan Dearg, on the other side of the Mull axis of symmetry, repeat many of the conditions of Beinn Talaidh and the Gaodhail River. It has been stated in Chapter XXII. that the Corra-bheinn Gabbro, for 100 or 200 yds. inwards from its southern and western margins, is practically free from cone-sheets, except where such are included as xenoliths. Farther in, cone-sheets of Late Basic types begin to appear, at first sparsely, but afterwards more abundantly and in thicker belts, until, at the summit of Corra-bheinn, they form more than half the rock and make up multiple sheets 30 or 40 yds. wide at the outcrop. Even here, they have not reached their full development, for, on Cruachan Dearg, there is evidence for a thickness of at least 1000 ft. of cone-sheets with only here and there a thin intercalated wedge of the gabbro country-rock still happily preserved. Where the rock, just south-west of the cairn, is quite bare, we can never pass a greater thickness of rock than 8 or 9 ft. without coming to a chilled margin; but it is possible that many of these margins belong to thin sheets which have split thicker ones. (C.T.C.)

Further information will now be considered under three headings —Distribution, Time-Relations, and Petrology.

### Distribution.

The one-inch Map, owing to its small scale, has to indicate the distribution of the Late Basic Cone-Sheets in a somewhat diagrammatic fashion. Where lenticles of country-rock are shown on this Map in the midst of a ground-colour of Late Basic Cone-Sheets, one must realize that the lenticles are generally much smaller than represented, and at the same time much more numerous. Apart from size and number, as little as possible has been sacrificed; but, in certain cases, it has been found impracticable to convey on the one-inch Map any idea of the time-relations of country-rock to associated Late Basic Cone-Sheets. For instance, near Gaodhail Cottage, in Glen Forsa, southwards to the tributary Allt nan Clàr, an important intermediate cone-sheet (aI) is shown as having a fairly continuous outcrop: in point of fact, this acid sheet, wherever seen, is cut into thin slices separated by Late Basic Cone-Sheets. E.B.B.

As a first approximation, it may be stated that the assemblage-outcrop of the Late Basic Cone-Sheets centres on the head of Loch Bà ($C_2$ of Fig. 58, p. 338). For a mile or two from this centre (according to direction), few or no sheets are to be met with, even in rocks like basalt-lava, which are undoubtedly older than the sheets. Beyond this limit, for the next mile or two, crowded sheet-assemblages are encountered, except where obliterated by some later intrusion, as, for instance, near the foot of Loch Bà. Beyond this again, the sheet-assemblage fails, though for some distance stragglers are encountered. W.B.W., J.E.R.

The main gaps in the assemblage-outcrop are due to interruption by later intrusions, and, as such, will be dealt with incidentally in the sequel. In Glen More, above Ishriff, there is evidence of a fairly important partial gap of an original character. There are quite

considerable areas of basalt-lava in this district (B and pB, Fig. 52, p. 308), and these are very sparingly cut by cone-sheets, although they lie where one might expect a continuation of the Late Basic Cone-Sheets that figure prominently in Cruach Choireadail farther west.
E.B.B.

## Time-Relations.

Enough has been stated, in the present and foregoing chapters (XXI. and XXII.), to indicate a time-scale running as follows :— (1) Early Basic Cone-Sheets; (2) Corra-bheinn Gabbro; (3) Late Basic Cone-Sheets.

The injection of Late Basic Cone-Sheets continued intermittently through a very long period, in fact, nearly to the close of igneous activity. Thus it happens in several cases that an intrusion belonging to a different category may cut some of the Late Basic Cone-Sheets, and in turn be cut by others. The quartz-Gabbro of Coir' a' Mhàim (south-east of Corra-bheinn) supplied the first-recognized example of this, which is now accepted as a commonplace of Central Mull. (C.T.C.)

It has been pointed out that, in an approximate sense, the Early Basic Cone-Sheets centre about $C_1$ of Fig. 58, and the Late Basic Cone-Sheets about $C_2$. There is evidence to suggest that the migration of the centre of sheet-activity was somewhat gradual. Two instances supporting this contention may be cited :—

(1) A Ring-Dyke of quartz-gabbro, with subordinate associated granophyre, runs from the east slope of Beinn Talaidh through Loch Sguabain, in Glen More (1 of Figs. 52 and 53, pp. 308, 312), to Cruach Choireadail (*cf.* Pl. VI., p. 307, and one-inch Map). It is known as the Glen More Ring-Dyke, Chap. XXIX., and, during this long continuous part of its outcrop, it cuts almost all the Late Basic Cone-Sheets with which it comes in contact—and these locally are very numerous. Beyond Cruach Choireadail, the Glen More Ring-Dyke has a discontinuous outcrop leading through Coir' a' Mhàim and Coir' an t-Sailein to Tòrr na h-Uamha. If one compare the Coir' a' Mhàim and Coir' an' t-Sailein exposures with those of Cruach Choriedail and Glen More, one finds the number of sheets cutting the ring-dyke distinctly increased in relative importance, though they still remain far inferior to the number which the ring-dyke cuts. Towards the western end of the Tòrr na h-Uamha outcrop, this change is accentuated ; in fact, the Glen More Ring-Dyke, as represented in Tòrr na h-Uamha, comes to be freely cut by Late Basic Cone-Sheets ; and these, from their position, seem referable to a more north-westerly centre than those of Cruach Choiredail. The only criticism, that might be offered in this connexion, would be a suggestion that the Tòrr na h-Uamha outcrop should not be referred to the Glen More Ring-Dyke. (C.T.C.), E.B.B., E.M.A.

(2) Just inside the Glen More Ring-Dyke, lies the Ishriff Ring-Dyke, composed of granophyre. Throughout most of its course, the Ishriff Ring-Dyke cuts the great majority of the Late Basic Cone-Sheets which it encounters ; but where it crosses Maol nam Fiadh (2, Fig. 53, p. 312), and takes its place at the back of Coire Ghaibhre, it is cut to pieces by Late Basic Cone-Sheets. Here again the cutting sheets seem, from their position, to belong to a more north-westerly centre than the ones that are cut.
E.B.B.

In the light of these two examples, it is apparent how carefully one has to proceed in matters of this kind. Thus, half-way between Glen More and Beinn Chàisgidle, Late Basic-Cone-Sheets cut very freely through a whole series of ring-dykes. It would be unsafe

300    *Chapter XXVIII.—Late Basic Cone-Sheets.*

to infer from this that the Ring-Dykes of Beinn Chàisgidle are earlier than those of Glen More, for it is quite probable that the Cone-Sheets themselves are of materially different dates in the two districts. Again, a first glance at the one-inch Map might suggest a comparison between the relative immunity from cone-sheets of the ring-dykes of the Glen More district with that of the Knock and Beinn a' Ghràig Granophyres, which belong to the ring-dyke system as developed near the foot of Loch Bà (Chapter XXXII.). But any such comparison is misleading. The Glen More and Ishriff Ring-Dykes are, in all probability, separated from the Knock and Beinn a' Ghràig Ring-Dykes by an interval of time that admitted of extensive cone-sheet injections. As already pointed out, what is taken as a continuation of the Glen More Ring-Dyke through Tòrr na h-Uamha is freely cut by Late Basic Cone-Sheets in its western part. In the same neighbourhood, the Beinn a' Ghràig Granophyre cuts every sheet it encounters.

The manner in which the Beinn a' Ghràig and Knock Granophyres cut through great numbers of Late Basic Cone-Sheets will be treated in greater detail in Chapter XXXII. All that need be emphasized at this juncture is that apparently one at least of these granophyres was followed by a minor revival of cone-sheet injection (p. 345).                                J.E.R.

In comparing the dates of ring-dykes and cone-sheets, it is probably safe to refer the Glen More and Ishriff Ring-Dykes of Chapter XXIX. to the latter part of the first half of the Late Basic Cone-Sheet period, and the Knock and Beinn a' Ghràig Ring-Dykes of Chapter XXXII. to a similar stage of the second half of the same period. The wonderfully continuous felsite of the Loch Bà Ring-Dyke of Chapter XXXII., so far as can be determined, is later than all the cone-sheets of Mull.        E.B.B., J.E.R.


PETROLOGY.

The Late Basic Cone-Sheets may be divided petrographically into two main types, namely, the quartz-dolerites and tholeiites that compose the greater part of Beinn Talaidh, and the variolites and tachylytes best seen in Cruachan Dearg. These types are often intimately associated in the field, and are sometimes linked by examples of a transitional character. Other variants from the normal representatives have affinities with such types as the craignurites (p. 227), leidleites (p. 281), and certain central types of lava (Chapter X.).

Situated well within the pneumatolytic zone of Central Mull (Plate III., p. 91), the Late Basic Cone-Sheets have shared to a varying degree in the general albitization and propylitization that have affected the rocks of this region ; but, in addition, the alkalinity of their own residual magma seems to have produced in them albitization and other changes, before or during the last stages of consolidation. In the case of these sheets, it is thought that a large proportion of the mineral-changes which are characteristic of them as a group are referable to auto-pneumatolysis, dependent partly upon composition and partly upon locality. The influence of composition is shown by the much more general albitization of these late sheets as compared with their predecessors, the Early Basic Cone-Sheets of Chapter XXI. The influence of position on their final state is suggested by the fact that the intensity of albitization among the Late Basic Cone-Sheets varies somewhat according to locality, and, in the case of the quartz-dolerites, appears to have progressed to a less extent in the sheets of Cruachan Dearg, than in those of the Gaodhail River. This difference is still further emphasized by a comparison with the Tertiary region of Ardnamurchan, where sheets, otherwise identical with those of Mull, are often not albitized to any appreciable extent. It can be readily understood that the conditions of Central Mull during the intrusion-period of the Late Basic Cone-Sheets may have been such as to favour the process of auto-pneumatolysis that appears to be an innate tendency of the sub-basic magma. But, whatever the conditions that controlled the albitization, it is certain they allowed the individual sheets to chill at their margins, and to develop a type of crystallization in keeping with their hypabyssal nature.

### TALAIDH TYPE OF QUARTZ-DOLERITE.

(ANAL. VIII.; TABLE II., p. 17)

The most prevalent variety of Late Basic Cone-Sheet is a moderately basic rock of quartz-dolerite affinities (18467). Texturally, the sheets are fine-grained rocks of a dark brownish-grey colour. They are usually devoid of porphyritic crystals, and seldom carry olivine as one of their constituent minerals. They are composed of a moderately basic plagioclase, augite, titaniferous magnetite and ilmenite, alkali-felspar of predominating albitic character, and quartz. One of their chief characteristics, however, is the seemingly early separation of the more basic and larger individual crystals, and crystal-groups, from an acid residuum, and the segregation of this acid mesostasis into well-defined regions, where it is consolidated with a type of crystallization differing from that of the rest of the rock, and more or less peculiar to itself. The amount of mesostatic, or intersertal, material present in any portion may vary within fairly wide limits, and is naturally greater in the more acid varieties that approximate to the craignurites in composition.

For the purposes of description, it will be convenient, as well as logical, to consider the mineralogical and structural features of the type in two sections—one dealing with the coarser and earlier crystalline constituents, and the other with the fine-grained acid mesostasis. Further, we shall take for the basis of our description the comparatively unaltered representatives of the type, that occur mainly on Cruachan Dearg, and to a less extent in the Gaodhail River.

*Early Constituents.*—Most of the augite appears to have separated from the magma during the period occupied by the crystallization of the relatively basic plagioclase, and may be in hypidiomorphic, ophitic, and seemingly eutectic, relationship with the latter within the same field of view. It is usually of a pale brownish tint, but occasionally shows a lilac tinge and slight pleochroism indicative of the presence of titanium.

The dominant type of crystallization of the augite is columnar. The columns are usually a little less than a millimetre in length and a fifth of a millimetre in breadth, and show a general lack of well-defined crystal-faces other than those of the prism-zone. They are elongated parallel to the C axis, and thus show traces of the prismatic cleavages parallel to their length. In cross-section, they frequently have a roughly octagonal outline due to the somewhat equal development of the prisms and pinacoids. They have the peculiarity of almost always carrying moderately large, and frequently well-formed, crystals of magnetite in their peripheral portions

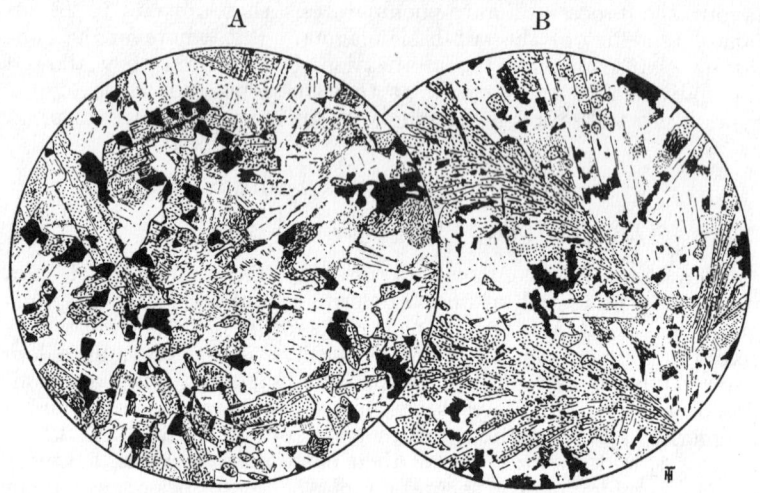

FIG. 50.—The Talaidh Type of Late Basic Cone-Sheets.

A. [14867] × 17. Quartz-dolerite. The section shows columnar augite associated with titaniferous magnetite, a colourless moderately basic and albitized plagioclase, and a mesostasis of alkali-felspar and quartz.

B. [14810] × 17. Quartz-dolerite. Mineralogically similar to the above, but with a highly characteristic cervicorn development of its augite (p. 303).

—a peculiarity also met with among the quartz-dolerites of West Lothian.[1]

Frequently, the crystals are simply twinned parallel to the orthopinacoid, and, almost invariably, show traces of salitic striation. This striation sometimes occurs throughout the crystals, but, at other times, is developed locally, more particularly at the crystal-borders and in the neighbourhood of the twin composition-plane.

A character of these columnar crystals, as in the acicular augites of the craignurites (p. 226), is the occurrence of a slender central rod of serpentinous material that presumably replaces original rhombic pyroxene or enstatite-augite.

[1] J. D. Falconer, 'Igneous Geology of the Bathgate and Linlithgow Hills,' *Trans. Roy. Soc. Edin.*, vol. xlv., 1908, Pl. II., Fig. 2, p. 150.

As commonly happens with columnar augites, of which the crystallization from a sub-basic magma was accompanied by a marked separation of an acid residuum, the crystals are sometimes curved and otherwise distorted. This is due to intercrystal pressure through the movement or withdrawal of the liquid mesostasis from the immediate neighbourhood. Similar curving of columnar augites has been noted by us in the less basic portions of the differentiated mass of Cruach Choireadail (p. 325), and was described and figured by Dr. Falconer[1] from the quartz-dolerites of Linlithgow.

In association with felspar, the columnar augites occasionally take on a stellate grouping, both augite and felspar crystals radiating from a common centre. This would appear to be the first stage of a structure which is highly characteristic of certain varieties of the Talaidh Sheets. In such rocks, the augite, commencing from a centre, spreads out sectorially by a continuous branching of several stocks in a cervicorn (antler-like) growth. Between crossed-nicols, it appears that this structure is not that of an ophitic individual with simultaneous extinction of its component parts, but that each stock has its own orientation, and that the extension of each branch is in the general direction of the prismatic zone-axis. In some of the fine-grained Talaidh Sheets that have tholeiitic affinities, this cervicorn structure of the augite is most conspicuously developed (14810, Fig. 50B).

The main felspar of these rocks builds columnar once-twinned, somewhat narrow crystals, that frequently attain greater lengths than the columnar augites with which they are associated. Often, they adopt a stellate grouping, and show a remarkable constricted development of their initial portions.

The extinctions vary somewhat from their central to marginal parts, and indicate a range of composition from acid labradorite, through andesine, to oligoclase. Like the columnar augites, they have frequently suffered deformation and have become curved (16558), or even broken, when the bending was too severe.

Cervicorn association of felspar and augite most commonly occurs within the limits of a single felspar-crystal of somewhat tabular form, and may possibly represent a eutectic crystallization. Less frequently, however, the augite growths may pass from one individual felspar to another, and thus indicate the independent and earlier crystallization of the pyroxene.

Iron-ore is usually an abundant constituent. That, which occurs in intimate association with the columnar augite, appears from its form to be either magnetite or titanomagnetite; but, scattered indiscriminately through the rock, are moderately large grains and patches which, from their manner of alteration, seem to be ilmenite.

*Mesostasis.*—The mesostasis is mainly collected into well-defined areas and varies considerably in amount. It appears as a very fine-grained crystalline mass, usually turbid, characterized by the acicular habit of its ferromagnesian constituent, and, to a less extent, of its felspar. It is composed of acicular augite and long narrow crystals of oligoclase in a somewhat chloritized microcrystalline matrix of alkali-felspar and quartz. Occasionally, quartz-patches

[1] J. D. Falconer, *op. cit.*, p. 140, and Pl. II., Fig. 3

reach moderately large dimensoins, and, sometimes, it is possible to detect a granophyric relation between quartz and alkali-felspar in the finer portions of the matrix. Apatite, as slender needles, is frequently abundant (14806). The acicular augite has usually been replaced by green fibrous hornblende or chlorite in a pseudomorphous manner. In rocks of this type, with a more than usually acid composition, the amount of mesostatic material becomes proportionately greater, and, at the same time, the ferro-magnesian mineral throughout the rock tends to assume the acicular habit. Such sheets reproduce, in a great measure, the essential characters of the more basic varieties of craignurite, and we may regard them as connecting links between the craignurites and the quartz-dolerites of Talaidh Type (17225, 17890).

*Albitization and other Secondary Changes.* As was stated above, these rocks of quartz-dolerite character have the inherent faculty of auto-albitization which they exercise to a variable degree. Where the acid residuum (mesostasis) is in contact with the earlier products of consolidation, it has frequently exerted upon them a local corrosive action. The original relatively basic felspars of the rock have been eaten into and replaced by a fresh growth of felspar of more alkaline character, and the columnar augites have developed on their exposed surfaces a fringe of fibrous green hornblende. These replacements may, in extreme cases, extend throughout the respective crystals.

Passing now to the more general type of alteration, we find it fairly well exemplified by sheets of Talaidh Type in Cruachan Dearg (16561-2), but more particularly by those of Gaodhail River (14792-5). Here, all the larger and relatively basic felspars show albitization in every stage of development, from strings of albite traversing the crystals to practically complete conversion into albite or albite-oligoclase. Similarly, we meet with the general change of augite into green fibrous hornblende.

### VARIOLITIC ROCKS OF CRUACHAN DEARG.

A series of fine-textured more or less tachylytic sheets, occurring on Cruachan Dearg, and, to a more limited extent, at other localities, possess an extremely beautiful variolitic structure (Fig. 51). On occasion, dark spherules, measuring up to a quarter of an inch in diameter, are discernible in the hand-specimen, but this is the exception rather than the rule. A beautiful rock of this character (17909) was found by Mr. E. G. Radley to contain 50·66 per cent. of silica.

These variolitic rocks consist of slender radiating and branching prismatic crystals of augite, sometimes showing titaniferous colouration, joined together transversely by numerous short rods of magnetite, and set in a colourless glassy matrix that makes but a small proportion of the rock. This base has sometimes devitrified (16553, Fig. 51A) to a variable extent, giving rise to an indefinite crypto-crystalline, or feathery, felspathic mass in which are included small definite patches of clear quartz. Felspar, in elongated form, occasionally accompanies the augite and contributes to the variolitic structure (16557, Fig. 51B); but, at other times, it forms small porphyritic and skeletal individuals (14808) that have no definite relation to the augite. A silica-determination by Mr. Radley gave 53·65 per cent. for the figured specimen (16553).

With the development of a cervicorn, rather than a delicate variolitic, structure on the part of the augite, and an increase in the number of recognizable felspar-crystals, these rocks pass over into variolitic and tholeiitic varieties of the Talaidh Type. In all these transitional varieties (16408, 18963, 17917), the cervicorn structure of the augite is the most distinctive feature, and shows clearly that this growth is an independent expression of the crystallization of

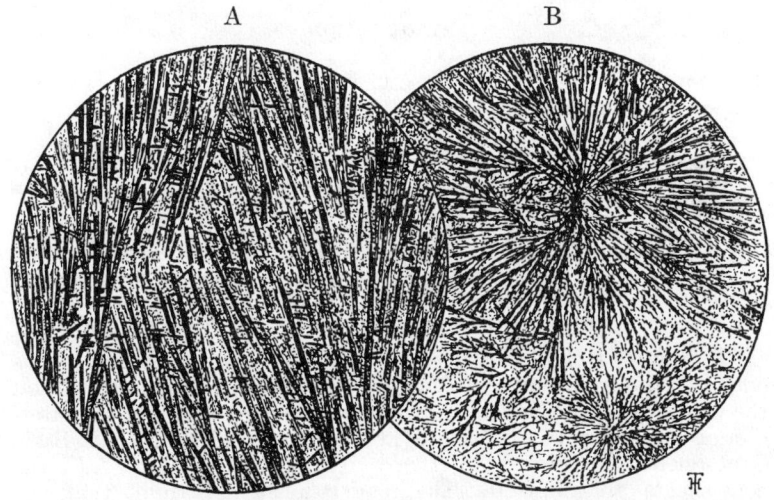

FIG. 51.—Variolite-Sheets of Cruachan Dearg.

A. [16553] × 17. Radiating and branching prisms of augite traversed transversely by short rods of magnetite, and set in a colourless devitrified matrix of indefinite felspathic material in which are small definite areas of clear quartz.

B. [16557]. Augite with attendant magnetite, and accompanied by elongated crystals of felspar, giving rise to a sub-spherulitic variolitic structure. The section shows two definite centres of radiation.

the augite, that is to say, it can take place without the intervention of any other crystalline phase.

A word may be said in conclusion concerning a beautiful variant of the variolitic type collected from the Gaodhail River (14805). It contains small well-shaped phenocrysts of partly albitized acid labradorite in a sub-variolitic ground-mass. The latter consists of cervicorn augite with abundant rods of magnetite, and elongated crystals of oligoclase, in a clear but chloritized base of alkali-felspars and quartz. H.H.T.; E.B.B.

## CHAPTER XXIX.

### RING-DYKES BETWEEN GLEN MORE AND BEINN CHÀISGIDLE.

#### INTRODUCTION.

Two terms, constantly made use of in the present chapter, may be defined as follows :—

> A ring-dyke is a dyke of arcuate outcrop, where there is good reason to believe that its arcuate form is significant rather than accidental. Only in rare instances are ring-dykes so completely developed as to show an entire ring-outcrop (Pl. VI.)
>
> A screen is a narrow partition of older rock separating two neighbouring steeply bounded intrusions. Screens separating ring-dykes have arcuate outcrops. Often in Mull, a screen is formed in large measure of a ring-dyke; but the screens which have proved of most assistance in unravelling the history of the island consist of rock-masses other than ring-dykes (*cf.* Fig. 52, p. 308).

The number of ring-dykes has already been commented upon as one of the main features of Mull geology (p. 6). In nature, and on the one-inch Map, this feature is much obscured owing to the intersection of many of the ring-dykes by multitudes of Late Basic Cone-Sheets (Chapter XXVIII.). Accordingly, in Plate VI., the cone-sheets are omitted. The reader must not think that the difficulties introduced by the presence of these cone-sheets are such as to render the interpretation of the ring-dykes uncertain. In the first place, there are several instances of ring-dykes, which are scarcely cut by cone-sheets—the Loch Bà Felsite seems in fact to be cut by none at all. In the second place, even where the ring-dykes are freely cut, it is still quite possible to draw boundaries for them: the problem presents itself to the field-geologist in much the same light as the mapping of a band through grass-covered country, where exposures, though numerous, are discontinuous. In using the one-inch Map, Sheet 44, the reader must understand that heavy black lines have been printed to enable him to follow boundaries, which have been rendered inconveniently discontinuous by cone-sheet intrusion. He will also find that the more basic ring-dykes are lettered qD or qE, according as they are quartz-dolerite or quartz-gabbro; and that the more acid are lettered (along with several other intrusions) F or G, denoting felsite or granophyre as the case may be. W.B.W., E.B.B., J.E.R.

Plate VI. shows clearly that the ring-dykes of Mull group themselves roughly about two centres ($C_1$ and $C_2$, Fig. 58, p. 338), the one situated in Beinn Chàisgidle, the other near the head of Loch Bà. In the present chapter, only the ring-dykes belonging to the former centre are dealt with; three important ring-dykes referable to the Loch Bà centre are reserved for Chapter XXXII. J.E.R.

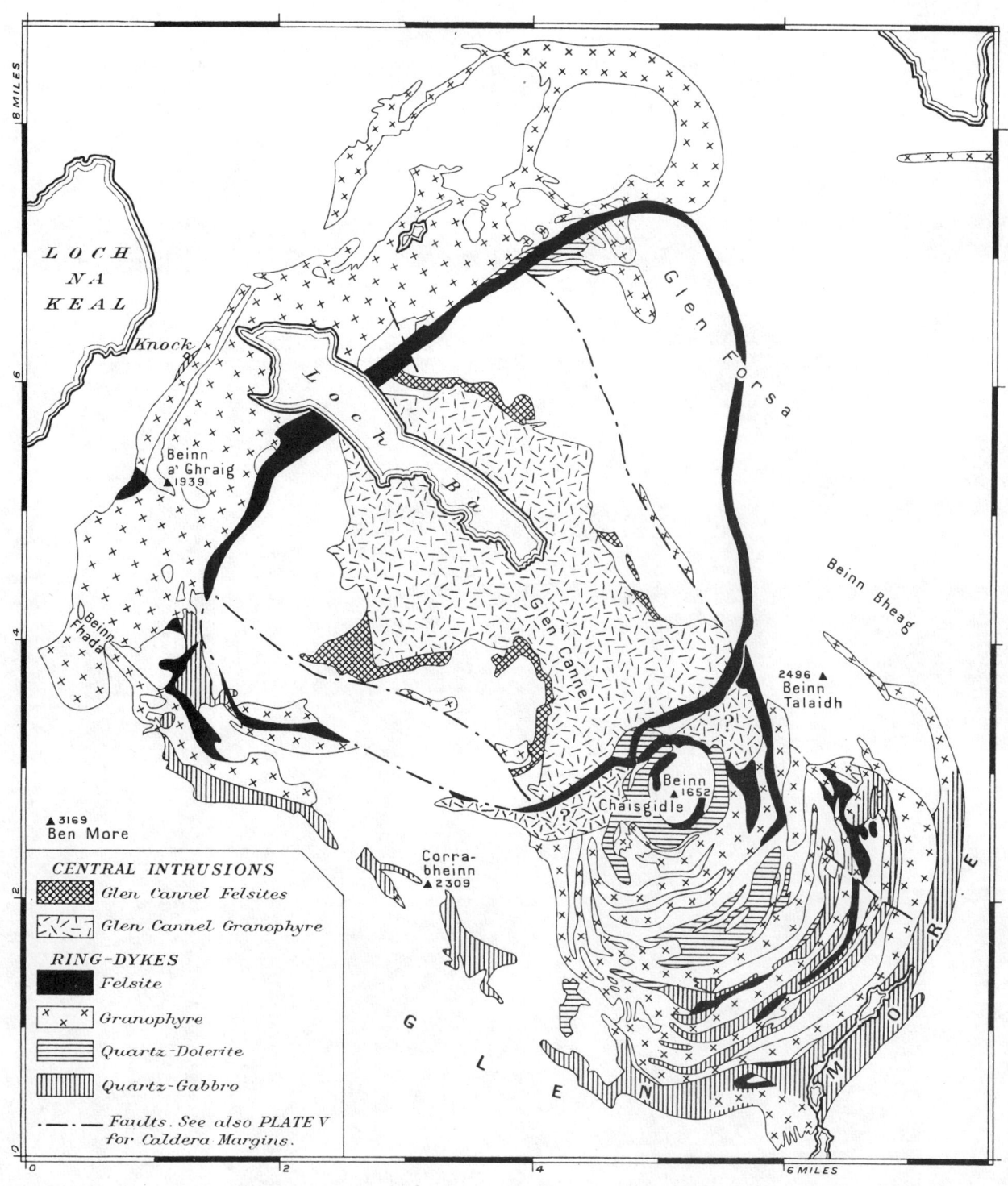

PLATE VI. RING-DYKES AND CENTRAL INTRUSIONS OF MULL

The two centres of ring-dyke activity do not seem to have behaved quite independently, for it is easy to recognize in Plate VI. a certain measure of dual control. In so far as they acted successively, the Beinn Chàisgidle centre functioned at an earlier date than that of Loch Bà. Probably, all the ring-dykes, included in the present chapter, date from the earlier part of the long period which saw the introduction of the Late Basic Cone-Sheets. This matter has already been discussed in some measure in the Time-Relations section of Chapter XXVIII. It is furnished with many additional illustrations in the detailed account of Field-Relations given below.

Unlike the cone-sheets, the ring-dykes of Mull very seldom show chilling at their margins. Instead, one commonly finds a little marginal assimilation of country-rock. The ring-dykes, too, are on the average coarser in crystallization than the cone-sheets. Evidently their cooling was slow; and this, combined with their erect posture, has given favourable conditions for gravitational differentiation, of which several examples are discussed in Chapter XXX.

The discussion of Field-Relations in the sequel is followed by a section on Petrology.

## Field-Relations.

The ring-dykes of Mull are much more elusive than the cone-sheets. This results, in large measure, from the relatively great bulk of the ring-dykes, and the rarity of smooth chilled margins, since both of these characteristics render it impossible to learn much from examination of isolated exposures. In general, it has been found necessary to map the ring-dykes before it was possible to visualize their behaviour. The country in which they occur affords many excellent stream and hillside exposures, and where, as in much of the Glen More district (Fig. 52, p. 308), there is no great complication due to later cone-sheets, the mapping presents no particular difficulty. Great arcuate outcrops of gabbro, dolerite, granophyre, and felsite are revealed, sometimes running side by side, sometimes separated by screens of more or less complex constitution. The boundaries of these belts can be laid down on the map with precision, and, as they happen to cross important ridges and valleys, it is possible to realize that they are very steep. It has not been established whether there is any significant departure from verticality, but this point will be returned to in the description of the Loch Bà Felsite (Chapter XXXII.). Anyone wishing to satisfy himself of the steep dyke-like character of the intrusions here considered would be well-advised to follow for a mile or so the felsite-dyke marked **6** (Fig. 52) in its course across Allt Molach, or, what is equally striking, the south-eastern margin of the granophyre marked **11**.

### ALLT MOLACH (Fig. 52.)

The Allt Molach District has just been mentioned. It was in

308   Chapter XXIX.—*Ring-Dykes of Glen More & Beinn Chàisgidle.*

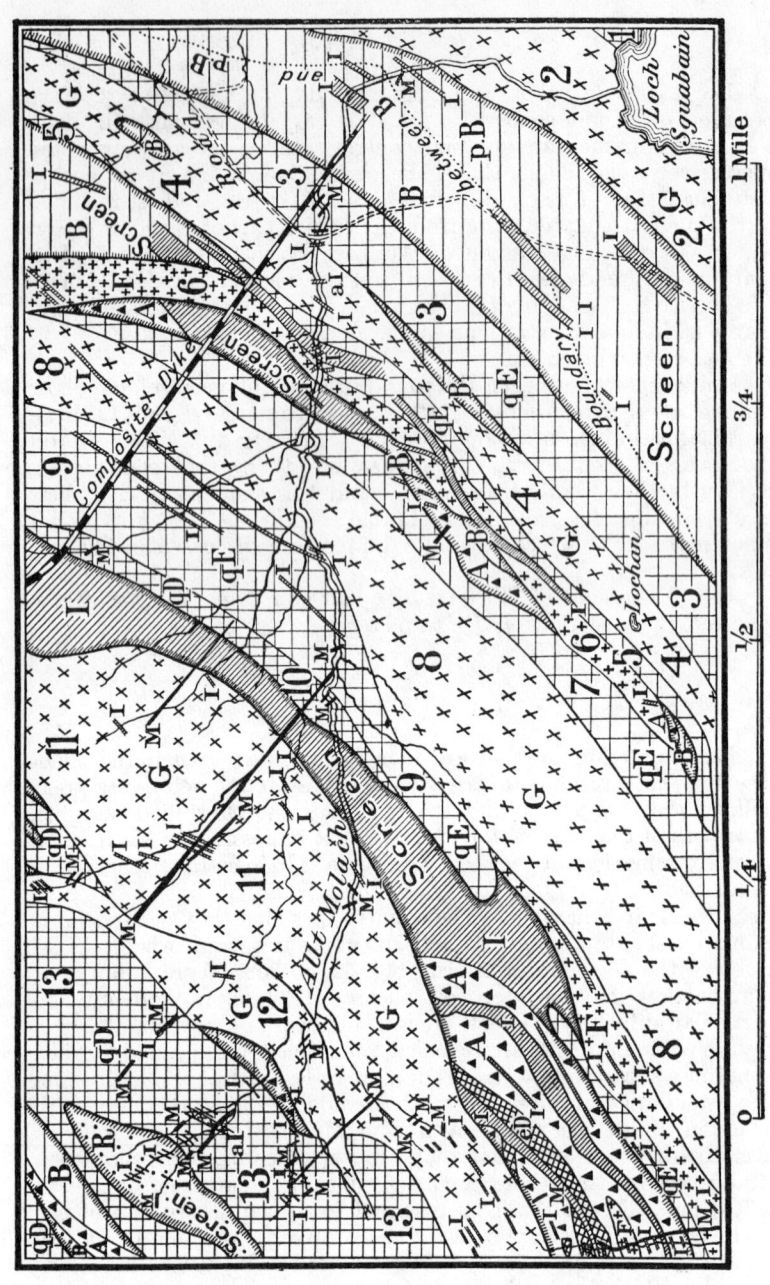

Fig. 52.—Ring-Dykes, Allt Molach.

EXPLANATION OF LETTERS.

DYKES.
  M basalt.

CONE-SHEETS.
  I basic.
  aI acid

RING-DYKES.
  F felsite.
  G granophyre.
  qD quartz-dolerite.
  qE quartz-gabbro.

SCREENS.
  R rhyolite.
  A agglomerate.
  B compact basalt-lava
  pB porphyritic basalt-lava.
  eD dolerite.
    also
  many of the cone-sheets (I).

Numbers explained in Text.
Margins of Screens shaded.

this district that the number and importance of the Mull ring-dykes first became apparent ; and it is hither that the geologist should still turn if he wants to investigate the subject to full advantage. Fortunately, no hard physical work is entailed, and the stream-section is crossed at one point by a driving-road, which renders it easy of access. The description, which follows, supplies a detailed guide to much of the geology illustrated in Fig. 52, for even the complex internal relationships of the various screens bear indirectly— if only by way of contrast—upon the subject matter of this chapter.

Intrusions 1 and 2 of Fig. 52 stand for the Glen More and Ishriff Ring-Dykes respectively, both of which make prominent features on the one-inch Map. Their further consideration is postponed, as there are no exposures of either within Fig 52.

Between 2 and 3 lies a Screen of basalt-lavas, poorly though sufficiently exposed. A fair number of Late-Basic Cone-Sheets (Chap. XXVIII) traverse these lavas. In Allt Molach, compact non-porphyritic basalt-lavas, along with neighbouring gabbro 3, are cut by a 15 ft. felsite-dyke with 1 ft. basalt-margins (composite). This dyke figures prominently again in a tributary stream, half a mile farther north-west.

Intrusion 3 is a Ring-Dyke of quartz-gabbro. Between the out-crop of the composite dyke, mentioned above, and the bridge, by which the road crosses Allt Molach, this gabbro is cut by three basalt-dykes. Just west of the bridge, it is cut by a Late Basic Cone-Sheet There is no evidence as to whether this cone-sheet cuts the neighbouring granophyre 4. The junction of 3 and 4 is exposed ; there is no chilling, but, instead, a very narrow belt of merging, such as is common at a contact of plutonic rocks. There is no evidence whether 3 is later than 4, or *vice versa*.

South of Allt Molach, a narrow Screen of compact non-porphyritic basalt-lava locally separates 3 and 4. Perhaps the outcrop of lava shown, north of Allt Molach, as surrounded by 4, may really act as a screen between 3 and 4, but exposures are not full enough to decide this point.

Intrusion 4 is a granophyre-felsite Ring-Dyke. It is granophyre with acicular crystallization as exposed in Allt Molach ; it is basic granophyre, but not acicular, in the stream-exposure between lava-outcrops 400 yds. farther north-east ; it is granophyre in most of its exposures south of Allt Molach, but locally, where its outcrop narrows west of the lava-screen, 200 yds. south of Allt Molach, it assumes the character of felsite with a small felspar-phenocrysts. In Allt Molach, the granophyre is cut by a thin sub-acid cone-sheet, and also by a 12 ft. Late Basic Cone-Sheet, of which the chilled top is well-exposed. The junction of 4 and 5, though exposed, gives no clue to age-relations.

Intrusion 5 is a Ring-Dyke of acid quartz-gabbro merging into granophyre. In Allt Molach, it is cut by a big Late Basic Cone-Sheet of medium grain with well-exposed chilled top. This cone-sheet is in turn cut by two minor Late Basic Cone-Sheets, and also by acid veins, the latter probably emanating from 6. Upstream from the big cone-sheet, the Ring-Dyke 5 is cut by two thin Late Basic Cone-Sheets, which are seen close to the junction of 5 and 6 ; the first of these runs north-west, and cuts veins from 6 ; the second runs north and south, and is cut by veins from 6. Thus Allt Molach supplies a local time-scale :— 5 followed by Late Basic Cone-Sheets, followed by 6, followed by more Late Basic Cone-Sheets.

North of Allt Molach, an important Screen of compact non-porphyritic basalt-lava wedges in at the outcrop between 5 and 6, or, where 5 locally fails, between 4 and 6. Exposures of these lavas are met with in a small stream 400 yds. north of Allt Molach, and continue northwards along the hillside well beyond the limit of Fig. 52. Near the southern edge of Fig. 52, a much less extensive screen of agglomerate and compact non-porphyritic basalt separates 5, on the one side, from 6 and 7, on the other. The agglomerate of this screen lies flatly on the lava (p. 197).

Intrusion 6 is a felsite Ring-Dyke carrying small felspar-phenocrysts. The Allt Molach exposure is so full of basaltic and doleritic xenoliths that it does not give a fair sample. A little tributary a few yards north of Allt Molach, and crags either side of the valley, are more representative. In the paragraph relating to Ring-Dyke 5, it has been shown that 6 cuts some Late Basic Cone-Sheets, and is cut by others. This relationship will be further illustrated in the succeeding paragraph on the Screen separating **6** and **7**. Here, it is only necessary to draw attention to five Late Basic Cone-Sheets shown in Fig. 52 as cutting **6**. The most prominent of these lies south of Allt Molach, and has been traced for about 500 yds. Its chilled top is well-exposed.

A Screen has been mapped for two thirds of a mile between **6** and **7**. In Allt Molach, it consists of fine dolerite, veined by **6** on the one side, and merging rapidly through digestion into **7** on the other. Probably, this dolerite consists of one or more basic cone-sheets greatly altered by the Ring-Dykes, which it separates. Half-way along its stream-exposure, this fine dolerite is cut by a thin dolerite cone-sheet, which is of later date than the veins from **6**. South of Allt Molach, the dolerite of the Screen gives place to basalt-lava and agglomerate. The basalt is of the compact non-porphyritic type; it has two isolated out-crops, and in the more southerly of these is seen passing steeply north-west under agglomerate (p. 197). The field-relations of the agglomerate to the Ring-Dykes (**6** and **7**) are full of interest: acid material from **6**, which merely veins the associated lava, floods the agglomerate; marginally, basic material from **7** also invades the agglomerate, but not to the same extent. The agglomerate contains many fragments of a type identical with the intruding felsite; and, at first sight, it is difficult to believe that agglomerate and felsite are not vitally connected in origin. However, the evidence against this view is complete; the agglomerate is seen, in bare rock-exposures, to be cut by basic cone-sheets with chilled margins; and these basic cone-sheets are themselves freely veined by the acid material from **6**, though not to the same extent as the agglomerate, which latter has in this respect reacted as an incoherent mass. Two of these interesting cone-sheets are picked out in Fig. 52, north-east of a basalt-dyke lettered M.

Intrusion **7** is a quartz-gabbro Ring-Dyke, no more than sufficiently exposed for mapping purposes. Its main interest lie in its relationship to the screen on the south-east; and this has been mentioned in the preceding paragraph. Only one Late Basic Cone-Sheet has been noticed traversing **7**.

Intrusion **8** is a Ring-Dyke of granophyre with a acicular crystallization. It is poorly exposed for the most part, though sufficiently to allow of its being mapped with certainty. It is cut by a few thin Late Basic Cone-Sheets. Allt Molach furnishes a section of the contact **8–9**, where it is fairly evident that **9** is earlier than **8**.

Intrusion **9** is a Ring-Dyke of quartz-gabbro, which, north of Fig. 52, passes uphill into granophyre (Chap. XXX.). It is cut by a few basalt dykes and Late Basic Cone-Sheets. The Allt Molach exposure of its junction with **10** shows that **9** cuts **10**. Thus Allt Molach supplies a fairly certain time-scale:—**10** followed by **9**, followed by **8**. E.B.B.

A Ring-Dyke of quartz-gabbro, which may be a discontinuous portion of **9**, reaches down to the south-west corner of Fig. 52, where it is separated from **8** by a felsite Ring-Dyke, that is unrepresented in the numbered sequence of Allt Molach. (C.T.C.)

Intrusion **10** is a Ring-Dyke of quartz-dolerite. It affords an additional, but rather poor, example of uphill passage into granophyre. Its time-relations with **9** are stated above. E.B.B.

Between **10** and **11**, and where **10** fails, between **9** and **11**, there stands an important Screen, poorly exposed in Allt Molach, but conspicuous on either side. North of Allt Molach, this screen consists of very steep dolerite cone-sheets with chilled margins. South of the valley, it is largely made of agglomerate and associated brecciated early dolerites, both cut by big dolerite cone-sheets with chilled margins. In the southern exposures, there are also many thin basic cone-sheets, which may be of later date than **11**. The massive dolerite cone-sheets, which figure so prominently in this Screen, are more like Early (Chap. XXI.), than Late, Basic Cone-Sheets in their field-appearance, and are grouped accordingly on the one-inch Map. (C.T.C.), E.B.B.

Intrusion **11** is a Ring-Dyke of non-porphyritic granophyre. The manner in which this granophyre truncates the more important of the cone-sheets occurring in the screen just described is strikingly exhibited both north and south of Allt Molach. The granophyre gives rise to soft grassy slopes, against which the crags, due to the dolerite-sheets, terminate abruptly. Several dykes, one of them composite, and also thin Late Basic Cone-Sheets, have been noted cutting **11**.

## Field-Relations.

Intrusion **12** is a Ring-Dyke of granophyre, distinguished from **11** by small phenocrysts of felspar and needles of augite. Probably **12** cuts **13**, but this is not quite clear.

In Allt Molach, a Screen of doubtful tuff and lava separates **12** and **13**.

Intrusion **13** is a great Ring-Dyke of quartz-dolerite, sometimes vesicular. It is freely cut by thin Late Basic Cone-Sheets.

### HILL-SLOPES BETWEEN FIGS. 52 AND 53.

The main features of interest illustrated on the hill-slopes connecting Figs. 52 and 53 are :—

(a) The horizontal shifting of the ring-dykes by a fault which runs just south-west of the stream that rises on Monadh Beag. The fault and its effects are clearly shown on the one-inch Map.

(b) The upward bifurcation of the Ring-Dyke **9** of Fig. 52 a little south of the fault just mentioned. This interesting phenomenon is due to the contemporaneous employment of two ring-fissures by a single intrusion. Another example will be described presently in the case of the Glen More Ring-Dyke (Fig. 54, p. 322).

(c) The upward transition of quartz-gabbro into granophyre in the two arms of the ring-dyke **9** (Chap. XXX.).

Anyone wishing to realize, without much expenditure of time, the continuity of the various ring-dykes, represented on the one-inch Map, will find very favourable exposures of the quartz-gabbro **3**, where it shows through the grass as a long narrow ridge north of the fault referred to above. The felsite **6** is also admirably exposed, both north and south of the fault.

### MAOL NAM FIADH (FIG. 53).

Fig 53 illustrates the geology of an interesting district south-east of Beinn Talaidh. The numbering of the ring-dykes is adopted from Fig. 52, already described.

Intrusion **1** is the Glen More Ring-Dyke. It enters the south-east corner of Fig. 53 as quartz-gabbro. Northwards, it merges gradually into granophyre—at first, coarse-grained and acicular ; farther on, fine-grained and acicular ; and, finally, in the north slope of Coire Ghaibhre, fine-grained and non-acicular. This change is dealt with again in Chap. XXX. ; here it is only necessary to point out that, though exposures are often poor, they are connected by a fairly continuous depression with rock rising on either side. In Coire Ghaibhre, the course of the Glen More Ring-Dyke is marked out in striking fashion. If, for instance, one looks from Torness on the Glen More road, one is impressed by the abrupt termination of crags of Beinn Bheag Gabbro (Chap. XXII.) against a grassy slope due to Glen More Granophyre. The crags just mentioned are themselves separated by lanes of grass following the outcrops of Late Basic Cone-Sheets. Of course, the distant view does not show whether these sheets are cut by the granophyre, but, fortunately, clear stream-sections in Coire Ghaibhre prove that the granophyre is here untraversed by sheets ; there can be no doubt, in fact, that the Glen More Granophyre cuts the host of Late Basic Cone-Sheets, which intersect the Beinn Bheag Gabbro in neighbouring exposures, on either side.

Just west of the termination of the Glen More Ring-Dyke, granophyre of a similar type is met with, but may be distinguished by the fact that it is freely cut by Late Basic Cone-Sheets. Moreover, this granophyre is traceable south-westwards for some distance past the Glen More Ring-Dyke, so that there can be little doubt that the two are quite different intrusions.

A large part of Fig. 53 illustrates a Screen which separates the northern terminations of **1** and **2**. The main element in this screen is afforded by Late Basic Cone-Sheets, separated, here and there, by lenticles of lavas, dolerite, gabbro, etc. :

312 *Chapter XXIX.—Ring-Dykes of Glen More & Beinn Chàisgidle.*

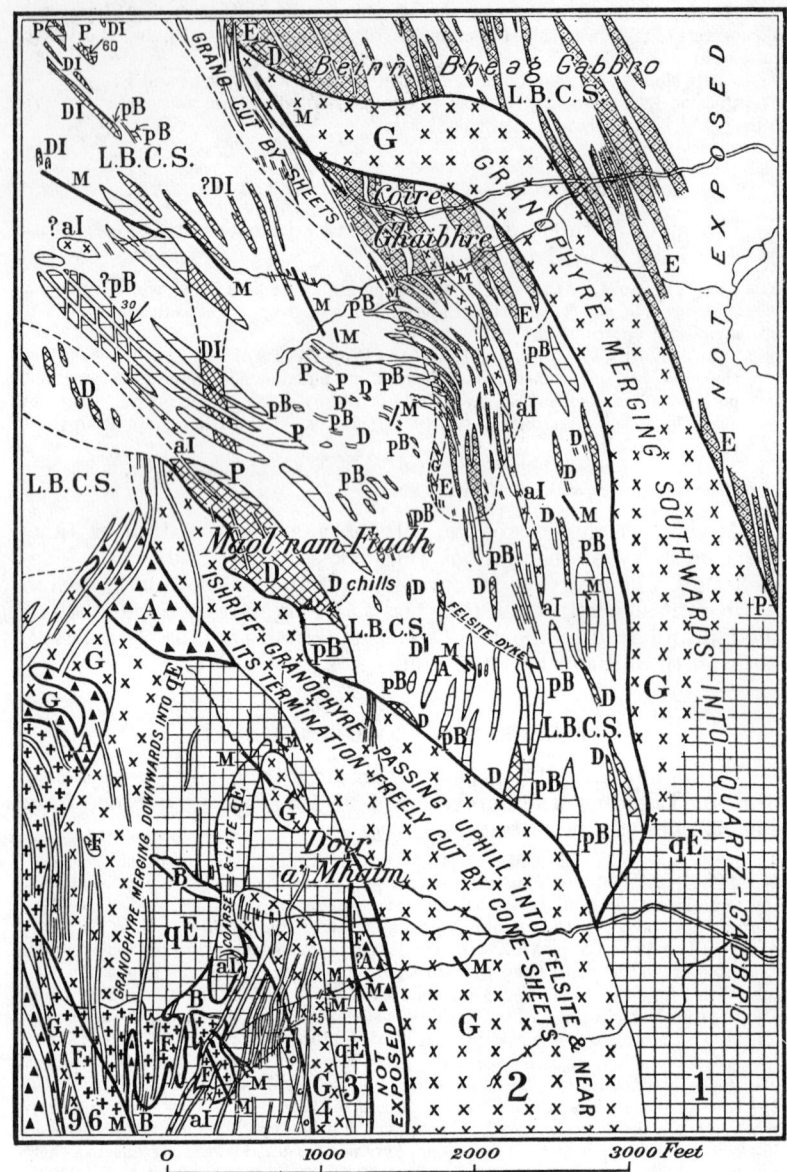

FIG. 53.—Ring-Dykes, Maol nam Fiadh.

**Dykes**: M basalt.
**Cone-Sheets**: DI dolerite; aI acid; L.B.C.S. Late Basic Cone-Sheets (shown without ornament).
**Ring-Dykes**: F felsite; G granophyre; qE quartz-gabbro.
**Screens**: A agglomerate; B compact basalt-lava; pB porphyritic basalt-lava; P pillow-lava; D dolerite; E gabbro; also many of the cone-sheets.
Numbers as in Fig. 52, *see* Text.

the lavas are porphyritic basalt, sometimes with pillow-structure; the dolerite includes representatives of Early Basic Cone-Sheets; the gabbro in Coire Ghaibhre belongs to a severed continuation of the Beinn Bheag mass.

Intrusion **2** is the Ishriff Ring-Dyke of granophyre passing northwards uphill into felsite. In the stream-sections of Doir' a' Mhàim, it is seen to be free of the numberless Late Basic Cone-Sheets that figure so largely in the Screen just described. Traced towards its termination in Maol nam Fiadh, it is found to be cut by many Late Basic Cone-Sheets. The inference is that the more westerly cone-sheets of the Late Basic Group are here of later date than the more easterly (p. 229).

Between **2** and **3**, a poorly exposed Screen intervenes. Tuff, and perhaps lava, are seen in the first stream north of the southern margin of Fig. 53; and some felsite in the second. Farther north, **2** comes against **3**, only to part company again and expose agglomerate, which is eventually lost sight of among Late Basic Cone-Sheets.

Intrusion **3** is a Ring-Dyke of quartz-gabbro, which merges up-hill into granophyre (Chap. XXX.). The acid western part of **3** is freely cut by Late Basic Cone-Sheets. The junction of **3** and **6** was found exposed at two localities, but age-relations were not clear.

Intrusion **4** is a Ring-Dyke of granophyre. It is freely cut by Late Basic Cone-Sheets.

Between **4** and **6**, stands a Screen largely composed of compact basalt-lavas.

Intrusion **6** is a Ring-Dyke of Felsite with small felspar-phenocrysts. An isolated protrusion of **6** into the lavas of the Screen, just referred to, is interesting for its marginal development of a very perfect fluxion-breccia consisting of fragments of flow-banded felsite enclosed in a felsitic base. **6** is freely cut by Late Basic Cone-Sheets.

For a quarter of a mile, a coarse quartz-gabbro can be traced north and south almost across the outcrop of **3**. It seems to be later than almost all the cone-sheets traversing **3**, **4**, and **6** in its vicinity.

Intrusion **9** is a Ring-Dyke of sub-acid granophyre, which, in its down-hill continuation in Fig. 52, is represented by quartz-gabbro (Chap. XXX.). E.B.B.

### GLEN MORE RING-DYKE: **1** OF FIGS. 52 AND 53.

The Glen More Ring-Dyke, owing to its exterior position, is easily identifiable on Plate VI. (p. 307). It consists essentially of quartz-gabbro with subordinate associated granophyre. Starting near Beinn Talaidh, the Ring-Dyke has a six-mile continuous arcuate outcrop through Glen More to Cruach Choireadail, where it branches, and, for a space, terminates (Fig. 54, p. 322). It is brought to light by erosion, again, in two valleys, Coir' a' Mhàim and Coir' an t-Sailein, situated either side of Corra-bheinn. It is hidden for a little beyond Coir' an t-Sailein, but then is traceable for a couple of miles north-west through Tòrr na h'Uamha; and, very probably, a northward continuation is to be recognized in Plate VI., east of Beinn Fhada.

The complex relationships of the Glen More Ring-Dyke with the Late Basic Cone-Sheets have already been dealt with in Chap. XXVIII. (p. 299), and the association of gabbro and granophyre at various points of the dyke supplies the main subject-matter of Chap. XXX. Accordingly, the present account is mainly devoted to two other features, namely, the **exposures** which enable the intrusion to be followed in the field, and the very interesting **branching phenomena** evident in Cruach Choireadail, Coir' a' Mhàim, and Coir' an t-Sailein. (C.T.C.), E.B.B.

*Exposures.* The northern termination of the Glen More Dyke has been sufficiently dealt with in the description of Fig. 53. South of this, along Glen More till Loch an Ellen is past, the exposures of the gabbro are scattered owing to their separation by spreads of moraine, peat, and alluvuim; but they are sufficiently numerous to introduce no significant uncertainty in drawing boundaries on the map. The most extensive of these exposures are met with :—

(1) About 100 yds. up from the Glen More road, on either side of an eastward flowing stream which crosses the road, 500 yds. south of the big bend near Torness.

(2) Between the road and the valley-bottom, from Loch Sguabain to Loch an Eilein.

(3) Above the morainic cover of the south-east slope between Loch Sguabain and Loch an Eilein. A junction with lavas is seen along this line, and the lavas are conspicuously traversed by white granophyric strings which likely originated, in some sense, from the Glen More intrusion. E.B.B., G.V.W.

West of Loch an Ellen, the Glen More Ring-Dyke consists in part of gabbro, in part of granophyre (*see* one-inch Map and Pl. VI.) The boundary between the two types is rather sharply marked, and does not show any definite tendency to follow the contours of the hill-slope. It is, therefore, an association which cannot be attributed simply to gravitational differentiation (*see* Chap XXX.). (C.T.C.), E.B.B.

The granophyre of this complex part of the Glen More Ring-Dyke extends in a marked southwards bulge to the foot of Loch Airdeglais. A stream entering the loch from the west gives a very interesting exposure of the southern end of the bulge, and shows it to have a unique type of junction with the adjoining rocks—probably lavas. Instead of, as elsewhere, presenting an almost smooth boundary, the intrusion penetrates the country-rock in all directions as irregular veins. Perhaps, the acid strings, already commented upon, in the border-zone of lavas above Loch Sguabain and Loch an Eilein are part of this phenomenon, less strongly developed; Unfortunately, the connecting exposures east of Loch an Ellen are too poor to furnish detailed information. G.V.W.

From Loch an Ellen westwards and northwards, exposures of the discontinuous outcrop assigned to the Glen More Ring-Dyke are so numerous that they need not detain us in their enumeration.

*Upward Branching.* The type-example of upward branching is displayed on the slopes of Cruach Choireadail, and is illustrated in Fig. 54 (p. 322). The Glen More Ring-Dyke at the bottom of the valley, roughly 500 ft. above sea-level, is 500 yds. wide. At 750 ft. above sea-level, it sends off a horizontal sheet in a south-westward direction. This horizontal sheet is roughly 500 yds. long and 400 ft. thick, and its horizontality is emphasized by its showing a vertical development of rude columnar jointing. At the north-east end of the sheet, the parent dyke continues upward, now reduced to 300 yds. in width; while, at the south-east end, as it were to compensate for this reduction, an additional dyke rises, measuring 200 yds. across. Both of these upward dykes thin out completely when followed for half a mile north-westwards. This tapering can, in part, be ascribed to an upward termination of the dykes; but, at the same time, it is clear that the dykes do not terminate upwards along horizontal lines, since nothing is seen of them in An Coireadail, west of the Cruach, and their next exposures, in Coir' a' Mhàim, are notably on lower ground.

Pl. VI. illustrates, sufficiently clearly, a tendency for both branches of the Cruach Choireadail exposure to manifest themselves in Coir' a' Mhàim and Coir' an t-Sailein, on the two sides of Corra-bheinn. Coir' a' Mhàim approximately reproduces the Cruach Choireadail phenomena of upward branching based upon a connecting horizontal sheet. In Coir' an t-Sailein, the two branches converge downwards upon one another.

It seems fair to claim these exposures as furnishing an occular demonstration of the simultaneous employment of two parallel fissures by the ascending magma of a ring-dyke. (C.T.C.)

*Tòrr na h-Uamha.* The Tòrr na h-Uamha outcrop does not show the branching phenomena just described; but it is grouped as part of the Glen More intrusion on account of its alignment and its similarity of lithological type. In Tòrr na h-Uamha, it is of basic character, whereas, in what appears to be its severed northward continuation, east of Beinn Fhada, it grades uphill towards granophyre (Chap. XXX.). In its more westerly (and northerly) parts, the Tòrr na h-Uamha outcrop is very freely cut by Late Basic Cone-Sheets. This, at first sight, distinguishes it from the greater part of the Glen More intrusion; but, as already stated, a strikingly dual relationship in regard to Late Basic Cone-Sheets can be adduced within the limits of the Coir a' Mhàim outcrop (pp. 299, 313), so that it is independently certain that a fair

number of Late Basic Cone-Sheets are later than the Glen More Ring-Dyke, although perhaps the majority are earlier.
E.M.A., J.E.R.

### ISHRIFF RING-DYKE (2 OF FIGS. 52 AND 53).

Several circumstances combine to emphasize the individuality of the Ishriff Granophyre. In fact, this ring-dyke is traceable for some four miles in arcuate outcrop from near Beinn Talaidh to the foot-slopes of Cruach Choireadail before it comes in contact with others of similar lithological type ; and then it no can longer be singled out. In most of this course, it has gabbro of the Glen More Ring-Dyke, on its outer side, and a screen largely composed of lava, on its inner. On the slopes of Cruach Choireadail, a very well-defined narrow screen of baked rock, probably in large measure Late Basic Cone-Sheets, is interposed between the Ishriff and Glen More Ring-Dykes.

Allt Molach and a very minor neighbouring burn, north of Ishriff, give no exposures of the Ishriff Granophyre ; otherwise, every little stream which crosses its outcrop—and there are four of them omitted from the one-inch Map between Ishriff and Maol Tobar Leac an t-Sagairt—reveals its presence. Hillside exposures are rare except on Maol Tobar Leac an t-Sagairt.

The Ishriff Ring-Dyke falls well within the period of Late Basic Cone-Sheets, as has been described already in relation to Fig. 53, and also in Chapter XXVIII., p. 299. The age-relations of the Ishriff and Glen More Ring-Dykes have not yet been determined.
(C.T.C.), E.B.B.

### BEINN CHÀISGIDLE (PL. VII., p. 307).

In the central region about Beinn Chàisgidle, all the ring-dykes indicated in Plate VI. are freely cut by Late Basic Cone-Sheets. In Socach a' Mhàim, the proportion of sheets is so overwhelming that the boundaries assigned to ring-dykes must be regarded as somewhat tentative. Elsewhere, they are sufficiently well-defined. From Màm an Tiompain along the Maol Uachdarach ridge, a feature, which helps greatly in the mapping of the ring-dykes, is the occurrence of conspicuous screens which are apt to weather out at the surface in low broken rocky ridges. These screens consist mostly of intrusive dolerites which have been to some extent brecciated by explosion before the advent of the ring-dykes. The screens also include a certain proportion of agglomerate and basalt-lava.

The contrasted relationships of ring-dykes and cone-sheets, in the Glen More and Beinn Chàisgidle districts, must not be taken as suggesting that the ring-dykes are of earlier date near the Beinn Chàisgidle centre, than they are towards the Glen More periphery. Very likely, it means no more than that the Late Basic Cone-Sheets, so prominently developed in Beinn Chàisgidle, are later than the great majority of those in Glen More (p. 300).
J.E.R.

### PETROLOGY.

The ring-dykes to be discussed in this chapter show, as an assemblage, a wide range of composition, passing from quartz-dolerite and gabbro, fairly rich in olivine-pseudomorphs in their more basic members, to felsite and granophyre, essentially composed of alkali-felspar and quartz. In several cases, such variations in composition are methodically linked together within the exposed limits of an intrusion in such a manner as to suggest differentiation of the mass in place. The detailed consideration of this side of the subject, is reserved for Chapter XXX. In other cases, there is well-established evidence of assimilation, or hybridization, of one mass by another at, and near, their mutual contacts.

The following petrological account is primarily intended as a supplement to the field-descriptions, that have already been given, of Figs. 52 and 53. The opportunity is taken at the end of the section to describe somewhat fully the Sgùlan Type of olivine-bearing quartz-dolerite (qD of one-inch Map), since this particular variety does not figure prominently in other chapters of the Memoir.

In the following summary of petrological characters, the numbers 1 to 13 are used as in Figs. 52 and 53 to distinguish various members of the ring-dyke-complex therein depicted.

### PETROLOGY OF FIGS. 52 AND 53.

Intrusion 1. The Glen More Ring-Dyke, in its extension to the west of the region shown in Fig. 52, has supplied most of the material for the petrological descriptions given in Chap. XXX. Elsewhere, its intermediate and acid phases are represented by three specimens (18673-5) obtained from the more southerly stream of Coire Ghaibhre (Fig. 53). Two of these were collected at a point where a wall impinges on the stream from the south, and in them we have a reproduction of the augite-diorite type described in Chap. XVIII.

The other specimen, taken from a point further up stream, is a granophyre rich in oligoclase and soda-felspar.

Intrusion 2. The Ishriff Ring-Dyke has not been sliced.

Intrusion 3. A doleritic rock (17428), collected from Allt Molach (Fig. 52), shows a hypidiomorphic medium-grained type of crystallization of its constituent olivine, augite, and labradorite, with conspicuous early-formed magnetite, and a fair proportion of late-formed quartz. Olivine is entirely replaced by chlorite and serpentine, and the basic felspar is in large measure albitized. The rock, as a whole, agrees closely with the type of quartz-gabbro that forms the lower portion of the Glen More Ring-Dyke (Chap. XXX.). Followed to the north (Fig. 53), it passes upwards, through an augite-diorite facies, into granophyre, but representatives of these intermediate and acid phases have not been examined microscopically.

Intrusion 4. A specimen (17432) from this mass, as exposed in Allt Molach (Fig. 52), is composed mainly of elongated crystals of albite, fringed with orthoclase and micropegmatitic growths. There is a small proportion of definitely individualized quartz, which generally occurs as irregular patches, but occasionally exhibits crystal-boundaries. The original ferromagnesian mineral was not abundant and is now entirely decomposed. A feature of the section is the occasional bending and breaking of the felspar laths. Such deformation has been noted in the case of other rocks of similar character.

Intrusion 5. An example, collected from Allt Molach (17434) to the east of the cone-sheet lettered I in Fig. 52, is of a rock that approaches more nearly to augite-diorite (Chap. XVIII) than quartz-gabbro in composition, although mapped for convenience as qE. An early crystallization has led to a development of somewhat stumpy crystals of oligoclase with associated hypidiomorphic augite and magnetite, together with a few pseudomorphs after olivine. The oligoclase is zoned marginally by, and converted locally into, albite, and there is a residium of alkali-felspar and quartz. Apatite is an important accessory. The acicular type of crystallization, so frequently developed in rocks of this composition, is wanting in the slice (17434); but, in the stream-section, the rock is seen to be variable in structure and sometimes acicular.

Contact 5-6. The Ring-Dyke 5 is represented by a specimen (17437) collected from Allt Molach, where it is veined by Ring-Dyke 6. This in many respects reproduces features of the augite-diorites, such as cervicorn augite (p. 303), zoning and corrosion of andesine by alkali-felspar, and a fairly abundant acid mesostasis; but some of this acid mesostasis may have penetrated from 6, and one notes the influence of the latter intrusion in a development of granulitic clusters of augite and rhombic pyroxene, which, it is clear, in some instances have formed in the place of pseudomorphs after olivine.

Intrusion 6. A specimen (17439), collected from the small tributary stream immediately north of Allt Molach (Fig. 52), represents the small-felspar type of Ring-Dyke 6. It is a felsite in which the chief phenocrysts are little rectangular crystals of orthoclase and albite. These are accompanied by subordinate oligoclase and corroded augite, both of which may be cognate xenocrysts rather than porphyritic constituents. The matrix is composed of crowded minute equidimensional crystals of alkali-felspar set in a quartz-residuum that has little or no tendency towards micropegmatitic intergrowth with the neighbouring felspar. The section shows the rock to be veined with felsitic matter, and, on the field-evidence, it would appear that these veins are auto-genetic as far as the enclosing rock is concerned. Another specimen (17438), taken as representing the xenolithic facies of this ring-dyke as exposed in Allt Molach, is closely comparable to the felsite (17439) described above, except that it is of slightly coarser texture and is crowded with xenoliths. Many of the xenoliths appear to have been derived from olivine-free basalt-lavas similar to those lettered B in Fig. 52. They exhibit a beautiful micro-granulitic structure of their component minerals—augite, felspar, and magnetite. Considerable assimilation has been effected by the felsite, and all stages of contamination may be studied. One of the chief effects is the production in the felsite of a pyrogenetic greenish hornblende as minute crystals (Chap XXXIII.). Specimens (18678-82) taken from the eastern side of the small detached mass of the Ring-Dyke 6, where it is exposed in the second streamlet to the north of the southern margin of Fig. 53, illustrate the intrusion-breccia which here constitutes the margin of the felsite. The chief fragments are of a fluxional rhyolite, such as might have formed on the rapid early consolidation of the ring-dyke felsite. There are also a few enclosures of olivine-free lava of basaltic composition. The matrix is generally of a felsitic character, but, exceptionally (18681), quartz, in equidimensional grains, predominates over the alkali-felspar that functions as a base.

Screen 6-7. The screen, that separates the Ring-Dykes 6 and 7 in the Allt Molach section (Fig. 52), has been so altered by these intrusions that its original nature is obscure. It is probably, however, of complex nature, in large measure consisting of Late Basic Cone-Sheets. A specimen (17460) was collected at a point where the basic material of the screen was manifestly veined by more acid material, that appeared, in the field, to be attributable to the felsite 6. The microscope-section shows a medium-grained basic rock in which a porphyritic tendency of the original felspars can easily be detected. The structure of the rock as a whole is decidedly granulitic with augite, rhombic pyroxene (pseudomorphed), plagioclase, and magnetite, all well represented. There are areas, however, that simulate closely the ocellar structure of camptonites. These areas have a relatively coarse texture, and the minerals present are hypersthene, biotite, augite, quartz, and alkali-felspar. The hypersthene is restricted to the margins of the areas and occurs as stumpy crystals, occasionally unaltered, but more frequently replaced by fibrous hornblende. The biotite usually builds poecilitic plates that occur either at the margins of the areas, or in the adjoining granulite; it also appears as smaller crystals within the areas, and is always marginally corroded. Hornblende occurs in various forms. It is usually grown about the corroded augite, and is then of a deeply coloured, well-formed pyrogenetic type. The paler varieties of hornblende, also replacing augite, are clearly of later formation than the darker varieties where the two are seen in contact. Quartz and alkali-felspar show micrographic intergrowth. Altogether, these areas, which are interpreted as the result of the introduction of acid material from Ring-Dyke 6, reproduce characters described in connexion with the hybrid types dealt with in Chap. XXXIII.

Another specimen (17440), taken from the assimilation-zone that connects the screen with Ring-Dyke 7, reminds one of certain rocks entering into the composition of the granulitic zone that borders the Ben Buie Gabbro near Loch Fuaran (Chap.

XXII.). It does not, however, permit of interpretation without the study of additional material.

Intrusion 7. The Ring-Dyke 7, as represented by a specimen (17441) from Allt Molach (Fig. 52), is of quartz-gabbro type, containing a few pseudomorphs after olivine.

Intrusion 8. Ring-Dyke 8, as exposed in Allt Molach (Fig. 52), is a felsit (17444) composed of crystals of microperthite and albite in a matrix of micropegmatite and free quartz. Groups of magnetite and augite crystals play a subordinate part. The augite is much corroded and frequently edged by an alkaline variety (aegerine-augite), and relics of a more basic plagioclase (oligoclase) may be detected as survivals in the centre of some of the more alakine felspars.

Intrusion 9. The Ring-Dyke 9 is represented by three specimens which were chosen to show the upward passage of quartz-gabbro into granophyre. The most basic of these (17442), collected from Allt Molach (Fig. 52) about 700 feet above sea-level, is an olivine-bearing quartz-gabbro. The intermediate variety (17443), from Monadh Beag (one-inch Map), at a height of about 1550 feet above the sea, has the usual characters or these rocks, and is allied to the augite-diorites of Chap. XVIII. The highest specimen (17461), from about the 1600 foot level on Monadh Beag, is a basic granophyre or craignurite. The types resemble those fully described in Chap. XXX. Here it is only necessary to mention, that the sliced intermediate specimen (17443) contains a xenolith of a type similar to the Late Basic Cone-Sheets, and that, by reason of assimilation, there has been a development of pyrogenetic hornblende as in the hybrids described in Chap. XXXIII.

Intrusion 10. The Ring-Dyke 10 has not been sliced.

Intrusion 11. A specimen (17445) of the Ring-Dyke 11, as exposed in Allt Molach (Fig. 52), is a granophyre. It is made up largely of oblong crystals of albite and microperthite in a micrographic matrix that contains a fair proportion of free quartz. Augite and magnetite, never abundant, have suffered corrosion, and the augite is frequently replaced by green hornblende. The quartz sometimes shows crystal-outline, while epidote, chlorite, and acicular hornblende occur together in what, presumably, were drusy cavities.

Intrusion 12. The Ring-Dyke 12, near the head-waters of Allt Molach (Fig. 52), is also a granophyre (17446) which consists mainly of alkali-felspar with subordinate quartz, and contains pseudomorphs after acicular augite with associated magnetite. The felspar is albite and microperthite, and occurs as rectangular phenocrysts of various sizes, which sometimes have skeletal outgrowths. The matrix contains irregular and acicular crystals of alkali-felspar and a fair proportion of micrographic material, that both edges the phenocrysts and occurs interstitially.

Intrusion 13. This Ring-Dyke introduces us to a somewhat special type of quartz-dolerite which is made the subject of the following description.

### SGÙLAN TYPE OF QUARTZ-DOLERITE.

Most of the intrusions lettered qD on the one-inch Map, belong to this type, and material is available for study and comparison from several occurrences:—

(1) The Sgùlan Mòr Ring-Dyke, numbered 13 on Fig. 52 (17218, 17448, 17449, 17924).

(2) The southern projection of the Beinn Chàisgidle quartz-dolerite (17925) into agglomerate, and the main outcrop of this dolerite (17921-3) on the southern face of Beinn Chàisgidle.

(3) Beinn nan Lus (17980) within the arc of the Loch Bà Felsite Ring-Dyke of Chap. XXXII, and Beinn na h-Uamha (17980) outside the same.

The Sgùlan Type of quartz-dolerite is distinguished from the quartz-gabbros described in Chapter XXX. primarily by its finer texture. On this account, the dolerites in the hand-specimen are dark grey rocks, and do not show the pronounced black and white character of the gabbros. Texture is wonderfully uniform throughout intrusions of this type, even though some of the masses are of large dimensions. Another characteristic is a tendency to possess drusy or vesicular cavities that are usually invaded, at any rate marginally, by mesostatic material. This invasion of vesicular cavities by mesostasis has been recognized to be a common characteristic of quartz-dolerites and tholeiites ever since Sir Jethro Teall[1] drew attention to it in connexion with the Tynemouth Dyke of the North of England.

Under the microscope, sections in most respects show the same general characters as are met with in the quartz-gabbros and associated differentiates (Chapter XXX.). Some are definitely basic rocks with conspicuous small pseudomorphs after olivine, and contain only a subordinate amount of acid mesostasis (17449). Others are on the border-line between basic and intermediate (17921), and thus lead to varieties (17922) showing affinities with the basic members of the craignurites (Chapter XIX).

The distinguishing fine texture of the Sgùlan Type, as compared with the quartz-gabbros, does not carry with it any essential difference in habit of the crystalline constituents.

The vesicles and druses alluded to above are frequently filled with quartz and chlorite in their more central portions (17449, 17921).

Where the rocks have been altered by the thermal action of neighbouring intrusions, the material filling the druses has been replaced in large measure by fibrous hornblende (17925) or small crystals of augite (17979, 17980). As a whole, the dolerites have suffered considerable alteration, especially in the replacement of olivine by chlorite and serpentine, and in the albitization of the basic felspars.   E.B.B.

[1] J. J. H. Teall, 'The Amygdaloids of the Tynemouth Dyke,' *Geol. Mag.*, 1889, p. 481.

# CHAPTER XXX.

## GRAVITATIONAL DIFFERENTIATION.

### INTRODUCTION.

A GRADUAL uphill change from gabbro (of the quartz-gabbro type, but often carrying olivine in its more basic portions) to granophyre [1] is sufficiently often encountered in Mull to leave no doubt as to the potency of gravitational differentiation. The subject is systematically treated in this chapter under three headings :—

(a) A series of examples is enumerated with a few critical remarks. It is hoped that this enumeration will convince the reader that he is faced with a recurrent phenomenon and not a mere local coincidence.

(b) The most conveniently situated example is illustrated with special reference to specific gravity determinations. The object in this case is to show how definitely the field-appearance of upward transitions from basic to acid can be confirmed by recourse to simple quantitative examination.

(c) The petrology of the transition is discussed in the light of microscopical and chemical analyses.

As a general remark, it may be added that, in every case, the instances of gravitational differentiation, here considered, come from massive, slowly cooled, more or less vertical intrusions. The basic portions do not show chilling where they come into contact with country-rock, and the same remark often applies to the acid. No similar stratification of acid, intermediate, and basic material has been met with among the numberless relatively thin and quickly cooled sheets of Mull, almost all of which show clearly chilled margins. (C.T.C.), E.B.B.

### LOCALITIES.

Eleven examples, cited below under eight localities, furnish good evidence of density-stratification ; one of them (Locality 6) is admittedly ambiguous ; and two (Localities 7 and 8) are somewhat obscured by later intrusions. Only in one noteworthy case, where gabbro and granophyre occur linked by transitional types, is there any marked failure of density-stratification (Locality 9). It is interesting to note that the discontinuous Glen More Ring-Dyke (Chapter XXIX.) supplies a series of examples considered separately under Localities 1, 2, 6, and 9, and perhaps also 7.

*Locality 1—one example.* The more northerly of the two western terminations of the main outcrop of the Glen More Ring-Dyke furnishes the type-example discussed below. With the help of Fig. 54, and Pl. VI. (p. 307), one can easily fix

---

[1] Granophyre is used in a rather broad sense in the present chapter and includes, in some instances, acid craignurites (Chap. XIX.), with rather less than 70 per cent. $SiO_2$.

the locality on the one-inch Map, Sheet 44. The Ring-Dyke crosses Glen More road rather more than a mile east of Craig Cottage, and the branch that concerns us strikes north-east to the summit of Cruach Choireadail. The upward transition from gabbro to granophyre is very gradual and complete. There are practically no later intrusions to complicate the exposures. The vertical range is about 1500 ft.

*Locality 2—two examples.* Both the south-easterly terminations of the Coir' an t-Sailean mass reaching up into Corra-bheinn yield good examples of upward gradual passage from gabbro to granophyre. The locality is easily recognized with reference to Corra-bheinn on Pl. VI. and the one-inch Map. There are a few later intrusions, but they do not lead to any confusion. The vertical range is about 1000 ft.   (C.T.C.)

*Locality 3—three examples.* Similar uphill passage from gabbro to granophyre can be traced in both limbs of a forking ring-dyke which runs up Monadh Beag from Allt Molach. Comparison with Fig. 52 (p. 308), where the unbranched gabbro stem is shown numbered **9**, will enable the reader to identify the intrusion on the one-inch Map. A very pretty complication is introduced by the horizontal faulting of this ring-dyke (p. 312). As one mounts the hill from the south, one finds that gabbro has given rise to granophyre in both arms of the intrusion before a fault, shown on Pl. VI., is reached. On crossing the fault, one finds both branches displaced somewhat to the south-east. This happens to be the direction in which the hill-face slopes, so that the severed arms are more deeply eroded north-east of the fault than south-west of the same. The result of this more searching erosion has been, in the case of the easterly branch of the ring-dyke, to re-expose its basic (or rather its sub-basic) lower part. Very soon, this sub-basic rock traced uphill northwards, away from the fault, gives place once more to acid. As in previous instances, later intrusions are very subordinate and do not trouble the observer. The vertical range is about 800 ft.

*Locality 4—one example.* Immediately west of the last-described gabbro ring-dyke in Allt Molach, is another ring-dyke numbered **10** in Fig. 52. It can be seen to grade upwards into granophyre. The exposures of this ring-dyke are intermittent. The vertical range is about 300 ft.

*Locality 5—one example.* A very definite and well-exposed case of upward passage from gabbro to granophyre is furnished by a ring-dyke (numbered **3** in Fig. 53) at the head of the Doir' a Mhàim corrie, 1 mile south-south-east of Beinn Talaidh. The locality is shown in the south-west corner of Fig 53, and is easily found on the one-inch Map. There are several thin cone-sheets in the western (granophyric) part of the exposures, but these do not confuse the issue, and the transition can be followed step by step. The gabbro portion of the ring-dyke continues southwards past the bridge across Allt Molach (**3** in Fig 52). The vertical range is about 1000 ft.

*Locality 6—one ambiguous example.* Merging of gabbro into granophyre is clearly demonstrable at the northern end of the Glen More Ring-Dyke, numbered **1** in Fig. 53. It takes place in the hollow connecting Glen More with Glen Forsa, at the east edge of Fig. 53. The locality is easily fixed on Pl. VI., and also on the one-inch Map, for there is no confusion due to outcrops of later intrusions. What renders the evidence ambiguous is the fact that the transition takes place in such flat ground that it can scarcely be trusted as evidence of gravitational stratification. All that is clear is that the ring-dyke, as exposed for three miles along the hollow of Glen More, southwards from the transition zone, consists of gabbro, whereas, in the high ground rising in the reverse direction through Coire Gaibhre towards the summit of Beinn Talaidh, it consists of granophyre. The vertical range is about 800 ft.

*Locality 7—one example.* East of Beinn Fhada and just outside the fault which has guided the Loch Bà Felsite Ring-Dyke, one can recognize a mass of quartz-gabbro shown on Pl. VI. as running north and south. Reference to the one-inch Map brings out the fact that this mass spans the summit of a ridge connecting Beinn Fhada with An Cruachan. At the summit of the ridge, the intrusion is much more acid than on the slopes on either side. A multitude of cone-sheets renders the evidence difficult to read. The vertical range is about 500 ft.

*Locality 8—one example.* The augite-diorite of An Cruachan (Pl. V., p. 165) is a variable mass passing at the top of the hill into granophyre and felsite, while on the slopes leading down to the River Clachaig it becomes gradually more basic. Again, there are too many cone-sheets present to allow an observer to trace the variation in detail. The vertical range is about 1500 ft.   J.E.R.

x

322    Chapter XXX.—Gravitational Differentiation.

*Locality 9—irregular association, one example.* Granophyre enters into the constitution of the main outcrop of the Glen More Ring-Dyke in three exposures. Two of these have already been mentioned (Localities 1 and 6). The third is in the southernmost extension of the mass west of Loch an Ellen. Here, while transitional types again occur, and the various types clearly form part of a single intrusion, there are marked departures from density-stratification. In fact, gabbro, in a marginal

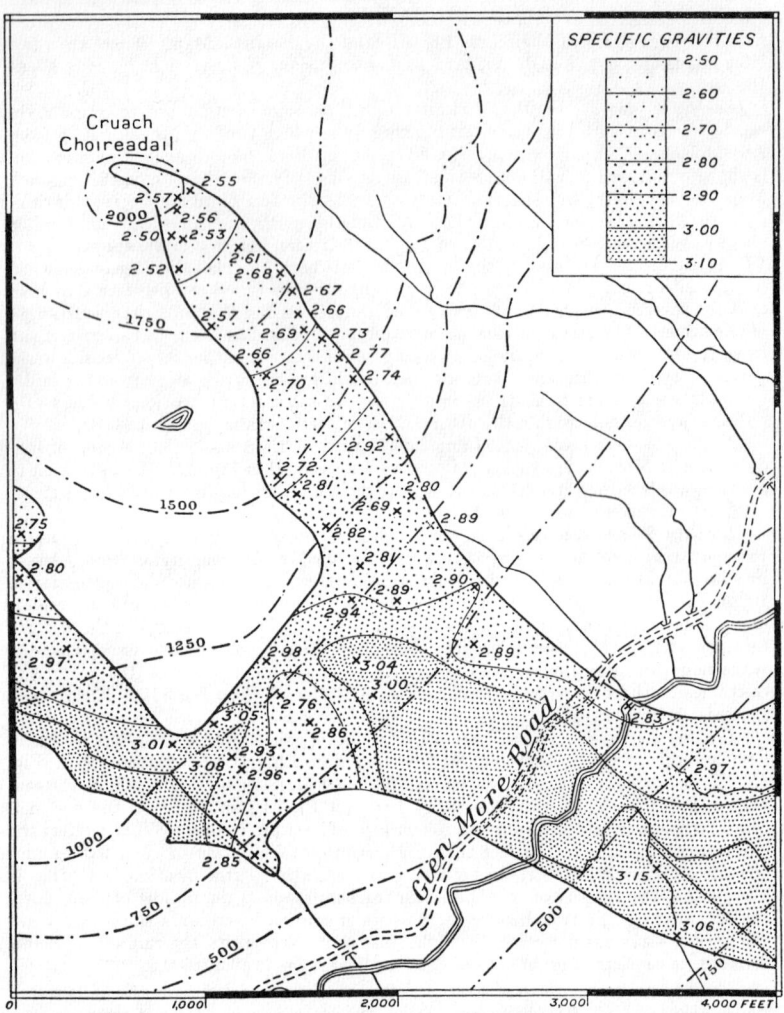

Fig. 54.—Map showing Density-Stratification in differentiated Ring-Dyke, Cruach Choireadail, Glen More (Locality 1). The extreme products exposed are olivine-bearing quartz-gabbro and granophyre.

strip, actually occurs in the highest exposures ; while, south-eastward, granophyre extends into the lowest. If, as one naturally supposes in the light of the other occurrences, the granophyre has resulted through gravitational differentiation, it would seem to have moved somewhat after it formed, and thus upset the regular stratification of the reservoir. The vertical range is 1000 ft., so that the anomalies are developed on a considerable scale.           (C.T.C.), E.B.B., G.V.W.

## Type-Example: One Mile West of Craig, Glen More.

A geologist, ascending the slopes of Cruach Choireadail (Locality 1, p. 320), from the Glen More road by way of the outcrop of the Glen More Ring-dyke, realizes very readily the upward transition from gabbro through rocks of augite-diorite affinities to granophyre (with sometimes felsitic margins). The gabbro is a dark rock with stumpy crystals of felspar bound together by ophitic augite. The diorite has a considerable proportion of pink felspathic base, in which early-formed grey felspar and dark augite-crystals abound, both characterized by a needle-habit. The granophyre is essentially a pink felspathic rock with subordinate quartz. The change of colour, of mineral-content, and of crystal-habit, all manifest them-

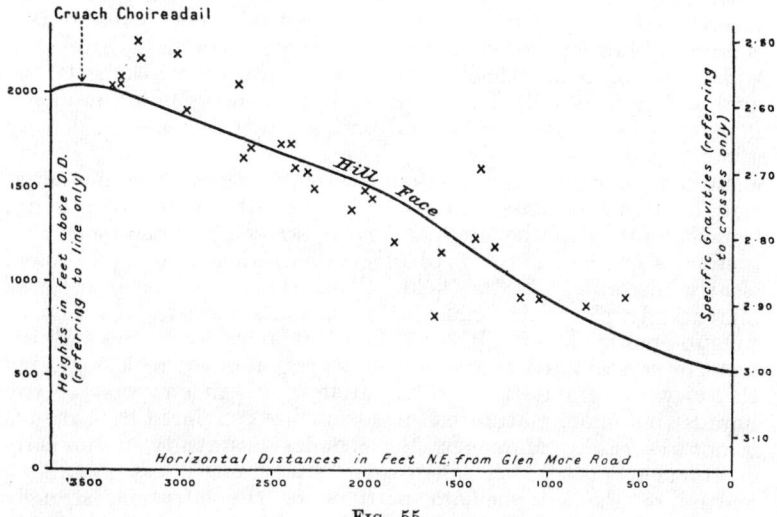

Fig. 55.

Graph showing relation of Specific Gravity to Altitude in gravitationally differentiated Ring-Dyke, Cruach Choireadail, Glen More.

selves quite gradually during the ascent. Above the 750 ft. contour, the hill-slope is free of drift; and, though much of the surface is covered by blocks loosened by frost, these have travelled so little that they do not obscure the issue.

The only important divergence from gradual upward transition from basic to acid is in the case of several horizontal acid veins which intersect the more basic lower portion of the mass. These veins are sometimes 3 or 4 ft. thick, and their edges are unchilled, though their margins are well-defined. They are clearly the infilling of fissures which developed in the almost completely consolidated gabbro.

To be able to portray on a map the general gradual upward passage from gabbro to granophyre, a suite of specimens was collected, and their specific gravities measured. The result is shown in Fig. 54. The same data have been employed in the construction of Fig. 55. It is obvious that an observer supplied with either of these two diagrams could, if he so chose, determine his height on the slope of Cruach Choireadail by taking specific gravity determinations of the rocks, and could attain something of the same degree of accuracy as by taking barometric readings of atmospheric pressure. (C.T.C.)

## PETROLOGY.

### TYPE-LOCALITY: ONE MILE WEST OF CRAIG, GLEN MORE.

(ANALS. I-IV; TABLE VIII, p. 29)

The almost complete exposure of the Glen More Ring-Dyke on the slopes of Cruach Choireadail (Fig. 54) presents features of great petrological interest. At the summit, we have a pale buff granophyric rock, and, in the valley, some 1500 ft. below, a moderately coarse gabbro or dolerite. Between these extremes, there is a wide zone of rock, which, with evident transition, connects the acid and basic types. The lower position of the gabbro, considered in conjunction with the general and gradual increase in density as we pass downward, as well as the practical absence of any sudden changes in texture or composition, invitingly suggests the differentiation of a magma aided, or controlled, by the action of gravity. Concomitant with the gradual change in composition there is a gradual variation in the type of crystallization of the rock-mass, clearly discernible in the field. Thus the augite, as it becomes increasingly plentiful, exchanges an acicular form for a columnar to sub-ophitic habit. It was first[1] thought that this adjustment of crystal-habit to the bulk-composition of the rock supported the view of gravitational differentiation of an emulsion of two liquids, but more mature consideration has convinced us that such is not the case. Moreover, the chemical instability of the early crystal-elements, which microscopic study reveals as a striking feature of the intermediate portions of the intrusion, strongly supports the view that crystallization was the chief agent in preparing the way for the separation of acid from basic material While the granophyre above, and the gabbro below, are both normal rocks, the transitional zone has a decidedly mixed aspect, as viewed under the microscope, and exhibits on a small scale many of the properties of h y b r i d s.[2] The observed characters are in keeping with the idea that there has been pronounced relative migration of early-formed crystals and residual magma within the limits of the differentiating intrusion (p. 355).

Only one alternative to some type of gravitational differentiation of the mass seems worthy of mention, and that is the injection and gradual assimilation of a basic rock by an acid magma. Dr. Harker[3] has found instances in the Island of Skye where acid magma, intruded into, or alongside, hot newly-consolidated basic rock, has

[1] E. B. Bailey, in 'Summary for Progress for 1913,' *Mem. Geol. Surv.*, 1914, p. 51.
[2] A. Harker, 'Natural History of Igneous Rocks,' p 333.
[3] A. Harker, 'Tertiary Igneous Rocks of Skye,' *Mem. Geol. Surv.*, 1904, pp. 169 *et seq*.

developed a hybrid zone by reason of a partial dissolution of the basic material. Hybridization is a common phenomenon in Mull (Chapter XXXIII.), but does not seem to us to have operated in the development of the variable mass of Cruach Choireadail (Fig. 54). The form of the intrusion, as determined in the field, renders it practically impossible for the granophyre to occupy the position it does if it be a separate injection ; and it must be remembered that the Cruach Choireadail example is but one of eleven of like character. Further, in contradistinction to the occurrences in Skye, there is a complete absence of all traces of xenolithic, as opposed to xenocrystal, structure ; in this and other respects, such as the absence of rhombic pyroxene, of biotite, and of pyrogenetic hornblende, the intermediate zone of the present intrusion differs from the undoubted hybrid zones described in Chapter XXXIII. We, therefore, feel that the true explanation of the observed variation of the mass must be differentiation *in situ*. As stated above, we have good reason to attribute the differentiation to progressive crystallization that allowed of a partial separation of solid and liquid phases, rather than to the immiscibility of two liquids. One's deductions in this respect are strongly reinforced by the fact that all recent investigations as to the behaviour of silicate-melts go to prove that such fluids are miscible with each other in every proportion.

*Serial Micro-Sections.*—A series of specimens (17625-36) was taken in sequence by Dr. Clough to illustrate the increase in acidity of the Glen More Ring-Dyke when followed from the bottom of Glen More to the summit of Cruach Choireadail.

The gabbro or dolerite of the Glen More Mass (17636), as exposed on the lower slopes, is a rock of moderately coarse grain and black and white aspect. The dark ferromagnesian component stands out in striking contrast to the white felspathic constituents which show distinct signs of unequal distribution.

Under the microscope the rock (Fig. 56A), is seen to be composed essentially of augite, labradorite, and titaniferous iron-ore. The augite in section shows a pale greenish-brown coloration and has a sub-ophitic habit. It occurs in moderately large individuals, that are ophitic towards the more basic felspars, but frequently exhibit crystal-boundaries and sub-idiomorphic habit when in contact with more acid material. Occasionally, the augite betrays signs of movement in the strongly curved form of its cleavage-traces. This bending of augite-crystals is frequently noticeable in other examples of these rocks, and was referred to by Dr. Falconer as a feature of the columnar augites of the quartz-dolerites of West-Lothian (p. 303).

The felspars are essentially labrodorite of normal composition, and have a tabular development. They are usually twinned and, unless completely enveloped by ophitic augite, show marked zoning, corrosion, and replacement by less basic plagioclase, culminating in a peripheral development of albite. This alteration of the first-formed basic felspars is clearly attributable to the action of acid material of later consolidation, and has taken place on all exposed surfaces of the basic felspar, and along all cracks, junctions of crystals, and other lines of weakness that rendered passage possible to the acidifying medium. In addition, narrow anastomosing veins of albite traverse the early-formed crystals of plagioclase in all directions. This veining with albite is probably part of the general albitization to which most rocks within the central region of Mull have been subjected, and is often clearly distinguishable from the albitizing effect produced by the acid residual magma.

The acid residuum shows a tendency to collect into ill-defined areas practically devoid of ferromagnesian minerals, and consists of a fine-grained turbid aggregate of alkali-felspar and quartz.

The iron-ore appears to be ilmenite. It occurs in moderately large patches and crystals that are frequently skeletal in form, but often show trigonal outline. Chlorite forms somewhat indefinite patches, but occasionally it retains the form of

# 326 Chapter XXX.—Gravitational Differentiation.

olivine, which was undoubtedly an original constituent of the rock, though now completely replaced. Apatite is practically absent except for a few slender prisms that are more or less confined to the acid portions of the rock.

Higher in the sequence (17635), although the rock shows but little change in appearance when studied in the hand-specimen, except perhaps in the more irregular distribution of the more purely felspathic matter, the microscope reveals differences of considerable importance. First we may note a distinct increase in the amount of late-consolidated acid residuum, which, as before, consists of alkali-

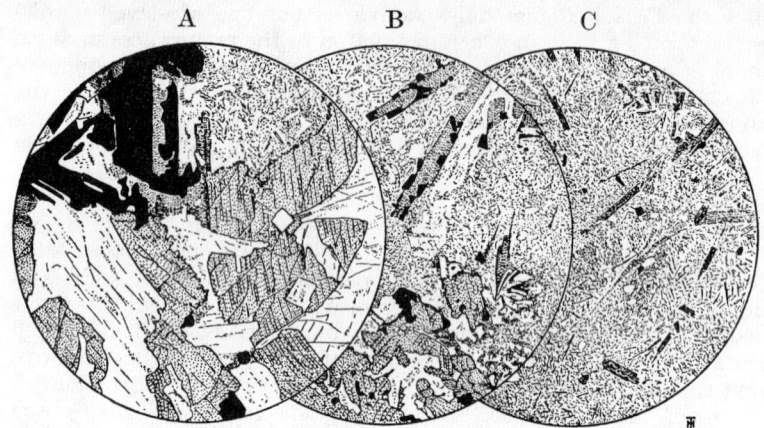

FIG. 56.—The Glen More Differentiated Ring-Dyke.

A. [17636] × 15. Lower Basic Portion. Quartz-gabbro. The rock is composed of labradorite, ophitic augite and large plates of ilmenite, with a variable amount of finely crystalline acid mesostasis (top). Where in contact with the acid residuum, the augite shows signs of resorption. Movement of the mass after partial consolidation has frequently resulted in the bending and breaking of crystals—note the curved cleavage-traces in the large crystal of augite.

B. [17632] × 15. Intermediate Portion. The figure shows a rock in which there is an increased proportion of acid mesostatic matter with characteristic acicular crystallization of its components. It has developed columnar crystals of augite (top) with their usual association of magnetite, and it encloses small patches of more doleritic material (bottom) which show signs of resorption and of being out of equilibrium with their surroundings.

C. [17626] × 15. Higher Acid Portion. Acicular type of crystallization is a characteristic feature. The rock is composed of elongated crystals of greenish hornblende, pseudomorphous after augite, in a feathery base of alkali-felspar and quartz, frequently in micrographic relationship to each other.

felspar and quartz, and takes the form of orthoclase, perthite, micropegmatite, and quartz. The augite, as far as can be judged, is constitutionally similar to that in the more basic rock described above, but in sympathy with the increase of acid residual matter, it tends to adopt a columnar prismatic habit, although it retains its ophitic development locally in the more basic portions of the section. Corrosion of the original felspars (labradorite), and their replacement by plagioclase of less basic composition, are marked characters of the rock. Apatite is still a poorly represented accessory, and is confined almost entirely to the replaced portions of basic felspars and to the acid residuum. Olivine is no longer represented.

The more basic portions of the Glen More Mass are occasionally traversed by segregation-veins of more acid character, usually with ill-defined boundaries. In such veins (17634), which are noticeably of a lighter colour and finer grain, the bulk of the felspar is of a variety more acid than labradorite, and is in all cases edged with albite, perthite, or orthoclase. The augite-crystals are columnar and intimately intergrown with magnetite near, or at, their surfaces. The residual material is plentiful, and consists of orthoclase, albite, perthite, and quartz, with apatite as an abundant accessory. Iron ore, which occurs plentifully as small crystals, appears to be either magnetite or titano-magnetite.

Passing upwards in the series (17633-17631), into that portion of the mass with a specific gravity of between 2·7 and 2·8, the rock becomes more finely grained, and medium-grey in colour. Viewed with a lens, it has a finely speckled character due to the alternation of ferromagnesian and felspathic constituents. Examined microscopically (Fig. 56B), the outstanding features are the reduced basicity of the larger felspars, the almost complete suppression of ophitic structure in favour of a columnar development of augite, and the collection of acid mesostasis into better defined patches and strings (17631, 17633). Further, the acid mesostasis has a pronounced acicular type of crystallization, which recalls that of the craignurites (Chap. XIX). It consists of skeletal and perthitic growths of alkali-felspar with quartz, either free or in micrographic intergrowth with orthoclase, and slender needles of fibrous green hornblende, presumably pseudomorphous after augite.

Apatite is moderately abundant in the acid portions and in the acidified parts of the early-formed felspars.

The augite of the basic portions of the rock, where in contact with, or surrounded by, acid material, has been converted into green fibrous hornblende.

Occasionally (17632), the grain becomes somewhat coarser, and the acicular type of crystallization of the rock as a whole more pronounced. We here may notice the marked columnar and elongated character of the augite crystals with their peripheral intergrowths of magnetite, and also a cervicorn development of this mineral in crystals of plagioclase. This cervicorn growth of augite is a feature characteristic of certain examples of the Talaidh Type of quartz-dolerites described in Chap. XXVIII. (p. 303), and also to some extent of the augite-diorites (Chap. XVIII.).

Another somewhat coarser variety (17630) of the rocks forming the intermediate zone shows well the genetic relation of the columnar augites to the acid mesostatic material which now forms a considerable portion of the mass. Also, it is interesting to note the incoming of a rudimentary salitic striation in the columnar augites and the passage of this mineral into green hornblende as the result of interaction with the acid mesostasis.

From here upwards the mass (17629-17627), although slightly variable in grain, becomes gradually paler in colour. The change is due primarily to the increased proportion of acid material and the more complete isolation of the early-formed and more basic constitutents. It is particularly interesting that this acid rock sometimes clearly chills against its containing-walls as a microporphyritic spherulitic felsite (17627). The more basic rocks at lower level do not exhibit chilled margins.

The more acid rocks of the series are compact, pale-grey to buff in colour, and are best described as soda granophyre. Their more basic representatives (17626) have craignurite-affinities, and are characterized by a finely acicular crystallization of their constituent felspar and augite, and a feathery base composed largely of alkali-felspar and quartz, frequently in micrographic relationship to each other (Fig. 56C). The early-formed felspar and augite, abundantly represented in the more basic portions of the transitional zone, are reduced to a minimum, and have suffered marked alteration.

### COIR' AN T-SAILEAN.

(ANALS. V. AND VI.; TABLE VIII., p. 29.)

The neighbouring exposures of Coir' an t-Sailean (Locality 2, p. 321) have also furnished interesting material of similar character (16530-8, 18455-9), which shows the same gradual transition from basic to acid types.

### DISCUSSION OF ANALYSES (TABLE VIII., p. 29).

For a more complete realization of the compositional changes that have taken place in these masses, five analyses of the various grades exposed have been made (p. 29). Of these, two from Glen

More (18463, 18462) and one from Coir an t-Sailean (18455), are of moderately basic character. These rocks are best described as olivine-bearing quartz-gabbro or quartz-dolerite. They contain only a small amount of residual acid material, and compare closely with the basic rock already described (17636). The granophyric part (18460) of the Glen More Ring-Dyke, as exposed at the summit of Cruach Choireadail, is a fine-grained soda-rich rock, poor in ferromagnesian constitutents, with an orthoclase to albite ratio of 1 to 2. It is composed of albite, orthoclase, microperthite, and quartz, with a little fibrous green hornblende after acicular augite. It is perhaps best classed as an acid craignurite unusually rich in soda.

The analysed rocks of the transitional zone of the Glen More Ring-Dyke (18461) and the Coir' an t-Sailean Mass (18456) may be regarded as representative.

Examined microscopically, these rocks present characters similar to those already described in connexion with Dr. Clough's specimens. They show the same collection of varying amounts of acid material into lakes and strings, the corrosion and replacement of early formed plagioclase by felspar of more acid character, the columnar and cervicorn development of augite, and a distinct concentration of apatite when compared with the basic and acid extreme types. They are fairly representative coarse craignurites, and differ little from some of the augite-diorites (Chapter XVIII.), except in their less mature type of crystallization.

It has already been pointed out (p. 28) that the differentiation series of Glen More corresponds remarkably closely in composition with the variation-diagram for the Normal Mull Magma-Series, illustrated in Fig. 2 (p. 14). Between the limits set by 50 and 70 per cent. $SiO_2$, this variation-diagram is approximately of straight-line character. The following table demonstrates the predominant linear tendency met with in the variable complex of Glen More.[1]

|  | $SiO_2$ | MgO. | CaO. | Alkalies. | $P_2O_5$. |
|---|---|---|---|---|---|
| Gabbro (18463) | 49·90 | 5·88 | 10·39 | 3·81 | 0·20 |
| Granophyre (18460) | 68·12 | 0·71 | 1·81 | 8·62 | 0·22 |
| Mixture 1 : 1 of above | 58·5 | 3·3 | 6·1 | 6·2 | 0·21 |
| Average of transition-zone (18456 and 18461) | 56·7 | 2·5 | 5·8 | 6·9 | 0·48 |

Viewed from the standpoint of the differentiation theory outlined above, linear variation bespeaks an approximate constancy of composition of the acid residuum during its migrations. This emphasizes, once again, the comparative pause that occurred in

---

[1] Straight-line variation is a characteristic of differentiation of quartz-dolerite magma in West Lothian, as is shown by curves drawn by E. B. Bailey in 'Geology of the Glasgow District,' *Mem. Geol. Surv.*, 1911, p. 147 to illustrate analyses by G. S. Blake, quoted by J. D. Falconer in 'Igneous Geology of the Bathgate and Linlithgow Hills,' *Trans. Roy. Soc. Edin.*, vol. xlv., 1906, p. 147.

the crystallization-history of the magma after the separation of its gabroic elements. Of course, the pause was not absolute. For instance, the microscope shows that a relative richness in apatite, to some extent, distinguishes the acid residuum, entangled in the basic and intermediate portions of the ring-dyke, from the fairly pure acid concentrate encountered at higher levels; so that it is clear that much of the crystallization of the apatite occurred during the migration of the acid residuum. Apatite is a very minor constituent of the Glen More series; but its content of phosphorus allows us to follow its distribution, even in bulk-analyses: here, the findings of the microscope are confirmed by a marked relative concentration of $P_2O_5$ in rocks that are otherwise of intermediate composition.

## Conclusions.

In these differentiated masses, we have as extreme types an olivine-bearing quartz-gabbro, or quartz-dolerite, and a soda-granophyre. The wide transitional zone between these types gradually varies in composition, but retains a fine-grained texture throughout, and shows no trace of xenolithic structure on a macroscopic scale. The basic rock, after the manner of quartz-gabbros, invariably contains a small, but inconstant, proportion of acid material in the form of strings and patches of alkali-felspar and quartz. Upwards in the succession, the ratio of light to dark-coloured constituents undergoes a progressive change. Under the microscope, we encounter increasingly frequent patches and strings, composed of albite, perthite, and micropegmatite, with little or no ferro-magnesian minerals; and these patches and strings separate earlier-formed crystals of augite and zoned labradorite (17872). The acid material has, wherever favourable opportunity offered itself, produced corrosion and substitution of the more basic felspars (18462), and, at its contacts with augite, has effected a change of this mineral into fibrous green hornblende. Further, there is the tendency, as the amount of acid matter increases, for the ophitic augite of the basic rock to assume a sub-idiomorphic columnar habit and to develop an incipient salitic striation. This columnar augite is highly characteristic of the augite-diorites (Chapter XVIII.), and of the quartz-dolerites of Talaidh Type (Chapter XXVIII.), and is almost always associated and intergrown with abundant magnetite. With further increase in the amount of acid base, acidification of the basic felspars of early formation is a marked feature, and progresses to the ultimate production of oligoclase-albite which may, or may not, retain a core of more basic felspar. It is obvious in the transitional zone that, although no xenolithic structure on a large scale is to be detected, the whole fabric of the rock is composed of minute patches and crystals that are of the nature of xenocrysts, and out of equilibrium with their surroundings. As was stated previously, the change in type of crystallization noticed on passing from the basic to acid portions of these ring-dykes is one primarily controlled by composition. The increase in acidity is accompanied by a greater tendency to acicular crystallization of the component minerals, a feature noticed in the craignurites

(Chapter XIX.), which in their acid representatives are structurally similar to the acid mesostasis of the transitional zone; the same feature is also characteristic of the acid differentiates of the quartz-dolerites of West Lothian.

There is little doubt that we have, in this transitional zone, a commingling and interaction of a fluid magma with solid material; but exactly similar phenomena can be studied in the quartz-dolerite sills of West Lothian, where it can be demonstrated that acid residual fluid, that originally existed more or less evenly distributed throughout a basic rock, has been squeezed out and segregated in certain portions of the mass in variable quantity. In connexion with this assumed application of external force, attention may be directed to the sharply flexed condition of many of the augite-crystals (Fig. 56A), both in the differentiated ring-dykes of Mull and in the quartz-dolerites of West Lothian. In the case of the Mull ring-dykes, the force responsible for the squeezing out of the acid mesostasis and the deformation of crystals appears in the main to have been gravity, for otherwise it is difficult to account for the almost universal and gradual downward increase in density.

The upward passage of the acid residual material, and its progressive concentration in higher portions of the intrusions, are naturally accompanied by a new set of physical and chemical conditions, whereby previously existing equilibria are disturbed. This accounts for the reproduction on a microscopic scale in these rocks of some of the characters of hybrids as discussed in Chapter XXXIII.; but, as there indicated, other characteristic features of hybrid rocks are unrepresented. There is all through the transitional zone of these differentiated ring-dykes abundant evidence of crystals, and groups of crystals, being out of sympathy with their environment, and of efforts made to stabilize the conditions through resorption, recrystallization, and replacement of earlier-formed crystalline phases. Absolutely analogous disturbances and readjustments are to be seen in the partially differentiated quartz-dolerites of West Lothian, where a migration of acid reiduum has locally modified the bulk-composition of the rock.

Our conclusions regarding the conditions, that in certain Mull Ring-Dykes favoured the migration of a partial magma, and its ultimate more or less complete separation from the early crystalline phases, are as follows. After crystallization had proceeded for some time, and had practically exhausted the magma as regards lime and magnesia, there still remained a residuum which retained its fluidity over a considerable range of temperature, so that there was, at this stage, a marked pause in the process of crystallization. Under such conditions, ample opportunity was afforded for migration of the fluid residuum under stress during the extended period that elapsed before crystallization was completed. In such a high column of magma as these ring-dykes represent, the gravitational pressure of the early crystalline phases on the still-liquid residuum would be considerable, and would in our opinion be quite sufficient to cause an upward filtration of the still-fluid portion of the mass, and its ultimate intrusion as an entity into the uppermost part of the dyke-fissure, where it chilled marginally against the country-rock.　　　　　　　　　　　　　　　　　　　　　　　　H.H.T., E.B.B.

# CHAPTER XXXI.

## GLEN CANNEL GRANOPHYRE AND ASSOCIATED FELSITES.

### Introduction.

The Glen Cannel Granophyre can easily be distinguished from the other granophyres, along with which it is lettered G on the one-inch Map, Sheet 44, by comparison with Plate VI. (p. 307), where it is separately ornamented. It lies surrounded by a country where geological complexity may be said to have reached a climax, and yet, in its extensive exposures, its relationships are relatively so simple as to afford a sensation of positive relief. A number of basalt-dykes can be seen traversing it in the many available stream-sections, but, except in its south-eastern portion, it is not known to be cut by a single cone-sheet.

The felsites which are dealt with in this chapter are cross-hatched on Plate VI. They are a rather heterogeneous assemblage; and some, perhaps, should have found a place in Chapter XVI., where similar acid rocks have already been discussed, along with intimately associated agglomerate. On the other hand, their distribution, illustrated in Plate VI., certainly suggests some causal connexion between themselves and the Glen Cannel Granophyre; it must, however, be confessed that such connexion, if it exist, is ill-understood. The idea, that the felsites might represent a local chilling of the Glen Cannel granophyre magma against the country-rock, is negatived by the fact that a number of excellent exposures demonstrate an absence of transition from felsite to granophyre; instead, the granophyre cuts the felsite abruptly, and is for the most part quite coarse-grained up to its margin. The relations of the two rocks can be well-seen on the east side of Bìth-bheinn and Creag Dhubh, and less clearly on Na Binneinean. Any doubt as to which is the earlier has been removed by microscopic examination, which shows the felsite to have been considerably altered by contact with the granophyre.

The granophyre is of tolerably uniform type throughout, a pale-pink well-crystallized rock, rich in felspar and quartz. The felsites are for the most part characterized by small phenocrysts of felspar. The strip of felsite along the slopes of Bìth Bheinn, west of Glen Cannel, is markedly vesicular in its northern portion; but this is unusual.

The further discussion is divided under three headings F o r m of I n t r u s i o n, T i m e-R e l a t i o n s, and P e t r o l o g y.

### Form of Intrusion.

The Glen Cannel Granophyre, towards the south-east, seems to have suffered displacement by the fault, along which the Loch

Bà Ring-Dyke of felsite subsequently arose (Chapter XXXII.). Elsewhere, its exposures reveal the gently domed roof of a major intrusion, locally cut through by erosion to a depth of a 1000 ft. without any indication of a base. The dome-roof roughly corresponds in shape with the upper surface of an ellipsoid, culminating above a line along Loch Bà and Glen Cannel—the familiar north-west axis of symmetry for Mull geology.

The general outward inclination of the granophyre-roof can easily be appreciated on inspection of the one-inch Map; for it is responsible for notable extensions of the granophyre outcrop into Glen Clachaig, the head-waters of Glen Cannel, and the pass leading eastwards to the Gaodhail River. To an observer on the ground, the relationship is doubly clear, for the rocks of the roof often rise above the granophyre in a little escarpment, so that the eye can follow on the hillside the boundary-line which has been drawn upon the map, and trace its gradual descent on all sides. Towards the north-west, the roof reaches lake-level just at the outcrop of the Loch Bà Felsite Ring, and there is no opportunity to determine the relation of the two intrusions at this point. The roof is, on the whole, remarkably evenly domed, but close observation shows a number of minor irregularities, with the inclination varying from the horizontal up to 50°, and, in places, interruptions are found, due to minor offshoots. There is no approach to the degree of breaking up and penetration of the roof observable in the case of the Beinn a' Ghràig Granophyre to the north-west (Chapter XXXII.).

At the southern end of the Glen Cannel Mass, there occur the only exposures which give any indication of its relations to the Loch Bà Felsite Ring and to the fault, which this latter accompanies. The junction is to be seen some distance above the confluence of the streams which flow north on either side of Coill' an Aodainn. In both streams, the granophyre on the north is seen to be brecciated in contact with the vertical edge of the felsite, which is itself intrusive into the resultant breccia and chilled against it. In one of the small streams descending from Beinn Chàisgidle, there is an exposure showing an irregular junction between the felsite and its northerly neighbour, the granophyre, without the intervention of any breccia. These relations, taken together, are very similar to those described later between the Beinn a Ghràig Granophyre and Loch Bà Felsite farther west (Chapter XXXII.), and clearly point to the disruption of the Glen Cannel Granophyre by the ring-fault along which the felsite-dyke has subsequently risen, guided by the line of weakness afforded by the breccia.

On crossing the Loch Bà Felsite in the Coill' an Aodainn district, one comes to granophyre on the southern side, which agrees in character with that on the north, and is similarly related to the felsite; that is to say, the felsite is intruded into the brecciated margin of the granophyre. It is suggested that this southern granophyre is merely a faulted continuation of the main Glen Cannel Mass (*see* western ?, Plate VI.). Its outcrop occupies a crescentic area extending along the dyke, and does not fit exactly with that of the main part of the Glen Cannel Granophyre just opposite, for it reaches considerably farther west; but this is easily interpreted on the hypothesis that the outlying part, remaining outside the cauldron-subsidence of Loch Bà (p. 340), has been subjected to relatively deep erosion with resultant wide exposure.

In this connexion, we must also consider the symmetrically disposed granophyre-area lying outside the Loch Bà Felsite on the other side of Beinn Chàisgidle. Here, the relations are not quite so plain, but they probably bear a similar interpretation (*see* eastern ?, Plate VI.).

Both these doubtful areas are composed of coarser-grained granophyre than that of the ring-dykes of the early Beinn Chàisgidle centre (Chapter XXIX.), with which they come into contact ; and this affords a ready means of separating them from the latter, and, at the same time, an argument for regarding them as portions of the Glen Cannel Mass. Hence, it seems that the latter is not strictly confined within the limits of the Loch Bà Felsite, but extends beyond to the south-east.

## TIME-RELATIONS.

The age of the Glen Cannel Granophyre, like that of many other of the intrusions of Mull, can only be roughly defined. The granophyre is earlier than a fair number of north-west dykes (Chapter XXXIV.). It is also earlier than the Loch Bà Felsite (Chapter XXXII.), and the associated ring-fault bounding the north-western caldera of Plate V. (p. 165). Those portions which lie within the Loch Bà Ring-Dyke are almost, but not quite, free from Late Basic Cone-Sheets (Chapter XXVIII.), except towards the south-east. This relative immunity, however, is shared by the roof-rocks of the granophyre, and is due to position rather than to date, for it is obvious that the main granophyre-exposure lies in a central area avoided by cone-sheets. The supposed extension of the granophyre to the south-east, beyond the Loch Bà Ring-Dyke, is freely cut by the Late Basic Cone-Sheets, though not perhaps to quite the same extent as other similarly situated rocks. The granophyre may accordingly date from the latter portion of the period of intrusion of the Late Basic Cone-Sheets. It seems distinctly earlier than the Beinn a' Ghràig Granophyre of Chapter XXXII., which truncates a very large number of Late Basic Cone-Sheets, and is only cut by one or two doubtful examples.

Another point about the Glen Cannel Granophyre is that it is clearly later than the intrusive felsites which fringe its roof ; but the age of these latter is not otherwise fixed. Again, on the presumption that the granophyre extends to the south-east beyond the Loch Bà Felsite, it appears to truncate the ring-dykes which centre round Beinn Chàisgidle (Chapter XXIX.) ; but this is only a deduction from the mapping, and has not been confirmed in exposures.

Finally, as the centre of intrusion for ring-dykes and cone-sheets seems to have migrated north-west during the long period of intrusive activity, there is a presumption in favour of placing the date of the Glen Cannel Granophyre towards the latter part of this period, as its centre lies well to the north-west, though apparently not quite so far in this direction as that of the Loch Bà Felsite. W.B.W.

## PETROLOGY.

### GLEN CANNEL GRANOPHYRE.

The Glen Cannel Granophyre (14844) is a moderately coarse rock composed of rectangular crystals and patches of turbid perthitic orthoclase in a matrix of somewhat coarse micropegmatite (Fig. 57A). Like the Beinn a' Ghràig Granophyre, it is characterized by the prevalence of a green pleochroic augite of alkaline characters (aegerine-augite), with which are commonly associated irregular patches and grains of magnetite, and occasionally a little sphene. The granophyre appears to be fairly constant in type, but a tendency to spherulitic structure and an assumption of an acicular type

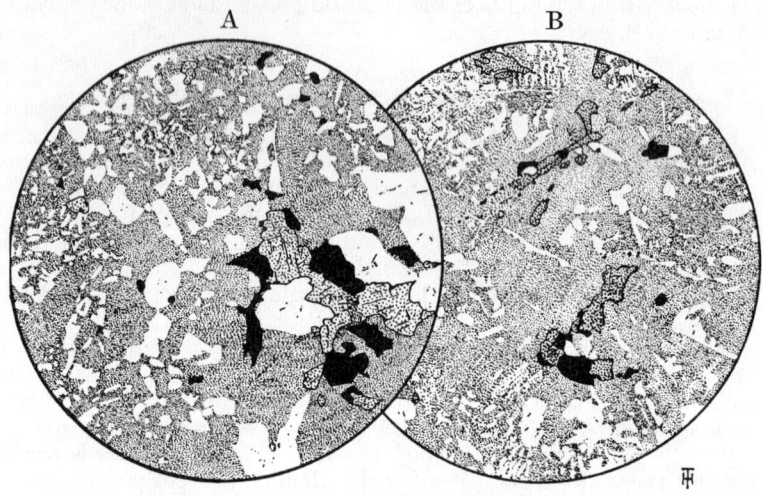

FIG. 57.

A. [14844] × 17. Granophyre of Glen Cannel. Green pleochroic augite (aegerine-augite) associated with magnetite, perthitic orthoclase and quartz in a somewhat coarse micrographic matrix.

B. [2146] × 17. Granophyre of Beinn a' Ghràig. Green pleochroic aegerine-augite with magnetite, in a moderately coarse matrix of quartz and turbid alkali-felspar in micrographic intergrowth.

of crystallization may be noted in certain instances. Specimens, collected from near the old burial ground to the east of Gortenbuie (14845), and about half a mile further up stream to the north-east (14788), are of more compact rock, in which the granophyric structure is developed on a more delicate scale. There is a tendency to a radiate and spherulitic grouping of the microgranophyric and felspathic material (14788), and the pyroxene is represented by slender acicular pseudomorphs of fibrous hornblende and magnetite. This spherulitic structure (14845), when most complete, is accompanied by a more definite crystallization of the felspar as small well-formed individuals, usually of perthite, edged with micropegmatite, but frequently composed throughout of cryptographic quartz and orthoclase.

## FELSITES ABOVE GLEN CANNEL GRANOPHYRE.

These felsites, regarded as possible earlier portions of the Glen Cannel Granophyre, have in all cases suffered intense thermal alteration by the granophyre-intrusion. Where least affected, they present characters similar to those of the so-called small-felspar felsites of the Loch Bà Ring-Dyke (Chapter XXXII.) and the Late Acid Cone-Sheets (p. 228). In many cases, they are of highly xenolithic character, but the original nature of the included xenoliths has been destroyed by contact-metamorphism following, probably, on earlier partial assimilation.

Judged by specimens from Na Binneinean (14571), Lochan nam Ban Uaine (14691-3), and Bìth-bheinn (15564-5), these felsites are all microporphyritic. The phenocrysts are usually of felspar with subordinate augite. The felspar is most commonly of a perthitic nature (14571, 14691), but, in what appear to be more basic variants (15564, 15562), it forms somewhat larger crystals and is oligoclase or oligoclase-andesine. Augite may be present as isolated hypidiomorphic brownish crystals, but commonly occurs in association with oligoclase-andesine felspar, magnetite, and apatite, with which it forms glomero-porphyritic groups (15565), that are probably of a xenolithic, but cognate, nature.

The matrix is felsitic to rhyolitic. Sometimes, it is completely granular (14691), but this is probably a structure of secondary origin. At other times, traces of original perlitic structure and fluxional banding can be clearly distinguished (14693, 15562). A somewhat unusual character emphasizes the rhyolitic nature, of the felsites that occur in the neighbourhood of Bìth-bheinn. The rocks show phenocrysts of oligoclase, with subordinate augite, in a devitrified, originally banded, matrix. In addition, they have a well-developed amygdaloidal structure. The elliptical vesicular cavities, lined with alkali-felspar and filled with quartz, are extremely numerous and reach three or four millimetres in greatest diameter (17128). A felsite with similar small amygdales occurs near Lochan nam Ban Uaine (14693).

## CONTACT-ALTERATION BY THE GLEN CANNEL GRANOPHYRE.

The effects of contact-alteration by the Glen Cannel Granophyre, as shown in the gabbro, rhyolitic lavas, and tuffs of Beinn na Duatharach, and in the dolerite-masses to the west of Clachaig Cottage, have already been discussed in Chapters XI. and XVI. We shall, however, consider here the contact-metamorphism of the felsites that occur above the granophyre, and also of some basic lavas that come within its influence.

The metamorphism of the felsites is best studied in material from Sròn nam Boc and Na Binneinean (14569-73). A specimen from Sròn nam Boc (14678) is of a xenolithic felsite that contained phenocrysts of oligoclase-andesine and augite in a patchy felsitic matrix. The augite has been resorbed and coated with a greenish-brown hornblende in crystallographic continuity. The matrix is charged with partially digested xenolithic basic material that, under the metamorphosing influence of the granophyre, has given rise to abundant little prisms of brownish hornblende; frequently, twinned and irregular grains of the same mineral are scattered throughout the mass. Granulitized basic material occurs as streaks and patches, and, where most obviously xenolithic, is composed of minute closely packed prisms of hornblende and grains of magnetite in a felspathic base.

The actual junction of the granophyre with felsite at Na Binneinean (14572) show a moderately coarse granophyre composed of perthite, quartz, and micropegmatite having an unchilled, but fairly sharp margin, against the felsite, into which it sends veins and tongues. The veins have been basified by resorption of material from the xenolithic felsite, and a green pyrogenetic hornblende has crystallized from them in moderately large individuals. The felsite has suffered more or less complete granulitization of its basic material, and is now charged with green hornblende, granulitic augite, and magnetite. A rock from the same locality, two feet above the junction, is a beautiful microgranulite (14573) composed of granulitic augite and magnetite in a granular base of oligoclase and alkali-felspar. There occur also frequent small poecilitic plates of red-brown biotite in association with granulitic areas of alkali-felspar. These areas are possibly due to invading acid material connected with the granophyre (*cf.* 17460, p. 317). At a height of 30 ft. above the junction, the felsite (14569), still charged with xenolithic basic material, shows signs of granulitization accompanied by assimilation of the foreign matter. The rock is full of little crystals and granules of greenish hornblende and magnetite. Some of the original perthitic phenocrysts have suffered corrosion, and augite is granulitized and partially converted into hornblende. Parts of the felsite have their original perlitic cracks now marked out by a development of strings of granulitized basic material along their course.

Farther away (14570-1), the original perthitic patches and phenocrysts are less uncomfortable in their surroundings, and we meet with a green alkali-augite that occasionally shows good outline. There is a tendency for this augite to change colour and seemingly become less alkaline towards the interior of the crystals, but it is uncertain whether this is an original or superimposed feature.

The felsitic matrix, here again, has suffered granulitization, but there appears now to be less xenolithic basic material present, and therefore a less abundant development of granulitic ferromagnesian minerals.

A basic lava (14574), from a point about 50 yds. west of the summit of Na Binneinean, is a microporphyritic olivine-free doleritic rock consisting of elongated crystals of oligoclase-andesine to acid labradorite, with ophitic to hypidiomorphic augite. The chloritic base has been altered with a development of scales of biotite, and there has been local granulitization of the augite. The rock is traversed by veins from the granophyre (14575), which show the effects of assimilation of basic material in the crystallization of a pyrogenetic deep-green hornblende. They also contain small irregular crystals of oligoclase-andesine, corroded and encased by perthite, which have, in all probability, been derived from the basic rock and only partly resorbed.  H.H.T.

## CHAPTER XXXII.

### THE RING-DYKES OF LOCH BÀ, BEINN A' GHRÀIG, AND KNOCK.

#### INTRODUCTION.

THE subject of r i n g - d y k e s has already been introduced in Chapter XXIX. of this Memoir, and the reader is referred to that Chapter for a definition of terms, and a general statement as to character. It is there pointed out that the ring-dykes of Mull group themselves about two centres, the earlier of which is situated on Beinn Chàisgidle, and the later at the head of Loch Bà ($C_1$ and $C_2$ of Fig. 58, p. 338). It has also been shown that the outermost of the earlier series, the great Glen More Ring-Dyke, shows indications of a dual control from both centres. The intrusion of the Loch Bà Ring-Dyke, dealt with in the present chapter is entirely controlled by the second centre. The Beinn a' Ghràig and Knock Dykes scarcely show any tendency to circulate, but are included in the category of ring-dykes because of their dyke-like form, and their general parallelism to the north-western segment of the Loch Bà Ring-Dyke.

The Loch Bà Ring-Dyke, illustrated in Fig. 58, is the most perfect example of a ring-dyke known to science. It is composed of compact porphyritic felsite, and has a fairly continuous outcrop, with a maximum diameter of five miles. The dyke itself ranges up to a quarter of a mile in width, but more often measures about 100 yds.
W.B.W., J.E.R.

The Beinn a' Ghràig and Knock Ring-Dykes are composed of pale well-crystallized granophyre. Their main interest attaches to the beautifully developed and well-exposed s c r e e n which separates them, and the fact that portions of the roof of the former are preserved and can be easily studied.

The subject-matter of the present chapter will be dealt with under the following headings :—

(1) Form of Intrusion.
(2) Age-Relations.
(3) Late Explosions.
(4) Petrology.

Throughout, the Loch Bà Felsite is taken in advance of its associates on account of the dominating interest of its ring-form. From the point of view of age-relations, it would have been preferable to consider the granophyres first, because, at any rate, the Beinn a' Ghràig Granpohyre is clearly of earlier date.

Certain special subjects are reserved for separate discussion. Thus the contact-alteration produced by the granophyres is described in Chapters VIII. and X., and the question of the assimilation of early basic plutonic rocks will be dealt with in Chapter XXXIII.
W.B.W.

338 Ch. XXXII.—*Ring-Dykes of Loch Bà, Beinn a' Ghràig, & Knock.*

INTRUSION-FORM.

LOCH BÀ FELSITE.

The Loch Bà intrusion may be described as an almost complete ring-dyke about the centre $C_2$, of Fig. 58. Considered with reference

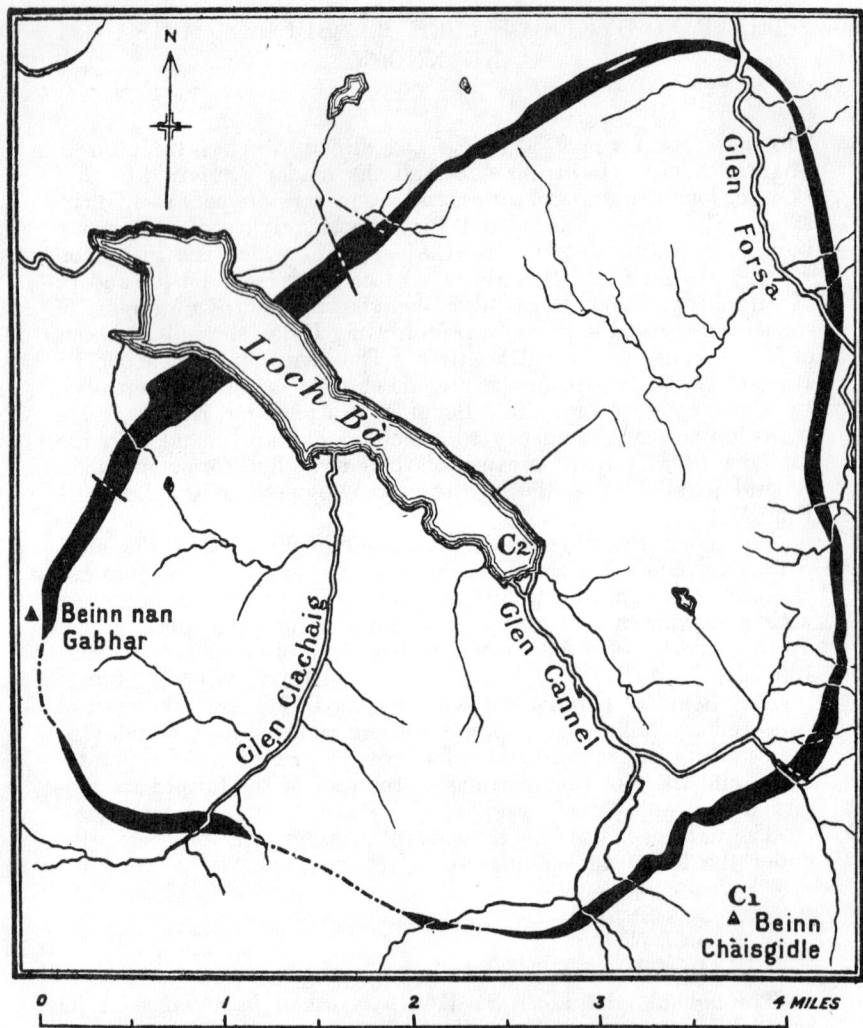

FIG. 58.—Loch Bà Felsitic Ring-Dyke along Fault.
$C_1$ and $C_2$ show two chief centres of ring-dykes and cone-sheets.
Quoted from *Summary of Progress for* 1914, p. 36.

to the centre of the whole plutonic complex, the ring-dyke assumes a sectorial aspect: the north-west segment of the sector, crossing Loch Bà, is nearly parallel to the circumference of the plutonic complex, though somewhat less curved; the apex near Beinn

Chàisgidle is greatly blunted; and the radii crossing Glen Forsa and Glen Clachaig, are characterized by a curious sigmoidal curvature. The dyke as a whole displays, more strikingly than any of the Mull intrusions, the symmetry about a north-west axis which is an outstanding feature of Mull geology. Even small details, such as the curvature of the radii, have been found repeated on either side of the line of symmetry.

A glance at Plates V. and VI. will convince the reader that this ring-dyke, apart from its beautiful development, marks a line of structural importance, for it forms the boundary of various rock-divisions. As a matter of fact, there is considerable evidence that it follows a line of faulting, and approximately, at any rate, outlines an area of central subsidence, the north-west caldera of Mull.

W.B.W., J.E.R.

In what follows, attention will be given more particularly to the following points :—

(1) Exposures upon which the mapping of the almost continuous ring-outcrop is based.
(2) Suggestions of a general outward inclination of the dyke.
(3) Evidence that the dyke has followed, or accompanied, powerful faulting.

### (1) Track of the Ring-Dyke.

In its course from Beinn nan Gabhar, north-east to Loch Bà, and again from Loch Bà to the 500 ft. contour on the west face of Glen Forsa, the Loch Bà Felsite is almost continuously exposed, and often gives rise to conspicuous crags. It is to this five miles of outcrop that anyone should go who wishes to realize the dyke-form of the intrusion. The hills flanking Loch Bà, from which the dyke takes its name, show it in strong relief, ascending from lake-level to a height of 1500 ft. On the upland, its northern face generally stands out boldly above the granophyre with which it comes in contact, but its southerly margin against the lavas and agglomerate is everywhere much less marked. Its scenic importance in this area is partly attributable to its frequent juxtaposition with granophyre free from minor basic intrusions, and therefore readily denuded. A number of north-west dykes cut across the felsite in this north-west sector, and erosion along certain of these makes possible a close study of its contacts.

W.B.W.

For half-a mile, just where the dyke is mapped as taking a rather abrupt turn in the bottom of Glen Forsa, there are no exposures. East of the River Forsa, however, the dyke is seen in four small streams, of which the first is Allt an Eas Dhuibh and the last Allt Mòr. There is also an exposure at the south end of the Còrrachadh ruins. Moreover, from a little north of Allt an Eas Dhuibh to beyond Còrrachadh, a distance of three-quarters of a mile, there is an almost continuous low escarpment of basalt-lava, etc., which defines the eastern boundary of the ring-dyke in agreement with the evidence derived from the exposures which have just been cited. The felsite, there, and in the rest of the ground to be described, shows no tendency to yield crags. In this it differs from its behaviour in the Loch Bà exposures. There is reason to believe that its local weakness as a scenic element is partly determined by its degree of crystallization. Where it forms crags, the rock is often rhyolitic in texture, and elsewhere felsitic.

It should be put on record that, though it is convenient to conform with the general clock-wise plan of the Memoir, and describe the Glen Forsa exposures after those extending either side of Loch Bà, this does not correspond with the development of our knowledge. It was Mr. Richey's completion of the circuit elsewhere that showed that the Glen Forsa outcrop belongs to the Loch Bà Ring-Dyke.

E.B.B.

For a mile south of Allt Mòr, superficial deposits cover the mapped course of the Loch Bà Ring-Dyke. Felsite of the characteristic type is seen in the most northerly of the little streams draining Liath Dhoire, and again, less than half a mile farther south, in the main stream at the foot of the slope. In the next mile, there are six stream-exposures, and, following a gap of half a mile, there is a capital section in

Allt a' Choire Bhàin. In the mile and a half separating the last-mentioned stream from Breapadail, the felsite is so often seen that there is no need for a detailed statement.

The Breapadail exposure of the Loch Bà Felsite is succeeded, three-quarters of a mile farther west, by another in Allt na h-Eiligeir. The outcrop, however, can be shown to be discontinuous in the interval. Another break in the outcrop is also demonstrable, west of Allt na h-Eiligeir, on the ridge connecting Cruachan Dearg with Maol Buidhe.

In the River Clachaig, the dyke reappears, and can be traced north-westwards for a mile, before it dies out within half a mile of the Beinn nan Gabhar exposures, which served as starting point in this description.

Thus, in the whole circuit of some fourteen miles, three gaps have been found, all situated on the south-west side, and these amount in aggregate to about a mile and a half. If exposures were available at every point, this figure might be slightly increased, but at most it would remain a small fraction of the whole.

In identifying the Loch Bà Felsite in its more isolated exposures, one is guided by several considerations, notably :— The character of the rock ; its late date—as shown by the absence of cutting intrusions other than dykes ; its tendency to occur along a line of crush, which may be traceable even where the felsite is absent ; and finally, the position of the exposures themselves relative to one another. It is to this last point that attention has been directed in the preceding paragraphs. J.E.R.

### (2) Inclination.

The Felsite Ring-Dyke has a steep north-north-westerly inclination along the whole of its well-exposed outcrop, either side of the Loch Bà. This feature is emphasized by the contrary inclination of a fairly well defined set of columnar joints that cross the dyke at right angles. Indeed, on this basis, the dyke appears to be inclined north-westwards at about 70°-80° on the two sides of Loch Bà, and at about 45° in Beinn nan Lus.

Elsewhere along its course, the Loch Bà Felsite is obviously steep, but the direction and degree of its inclination have generally defied detection. In the little stream mid-way between the summit of Beinn Talaidh and Beinn na Duatharach, there is, however, a clear exposure of the dyke's fluxion-edge (or top) inclined eastwards at a comparatively low angle.

There is thus a slight suggestion of a general outward inclination in the case of the Loch Bà Ring-Dyke, but far more evidence is required upon the point. E.B.B.

### (3) Ring-Fault.

There is abundant evidence for the existence of a ring-fault, along the course of which the Loch Bà Felsite has forced its way upward ; and it is substantially proved that the downthrow is towards the centre. The central subsidence may at one time have produced at the surface a caldera comparable with that which occupied the central area around Beinn Chàisigidle at the period of the extrusion of the Central Type of Lavas. It should, however, be noted that here we have no actual indication of a surface-basin, such as is so beautifully demonstrated by pillow-lavas in the case of the south-eastern caldera (Chap. V.) The evidence for cauldron-subsidence is as follows :—

(1) There are several fine sections of the north-west segment of the Loch Bà Ring-Dyke, where this dyke is clearly intruded along a line of crush. The best of these lie north of Sròn nam Boc, and exhibit the granophyre, sheared into a breccia, cut across by the felsite. A thin slice (14766), from the junction of the felsite with the mixed basic-acid intrusion lettered eD on Coille na Sròine, shows fine granophyric and ophitic debris invaded by beautifully spherulitic rhyolite constituting the edge of the Loch Bà Ring-Dyke.

(2) Poor exposures prevent our determining whether a similar crush-line continued along the eastern limb of the ring-dyke ; but, coming to the south-eastern and southern portion of the ring-dyke, we find everywhere, in the numerous streams which descend into the headwaters of Glen Cannel on both sides of Beinn Chàisgdle, evidence of crushing in the rocks along the inner and outer edges of the felsite. North-westwards the crush-zone continues across

Maol Buidhe, where felsite is missing. In Glen Clachaig, the outer margin of the Loch Bà Felsite makes a sharp junction against crushed granophyre. The crushing extends a few yards upstream from the felsite and involves cone-sheets which cut the granophyre. The belt of crush continues northwards, and serves to bridge the gap in the felsite-ring between An Cruachan and Beinn nan Gabhar.  W.B.W., J.E.R.

(3) Only small interrupted outcrops of basalt-lava have been preserved in the area ringed about by the Loch Bà Felsite. All the specimens collected and sliced have proved to be of Central Types (Chap. IX.). One of these specimens is the analysed rock (pp. 131, 149) from Na Bachdanan, north-east of Loch Bà. Here Central Type of Basalt reaches down to the 700 ft. level just inside the ring-dyke (the basalt-lavas at Coille na Sròine, south-west of Loch Bà, are better placed for our purpose, but have not been critically examined). Outside the Loch Bà Ring-Dyke, in this part of its course, the country consists of Plateau Basalts. The evidence can be followed on the one-inch Map: in Ben More (Area 8, Chap. VIII.), the Plateau Group reaches to 3169 ft. above O.D., with the base of the Ben More Mugearite at about 2100 ft., and the base of the Pale Group at about 1700 ft.; in Beinn Fhada, the base of the Mugearite, near the Loch Bà Ring-Fault, is at about 2000 ft; in Beinn a' Ghràig, the Mugearite has not been distinguished, but the base of the Pale Group is still about 1250. The lavas of Beinn Bhuidhe, north-east of Loch Bà, have not been very critically examined, but Maol Buidhe, across Glen Forsa (Area 9, Chap. VIII.), belongs to the Plateau Group with a mugearite-position—perhaps the Ben More Mugearite—toward its summit. It is certain that nothing but a marked subsidence could lead to the preservation of Central Types of lava in Na Bachdanan. It is possible that some proportion of the rhyolites and agglomerates of the region may belong to cones, rather than vents (p. 197); if this be the case, their preservation also requires subsidence to account for it, but this is more speculative.

(4) Near Gaodhail, in Glen Forsa, there is strong suggestive evidence of faulting of Early Basic Cone-Sheets at the line of the Loch Bà Ring-Dyke (Fig. 36, p. 238). The throw indicated is again one of central subsidence.

(5) Both sides of Beinn Chàisgidle, there is good reason to believe that the Glen Cannel Granophyre stands considerably higher outside the Loch Bà Ring-Dyke than it does inside (p. 332).

(6) There is also the contrast of gneiss-free vent-agglomerate, inside the ring-dyke, and gneiss-laden vent-agglomerate, outside (Fig. 29, p. 201). This contrast, like the rest of the evidence, suggests central subsidence, but quite possibly of a different type. The main constituent of the agglomerate within the ring is granophyre, obviously a much earlier granophyre than any exposed in the immediate vicinity, though quite possibly of the same age as the Glas Bheinn Granophyre of Chap. XII. In the intrusion of this hidden granophyre, the floor of gneiss must have either foundered piecemeal by stoping, or been let down *en bloc* in the manner suggested by the cauldron-subsidence of Glencoe.[1] We cannot, of course, decide which of these two modes of intrusion actually operated; but if the latter, the absence of gneiss-fragments in the agglomerate suggests subterranean cauldron-subsidence in this area at a date prior to that of the formation of the agglomerates.

W.B.W., E.B.B.

## KNOCK AND BEINN A' GHRÀIG RING-DYKES.

These, the most north-westerly of the plutonic masses of Mull, are classed as ring-dykes mainly on account of their elongated form, the fact that they are separated by an exceedingly well-preserved screen, and their parallelism with the neighbouring portion of the Loch Bà Felsite. Their curvature is that of the periphery of the Mull plutonic complex as a whole, and, as such, is a very

---

[1] *See* C. T. Clough, H. B. Maufe, and E. B. Bailey: 'The Cauldron-Subsidence of Glencoe,' *Quart. Journ. Geol. Soc.*, vol. lxv., 1909, p. 669.

inconspicuous feature. The belt of country occupied by their outcrops is characterized by marked relief, the result, to a large extent, of the contrast between the easily eroded character of the granophyre and the resistant nature of the baked lavas and intrusions which form the walls and roof. The highest summit is Beinn a' Ghràig, 1939 ft. O.D., while the two principal valleys, those of Loch Bà and Glen Forsa, cross the granophyre-outcrops at levels lower than the 50 ft. contour. The steep hill-slopes towards the west afford admirable exposures, and, on the whole, the conditions are very favourable for the investigation of intrusion-form. The subject will be dealt with under the following heads :—

(1) The form of the walls, screen, and roof.
(2) The ring-like intrusions of the north-eastern extremity.
(3) The general absence of disturbance outside the intrusions.

### (1) Form of the Walls, Screen, and Roof.

The Knock Granophyre, in a course of one and three-quarter miles, has a minimum width of 250 ft. and a maximum of 1000 ft. Its form is therefore distinctly dyke-like.

The Beinn a' Ghràig mass is much wider, ranging from about half a mile to a mile, but at the same time it is much longer, having a total length of seven miles from Beinn Fhada to Glen Forsa; on either flank, it extends a little beyond the gently curved north-western segment of the Loch Bà Ring-Dyke. The outer margin of its outcrop can be seen in places to be a nearly vertical wall, but in others shows a marked irregularity, due no doubt in part to the preservation of portions of the roof. Its inner or south-eastern margin is to a large extent defined by the fault accompanying the Loch Bà Dyke, but there are preserved, at this side also, segments of the roof, the form of which suggest that there is a gradual passage from roof to wall—although it is an open question whether the boundary on this side, previous to faulting, approached verticality in depth. The downward tendency of the roof towards the south-east is especially noticeable on Beinn Fhada and Beinn nan Gabhar, where the ring-fault swings south and leaves the granophyre.

It is impossible to say whether the Beinn a' Ghràig Granophyre extends beyond the fault within the area of central subsidence. Certain masses of hybrid basic rock outside the fault on Sròn nam Boc and Coille na Sroine suggest a limit to it in this direction, but the evidence on this point is very weak. A possible exposed extension of the granophyre lies just within the circuit of the Loch Bà Ring-Dyke on the slopes above Glen Forsa, between the 400 and 900 ft. contours, where its freedom from cone-sheets attracts attention.

The outer margin of thr Knock Granophyre is approximately vertical, and so is the screen of baked lava which separates it from the Beinn a' Ghràig Mass. This screen can be traced over hill and valley from near sea-level up to the 1750 ft. contour. It is apparently absolutely continuous for a mile and three-quarters, but varies considerably in thickness, being as little as 10 ft. in one place, and reaching 300 ft. in others. At both ends of the Knock Granophyre, the screen connects with the general mass of the lavas outside. At the south end, the connection is not quite clear on mere inspection of Plate VI. (p. 307), for the screen, as exposed on the west slope of Beinn a' Ghràig, locally consists of intrusive felsite instead of the usual basalt-lavas. In the field, however, it is obvious that the felsite concerned has this much in common with the basalt-lavas, that it is of earlier date than the two granophyres which it separates, for it is cut by many late Basic Cone-Sheets that are entirely absent from neighbouring exposures of the granophyres.

The course of the screen up the hill-face south of Knock can be readily traced by the eye even for a considerable distance. The caps on Beinn a' Ghràig and Beinn nan Gabhar, which can easily be recognized on the one-inch Map, are also scenic features obvious to the most casual observer. One of the caps on Beinn a' Ghràig joins up with the screen, forming with it an inverted letter L. At the south-western end of the Beinn a' Ghràig Mass, there is thus a marked contrast between the two sides: on the north-west, the arched roof joins up almost at a right angle with the vertical screen, whereas on the south-east it inclines rapidly downwards as far as erosion has brought it to light. North-east of Loch Bà, however, the asymmetry does not persist, and there seem to be two sides to the arch. Still farther in this direction, the conditions become so complex as to preclude any profitable speculation as to the general form of the roof.

The condition at the north-east end of the Beinn a' Ghràig granophyre, though comparatively speaking obscure, are sufficiently interesting to merit further discussion. It is a noteworthy feature of many of the cappings of the Beinn a' Ghràig interior that they frequently show steep-sided or vertical contacts with the granophyre, being apparently more of the nature of half-submerged blocks than true cappings. In this, they contrast with the remarkably evenly developed roof of the Glen Cannel Granophyre; and there arises a suggestion that the difference is fundamental, and connected with a clearly marked distinction between the ring-dyke and batholithic modes of intrusion. The form of the roof at the north-eastern end of the Beinn a' Ghràig Dyke indicates that subsidiary dykes, often with a ring-like character, may arise from the top of the main mass. The most remarkable of these is that to which the name Killbeg Ring-Dyke may be applied. It has an internal diameter of just over a mile; and is apparently a parasitic outgrowth of the Beinn a' Ghràig Dyke, of which it forms the north-eastern extremity.

## (2) Ring-like Intrusions of North-Eastern Extremity.

(2a) A well-marked sector of the Killbeg Ring-Dyke, introduced in the previous paragraph, crosses Maol Bhuidhe in a straight course, three-quarters of a mile long; its horizontal thickness here is 500 ft., and its exposed vertical depth 700 ft. In Allt nan Leòthdean, its south-east margin, as seen, is both vertical and chilled.

From Callachally to Pennygown, superficial deposits obscure the outcrop of the Killbeg Ring-Dyke; but south-eastwards from Pennygown, its arcuate track through An Càrnais is a conspicuous feature of the landscape. The outer margin of the granophyre is here well-exposed in contact with the truncated basalt-scars of Coire nam Muc. The inner margin is covered by alluvium, but a low ridge of baked lava, constituting Ceann an Tùir, clearly marks its approximate position.

At the south-east angle of the Killbeg Ring-Dyke, there are granophyre-exposures in two westward-flowing tributary burns, a little above their junction with the alluvium of Glen Forsa. In the River Forsa, no rock of any sort is seen in this neighbourhood. West of the river, however, as soon as the 50 ft contour is crossed, it is possible to draw a fairly well-defined southern limit for local exposures of basalt-lava; probably this limit corresponds closely with the northern edge of the granophyre, for between the 500 ft. and 1000 ft. contours a continuation of the same line separates basalt from exposed granophyre. The mapping of the granophyre above the 500 ft. contour towards the summit of Beinn nan Lus, and thence round the head of Coire an Ùruisge to Maol Bhuidhe, presents no difficulties.

(2b) Plates III., V., and VI. all show the irregular nature of the Beinn a Ghràig Granophyre at its north-eastern extremity, north of Salen, which we have just pointed out to be probably due to the extensive preservation of the uneven and fissured roof. They bring out very clearly, not only the parasitic ring of Killbeg, but also an extension westward from it, which takes a course in rough continuation of the Knock Dyke and of the general outer margin of the Beinn a' Ghràig mass farther west. This rather irregular outer dyke, known as the Toll Doire Dyke, seems, in Cnoc Maol Mhucaig, to be partially double, suggesting contemporaneous employment of two parallel fissures. Although shown as granophyre throughout in Plate VI., this intrusion includes subordinate outcrops of dioritic composition either side of Glac a' Chlaonain, and these are picked out by colour and lettered H on the one-inch Map.

A question naturally arises whether the Toll Doire Dyke is an offshoot of the Beinn a' Ghràig mass, an interrupted continuation of the Knock Granophyre, or an independent entity. Unfortunately, no positive answer can be given. This is the more to be regretted, since, as will be shown later (p. 345), it has a distinct bearing upon the interpretation of the age as well as the form of the Beinn a' Ghràig intrusion. As regards form, it may be pointed out that the width of the Beinn a' Ghràig Dyke, south of Loch Bà, suggests that the intrusion has there filled a vacancy left by subterranean foundering of a mass of country-rock bounded at the sides by parrallel fissures. If the Toll Doire Dyke is part of the Beinn a' Ghràig Granophyre, it may be taken as marking the continuation of one of the fissures concerned.

W.B.W.

### (3) Absence of Disturbance Outside the Intrusions.

The intense folding, which characterizes much of the periphery of the south-eastern caldera of Plate V. (p. 165), is unrepresented about the Loch Bà centre.

As Sir Archibald Geikie has emphasized (p. 48), there is a general lack of disturbance of any kind outside the Knock and Beinn a' Ghràig Granophyres: instead, the featuring of the hill-sides show these great intrusions abruptly truncating gently inclined Plateau Lavas. One gets an impression of an essentially tranquil process during which granophyre has replaced basalt, either by piecemeal stoping, or by contemporaneous infilling of a space prepared by large-scale subsidence *en bloc*.

The phenomenon is particularly clear in Beinn Fhada. On the western termination of the ridge, subdued, but none the less obvious, trap-features indicate that the lavas continue with a gentle and steady inclination until they are clearly terminated at the margin of the granophyre. This scenic evidence is reinforced by close examination. The one-inch Map, and Pl. III. (p. 91), show a mugearite-lava traced for many miles through the Ben More country. In Beinn Fhada, this mugearite is found on both sides of the granophyre, and its two separated portions have not suffered appreciable vertical displacement relatively to one another.

A local exception to the general undisturbed nature of the country-rock on this side of Mull should, however, not be passed over without notice. It occurs in Glac a' Chlaonain, at the north-west edge of the Toll Doire Dyke. The lavas are here bent suddenly upwards, and the underlying Liassic sediments are exposed to view.

## Age-Relations.

The intrusions dealt with in the present chapter are among the latest of all the igneous rocks of Mull, a circumstance consistent with their position towards the north-western end of the axis of symmetry. Of the three, however, the Loch Bà Felsite, which is clearly the most recent, does not occupy the most north-westerly position.

The only rocks that cut the Loch Bà Felsite are a number of north-west dykes (Chapter XXXIV.). It is almost certainly later than all categories of cone-sheets, a claim that can be put forward for no other large-scale intrusion. Even the Beinn a' Ghràig Granophyre, which is obviously later than the majority of the Late Basic Cone-Sheets of Chapter XXVIII., is found on close examination to be intersected by what appear to be a few late stragglers. These, however, are extremely rare in spite of the fact that the granophyre lies right in the track of the main belt of Late Basic Cone-Sheets, which constitute much of its walls and roof.

The Glen Cannel (Chapter XXXI.) and Beinn a' Ghràig Granophyres had been intruded, solidified, and brecciated along the Loch Bà Ring-Fault before the uprise of the felsite guided by this fracture. The relation is everywhere so obvious as hardly to need further comment. An examination of almost any of the exposures mentioned in connexion with the brecciation along the fault (p. 340) will convince the reader of its truth.

If the conclusion arrived at in Chapter XXXI. (p. 332), to the effect that the Glen Cannel Granophyre extends outside the Loch Bà Felsite to the south and south-east, be correct, then a comparison is made possible between the Glen Cannel and Beinn a'

Ghràig Granophyres in respect of their behaviour towards the Late Basic Cone-Sheets; for these extended areas of the great Glen Cannel mass lie in the belt of the main development of these sheets, and are cut freely by a fairly large proportion of them. It is therefore highly probable that the Glen Cannel Granophyre is distinctly earlier than that of Beinn a' Ghràig.  W.B.W., J.E.R.

Good exposures for studying the relations of the Beinn a' Ghràig Granophyre to the Late Basic Cone-Sheets are met with west of Loch Bà. A belt of late Basic Cone-Sheets, crossing the south-east summit of Beinn Fhada, is interrupted for a space by the Beinn a' Ghràig Granophyre, only to reappear in the lava-cappings of Beinn nan Gabhar and Beinn a' Ghràig, and in the country-rock beyond, which consists in part of basalt-lavas (cf. one-inch Map). The Cone-sheets in Beinn Fhada are locally much traversed by granophyre-veins, and in many cases are demonstrably cut off by granophyre in mass. In the Beinn a' Ghràig caps, a felsitic modification of the granophyre occurs as tongues crossing the lavas with the same inclination as characterizes the cone-sheets, and sometimes splits these latter up the middle.

The Knock Granophyre, north of Beinn a' Ghràig, behaves in the same manner to the Late Basic Cone-Sheets as does its bigger companion just described. The two granophyres offer a very marked contrast, in this respect, to the felsite that forms the southern end of the separating screen; for this felsite is cut by numerous Late Basic Cone-Sheets which are entirely absent from the granophyres on either side. Another good exposure showing the Knock Granophyre cutting Late Basic Cone-Sheets is afforded by the hill-face above Gruline House.  J.E.R.

Only two exposures can be quoted where inclined sheets, apparently indistinguishable in any way from the general group of Late Basic Cone-Sheets, can be seen cutting the Beinn a' Ghràig Granophyre. One of these two exposures is afforded by the stream descending from An Coire, nearly half-a-mile west-south-west of Benmore Lodge, at the lower end of Loch Bà. It shows the granophyre traversed by a pale basaltic sheet, or dyke, with columnar structure: this sheet runs north-east along the stream and dips south-east at 60°. The other exposure is not indicated on the one-inch Map, but occurs in the stream directly above Gruline House, where the granophyre is cut by a sheet of Talaidh Type (14345) which dips E. 25° S. In this case the angle of inclination has not been recorded.

Other sheets which probably belong to the Late Basic Cone-Sheets are seen to cut the Toll Doire Dyke (p. 343) in Glac a' Chlaon-ain. Their dips have not been recorded. There may be one or two other cases east of this, but, where the sheets swing into a direction nearly coincident with that of many of the dykes, it is much harder to pick them out unless they are very well-exposed.  W.B.W.

## Late Explosions.

In places along its south-eastern exposures, the Loch Bà Felsite seems to pass into compact tuff containing fragments of granophyre, felsite, etc. Good stream-sections occur in two burns at the foot of Coill' an Aodainn. A specimen from Allt a' Choire Bhàin (15569) shows a fine-textured breccia composed of fluxional rhyolite, well-crystallized granophyre, and some more basic rock, the whole invaded by tongues of spherulitic rhyolite. It certainly looks as though the felsite has locally shattered itself by explosion. W.B.W., E.B.B.

## PETROLOGY.

### LOCH BÀ FELSITE.

(ANAL III.; TABLE IV., p. 20).

The Loch Bà Felsite is for the most part a fine-grained rock that may be termed either rhyolite, or felsite according to the state of its devitrification and the presence or absence of dominant fluxion-structure.

A series of specimens, taken in succession along the dyke from the hill-top west of Loch Bà by way of Glen Forsa and Beinn Chàisgidle to Glen Clachaig, shows that in the more northerly portions of its course it is a beautifully banded rhyolite (2121) with delicately developed fluxion-structure (Fig. 59B), while towards the south and east it becomes more definitely felsitic. It contains small phenocrysts of felspar and well-formed little rectangular crystals of yellowish augite (14322), in a matrix that exhibits various degrees of banding and devitrification. The felspar-phenocrysts are usually of albite or perthite, and occasionally are edged with turbid orthoclase (14697). There is a general absence of granophyric structure, but it has been noted in some of the more felsitic varieties of the rock where microgranophyric matter fringes turbid phenocrysts of perthite (14969). The augite, for the most part, appears to be an ordinary aluminous species, but, in a few instances, both in the rhyolitic (14321) and felsitic varieties, an augite, replaced or partly replaced by serpentine, would appear to be of enstatite-composition (enstatite-augite, Chapter XXV.). The main differences presented by the various portions of the dyke lie in the state of devitrification and the nature of the resulting crystalline matrix. Frequently, the matrix has either assumed a minutely patchy character, due to the separation of alkali-felspar and quartz into small mutually interfering areas (2119, 14319), or has become microcrystalline with a definite separation of felspar (oligoclase) microlites (14362). Acicular augite, converted into greenish hornblende and magnetite, is a feature of some varieties (15570), and may possibly indicate some slight compositional change in the rock, especially as it seems to be more frequently associated with the occurrence of microporphyritic enstatite-augite (17104). This type of rock connects naturally with the acid varieties of the inninmorites (Chapter XXV.).

Very occasionally, a still more pronounced type of crystallization is encountered, and then the matrix consists largely of minute short rectangular crystals of orthoclase, free-quartz, and subordinate plagioclase (17969). This type of matrix is most ususally present when the rock has a more definitely granophyric character. In the felsitic portions of the mass, the rock frequently shows the results of auto-brecciation and the incorporation of darker and more vitreous angular patches of banded rhyolite. These patches probably represent ribbons of the more quickly cooled rock broken up and carried forward in the course of the intrusion. In one case only, a pyrogenetic hornblende, occurring in small irregular grains, appears to be the dominant ferro-magnesian mineral (17970); but the rock, in this instance, seems to have an abnormally basic character, and may owe its unusual character to assimilation of basic material.

The analysed specimen (14825 Anal. III., p. 20) is characteristic of the mass. It shows good fluxion-structure, and has small

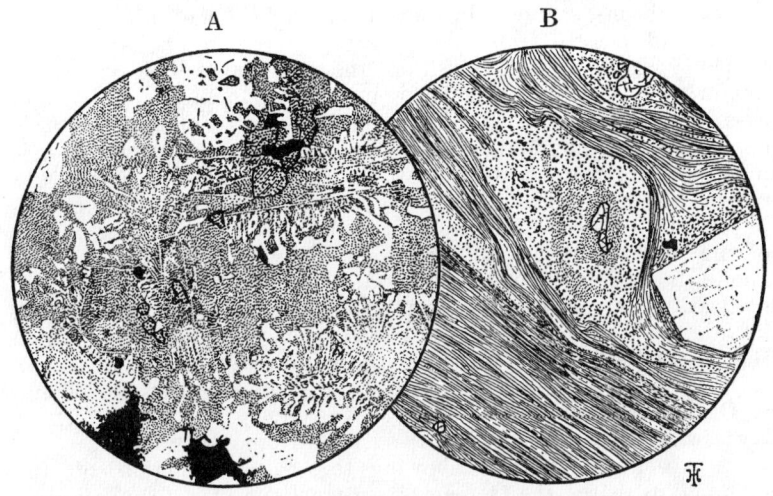

Fig. 59.

A. [14841] × 17. Knock Granophyre. Brownish-green augite and crystals of oligoclase edged with perthite, enveloped in a typically granophyric matrix of which the structure is emphasized by the turbidity of the alkali-felspar.

B. [14825] × 17. Felsite of Loch Bà. Rhyolitic type with well-developed fluxion-structure. The phenocrysts are of yellowish augite and albite. Areas devoid of banding have suffered a more pronounced devitrification.

phenocrysts of albite and aluminous augite. The matrix is devitrified to a considerable extent, but the degree of devitrification varies greatly from band to band. It is interesting as containing relatively basic glomeroporphyritic groups of oligoclase-albite and serpentinized enstatite-augite.

### BEINN A' GHRÀIG GRANOPHYRE.

(ANAL. IV.; TABLE IV., p. 20.)

The greater part of this rock consists of quartz and orthoclase in granophyric intergrowth of varying coarseness, which furnishes a matrix to, or fringes, small phenocrysts of albite, perthite, orthoclase, or even microgranophyre (Fig. 57B, p. 334). The chief characteristic of the rock, however, is a green pleochroic augite that occurs in some abundance as slightly elongated irregular crystals and grains (2146, 13857). It probably has an alkaline composition, and its optical properties would suggest its inclusion in the series of aegerine-augites (p. 350, also Chapters XIV., XXXI., etc.). It commonly is of uniform composition throughout, but occasionally it may form a peripheral coating to an augite of brownish tint and presumably of less alkaline character (13859). It is usually associated closely with magnetite, but without that regularity of arrangement observed in the columnar or acicular augites of such rocks as the augite-diorites, leidleites, and quartz-dolerites of Chapters XVIII., XXV., and XXVIII.

Viewed in section by ordinary light, the alkali-felspar stands out clearly from the quartz by reason of its extreme turbidity (2094). In composition, the rock shows little variation, but in texture it may become more finely granophyric (2095) and assume a subspherulitic structure, or its constituent minerals may crystallize as equigranular individuals (13857). In rarer cases, it may assume characters that would indicate affinities with the craignurites of Chapter XIX.; the texture being of a finely granophyric nature, with a tendency towards a spherulitic grouping of its microperthitic and microgranophyric constituents. The augite is mainly acicular in form, and represented by narrow magnetite-charged hornblende-pseudomorphs (13858, 13860), and, in addition, there are a few well-shaped pseudomorphs (13860) that recall the enstatite-augite of the inmorites (Chapter XXV.).

Very interesting modifications have been noted in the Beinn a' Ghràig Granophyre at and near its contact with the screen of lavas which separates it from the Knock Granophyre.

A specimen (14316), showing the actual junction of the granophyre with the screen, is instructive. The country-rock, much broken up by the invading granophyre, is now an extremely fine-grained granulite composed of recrystallized felspar, and minute granules of greenish augite, pale-brown hornblende, and magnetite. The granophyre, though finer-grained than usual, continues moderately coarse to its margin, without any sign of rapid chilling. It has resorbed a considerable amount of basic material with a resultant separation of fairly large crystals of zoned labradorite and oligoclase-andesine felspars. The more alkaline portions are strung through with granulitic augite, hornblende, and magnetite, with a crystallization similar to that of the granulitized country-rock. A few feet away from the screen (14315), the granulitization and assimilation of basic material are equally well-marked, though the resultant greenish augite and brownish hornblende have coalesced to a greater degree. A little biotite is also present, mainly in the hornblendic areas and in association with magnetite.

### KNOCK GRANOPHYRE.
(ANAL. V.; TABLE IV., p. 20).

The Knock Granophyre, like the Beinn a' Ghràig rock, has

typically a granophyric structure (14841, Fig. 59A). The dominant ferromagnesian mineral, however, is a brownish-green, presumably aluminous, augite that forms sub-idiomorphic to irregularly bounded crystals. Another point of difference is the occurrence of a more obvious quantity of plagioclase. Augite occasionally assumes an acicular habit, but only to a limited extent.

A specimen from Beinn Bheag (14840), much more basic than the normal granophyre, has a relative abundance of basic plagioclase, together with a fair quantity of augite that has assumed a columnar habit and is intimately intergrown with magnetite. Indefinite chloritic and magnetite-bearing patches probably represent imperfectly assimilated basic material.

Close to the screen of lavas, which borders the granophyre to the south-east, the rock (14317) has a distinctly abnormal facies, more striking than that of the hybrid portions mentioned above. It may be possible that we are here dealing with a normal basic edge of the granophyre, but it appears from the general character of the rock, especially when considered in connexion with the similarly placed portion of the Beinn a' Ghràig Granophyre (*see* 14315, 14316, p. 348), that its peculiarities are due to an assimilation of basic material.

### TOLL DOIRE DYKE.

This mass presents characters, in different parts, that would justify the application of the name either of granophyre or felsite. In the granophyric portions (13855), the rock consists of fairly large individuals of microperthite strung with quartz, with which the felspar is in micrographic relation. The perthitic areas, when showing definite crystal-form, are usually fringed with granophyric material, of which the structure becomes increasingly coarse in an outward direction. There are moderately large areas of free quartz, and a fair abundance of bright green pleochroic alkaline augite of sub-ophitic habit and intimately associated with magnetite.

The felsitic types (13879, 13886) are somewhat poor in ferro-magnesian minerals, but, in addition to small phenocrysts of albite and perthite, a few microporphyritic crystals, replaced by serpentine and magnetite, appear to have been augite.

The phenocrysts of felspar are frequently surrounded by radiating sheaves of fibrous felspathic matter, with which are associated pseudomorphs after a finely acicular augite. With a more pronounced development of this radial growth, the rock may take on a definite spherulitic structure (13886), in which small regular round or ovoid patches of quartz occasionally act as nuclei.

Rocks of a somewhat abnormal character, and mapped variously as granophyre and augite-diorite, present features that appear to indicate considerable assimilation of basic material by the acid magma. A specimen (15590), from the more easterly outcrop of augite-diorite, as shown on the one-inch Map, is a granophyric rock that contains a large amount of corroded acid-labradorite felspar. Alkali-felspar has penetrated cracks and replaced weak spots, showing that the original felspar was out of sympathy with the magma by which it was ultimately enveloped. Augite is abundant, and occurs mainly in long irregular columnar crystals crossed by a salite-structure. This rock, in its main characters, is extremely like others to which the name augite-diorite has been applied, and which are dealt with in Chapter XVIII.

Another rock, mapped as granophyre (13880), has certainly much in common with the variant described above, but there is a greater proportion of granophyric matter in its constitution, suggesting assimilation of rather a different order. The augite, as in most Mull rocks of intermediate to sub-acid composition, occurs mainly as irregular columnar crystals; but an original ophitic augite can be recognized by its external form, though now completely granulitized by re-heating into a fine mozaic of augite and magnetite. A still more advanced stage in the assimilation of basic material by the Toll Doire Dyke may be seen in a rock (13877) which contains irregular aggregates of bright green pleochroic alkaline augite, with granular strings of the same mineral in association with magnetite and a little granular sphene, in a matrix mainly composed of small irregular grains and patches of alkali-felspar.

### DISCUSSION OF ANALYSES (TABLE IV., p. 20).

It will be seen, from the analyses of the rocks described above, that there is considerable uniformity in the percentages of silica and alumina, but a noteworthy variation in the percentages of ferric iron, lime, and the alkalies.

Allowing for such accessory minerals as magnetite and apatite, it is found that, in the case of the Knock Granophyre, orthoclase and albite make up respectively about 14 and 33 per cent. of the total rock, and that sufficient alumina remains available to produce an aluminous augite with the magnesia, ferrous-iron, and lime. In the case of the Beinn a' Ghràig granophyre, however, we note an increase in the percentage of ferric iron and potash, with a decrease in the relative amount of lime. In this case, orthoclase will form about 27 per cent., and albite about 30 per cent. of the rock, and, at the same time, the amount of available alumina and lime is too small to allow the formation of a normal aluminous augite. It is probable, therefore, as is confirmed by the green colour and optical characters of the contained augite, that in this rock we have a pyroxene (aegerine-augite) in which ferric iron partly replaces alumina, and in which soda enters into the composition.

Similarly, the high percentage of alkalies in the Loch Bà Felsite, and the percentages of alumina, lime, magnesia, and iron, when compared with the percentages of the same constituents in the Knock Granophyre, demand a high felspathic content, and the consequent using up of the available alumina in the formation of these minerals. It is natural to expect, therefore, that the lime and magnesia would tend to separate in the form of a non-aluminous pyroxene, and this to a limited extent has been found to be the case from actual observation.   H.H.T.

# CHAPTER XXXIII.
## HYBRIDS OF SRÒN NAM BOC AND COILLE NA SRÒINE.
## LOCH BÀ.

### Introduction.

The Sròn nam Boc and Coille na Sròine complexes of basic, intermediate, and acid rocks are grouped under the index-letter eD on the one-inch Map, Sheet 44. They both lie immediately outside the Loch Bà Felsite of Chapter XXXII., between which and the Beinn a' Ghràig Granophyre they constitute an interrupted local screen. Possibly they may have originated as a ring-dyke, but this is quite uncertain. Their chief interest depends upon the fact that they exhibit phenomena of intricate intrusion and partial assimilation of a basic rock by the Beinn a' Ghràig Granophyre.

Within the areas coloured as dolerite or gabbro at the two localities under consideration, there is to be met with every grade of composition from basic to acid. Sometimes, one type of rock merges gradually into another; but, often, the transition is rapid, with concomitant veining and enclosure of more basic by more acid material, so that angular blocks of undigested gabbro lie here and there immersed in granophyre.

The remainder of the chapter is mostly devoted to the microscopic Petrology of material collected from the two localities named above. Two concluding sections, however, look further afield, and are entitled Other Mull Examples and Internal Migration compared with true Hybridization.
W.B.W.

### Petrology.

The work of Dr. Harker on the gabbros and granophyres of Skye, and the subsequent elaboration of his views, have familiarized all students of petrology with the conception of hybridization as applied to igneous masses, and with the essential characters of hybrid rocks. It is interesting, therefore, to find in Mull further striking examples of the development of hybrids of intermediate composition characterized by special mineralogical and structural features.

Mr. Wright has provided us with a large number of excellently selected specimens, with carefully determined field-relations, to illustrate the phenomena of hybridization. His localities are Sròn nam Boc and Coille na Sròine on the slopes on either side of Loch Bà. The two suites of specimens show many resemblances, but it will be well to treat them separately at first, and then to summarize the essential features common to both.

*Sròn nam Boc.*—The gabbro (14724) of this locality is a moderately coarse, olivine-gabbro composed of olivine in well-formed crystals, zoned labradorite, coarsely ophitic pale augite, and iron-ore. The rock has suffered changes that may be attributed, partly to the thermal action, and partly to the chemical influence of the invading acid magma.

The olivine, which is full of dentritic magnitite, has been in some measure converted into rhombic pyroxene, actinolite, and serpentine, and shows good schiller-structure. In portions of the rock that appear to be albitized and acidified, olivine is surrounded by rhombic pyroxene, and independent rhombic pyroxene has developed, together with green hornblende.

In a specimen, collected to show the usual more pronounced type of acidification of the basic rock, we may detect three grades of alteration within the limits of a single slice (14311). Where least altered, the rock contains pseudomorphs after olivine, similar to those described above. This is succeeded by a zone in which the augite and felspar of the original rock have been largely recrystallized. The augite, locally, shows signs of granulitization, but more particularly of recrystallization in clots, where, along with slender radiating prisms of re-formed felspar, it gives rise to a sub-variolitic structure. In the most acidified portion, there is a hypidiomorphic, to panidiomorphic, development of the ferro-magnesian minerals and more basic felspars: augite has recrystallized as small idiomorphic crystals; rhombic pyroxene, represented by pseudomorphs in green fibrous hornblende, has developed as finely ophitic patches and isolated crystals, and is frequently intergrown with augite; biotite occurs in general association with iron-ores; and basic re-formed felspars are deeply zoned with oligoclase. Patches of turbid alkaline material, with a fair amount of free quartz, can be seen to have exerted a powerful corrosive effect on the felspars, producing a restricted, but intense, albitization.

In a similar specimen of acidified gabbro (14312), all trace of original olivine is lost. Augite has recrystallized as small crystals and grains with hypidiomorphic outline, but occasionally one may note a columnar development of this mineral in the form of large and definitely elongated prisms. The felspars are of two kinds, strongly corroded large individuals of acid labradorite, zoned by turbid alkali-felspar or micropegmatite, and smaller elongated oligoclase-crystals, of an evidently subsequent generation, edged with orthoclase and quartz. Rhombic pyroxene (pseudomorphed) in well-shaped crystals is a noticeable feature of the hybrid zone, as also is a general separation of small regular crystals of magnetite. The rock contains much quartz, and, locally, a moderately coarse graphic intergrowth of this mineral with alkali-felspar.

A specimen (14313), collected to show the nature of the obviously acid veins that traverse the basic rock, is of a beautiful granophyre. It consists of small extremely turbid crystals of plagioclase, possibly remnants from the basic rock, surrounded by granophyre-aggregates. The more obviously contaminated portions of the vein show hypidiomorphic augite, occasionally in columnar crystals, intergrown with magnetite,[1] and fibrous hornblendic pseudomorphs after rhombic pyroxene.

Other specimens (14721, 14725, 14726) were collected to show the characters of the doleritic assemblage and the acid rock by which it is veined. The acid rock in these cases is a coarse granophyre. The doleritic rock is similar to the acidified rocks described above (14312), but, adjacent to the granophyre, there is a zone characterized by coarseness of grain and a richness in augite and hornblendic and chloritic pseudomorphs after rhombic pyroxene. In this zone, there is a little brownish-green hornblende of pyrogenetic origin. We may also note the original augite of the dolerite being eaten into and developing a salitic striation (14725), or recrystallizing with columnar habit (14726). Rhombic pyroxene (pseudomorphed) has developed at the expense of olivine, and biotite is associated with the iron-ores (14725).

An actual xenolith of basic rock (14722) surrounded by granophyre shows well the nature of their reciprocal action. Internally, the xenolith has the usual

[1] A. Harker, 'Tertiary Igneous Rocks of Skye,' *Mem. Geol. Surv.*, 1904, Pl. XXI., Fig. 2.

characters of the acidified dolerite (*cf.* 14311). Olivine is replaced by serpentine, fibrous hornblende, and magnetite. Augite is in part recrystallized, and in part surrounded by a pseudomorph after rhombic pyroxene. In places, the original augite is curved and bent (p. 325), and shows an incipient development of salitic structure. Nearer the granophyre, augite has been resorbed with subsequent formation of greenish-brown hornblende; and all stages of dissolution may be studied. The granophyre itself, near the junction, contains columnar magnetite-bearing crystals of augite, now replaced by fibrous green hornblende.

A specimen (14723), collected as illustrative of the typical medium-grained hybrid, may be best regarded as a basified granophyre. Little is left in the way of remnants of the basic material. The felspar is almost all fresh oligoclase and perthite; and there is an abundance of quartz with a little micropegmatite. The ferromagnesian minerals are augite with columnar habit, and hypidiomorphic crystals of smaller but stouter form. There are the usual pseudomorphs after rhombic pyroxene, and, in addition, a good deal of greenish-brown pyrogenetic hornblende.

*Coille na Sròine.*—At the other locality, on the western slopes above Loch Bà, the more basic phases, acidified dolerites, are represented by four specimens (14309, 14310, 14719, 14720). They contain patches of micropegmatite; their basic plagioclase is much albitized; and their augite occurs in prominent twinned crystals showing herring-bone salitic striations. Olivine is represented by large chloritic pseudomorphs. Signs of contact-alteration, due to the proximity of the granophyre, are forthcoming in flecks of biotite developed in partially chloritized augite. In certain instances, the augite is strongly columnar with salitic striation (14720), and is edged with greenish-brown pyrogenetic hornblende (14719).

A rock (14318), collected to illustrate a variety into which the above-described basic phases rapidly pass, is of generally more acid character. It completely lacks pseudomorphs after olivine, but has a little pyrogenetic hornblende in its quartz-rich portions. The original augite appears to have been recrystallized with a tendency to assume the columnar habit. There are pseudomorphs after rhombic pyroxene. Basic felspars have been broken up, albitized, and fringed with perthite.

A rock of still more acid character (14712), which encloses one of the basic phases (14719) as a xenolith, is a basified granophyre with an acicular type of crystallization of its constituents. The slice itself contains no recognizable fragments of basic material, but consists of elongated crystals of augite pseudomorphed in fibrous hornblende, and elongated crystals of oligoclase-albite fringed with orthoclase and perthite, in a matrix of micrographic material and free quartz.

Three specimens (14713-5) were collected to show varieties of rock that, in the field, apparently pass gradually into each other. They are all characterized by a prevalence of pyrogenetic hornblende. The augite of the original rock has departed from the ophitic form, and has assumed a hypidiomorphic columnar habit, and is occasionally intergrown with the brown-green hornblende. They show a variable amount of true granophyric material, between which and the more basic portions, the usual rhombic pyroxenes have been developed. Felspars have suffered all changes from partial acidification to complete resorption.

Further specimens may be noted, of which one (14716), taken to illustrate an indefinite junction between more and less acid types, shows an abundance of pyrogenetic hornblende; while another (14717), of a type that sometimes presents definite boundaries against the more general and coarser material of its neighbourhood, is a felspathic rock rich in pseudomorphs after rhombic pyroxene.

Three specimens (14711, 14710, 14709) show the gradual transition of the hybrid rock to the granophyre (14708). They are well-mixed rocks, in which pseudomorphs after rhombic pyroxene are characteristic of the more basic material (14710), while pyrogenetic hornblende is a feature in the more acid regions (14709). The granophyre itself is a beautiful rock, and, at the point from which the specimen was taken, appears to be unmodified by the assimilation of basic material.

## SUMMARY.

To appreciate the significance of the mineralogical and structural peculiarities of the rocks described above, we must first endeavour to disentangle the effects of thermal alteration from those attributable to chemical activity. In the former category, we may safely place the development of actinolitic hornblende within the pseudomorphous representatives of olivine, the granulitization and recrystallization of augite, and the production of biotite from chlorite and in the neighbourhood of iron-ores.

Turning now to those changes which can be referred directly to mutual action between an acid magma and an already consolidated more basic rock, we note effects due to both local and general absorption. Acidification of the basic rock is responsible for the partial, or complete, replacement of olivine by rhombic pyroxene. Similarly, it accounts for the edging of original augite with brownish-green hornblende of pyrogenetic character, and also for the attack of original basic plagioclase crystals, their corrosion, and subsequent irregular replacement by felspar of more alkaline character. In like manner, basification of the acid magma has led to crystallization of newly-formed hypidiomorphic rhombic pyroxene, of independent grains and prisms of pyrogenetic hornblende, and of comparatively basic plagioclase felspar. Hydrothermal exchange has probably been involved in the alteration of augite to fibrous green hornblende, frequently noticeable in cases, where the acid magma has come in contact with pyroxene and yet has failed to produce a hornblendic fringe of typical pyrogenetic character.

The outstanding features, therefore, of these hybrid rocks of Mull are:—the acidification of basic felspars in irregular fashion; the replacement of olivine by rhombic pyroxene; the fringing of augite, frequently recrystallized, by pyrogenetic hornblende; the complete local assimilation of basic material followed by the independent crystallization of rhombic pyroxene; the basification of the acid magma with the production of pyrogenetic hornblende; and a general increase in the basicity of the early separating felspars. Referring to the independent crystallization of rhombic pyroxene in these rocks of hybrid character (acidified basic rocks), it must be remarked that, while this mineral is highly characteristic of the Mull occurrences, it must not necessarily be regarded elsewhere as an index of hybridization. It must be remembered that, even in Mull, it is a common constituent of the leidleites (Chapter XXV.).

The microscopic structures of these hybrid rocks are equally characteristic. In almost every case, rapid transitions from acid to basic types are met with, and a xenolithic structure is noticeable, even when the enclosed rock has been greatly modified by the surrounding magma. It is a noteworthy fact, pointing to selective assimilation and diffusion, that the shape of included xenolithic masses remains practically unaltered although their constitution is radically changed.

### Other Mull Examples.

There are many other examples of hybridization in Mull, where relatively basic and acid rocks are in contact; and perhaps the most noteworthy is that discussed under the heading of Craig Porphyrite in Chapter XXII. (see also pp. 210, 317).

Marked interaction between magma and its containing sedimentary walls is dealt with in connexion with the xenoliths of the

Loch Scridain sills (Chapter XXIV.), and the sandstone-granophyre assimilation-zone of the Glas Bheinn Granophyre (Chapter XII.).

### INTERNAL MIGRATION COMPARED WITH TRUE HYBRIDIZATION.

There is no positive evidence in Mull that any of the recurrent types of intrusions owe their distinctive characters to assimilation of foreign material, or commingling of magmas, prior to their arrival in their present position. In fact, there appears good ground for assuming that such divergent characters as they possess are more generally the outcome of normal processes of differentiation.

A word must be said here concerning many rocks, that, in some measure, present microscopic characters suggestive of hybridization in a restricted sense of the term. Such rocks are the Late Basic Cone-Sheets (Chapter XXVIII.), some of the Ring-Dykes (Chapter XXX.), and the so-called augite-diorites (Chapter XVIII.). We have, in Mull, numerous rocks of quartz-dolerite composition, derived from a magma, which, as crystallization progressed, clearly gave rise to an acid differentiate. This acid partial magma was of strikingly different composition to the early crystalline phases, and its temperature of complete consolidation was evidently far below that at which the larger and earlier-formed individuals had practically ceased to grow. It represents the original magma, almost depleted as regards lime and magnesia, but retaining abundant alkalies and dissolved water, and, with them, extreme fluidity. That it was capable of migration under gravitational, or other, forces from one portion of the mass to another is quite clear ; for, owing to such movement, it often achieved a local concentration, or traversed the earlier products of consolidation as strings and veins. Under these circumstances, it was brought into contact with early crystalline phases, with which it was no longer in equilibrium, and an appearance of injection of a relatively basic rock by a relatively acid magma has been produced. The disturbances of equilibria, moreover, have simultaneously engendered absorptive processes, similar to, but generally less active than those known to have operated in the formation of true hybrid rocks that resulted from invasion of magma from an external source.

The process of true hybridization involves the action of an independent magma upon an already consolidated rock, or the commingling of two independent magmas of different chemical composition ; but the process outlined above involves merely the reaction, without any considerable rise of temperature, of a differentiate, or partial magma, on already separated crystalline phases. It will be seen that the two processes have much in common, and in extreme cases the results may be indistinguishable. The main difference noted is that the disturbances of chemical equilibria are relatively reduced in the case of internal migration. In this connexion it is interesting to note the general absence of rhombic pyroxene and pyrogenetic hornblende from most of those rocks of Mull which have a mixed acid and basic character, except where definite xenolithic structure, observable in the field, furnishes, in itself, direct evidence of hybridization.  H.H.T., E.B.B.

# CHAPTER XXXIV.

## DYKES.

### Introduction.

Dykes of Tertiary age furnish one of the most striking features of West Highland igneous activity, and are characteristically exposed in the shore-sections of Mull, Lorne, and Morven. The basic dykes are lettered M on the one-inch Map (Sheet 44); the acid F; and the intermediates either M or F. As explained in the index, F is also used for certain other acid intrusions besides dykes. A few camptonitic dykes of very doubtful Tertiary date are discussed separately in Chapter XXXV.

Dykes, as here understood, are roughly vertical seams of igneous rock resulting from underground intrusion. Individual dykes may be traceable for miles along their outcrop; but, often, it would be hopeless to attempt the task of following them in this fashion where many close neighbours are of essentially similar lithological type. The breadth of a dyke is relatively small, and 5 ft. is a very common measurement.

Most of the Mull dykes are of basaltic types; but, whether basic, intermediate, or acid, they have their particular habit of jointing and weathering, controlled, in large measure, by composition. A rude approach to horizontal columnar jointing is often evident; while vertical joints, parallel to dyke-margins, are also well-represented. In many cases, jointing of any kind fails towards the centre of a dyke, where a median vesicular zone may be strongly developed.

Chilled edges, consisting of very compact crystallized rock, are typical, and glassy selvages occasionally occur (p. 47). There is a tendency for faulting and decomposition to follow the edge of a dyke, and this sometimes obscures an original chilled edge. In composite dykes, with more basic margins and more acid interiors (p. 8), it is customary to find marked external chilling of the basic flanking bands, but no chilling at all of basic against acid, or *vice-versa*, in the interior of the dyke; the explanation, of course, is that the acid followed the basic before this latter had cooled. A good instance of a composite dyke crosses Fig. 52 (p. 308), and others will be cited in the petrological section of this chapter. E.B.B.

Multiple dykes are at least as common as composite ones. In this memoir, the distinction between multiple and composite minor intrusions is based upon the presence or absence of chilling at the contacts of the participants. Coastal roads render the two shores of Loch na Keal (Sheet 44) exceptionally convenient for the examination of a great number of dykes; and anyone walking south-west from Eilean Feòir will find about twenty instances of multiple dykes in a distance of little more than a mile. There may be as many as four or five partners in a multiple dyke.

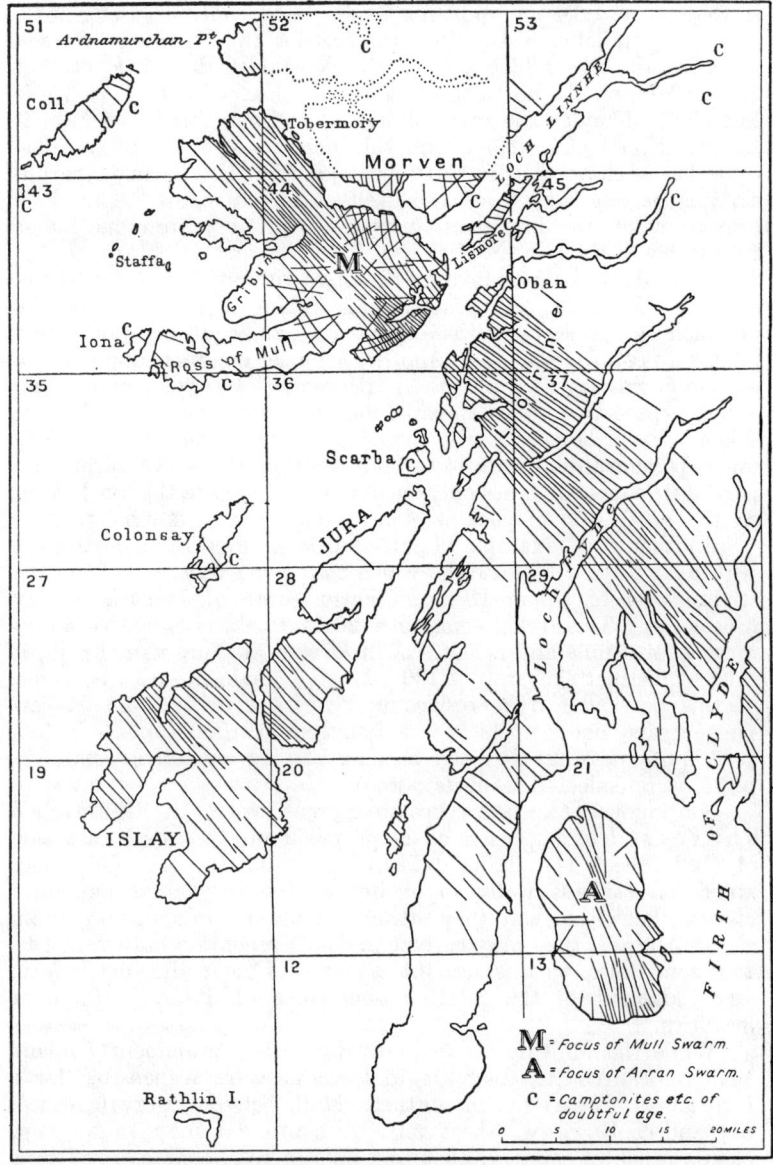

FIG. 60.—Tertiary Dykes of the South-West Highlands.
Only about one dyke in every ten or fifteen is shown. The mainland portion of Sheet 52 of the one-inch Map of Scotland has still to be surveyed.

Minor transgression is the rule, and it is rare to find two partners lying side by side with juxtaposed chilled edges, so that it is generally quite easy to arrive at a fairly complete time-scale for any particular multiple dyke. J.E.R.

Another familar incident of dyke-intrusion, illustrated in Mull, is that of a dyke s t e p p i n g aside so to continue along some adjacent parallel course. Much the best instance of this is afforded by a basalt-dyke which crosses the coastal road a little east of Kellan Mill, on the north shore of Loch na Keal (Sheet 44). Where lost sight of under the raised beach at the mouth of Allt Mòr, it has a thickness of 10 ft.; inland, this increases locally to 20 ft. The dyke has a characteristic platy cross-jointing, but its main peculiarity is its tendency to expose itself on inland moors. Thus it is easy to make sure of its identity, even when it steps from one course to another. From near Kellan Mill to the first tributary of Allt Mòr, the dyke strikes inland, running without break for three-quarters of a mile. At this tributary, where M is printed on the one-inch Map, it steps south-west 100 yds., and continues north-west on its new course for a mile. Another M on the one-inch Map marks a side-step, this time towards the north-east, of 100 yds.; and within a short space two minor steps in the same north-east direction are taken. From here onwards, the dyke is easily traceable for three miles, passing from Sheet 44 to Coire Bàn, in Sheet 43; during this part of its course, it takes only one minor side-step to the south-west, on the north side of Gleann Mhic Caraidh. E.B.B., J.E.R.

Another good example of side-step is afforded by a north-west p o r p h y r i t i c basalt-dyke which runs south-eastwards from the stream west of Upper Druimfin Farm, south of Tobermory Bay (Sheet 52). This dyke, sometimes 20 ft. thick, is traceable at intervals for a mile and a half. A little west of Coire nam Fiadh, it takes a distinct side-step of 100 yds. The main interest, however, of this particular dyke resides in its exceptionally large felspar-phenocrysts, one of which was found measuring 4 by 6 inches. It is quite possible that the phenocrysts represented in Anal. II., p. 34, were collected from its outcrop., G.V.W.

Continuously exposed dykes are a great rarity. A clean-washed s h o r e - s e c t i o n, such as is provided on the south-east side of Mull (Sheet 44), shows dykes to great advantage. They may stand as resistant walls, or shelter at the bottom of miniature chasms, but either way they attract attention. Inland, away from stream-courses, they may become invisible beneath a grassy mantle. It is sometimes, on this account, a matter of great difficulty to form any judgment of the relative abundance of dykes in different localities. E.B.B.

While the Mull dykes generally figure less prominently inland than on the coasts, they are in large measure responsible for a l i n e a t e d  s c e n e r y in northern Mull, between Dervaig (Sheet 51) and Tobermory (Sheet 52). A contoured map shows very clearly a belt of country, about six miles across, characterized by a north-westward direction of ridge and hollow. Erosion has been guided to a very large extent by dykes, and to a lesser degree by parallel crush-lines. Loch Frisa may be mentioned as a major element of this north-westwardly directed scenery. G.V.W.

Various matters of importance will now be dealt with under separate headings, more particularly, the M u l l  S w a r m  o f

North-West Dykes, Dykes of other Directions, Explosions along Dykes, the Relation of Dykes to the Central Field of Pneumatolysis, and Petrology.

## Mull Swarm.

We are now in a position to emphasize certain facts regarding the distribution of Mull dykes. The dykes are mostly of north-westerly trend, and are grouped in a swarm, rather more than 10 miles wide. They are known to extend from a centre in Mull, south-eastwards with continually diminishing numbers, across the Firth of Clyde, 50 miles distant, and thence, sorely depleted, into England. In a north-westerly direction, they pass under the sea, after having been traced for about 15 miles. The reality of the Mull swarm will be gathered on inspection of Fig. 60, where a careful generalization of the data supplied by Geological Survey six-inch Maps has been attempted for the whole south-west Highlands. Perhaps the most definite irregularity of the Mull swarm is the northward, or north-north-westward, trend of many dykes west of Scallastle Bay and Loch Aline. A similar tendency is manifested in the Applecross peninsula in regard to the Skye centre (Sheets 71 and 81), and also in Rum. In the latter case, especially, it seems that the northerly trend has a radial significance.[1]

The north-north-westward, or northward, trending dykes, extending from the north side of the Mull centre, are thoroughly representative of the Mull dyke-assemblage from the point of view of petrology. At any rate, they include typical Plateau and Central Types, as, for instance, crinanite and tholeiite. Accordingly, they are not easily attributable to any single epoch, and this fact, combined with their number, favours their interpretation as a radial offshoot from the main swarm.

The petrological variety of the more or less northward-trending dykes is well illustrated in the Fishnish Peninsula (Sheet 44). The dykes of this peninsula are divisible into two great groups:—those running more or less north and south are of various types, and those running more or less east and west are tholeiites. The latter seem to belong to a single relatively late episode, for, in seven exposures, east-and-west tholeiites cut their northwardly trending associates. The details are as follows:—

Talaidh Type of tholeiite (13898) cuts porphyritic basalt, like Central Lava, with phenocrysts of felspar and augite (13897).

Brunton Type of tholeiite (15677) cuts non-porphyritic olivine-basalt, like a Plateau Lava, with rich purple augite (15678).

Four dykes of Salen Type of tholeiite (15679, 15680, 15682, and 15683) cut dyke of unusual type, olivine-basalt approaching mugearite (15681).

Salen Type of tholeiite (15684) cuts small-felspar dolerite with purple augite (15685).

Other dykes, belonging to the north-by-west set in Fishnish Point, are represented by (15675), of character intermediate between the Plateau Lava and Salen Tholeiite Types, and (15676) of strongly marked Plateau Lava Type with very purple augite and interstitial analcite (crinanite). Anyone wishing to make sure of obtaining a crinanite-specimen from among Mull dykes can collect from this dyke.

[1] A. Harker, 'The Geology of the Small Isles of Inverness-shire,' *Mem. Geol. Surv.*, 1908, Pl. III., p. 144.

It is 5 ft. thick, and is the most southerly dyke shown on the one-inch Map on the east shore of Fishnish Bay. An unmapped parallel dyke, a few yards south of it on the shore, may be taken as a landmark, for it is cut through by a 3 ft. dolerite-sheet dipping east at 30°.                                                                                                                           E.B.B.

In Morven (Sheet 44), several tholeiites have been encountered running roughly north-north-westwards (14969, 14974, 15784, 15798, 15838).

It will also be noticed, from inspection of the map, that the main fault-direction in Morven is approximately north and south.                                                                   G.W.L.

Quite apart from the north-north-west dykes, just alluded to, it is open to anyone to see a radial tendency in the arrangement of Mull dykes, since Tertiary dykes of every direction are unusually abundant in Mull (Fig. 60). The view accepted here is that the Mull focus has served as a localizer and injector of dykes (p. 10). Most of the dykes it has localized run north-west and south-east; but occasionally a tendency to dyke-formation in other directions has developed, with the result, in the aggregate, of a very ill-defined radial assemblage. Under these circumstances, it seems advisable to lay stress upon the parallelism of the many, and to recognize a Mull Swarm of north-west dykes as the outstanding feature of the district.

### AGGREGATE BULK OF DYKES IN MULL SWARM.

The immense bulk of the swarm, as a whole, is brought home by a consideration of the number of dykes which enter into it, as well as of the great area through which these dykes are distributed, far outside the confines of Mull.

In the $12\frac{1}{2}$ miles, measured in a south-west direction, that separate Duart Point from Frank Lockwood's Island (Sheet 44), 375 more or less north-westerly dykes have been mapped in tolerably complete coastal exposures, only interrupted for 1000 ft. at the mouth of Loch Don. Two of the dykes are of quite exceptional thickness: one, a north-north-west felsite at Duart Point, is about 130 ft. thick; and another, which terminates upwards in the cliff of Rudha Tràigh Gheal, is, at sea level, 200 ft. thick. Of the remainder, 291 dykes have been roughly measured, and give an aggregate thickness of 1698 ft.—or on an average 5·8. If this average holds for the remaining unmeasured 82 dykes, towards the southern end of the swarm, then the total thickness of the north-westerly Mull dykes actually exposed on the south-east coast of the island is:— $330 + 1698 + 476 = 2504$ ft., or approximately half a mile.                                                                                   E.B.B., G.V.W.

The exposure of the swarm, afforded on the other side of the Mull centre by the south-east shores of Loch na Keal (Sheet 44), is much more partial; but, so far as it goes, it is in wonderfully close agreement with that considered above. Thus, in $1\frac{1}{4}$ miles, 142 dykes have been counted with an aggregate thickness of 817 ft., which again gives an average of 5·8 ft. per dyke.                                     J.E.R.

### AGE OF THE NORTH-WEST DYKES.

The north-west dykes of the Mull Swarm constitute a complex, built up during a long period of time. It is certain that many of the north-west dykes of Mull are the latest igneous products of the focus; but it is equally certain that others are of earlier date than associated intrusions of non-dyke habit. It is impossible to decide whether any of them are as early as the lavas still preserved.

Apart from normal dykes, such as furnish the subject-matter of the present chapter, the Loch Bà Felsite Ring-Dyke is the latest intrusion in Central Mull (Chap. XXXII.). Its exposures are particularly clear on either side of Loch Bà, and it is there seen traversed by 24 north-west basalt-dykes in a total distance of about four miles. Of course, 24 does not represent quite the full number of dykes, but it is established beyond doubt that there are not nearly so many here as one meets with in the middle of the Mull Swarm on the coast, where the concentration exceeds 100 per mile. The deficiency of dykes, whatever the cause, is particularly marked in the first 700 yds. north-east of the head of Scarrisdale River: here, in virtually complete exposures, not a single basalt-dyke has been observed to cross the Loch Bà Felsite.

A thoroughly representative collection has been made of north-west dykes cutting the Glen Cannel Granophyre, with the following results:—

(14784): felsite.

(14778, 14782-3): quartz-dolerite of a rather peculiar type, with affinity to porphyrite.

(14779, 14826-14839): olivine-basalt, generally with purple augite; (14833) and (14835) are particularly interesting as they carry a little of the deep brown hornblende often found in camptonites (Chap. XXXV.), and the former also shows ocellar structure.

Two pairs of composite dykes (14780-1): olivine-dolerite with pale augite bounding olivine-quartz-dolerite; and (14786-7) olivine-tholeiite bounding curious quartz-porphyry.

While it cannot be claimed for certain that all these dykes are of later date than the Loch Bà Felsite, it is unquestionable that most of them are. Accordingly, this microscopic examination demonstrates a reappearance of the Plateau Type of magma (p. 7), in post-Loch-Bà-Felsite times. W.B.W.

It will be pointed out in the section on pneumatolysis (p. 366) that there is reason to suspect that as many as a third of the north-west dykes exposed on the south shore of Loch na Keal may be earlier than the Loch Bà Felsite and the Knock and Beinn a' Ghràig granophyres. In a suite of slices from these supposed early dykes, the Plateau Type is again predominant. J.E.R.

Direct evidence that a north-west dyke of this district is earlier than the Knock and Beinn a' Ghràig granophyres is exposed on the hill face a quarter of a mile south of Knock burial-ground. A 7 ft. north-west basalt-dyke is here seen cut across by a 3 ft. north-east Late Basic Cone-Sheet, itself veined by granophyre. The exposure lies 200 yds. outside the Knock Granophyre. W.B.W.

Another very good example of a pre-granophyre north-west basic dyke can be easily recognized on the one-inch Map, Sheet 44. It is traceable for about two miles from the coast near Dishig almost to the base of the Ben More Mugearite. The dyke consists of dolerite; it is sometimes 40 ft. thick, and is excellently exposed. It is cut by at least four of the pre-granophyre basalt-sills of the district (Chapter XXVI.).

A shore-section showing a similar relationship, with a 2 ft. 6 in. north-west basalt-dyke cut by a 5 ft. basalt-sill, occurs a mile farther north-east, just beyond the mouth of Abhuinn na h-Uamha. Inland exposures, on much the same line, show a 21-ft. dolerite-dyke cut by a couple of thin sills in, and below, the level of the Ben More Mugearite, north of A'Chioch. J.E.R.

Intersections, comparable with the above, are very rarely seen. An example will now be cited from another part of the island, where the interest is enhanced, since one can deduce a lower as well as an upper time-limit for the dyke. A north-west basalt-dyke crosses some agglomerate-crags, 500 yds. north-east of Dùn da Ghaoithe summit (overlooking Scallastle Bay, Sheet 44), and is itself cut across by a massive Early Basic Cone-Sheet (Chapter XXI.). In this case, the dyke clearly belongs to some period following that of maximum explosive activity (Chapter XVI.), and, therefore, considering its composition, is probably a product of Early Basic Cone-Sheet times.

Two additional instances of relatively early north-west dykes are illustrated in Fig. 53 (p. 312). They occur on the east face of Maol nam Fiadh, and are seen traversing lavas of Central Type (Chapters IX. and X.), and cut to pieces by Late Basic Cone-Sheets (Chapter XXVIII). One of them is a felsite, and is named as such on Fig. 53. The other occurs about 60 yds. farther south-west and consists of rhyolite-breccia, apparently indicative of explosion. A north-west basalt-dyke, lettered M on Fig. 53, lies almost immediately on the south-west side of the breccia-dyke, and serves as a landmark; it is of late date, and cuts Late Basic Cone-Sheets as well as lavas. The age of these two dykes, felsite and rhyolitic breccia, cannot be precisely fixed, because the time-interval separating the Central Lavas from the Late Basic Cone-Sheets is a very long one.

Only in one case, can an intersection be pointed to that suggests so early a date for a north-west dyke as that, perhaps, of some of the lavas preserved in the island. For a mile south from the entrance of Loch Don, an Early Acid Cone-Sheet runs more or less along the shore-line. It is seen to be cut by eight north-west non-porphyritic basalt-dykes; but it almost certainly cuts another dyke, which, while it runs north-west, is distinguished from its fellows by porphyritic felspars. The crossing occurs just where *Exogyra*-sandstone is noted on the one-inch Map, Sheet 44, under flint-conglomerate.

E.B.B.

## DYKES OTHER THAN NORTH-WEST.

It has already been pointed out (p. 359) that the north-north-west, and north-and-south dykes of Mull, west of Scallastle Bay, and of Morven, west of Loch Aline, are probably an offshoot from the main north-west swarm; it is thought that they have been intruded intermittently through a very long interval of time.

It has also been shown that in the Fishnish district a set of roughly east-and-west tholeiite-dykes cuts across the more or less north-and-south assemblage (p. 359).

More towards the centre of Mull, two additional examples of east-and-west dykes cutting north-and-south dykes can be cited, and two other cases where this relationship is reversed :—

In Gleann Dubh, near the head of Loch Scridain, just above where a small tributary joins the main stream, 200 yds. south of an east-and-west fault shown on the one-inch Map, an east-and-west dyke cuts a north-and-south dyke.

So also, on the south shore of Loch Beg, another east-and-west dyke cuts a north-and-south dyke.

But, where the first stream west of Sròn Daraich crosses the Pre-Glacial coast above Loch Beg, a north-and-south dyke cuts an east-and-west dyke.

And also, a mile east-south-east of Ben More summit, a north-and-south dyke cuts an east-and-west dyke.

It will be seen, therefore, that, in the Ben More and Loch Beg district, the rule established for the Fishnish district seems to be as often broken as obeyed. It may be added that, while the east-and-west dykes of Ben More and Loch Beg have been observed cutting cone-sheets in three different exposures, the reverse has never been detected. E.M.A.

A group of north-north-east and north-and-south dykes on the south-west slopes of Corra-bheinn can be dated fairly accurately. They vary in composition from olivine-dolerite to felsite, but they agree in being cut across by both Early Basic Cone-Sheets (Chapter XXI.) and Corra-bheinn Gabbro (Chapter XXII.). Since the basic dykes are seen to cut the Derrynaculen Granophyre (Chapter XII.), it is almost certain that they belong to some interval in the earlier part of the Cone-Sheet Cycle of Mull chronology. As might have been expected, these basic dykes show manifest signs of contact-alteration with the production of biotite and hornblende (16542-4). (C.T.C.)

There are quite a number of north-east dykes in the country overlooking Scallastle Bay on the Sound of Mull. In one case, it is clear that a basalt-dyke of this direction is earlier than the Early Basic Cone-Sheets of its vicinity. It is exposed 800 yds. north-north-east of the summit of Beinn Chreagach Mhòr, just on the east side of a deer-fence. It cuts a post-lava agglomerate-vent, and is distinguished from neighbouring cone-sheets by its baked appearance. However, all the north-east dykes of this district are not similarly altered. The lavas of Beinn nam Meann (1½ miles farther east-south-east) are greatly affected in proximity to an agglomerate-vent cut through by the Beinn Mheadhon Felsite ; but the dykes traversing the lavas, whether running north-east, north and south, or north-north-west, are no more visibly changed than the cone-sheets near by. E.B.B.

That some of the north-east dykes of Mull are comparatively late is shown by a north-east basalt-dyke cutting the Glen Cannel Granophyre south of Lochan nam Ban Uaine. W.B.W.

North-east dykes are rather common in the south-east part of Mull, though far out-numbered by their north-west fellows. The north-east dykes have a tendency to incline steeply north-west. They are, on the whole, earlier than the north-west dykes ; for the latter have been seen to cut them fifteen times, whereas the reverse relation has only been noticed four times,

## Chapter XXXIV.—Dykes.

### Explosions along Dykes.

Much the best example of blow-holes along the course of Tertiary dykes, belonging to the Mull Swarm, falls a mile outside the east margin of Sheet 44 (Fig. 61), and has been described by Messrs. Peach, Symes, and Kynaston.[1] In this case, vents occur along a line of multiple dyke-intrusion in which both basalt and rhyolite

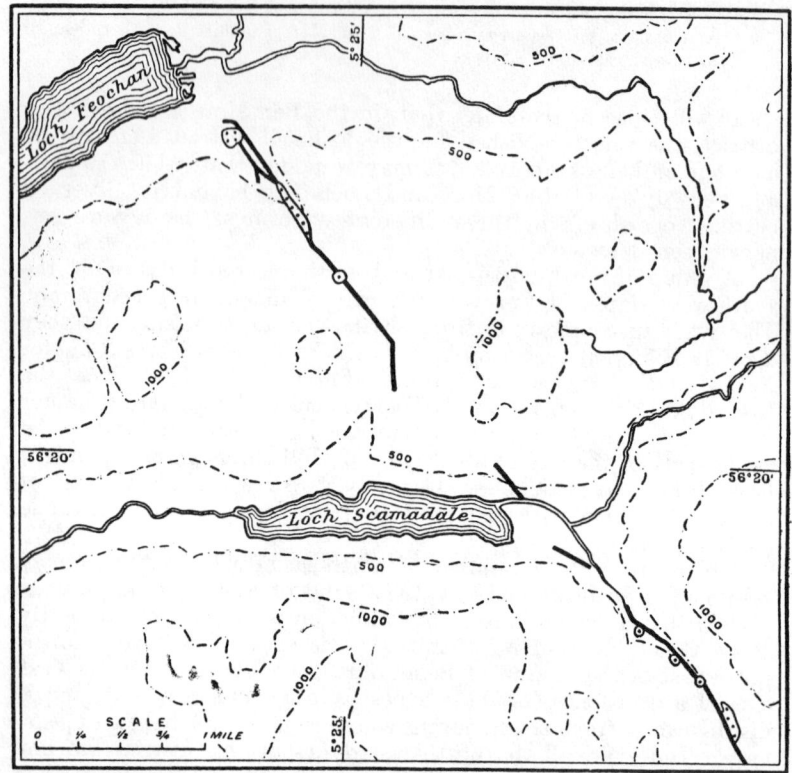

Fig. 61.

Agglomerate-Vents along multiple dyke (basalt and rhyolite) south-west from head of Loch Feochan (after B. N. Peach, R. G. Symes, and H. Kynaston).

play a part. Often, the dyke-rocks cut and include the agglomerate, but, in some exposures, this relationship is reversed. The agglomerate consists of fragments of basalt, rhyolite, and country-rock—the latter mostly lava of Old Red Sandstone age.

The neighbouring south shore of Loch Feochan affords an exposure of a similar and parallel multiple dyke, still half-a-mile outside Sheet 44. Here, basalt is cut by dolerite, and both of them are traversed by fissures filled in with agglomerate; which in its turn

[1] 'The Geology of the Country near Oban and Dalmally' (Sheet 45), *Mem. Geol. Surv.*, 1908, pp. 135-137.

is cut by rhyolite. Dr. Kynaston has described this example in the Memoir just referred to, and has illustrated it with a sketch-map drawn by Dr. Peach.[1]

The next instance of the sort to be noted lies half-a-dozen miles farther south-west. It is best exposed on an islet, north of Eilean Buidhe, to which a note on the one-inch Map (Sheet 44) directs attention. It consists of a basalt-, or dolerite-dyke, about 8 ft. thick, though irregular. The basalt veins and surrounds lenticles of breccia made up of the country-rock (black slate) along with a subordinate proportion of basalt, some of which is vesicular. Half-a-mile farther south-east, the same dyke is seen again, retaining its characters, in a small quarry in the face of the bluff behind the raised beach north-west of Clachan Bridge.

Perhaps, the Clachan Bridge dyke turns from its south-east course at the bridge. At any rate, showing through the raised beach, half-a-mile south-south-east of the bridge, there is exposed a north-north-west dyke of agglomerate some 12 ft. wide, separated from the slates on either side by a narrow basaltic border. The agglomerate consists of various types of basalt, and its junction with the marginal basalt is rather vague. E.B.B.

Three explosion-vents, which are interpreted with some hesitation as belonging to north-west dykes, are exposed on the south-east coast of Mull near Port a' Ghlinne ; and two of them are noted on the one-inch Map (Sheet 44). The first lies about 200 yds. east of Port of Ghlinne. As seen on the shore, it is a broad irregular north-westerly dyke, consisting of porphyritic basalt of Central Type (17374—a coarser specimen is a good example of small-felspar dolerite 17378). It is crowded with an assortment of basaltic and felsitic xenoliths (17368, 17377, 17381), and surrounds a central patch of agglomerate, full of fragments of variolites (17367, 17369, 17379, 17380). This complex dyke-vent is seen cutting an east-and-west dyke, and also a later acid intrusion.

Eleven hundred yards farther north-east, a small patch of agglomerate has been found on a north-west basalt-dyke, though not indicated on the one-inch Map. G.V.W.

Two hundred yards beyond this, at the second point noted on the one-inch Map, there is a good exposure of breccia belonging, perhaps, to a dyke-vent. The fragments include both igneous and sedimentary types (15859-15865), and some of the latter are intensely altered. E.B.B.

<small>The sedimentary fragments appear to be mainly quartzites or sandstones, and have suffered various degrees of alteration and magmatic digestion. A moderately large siliceous fragment (15862) shows, beautifully, the effect of continued heating. The original quartz has undergone almost complete digestion, accompanied by the separation of large plates and wedge-shaped twins of tridymite, giving rise to a structure similar to, and as coarse as, that observed in silica-bricks after continued use in a steel-furnace. The tridymite, as is usual in these rocks, has reverted to a granular mozaic of quartz, but its external form is sharply preserved. *Lacunae* of melt throughout the rock have developed crystals of yellowish to greenish augite which appear to be in eutectic relationship to the silica. The eutectic structure is in some instances of extremely delicate nature. The crystals of augite are relatively homogeneous at their centres, but towards their peripheries give evidence of rapid growth in the cooling magma, resulting in a fibrous or acicular type of crystallization.</small>

---

[1] *Op. cit.*, Fig. 2.

The residual glass contains abundant minute crystals and skeletal growths of magnetite. The chief interest in the rock lies in its original richness in tridymite, for this mineral constituted the greater part of its mass. H.H.T.

If, instead of going along the coast from Port a' Ghlinne, one turns up Glen Libidil, another dyke-vent is encountered. It traverses the lavas on the steep western face of the glen, a little north of the second fault marked on the one-inch Map. In this case, the vent shows as a narrow belt of fine breccia, or tuff, traceable in an east-north-east direction. Acid fragments are a feature of its constitution (17371-2). G.V.W.

The only other example of the kind in Mull is an early north-west dyke of rhyolite-breccia on the east face of Maol nam Fiadh. Attention has already been directed to this dyke on p. 362. Possibly it is a fluxion-breccia, and not a product of explosion.

## RELATION OF DYKES TO THE CENTRAL FIELD OF PNEUMATOLYSIS.

A considerable proportion of the Tertiary dykes of the region carry fresh olivine, where they have been collected outside the Limit of Pneumatolysis recognized for the Mull lavas of Chaps. V.-X. (Plate III., p. 91). A little distance inside this limit, fresh olivine seems to fail altogether. On the north-west side of the Pneumatolysis Field, the boundary, at which the fresh olivine of the dykes gives place to pseudomorphs, agrees approximately with the limit traced on Plate III.; on the south-east side, however, the limit for the dykes lies materially farther in than for the lavas, and fifteen examples retaining fresh olivine have been sliced from the south-east shores of Mull, on the two sides of the entrance to Loch Spelve. A dyke with fresh olivine has also been sliced from near Kinloch Inn at the head of Loch Scridain.

The dykes which traverse the plexus of intrusions characteristic of Central Mull are manifestly of very much later date than the lavas; and their comparatively unaltered appearance might in some cases lead one to expect a certain proportion of fresh olivine. The dykes cutting the Glen Cannel Granophyre have been carefully examined from this point of view. Sixteen among the sliced specimens show olivine-pseudomorphs, and not one shows the mineral fresh. Moreover, as in many cases there is a fairly free development of fibrous hornblende in amygdales, etc. (14779, 14786, 14827, 14831), it does not seem possible to account for this decomposition of olivine as a result of mere weathering; in fact, their condition is on the border-line where pneumatolysis gives place to contact-metamorphism.

It is established, therefore, that all, or almost all, the dykes cutting the Glen Cannel Granophyre have suffered some degree of pneumatolysis. The majority, at any rate, of these dykes are later than the adjacent Loch Bà Ring-Dyke—that is, they are the latest of Mull's igneous products. Accordingly, the phenomenon of central pneumatolysis continued to find expression in Mull until the very close of igneous activity. This is in keeping with the view that the concentration of a swarm of dykes, in itself, bespeaks the underground existence of a pipe filled with molten magma (p. 10). E.B.B.

We may note two further examples of definitely late, and yet much altered, dykes (16633-4), both of them cutting the Beinn a' Ghràig Granophyre. The first is a north-and-south dyke in Allt Coire nan Gabhar in the middle of the granophyre-outcrop. The other is the second north-west dyke shown on the one-inch Map in the Scarisdale River above the west margin of the granophyre. Both are tholeiites with an abundant development of fibrous hornblende. E.B.B., J.E.R.

It must not be supposed that the dykes of Mull have, as an assemblage, been uniformly affected. There is no place where this point can be better appreciated than within the Pneumatolysis Limit of Plate III. (p. 91) on the south shore of Loch na Keal. About 30 per cent. of the dykes here seem to have reached the same condition as the lavas which they traverse, while the rest are notably fresher; so much so, that one is justified in regarding the intrusion-periods of the two sets of dykes as separated by an interval of intense pneumatolysis, connected possibly with the introduction of Knock and Beinn a' Ghràig Granophyres. J.E.R.

Mr. Richey has collected a series of specimens (17064-17075) to illustrate the marked alteration of some of the dykes referred to above. Occasionally, their condition renders it difficult to assign them to their proper place in the petrological classification proposed in the succeeding section; but certainly most of them are Plateau Types, often with rich-purple augite, while one (17072-3) is an olivine-tholeiite (Salen Type). The alteration of these rocks is exactly on a par with that of neighbouring lavas. There has been a complete decomposition of olivine, and a considerable change in the felspars, and sometimes also in the augite. The resulting secondary minerals, mainly chlorite, albite, and epidote, with subordinate calcite, occupy vesicles and also veins (17065, 17074). The alteration is sufficiently intense to lead to the development of fibrous hornblende in many of the vesicles (17066, 17067, 17071). Occasionally, somewhat fibrous hornblende completely replaces the augite of the groundmass; and, in the two cases where this has been observed, garnet occurs in the vesicles (17068, 17070).

## Petrology.

Examples of the Tertiary dykes of Mull, and the surrounding district, are well-represented in the Survey collections, and abundant material is available for safe generalizations.

In the following account of the petrography of these rocks, only such dykes as cut Mesozoic sediments or Tertiary lavas are made use of, *except where slide-numbers are quoted in square brackets*. Camptonites and allied types, of doubtful age, are dealt with in Chapter XXXIV.

As has already been indicated, the majority of the dykes are of more or less basaltic character, but they exhibit considerable variety. They are dealt with in the sequel under the headings Plateau Basalt Type, Porphyritic Central Lava Types, Tholeiites, Leidleites, Felsites, and Composite Dykes.

# Chapter XXXIV.—Dykes.

## DYKES OF PLATEAU BASALT TYPE.

A very large proportion of the basaltic dykes are directly comparable, both as regards texture and composition, with the Plateau Basalt Lavas (Chapter X.), and it will be convenient to refer to them as dykes of Plateau Basalt Type. Olivine-rich varieties, of basaltic and doleritic character, are extremely common, often with an appreciable amount of analcite in the base—crinanites of Dr. Flett (p. 16). Rocks with mugearitic affinities (16270) occur very sparingly.

The dykes of Plateau Basalt Type, as a class, are characterized by a highly titaniferous augite of purplish tinge and distinct pleochroic character. Texturally, they range from coarsely to finely ophitic varieties, and porphyritic constituents, other than olivine, are usually wanting. The ultimate residual products of consolidation of these rocks consist mainly of analcite, which occurs together with a certain amount of alkali-felspar. Olivine is always present, frequently in an undecomposed condition (p. 366), and may occasionally be more abundant than augite (14870). Porphyritic constitutents, other than olivine, are usually wanting, but types are met with which have a multitude of small tabular sub-porphyritic felspars (14990) that may take on a linear arrangement according to the direction of flow. In rarer instances, large porphyritic felspars, ranging up to half-an-inch in length, may be noticed (15781) as a minor feature of a rock that otherwise is of normal Plateau Type.

A coarse type, forming a six-foot dyke in a stream on the western side of Clachaig (14836), is more than usually rich in analcite. For the most part the augite is ophitic, but when in contact with the analcitic areas it assumes a hypidiomorphic habit (p. 137). In this rock, olivine is entirely decomposed (p. 366), and areas, originally occupied by residual material, are frequently replaced by chlorite of a late secondary character.

A beautiful fine-grained crinanite, with abundant analcite, occurs as a north-and-south dyke, 500 yds. to the south of Fishnish Point (15676. See also p. 359).

A few somewhat unusual types are worthy of individual mention. In a dyke south-west of Beinn Chàisgidle (17942), there is a considerable falling off in the amount of felspar, and the principal minerals present are hypidiomorphic titaniferous augite and olivine (decomposed, p. 366). The whole rock shows a tendency to assume the characteristic panidiomorphic structure of the lamprophyres (camptonites, etc.). A neighbouring dyke, north-north-west of Sgulan Mòr, shows segregation of titaniferous augite around the vesicular cavities, with the augite-crystals sometimes interrupted at the surface of the vesicle (17943). In specimens from Loch na Keal, titaniferous augite may be seen to project, in a position of growth, into the vesicular cavities (17065)—a feature not uncommon in Mull lavas, as was first described by Dr. M'Lintock (p. 141).

In many instances, we may detect an approach of the Plateau Basalt magma to that of the camptonite-branch of the lamprophyres. One case (17942) has been cited above, but others may be noted in which biotite or hornblende makes an appearance (14217*a* and *b*, 14833, 14835) and a definite ocellar structure is discernible. The last two slides have already been dealt with (p. 361). The first two come from an easily located 10 ft. dyke, at a wall south of Port na Tairbeirt, on the east coast of Mull. Its amygdales, with ocellar linings, can be recognized in the field. In petrological character, the rock appears to stand about half-way between the basalts and the camptonites, and for this reason is of special interest. The ocelli (14217 *a* and *b*) are moderately coarse-grained, and are composed of somewhat tabular orthoclase in association with a deep red-brown intensely pleochroic biotite. The microscopic structure recalls vividly that of the Carsaig Alkaline Syenite (Chapter XIV.).

'*Cognate*' *Xenoliths.* — A particularly interesting dyke occurs on the north side of Loch na Keal, a third of a mile west by north of the cairn (769 ft.) on Cnoc na Dì-chuimhne (inside the angle of the stream shown on the one-inch Map, Sheet 44). It is only partially exposed, but is seen to consist of two lateral bands, each $1\frac{1}{2}$ ft. wide, marked by a great abundance of 'cognate' xenoliths, and separated by a two-foot central portion of olivine-basalt of Plateau Type (17031). The rock, as a whole, is exceedingly fresh, with its olivine showing little or no sign of decomposition. The xenoliths are gabbroic or peridotic in character; some consist of moderately coarse aggregates of bytownite and olivine with the local ocurrence of large (2-3 mm.) ill-formed crystals of sage-green spinel (17030); others of bytownite and bottle-green augite (17033), and others again of similar augite poecilitically enclosing olivine (17032). The augite shows evidence of re-heating and resorption. It is difficult to draw a valid distinction between these xenoliths and the well-known olivine-nodules, so conspicuously absent, as a rule, from the British Tertiary rocks.

### DYKES OF PORPHYRITIC CENTRAL LAVA TYPE.

Porphyritic rocks with phenocrysts of felspar are well-represented amongst the dykes, and approximate in character to lavas of Porphyritic Central Type (Chap. X.). The felspar is usually a moderately basic labradorite, frequently of surprisingly uniform composition, as is indicated by its insignificant zoning. The porphyritic individuals exhibit considerable variation in size, commonly from a few millimetres to a centimetre in greatest dimension (15745), and in one case ranging up to $6 \times 4$ inches across (p. 358); the average size may be taken as 4 to 5 mm. The matrix shows variation comparable to that exhibited by the corresponding lavas. It may be granular (14207) [19040], micro-ophitic (16666), or fluidal (17374), or show a tendency to intersertal structure (14954). Olivine may occur as additional phenocrysts (16666, 17374), but is more commonly restricted to the matrix in the form of small crystals and grains.

Augite is not particularly common as a porphyritic constituent, but a dyke from near the summit of Ben Buie contains a few badly formed crystals, comparable in size to the porphyritic felspar.

A highly porphyritic type, in which the phenocrysts of felspar are extremely abundant—so abundant that they practically touch each other—furnishes another example of the 's m a l l-f e l s p a r d o l e r i t e' (pp. 163, 286). This type of rock was previously met with by Dr. Harker[1] in Skye, and referred to by him as an olivine-dolerite of G h e a l G i l l e a n  T y p e (7366-7).

The dominant felspar is basic labradorite, much twinned and usually zoned. It commonly contains inclusions of the matrix which show a rudimentary type of crystallization (14956).

The ground-mass, which in many cases is reduced to a minimum, is invariably rich in augite relatively to felspar, and frequently contains granular olivine (14956, 14975). The augite usually has an ophitic development, and is commonly of a somewhat deeply coloured titaniferous variety. The ophitic structure varies in coarseness in different dykes, giving in some instances a normal basaltic character to the ground-mass (14537, 14889), but in others a much coarser, almost gabbroic texture (16312, 16332). The type is very definite, and, although not particularly prevalent, is widely distributed.

Olivine most commonly occurs in the ground-mass; but in the more coarsely crystalline rocks, in which the proportion of matrix is small, olivine builds large porphyritic individuals associated with the felspar (16662) [17451], and thus emphasizes the gabbroic character.

'*Cognate*' *Xenoliths.* — 'Cognate' xenoliths are occasionally met with in the porphyritic dykes; and a good example may be cited in the case of a dyke that occurs on the shore of Loch Spelve, at the first point south of Seanvaile. The xenolith-bearing rock is somewhat different to the common type, in that the augite tends to assume an idiomorphic, rather than an ophitic, form (15066).

The xenoliths are large masses of greenish augite, ophitically enclosing somewhat tabular crystals of bytownite (15065). Against the enclosing rock, the augite shows signs of corrosion, as also do isolated xenocrysts of augite which occur scattered throughout the ground-mass.

A similar dolerite, with xenocrystal development, occurs at Loch na Keal, 530 yds. S. 13° W. of Eilean Feòir (16235).

### THOLEIITE DYKES.

(ANALS. I., VI., AND VII.; TABLE II., p. 17).

Tholeiites have already been defined (Chapter XXV.), and it has been pointed out that, in Mull, they group themselves conveniently under three headings: (*a*) olivine-tholeiites of Salen Type; (*b*) tholeiites of Brunton Type; and (*c*) tholeiites of Talaidh Type. The Brunton and Talaidh Types agree in having little or no olivine, but are distinguishable on structural grounds.

#### OLIVINE-THOLEIITE OF SALEN TYPE (ANAL. I.; TABLE II., p. 17).

Rocks of this type are abundantly represented among the tholeiite-dykes of Mull. They are even-grained, finely crystalline dark-grey rocks, usually without any trace of porphyritic con-

---

[1] A. Harker, 'Tertiary Igneous Rocks of Skye,' *Mem. Geol. Surv.*, 1904, pp. 329, 330.

stituents. Microscopically, they are seen to be composed mainly of augite and labradorite felspar, with subordinate olivine, iron-ore, and a variable quantity of residual glassy matter (Fig. 62A) of relatively late consolidation (14218, 16807, 16808) [19035]. The augite is, usually, at least as prevalent as the felspar. It has a pale yellowish-brown colour, is sometimes pleochroic (15744), and occasionally somewhat titaniferous; it forms somewhat small crystals, that, while behaving ophitically towards the felspar [19034], show a distinct tendency towards idiomorphism, when in contact with later products of consolidation. Occasionally, acicular augite is present as a consolidation-product of the residual matter (14209);

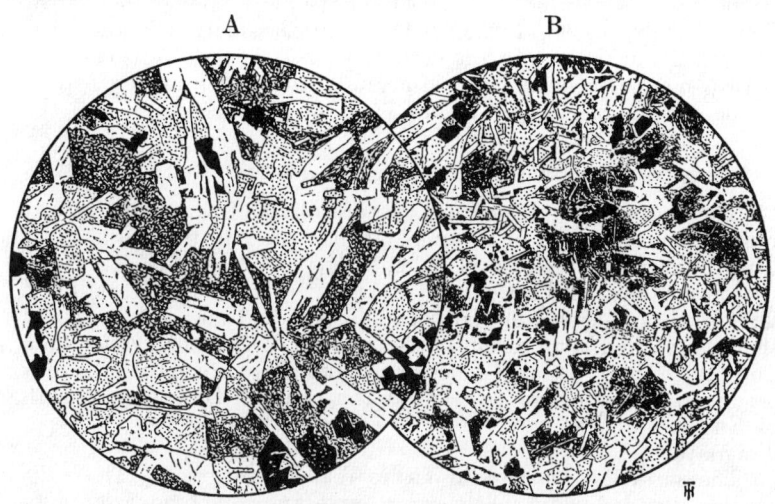

Fig. 62.—Tholeiites.

A. [16807] × 17. Tholeiite of Salen Type. Composed of augite, labradorite-felspar, subordinate olivine and iron-ore, and a variable quantity of residual glass.

B. [16809] × 90. Tholeiite of Brunton Type. Augite, labradorite, magnetite and glass. The well-marked intersertal structure, produced by the arrangement of the crystalline elements with regard to the glassy base, is a constant and characteristic feature.

and, now and again, this mineral may assume to a limited extent that cervicorn structure (14853) [15792], so characteristically developed in some of the sub-basic sheets (Chapter XXVIII.).

Olivine appears to be the earliest silicate to have separated; but it seldom reaches any considerable importance as a constituent. It builds small somewhat badly formed crystals, or ellipsoidal grains, on which both felspar and augite are moulded.

The felspar is usually a much-twinned and zoned labradorite, that occurs in somewhat irregular, and probably rapidly grown, lath-shaped crystals. The individuals frequently show a tendency to crystallize from common initial points, a fact that leads to a roughly radiate or stellate grouping (14853), such as is often found in rocks of the Brunton Type.

Iron-ore, comparable in size to the olivine, has the magnetite-habit, and commonly clings tenaciously to the augite, or is concentrated in the areas occupied by residual material. Less frequently, it builds large plates with narrow cross-sections (14209).

The residual matter shows a tendency to collect into definite areas, giving rise to intersertal structure. It is usually chloritized and full of skeletal growths of magnetite ; but its devitrification-products, and the increased alkalinity of the felspars in its immediate neighbourhood, suggest that it is much more siliceous and alkaline than the bulk of the rock.

Coarse-grained representatives of the Salen Type of tholeiite pass over to the quartz-dolerites [19017, 19021, 19037, from Lorne], and connect with somewhat similar rocks on the border-line of the Plateau Type (15068, 15866, 16252, 16348). In the more acid direction, we find the incoming of more definitely siliceous mesostatic material, with a marked acicular type of crystallization, such as is a feature of the quartz-dolerites of Talaidh Type (p. 303).

THOLEIITES OF BRUNTON TYPE (ANALS. VI. AND VII. ; TABLE II., p. 17).

The tholeiites of Brunton Type are olivine-free, or are, at any rate, definitely poorer in this mineral than those of the Salen Type. They differ from tholeiites of Talaidh Type in an absence of acicular or columnar crystallization of the augite. The general texture is finer, with a considerable proportion of residual glassy matter, and there is marked aggregation of the crystalline elements, other than iron-ore. The type, as developed in Mull, is identical with the rock described by Sir Jethro Teall[1] from the Brunton Dyke in the North of England, and we have therefore adopted this name as the type-designation.

There is little to add to Sir Jethro Teall's description. The rocks are usually devoid of phenocrysts, and are composed essentially of augite, labradorite, iron-ore, and glass: the iron-ore shows a definite tendency to restriction within the areas occupied by residual material (15724). The well-marked intersertal structure produced by the arrangement of the crystalline elements with reference to the glassy residuum is a constant and characteristic feature, and is illustrated in the appended figure (Fig. 62B). The type is uniform, because a little variation introduces the characters either of the Salen or Talaidh Type. The analysed rocks (16809, 16810) may be regarded as typical.

THOLEIITES OF TALAIDH TYPE.

When dealing with the petrology of the Late Basic Cone-Sheets (Chapter XXVIII.), we described numerous examples, with quartz-dolerite affinities, as belonging to the Talaidh Type, on account of the prevalence of such rocks in the region of Beinn Talaidh. The more finely crystallized varieties of the type are tholeiites, as already pointed out in Chapter XXV.

[1] J. J. H. Teall, 'Petrological Notes on some North of England Dykes,' *Quart. Journ. Geol. Soc.*, vol. xl., 1884, pp. 236-7, Pl. xii., Fig. 6 ; *see also* A. Harker, 'Petrology for Students,' 5th ed., 1919, Fig. 62, p. 180.

The Talaidh Types of tholeiite and quartz-dolerite, as represented by dykes, are olivine-free rocks, for the most part of moderately fine grain and dark colour. They vary more in their textural characters, than in the relative proportions of their constituent minerals. They carry augite, moderately basic, but much-zoned, plagioclase, and abundant magnetite, in an acid base, partially devitrified to quartz and alkali-felspar, with or without ferromagnesian constituents. This acid base, of a mesostatic character, is occasionally more or less evenly distributed as a discontinuous matrix to the larger crystalline elements (14353), but usually is collected into fairly definite irregularly-shaped areas (13898, 14341).

The augite of these rocks shows great variation in its manner of crystallization. Frequently, it is coarsely ophitic, though with a tendency to idiomorphism towards the mesostatic material [15841]; and, in this form, it is common to encounter a curving of the cleavage-planes [15793], as remarked upon in the case of the cone-sheets (p. 303). More frequently, the augite tends to assume a stoutly columnar habit [15838] (17614), with an incipient development of salitic striation [15840], a common habit of augite in fairly acid quartz-dolerites. We also notice those curious cervicorn growths of this mineral (14206, 14854), so characteristic of many of the cone-sheets of Bheinn Talaidh (Fig. 50B, p. 302); and all stages exist between this structure and a micro-ophitic development [15802]. The examples cited in square brackets in this paragraph cut Pre-Mesozoic rocks in Morven.

The mesostatic matter of late consolidation is a finely crystalline mass, in which practically all the quartz is segregated, along with most of the alkali-felspar, other than that which may fringe zonally the larger crystals of plagioclase. It is characterized by a marked acicular type of crystallization (16658), such as is the dominant feature of the mesostasis of the differentiated rocks of Glen More (Chapter XXX.), and the matrix of the craignurites.

INTERMEDIATE AND SUB-ACID DYKES.

LEIDLEITE.

A comparatively small proportion of the Mull dykes belong to the leidleite type as already defined (p. 281). They are dark grey in colour, with a variation in texture corresponding to their degree of devitrification (14534, 16663). They are composed of plagioclase, augite, and magnetite, with glass, or the products of its divitrification, in varying amounts. In those which approach to pitchstone, the glassy matter is abundant and the crystalline constituents are of small dimensions (14588) [19003, 19018, from Lorne], while, in the more stony varieties (14974) [15794], the glass is reduced to a minimum, with the development of very fine-grained devitrification-products, and the crystalline elements are of relatively greater size.

The felspars build narrow, somewhat ill-formed laths or skeletal growths, and have the composition of acid labradorite or andesine. The augite is a normal aluminous variety, that occurs commonly as blade-like crystals, frequently with stellate grouping. Magnetite exists as strings or rods, formed by adherent

374         *Chapter XXXIV.—Dykes.*

octahedra, and it is either closely associated with the augite or more or less restricted to the residual material. By the presence of uniaxial augite as small rounded phenocrysts, the rocks [15801] occasionally may show a leaning towards the ininmorites (p. 282); and, by the occurrence of an acid residuum with clearly individualized quartz, they connect with the craignurites of Chapter XIX. [19001]. Certain other varieties with a variolitic structure, on a microscopic scale (16317, 17379), recall in a measure the variolites of Cruachan Dearg (p. 304).

A dyke (16659), that occurs three-quarters of a mile to the east of Meall an Fhìar Mhàim, north of Loch na Keal, is a rather more exceptional variety. It is a fine-grained, compact, somewhat glassy rock, built up of hypidiomorphic columnar augite-crystals that are intimately associated with regular crystals and plates of magnetite. The felspar is andesine-labradorite, and forms irregularly bounded lath-shaped crystals. The larger crystalline elements are spaced without crowding in a micro-variolitic glassy base that is full of skeletal growths of augite and magnetite.

### INNINMORITE.

A few ininmorite dykes occur associated with sheets in the Inninmore district of Morven (p. 293).

### ACID DYKES.

The acid dykes of Mull are in a minority. They present, however, considerable variety amongst themselves, both in macroscopic and microscopic characters. Judged by the phenocrysts that are usually present, the composition, too, should vary within wide limits, but by reason of the fine texture of the matrix of most of these rocks, compositional differences are difficult to gauge without ultimate chemical analyses.

A microscopical examination shows that these dykes are mostly microporphyritic, and have frequently, in addition, some definite structure that enables us to describe them as granophyric, spherulitic, or trachytic. The greater number of the dykes are felsites, amongst which a few with quartz-phenocrysts may be distinguished as quartz-felsites. They are almost invariably microporphyritic, with phenocrysts of albite (13907), orthoclase (14773, 14789), or quartz (16622, 17269), in a microcrystalline base.

True granophyres are not well-represented, but, in a certain number of dykes, the base has a granophyric structure, either fairly coarse (14375), or more commonly of a micrographic character produced by the very fine and intimate intergrowth of alkali-felspar and quartz (14735, 18525).

The spherulitic rocks are mainly referable to the quartz-felsites, and two beautiful rocks may be cited. One is a dyke having a west-north-west direction and exposed 600 yds. to the south-west of Killbeg, Glen Forsa (14768); while the other is a six-foot north-north-west dyke that occurs 1580 yards S. 35 W. of Eilean Feòir, Loch na Keal (16244). In both cases, the quartzo-felspathic spherulitic and radiate growths have small quartz-phenocrysts as nuclei, and occur in a microcrystalline, presumably divitrified, matrix. The spherulites reach a few millimetres in diameter. In the second named example, small phenocrysts of alkali-felspar are present in addition to those of quartz.

Of the rocks that assume a trachytic, or orthophyric, structure we may mention the acid margin of a multiple dyke occurring on Speinne Beag, on the junction of Sheets 44 and 52 (15765). It is built up of narrow, elongated, indefinitely terminated crystals of orthoclase, having a sub-parallel arrangement.

A beautifully banded rhyolite (14789), reminiscent of the Loch Bà Felsite (Fig. 59B), occurs as a north-and-south dyke that cuts the Glen Cannel Granophyre, three-quarters of a mile south of Clachaig. It has small phenocrysts of albite in a micro-crystalline divitrified matrix, that has a somewhat patchy appearance and has undergone some secondary silicification. Occasionally, rocks, similar in other respects, may develop a spherulitic structure, together with an acicular type of crystallization of the ground mass (17034).

From these acid rocks, there is a gradation into several subacid and intermediate types that have been already discussed when dealing with the Cone-Sheets (Chapter XIX.).

Two exceptional acid dykes of alkaline character are worthy of mention, although their Tertiary age is not quite established, since they have only been found cutting Pre-Mesozoic Rocks. Both run north-west and south-east. One of them, a pale microporphyritic dyke, has been collected from the north-east of Rudh' an Fheurain, Sound of Kerrera [18722], Dùn Ormidale [18723], and Lerags House [18724]. It consists almost entirely of alkalifelspar with subordinate quartz The felspar occurs as two generations, seemingly of approximately the same composition : the first is represented by porphyritic crystals of anorthoclase, somewhat rounded in form and edged with orthoclase ; and the second, by small rectangular, often equilateral, crystals that have fairly uniform extinction, and are closely packed together to constitute the matrix in which the porphyritic crystals lie. Quartz occurs sparingly between the felspars of the matrix, but there is no attempt at micrographic intergrowth on the part of the two minerals. Apatite is present in small quantity only. A striking and unusual feature of the rock is the occurrence throughout the matrix of minute, but abundant, crystals and patches of chalybite. The origin of this carbonate is obscure, but there is some evidence that it occupies pseudomorphously the place of a mineral, now completely removed, possibly of some alkaline pyroxene or amphibole.

The other dyke [18721] was collected from the mainland shore, 200 yds. north-east of Kerrera Ferry. The matrix takes on a more bostonitic character with frequent radial grouping of the elongated felspars. As a whole, the rock presents many points of similarity to the bostonites of Rossal Type, and might well be regarded as the dyke-equivalent of such sills (Chapter XIV.).

### COMPOSITE DYKES.

An enumeration of a few sliced examples of composite dykes is all that is necessary in this connexion :—

(1) At Allt Molach, a conspicuous composite dyke, shown on the Map (Fig. 52, p. 308), has a margin (17427) of quartz-dolerite with columnar augite and acid mesostasis; the interior (17426) is a fine-grained soda-felsite.

(2) An east-north-east dyke, not shown on the one-inch Map, Sheet 44, 100 yds. north of the top tributary of Allt na Coille Mòire, north of Loch Scridain, has a margin (17274) of Salen Type of tholeiite, and an interior (17273) of felsite with small perthitic phenocrysts.

(3) A west-north-west dyke, traced for half a mile on the one-inch Map, Sheet 44, three-quarters of a mile west-south-west of Beinn Talaidh summit, has a margin (14372) of micro-ophitic basalt. Farther in (14373-4), it is more acid; while its interior (14375) is a felsite with quartz- and perthite-phenocrysts, the quartz showing micrographic fringes. The matrix of the felsite is an intricate intergrowth of quartz and alkali-felspar, often in the form of sub-radial sheaves. H.H.T., E.B.B.

## CHAPTER XXXV.

### CAMPTONITES, ETC., OF DOUBTFUL AGE.

#### Introduction.

TRUE camptonites were first distinguished in Britain by Dr. Flett,[1] when he described the camptonite-monchiquite dyke-assemblage of Orkney. The name was previously used in Scottish petrological literature for rocks of the spessartite class. The camptonites of Orkney have an east-north-east trend. When Dr. Flett described these dykes, he pointed out that they had been generally assumed to be of Tertiary age; and he added that they might well be of this date, although further research in the north-east of Scotland was required before it would " be possible to arrive at any definite conclusion." Later, after describing a few examples from Caithness, belonging to the Orcadian field of dyke-intrusion, he summed up as follows : " The whole question of the age of these rocks is still very obscure, and the possibility that they are really Tertiary cannot be definitely excluded."[2]

Dr. Flett[3] has also described a few east-north-east dykes of Orcadian facies and direction from the Glen Strath Farrar district of Ross; and has pointed out that others are known in Eastern Sutherland and the neighbourhood of Glenelg. As regards the Glenelg examples, the most definite reference that can be given is due to Dr. Clough,[4] who under the heading Pre-Triassic Igneous Rocks, refers to an east-north-east dyke of teschenitic affinity occurring towards the north end of the Sleat peninsula of Skye (south of Rudha na Caillich). This particular dyke has well-developed ocelli; and Dr. Clough has taught the geologists working in the Mull district that ocellar structure is of the utmost service in the field for distinguishing camptonites from other thoroughly basic types. It is true that one may be misled in applying this criterion, for felspathic segregations of ocellar type are found in some of the Mull basalt-lavas (p. 114), and also commonly among quartz-dolerite intrusions; but, in practice, it is comparatively seldom that one mistakes a camptonite. A further characteristic is the proneness of camptonites to a pustular or nodular weathering, so as to be covered on the surface by protruding spheres half an

---

[1] J. S. Flett, 'The Trap Dykes of the Orkneys,' *Trans. Roy. Soc. Edin.*, vol. xxxix., 1900, p. 865.
[2] J. S. Flett *in* 'The Geology of Caithness,' *Mem. Geol. Surv.*, 1914, p. 117.
[3] J. S. Flett *in* 'The Geology of Central Ross-shire,' *Mem. Geol. Surv.*, 1913, p. 79.
[4] C. T. Clough *in* 'The Geology of Glenelg, Lochalsh, and the South-east part of Skye,' *Mem. Geol. Surv.*, 1913, p. 79.

# Chapter XXXV.—Camptonite-Dykes of Doubtful Age.

inch or less in diameter. Dr. Peach[1] and Dr. Flett[2] seem to have been the first to draw attention to this feature in particular cases in Scarba and Caithness. It is an exaggeration of the pimply type of weathering characteristic of many of the Mull Plateau lavas (p. 138).

Dr. Flett's discussion of the age of the Scottish camptonites, as given in the Caithness Memoir, may be turned to by anyone wishing for an introduction to this difficult subject. Further assistance is afforded by Fig. 60 (p. 357), where localities for camptonite (or its allies, monchiquite, ouachitite, nepheline-basalt, etc.) are marked with a C. For descriptions of these occurrences the reader may turn to the Geological Survey 'Explanations' of the various one-inch Sheets indicated by numbers on Fig. 60. Explanations have not yet appeared in the case of Sheets 43, 51, and 52, but accounts of camptonites are given in Prof. Jehu's[3] description of Iona, and also in the *Summaries of Progress of the Geological Survey* for the years 1920 (p. 34), 1921 (pp. 96, 104), and 1922 (p. 94).

The camptonitic assemblage, recognized at several places in the West Highlands, agrees very closely in petrological type with what is characteristic of the Orkney-Glenelg country farther north. On the other hand, the predominant direction for camptonite-dykes in the West Highlands is north-west, and not east-north-east as it is farther north.

Thus the camptonites of the West Highlands agree in trend with the undoubtedly Tertiary dykes of the same region. Moreover, when a large collection of slides is examined, it becomes evident that the camptonites are petrologically united with basalts of a type such as undoubtedly figures largely among Tertiary lavas and intrusions of Mull. On these two grounds, it has been thought, with varying degrees of confidence, that the camptonites of the West Highlands may be of Tertiary age.[4]

At the same time, the fact remains that, while typical camptonites, monchiquites, etc., occur round about Mull, and even in Mull itself, they have not yet been found cutting Mesozoic sediments or Tertiary lavas. Furthermore, wherever camptonite-dykes are seen in contact with undoubted Tertiary intrusions in the West Highlands, the rule seems to be that the camptonite-dyke is cut by its companion. Dr. Harker,[5] with these two circumstances in view, has suggested a Permian age for the West Highland and Orkney camptonites. It will be seen that no one favours a Pre-Permian date; and, indeed, such a view would be difficult to maintain, for straggling examples of the north-west

---

[1] B. N. Peach *in* 'The Geology of the Seaboard of Mid Argyll,' *Mem. Geol. Surv.*, 1909, p. 90.

[2] J. S. Flett *in* 'The Geology of Caithness,' *Mem. Geol. Surv.*, 1914, p. 115.

[3] T. J. Jehu, ' The Archean and Torridonian Formations and the Late Intrusive Igneous Rocks of Iona,' *Trans. Roy. Soc. Edin.*, vol. liii, 1922, p. 186.

[4] J. S. Flett *in* 'The Geology of the Country near Oban and Dalmally,' (Sheet 45), *Mem. Geol. Surv.*, 1908, p. 124; C. T. Clough, W. B. Wright, and E. B. Bailey *in* 'The Geology of Colonsay and Oronsay with parts of the Ross of Mull' (Sheet 35), *Mem. Geol. Surv.*, 1911, pp. 41-88.

[5] Presidential Address, *Quart. Jour. Geol. Soc.*, vol. lxxxii., 1918 for 1917, p. lxxxvi.

camptonites are known cutting coal-measures in Ayrshire and Lanarkshire.

The evidence, as outlined above, does not permit of any certain conclusion. At the present time, however, the writer is inclined to adopt the Tertiary hypothesis, with the proviso that most of the West Highland camptonites are amongst the earliest manifestations of the Tertiary magma. Before elaborating these suggestions, it is well to give an account of the Ross of Mull and Morven occurrences, which seem, at first sight, to point in the opposite direction.

E.B.B.

*Ross of Mull.*—It has been shown in the 'Explanation' of Sheet 35 that most of the north-west dykes met with on the south coast of Mull beyond the Loch Assapol (Sheet 43) boundary-fault of the Tertiary lavas are of the camptonitic suite. Out of ten sliced specimens, four have been determined by Dr. Flett as camptonites, one as fourchite, two as monchiquites, and three as dolerites or basalts. One of the dolerites (14445) is certainly cut by the Tertiary xenolithic sill of Tràigh Bhàn na Sgurra (p. 266), and the same is in all probability true of the fourchite (14446); in addition, the latter is crossed by a number of small faults. The rest of the Ross of Mull evidence can be left to the 'Explanation' of Sheet 43. Here, it is enough to say that the camptonitic dykes are fairly well represented westwards right out into Iona.

(C.T.C.)

The Loch Assapol boundary-fault runs roughly parallel with the dykes of its neighbourhood. The Tertiary lavas on its north-eastern side, are, at first, markedly deficient in dykes of any sort. Even where, after some miles, dykes occur in profusion, typical camptonites, fourchites, or monchiquites are absent; or, at least, they have not been found in spite of free collection and slicing from all parts of the island.

E.B.B.

*Loch Aline, Morven.* The same absence of camptonites seems to characterize the Tertiary lava-area of Loch Aline. But, across the boundary-fault of Inninmore, (p. 181), camptonites are fairly well-represented on the mainland, and also, as Mr. Maufe found, in Lismore. The reappearance of camptonites is certainly striking, considering their absence throughout many intervening miles; but none of them is found, in the Loch Aline district, within 1½ miles of the lava-boundary, nor indeed aiming straight at it. Almost all the occurrences noted have been inserted on the one-inch Map and lettered C. Two of these are of particular interest, since, in them, camptonite is cut by normal Tertiary basaltic types: one shows an east-and-west camptonite (15844), on the coast south of Camas Lèim an Taghain, cut by a north-and-south basalt, similar in type to a Plateau Lava (15843); the other, a north-west camptonite (15785), in a stream north-east of Allt Dubh Dhoire Theàrnait, cut by a north-and-south tholeiite of Salen Type (15784).

It will be noticed, from the above, that the camptonites of the Loch Aline district vary considerably in direction; but their prevalent trend is roughly north-west. In Lismore, Mr. Maufe has found that they share with their companion-basalts a tendency to run west-north-west.

G.W.L.

If one had no more evidence in regard to the distribution of camptonites round about Mull, their occurrences in the schist and granite parts of Morven and the Ross of Mull, and their absence from the intervening lava-fields, would seem to point almost conclusively to a Pre-Tertiary date of intrusion. As a matter of fact, this evidence stands for nothing, since c a m p t o n i t e s a r e   u n k n o w n   i n   t h e   M u l l   S w a r m   f r o m   O b a n   t o t h e   C l y d e. To supplement earlier enquiries, Mr. Tait collected a suite of forty-eight specimens across the swarm from Sloc nan Uan on the west coast of Seil (near southern edge of Sheet 44), northwards to near Barrnacarry Bay (18993-19041); and these showed no greater camptonitic tendencies than the dykes across the water in Mull itself. Accordingly, the absence of camptonites in the greater part of Mull is not due to their concealment under Mesozoic and Tertiary cover.

Other evidence which has been collected bearing on the question is as follows :—

(1) Camptonites are unrepresented among the few dykes which cut the Moine Gneiss, exposed beneath Tertiary lavas for four miles in a north-easterly direction along the shores of Gribun and Inch Kenneth (Sheet 43). Visitors must be warned against a very deceptive appearance afforded by a 2 ft. west-north-west basalt-dyke on the west shore of Inch Kenneth. At a point noted on the one-inch Map, this dyke traverses the gneiss of the foreshore, but ends in the cliff 10 ft. below the base of the Trias. One has to climb up the cliff-face to realize that the dyke ends below the base of the Trias, and is not cut off by the same.

(2) The absence of camptonite-dykes from the Mull Swarm and its immediate neighbourhood no longer holds good in Coll, where Mr. V. A. Eyles and the writer have found north-west camptonites striking at the lava-field of Mull.

(3) On the other hand, camptonite-dykes do seem to be absent from the *western* half of the Arnamurchan peninsula. Messrs. Eyles, Simpson, and the writer have found them well-represented in the *eastern* half of the peninsula, where they continue the camptonitic belt of south-east Morven, Lismore, and Arnamucknish.

(4) Though typical camptonites are unknown from the lava-field of Mull, at least one camptonitic basalt (or dolerite) has been met with (p. 369), which might possibly be classed as camptonite. Then, too, one cannot lose sight of the somewhat camptonitic affinities of many of the Mull dykes and lavas—more especially in the case of lavas with ocellar segregations (Chap. X.). Not only so, but the occasional mugearites (Chap. X.), trachytes, bostonites, and syenite (Chap. XIV.), are just such associates as one expects to meet with camptonites.

(5) Though rocks of camptonitic affinity recur in the Mull petrological time-table, it is almost certain that the West Highland camptonites, if Tertiary at all, are of relatively early date; for they have often been seen cut by other Tertiary intrusions. Perhaps an early date in some small measure accounts for their not being found exposed in the region of depression occupied by the lava-field; but it is difficult to press this suggestion in view of the apparently continuous absence of camptonites along the track of the Mull Swarm south-eastwards to the Clyde.
E.B.B.

## PETROLOGY.

For the most part the dykes of the present chapter are fairly uniform in their microscopic characters, but vary amongst themselves to some extent in the relative abundance of their constituent minerals, among which first one bisilicate and then another gains preponderance. They are holocrystalline rocks with the typical panidiomorphic structure of the lamphrophyre-group; and are composed essentially of olivine, purple augite, deep-brown hornblende, soda-lime and alkali-felspars, and magnetite, with generally some analcite as the ultimate residual product. The felspar appears to be dominantly plagioclase; but orthoclase is usually, though not universally, present. The residual analcite occurs alike in the base, in vesicular cavities, and in ocelli. Quartz has not been detected.

Dr. Flett, whose writings have been quoted freely in the earlier part of this chapter, was the first to claim that Scottish dyke-rocks with the characters outlined above are true camptonites; and, at the same time, he pointed out that the name had previously been misapplied in Scottish literature. Dr. Flett's determination of the camptonites was largely based upon the colour of the augite and hornblende, and the presence of analcite, instead of quartz, as a residuum. He arrived at his conclusions after comparison of Scottish material with slides from continental areas ; and his results have been accepted by Sir Jethro Teall and Rosenbusch.[1]

The figured Lismore dyke (13744, Fig. 63A) contains moderately large crystals of augite and olivine in a matrix of well-formed elongated crystal of rich-brown hornblende, lath-shaped crystals of acid labradorite, and fairly abundant magnetite, in a base of what appears to be turbid orthoclase and analcite. In some cases (13748), olivine, as large crystals, is the dominant micro-porphyritic

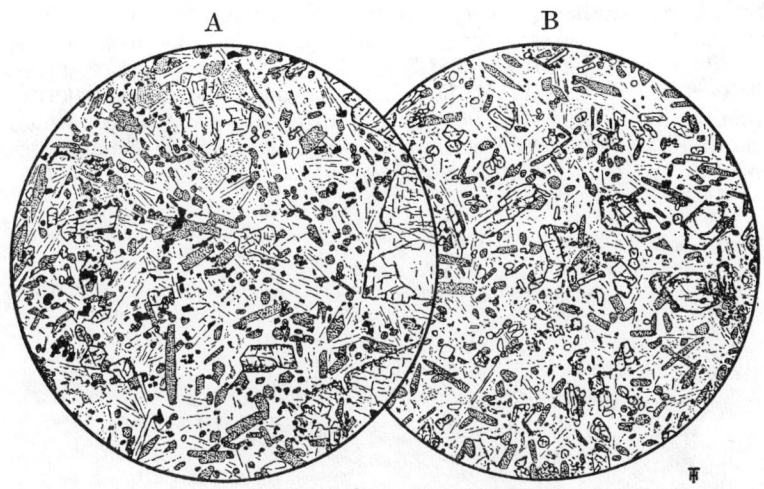

FIG. 63.—Camptonite-Dykes.

A. [13744] × 17. Camptonite from Lismore, showing moderately large crystals of augite and olivine in a matrix of well-formed elongated crystals of hornblende, labradorite, and magnetite, in a base of turbid orthoclase and analcite.

B. [15788] × 17. Camptonite from Morven. The phenocrysts are mainly olivine. Augite is a less prominent constituent than in A ; otherwise the rocks are similar.

constituent, and the place of hornblende in the matrix is wholly taken by augite. In other cases, olivine is practically wanting (13746), and hornblende almost unrepresented. Augite occurs in two generations, the one as larger somewhat-coloured individuals, frequently zoned towards the margin and showing hour-glass structure, and the other as hypidiomorphic prisms occurring in a matrix of plagioclase, orthoclase, and analcite.

The approach of certain dykes to the composition of vogesites is shown by the occasional increase in the amount of alkali-felspar (13751).

[1] H. Rosenbuch, 'Mik. Phys. d. Min. und Gesteine,' 1907, Bd. ii., 1, p. 170, also E. B. Bailey in 'The Geology of Ben Nevis and Glen Coe,' Mem. Geol. Surv., 1916, p. 157.

In Morven, the hornblende-rich rocks constitute the dominant variety, and the figured example (15788) may be regarded as characteristic. As before, olivine forms well-shaped phenocrysts (13744, 15834), and augite and hornblende figure prominently in the matrix. The felspar is largely a moderately acid plagioclase, but some alkali-felspar and analcite appear as ultimate products of consolidation. Occasionally, the crystals of olivine reach large dimensions. The hornblende is the usual deep red-brown variety, and occurs as elongated crystals showing strong pleochroism. Augite may occur as phenocrysts of considerable size (13744, 15842), but frequently it is more or less restricted to the matrix, and is comparable as regards dimensions with the hornblende. The amount of alkali-felspar and analcite in the base is a variable quantity. It seldom, if ever, reaches a value sufficient to cause the rocks to be classed as vogesites, but is readily detected (15790, 15812).

Certain of the dykes have, in the main, distinct doleritic affinities (15795, 15844), but present an ocellar structure of which the ocelli are lamprophyric both in composition and in the arrangement of the constituent minerals.   H.H.T.

## CHAPTER XXXVI.

### POST-VOLCANIC AND PRE-GLACIAL.

#### Introduction.

One of the most impressive lessons, which Macculloch, Sir Archibald Geikie, and Professsor Judd derived from their study of Mull, is the immensity of the erosion that has affected the district since the extinction of its volcanic fires. It is clear that its typical West Highland scenery has been fashioned during Tertiary and Post-Tertiary times; for its earliest lavas are of Eocene date. The testimony is all the more striking because the texture of many of the Tertiary rocks involved proclaims that they consolidated under considerable cover. The Ben Buie Gabbro, which can be examined through a vertical thickness of more than 2000 ft., is a case in point.

The maturity of the erosion is remarkably evident. The mountainous region of Central Mull stands up above the rest, because it is constituted of relatively resistant intrusions and indurated lavas, which in Ben More reach the highest altitude met with in the Hebrides, 3169 ft. Another feature of maturity is the failure of faults to step the scenery. The best example of this is afforded by the Inninmore Fault, of 1000 ft., or so, downthrow. This dislocation, as Judd realized, introduces a new type of scenery (Fig. 26, p. 182), without affecting the general surface-level of the ground. Other examples can be drawn from faults of the Gribun Peninsula, one of them with a downthrow of 400 ft. (p. 183). The exposures of the Loch Don Anticline, south of Loch Don (Fig. 25, p. 174) supply a variant of the same story. Instead of the Tertiary lavas of this anticline towering above their neighbours of the adjoining synclines, they have been entirely stripped away; the whole district has been reduced to an uneven surface, seldom more than 500 ft. above sea-level; and subsidiary ridges and hollows have been developed according to the strength or weakness of the various rock-groups with which erosion is at the present time confronted.

All this is in keeping with what has been established in the other Hebrides, Skye, Rum, etc. It is also at one with the well-known fact that the valley-system of the neighbouring mainland is cut through a great number of Tertiary dykes.

These general statements will now be followed by a consideration of relics of an Early Drainage-System, certain Obscure Plateaux, well-marked Notches and Caves of Marine Erosion between 100 and 160 ft. above High-Water-Mark, and the Gribun Landslip.

## EARLY DRAINAGE-SYSTEM.

Mr. Cadell, Sir H. J. Mackinder, Dr. Peach, Dr. Horne, Dr. Bremner and others, including the present writer, have attempted to piece together relics of an old drainage-system which constantly obtrude themselves upon observers in the Western Highlands and elsewhere in Scotland.[1] Such relics undoubtedly occur in Mull; but they present great difficulty of interpretation, owing to the island itself being a mere fragment, isolated by erosion combined with general subsidence. Under these circumstances, there is very little to gain from detailed discussion; but a brief statement of certain facts and hypotheses may be of interest.

A 'through-valley' is traceable from the mouth of Loch Scridain (Sheet 43) east-north-east across the valley-bottom-watershed of Glen More (650 ft.), where the road passes north-west of Loch an Ellen (Sheet 44); and thence, past Ishriff almost to Torness. Here, Glen More, followed by its road, takes a right-angled bend, and accompanies the Lussa River to Loch Spelve. Obviously, however, this Torness bend is a mere 'subsequent' development along the 'shatter-belt' of the Port Donain Fault (p. 183). The direct, and almost certainly original, continuation of the Ishriff segment of Glen More is northwards across an almost imperceptible valley-bottom-watershed, and so on into Glen Forsa. The connection between Glen More and Glen Forsa is with little doubt due to a river having once flowed across the site of the now intervening valley-bottom-watershed. The significance, however, of the continuity of Glen More from Loch Scridain to Ishriff is more uncertain. In this latter case, the valley-bottom-watershed may be interpreted, either as a watershed in course of development, interrupting a previously continuous river-bed, or as a watershed in course of demolition. Tentatively, we adopt the former alternative, and shall now sketch the tributary drainage-system attaching to our hypothetical Scridain-Forsa trunk-river. Even though the trunk may eventually prove to be double, instead of single, this will not seriously affect what is stated below in regard to the tributary system. E.B.B.

The Scridain segment receives an interesting set of 'beheaded' tributaries:—Gleann Seilisdeir from the north (Sheets 43, 44); the Beach River, Glen Leidle, and the Abhuinn Loch Fhuaran from the south (Sheet 44). It is quite clear that this tributary system has been restricted, since its initiation, through the retreat of the Gribun and Carsaig cliffs, and a diagrammatic example of beheading, with development of 'obsequent' streams, has resulted. The modern valley-bottom-watershed of Gleann Seilisdeir is a delta-watershed, or 'corrom,' as is so often found to be the case in beheaded river-valleys. E.B.B., E.M.A.

The Ishriff segment of our hypothetical trunk receives a prominent beheaded tributary from the valley of Loch Airdeglais.

The Glen Forsa segment is joined from the south-west by through-valleys passing Tomsléibhe and Goadhail Cottages respectively. It is quite possible that both these valleys originally

[1] For references and discussion see 'Geology of Ben Nevis and Glen Coe,' *Mem. Geol. Surv.*, 1916, p. 4.

headed on the south-east of the Loch Bà and Glen Cannel hollow, and that they have been subsequently cut across by this great scenic feature of Central Mull along the line where erosion first met the Glen Cannel Granophyre of Chapter XXXI.

A little beyond its present mouth, the Forsa valley joins with the Loch na Keal hollow and Glen Aros.

The old drainage-system of the West Highlands antedates any parcelling up of the country into 3000-ft., 1000-ft., etc. plateaux, such as is contemplated in the next section. For instance, if a river once flowed from Loch na Keal down Glen Forsa to Salen, it must have started at a time when Mull did not possess a central mountain-group.

OBSCURE PLATEAUX.

There can be little doubt that Mull was an island long before Glacial Times : on the one hand, Late-Glacial and Post-Glacial marine erosion has had but a trifling share in shaping the coast-line, and this little has almost been undone by a comparatively recent elevation of the district ; on the other hand, Mull stands boldly up from the Continental Platform, a thing apart. The contrast is evident when comparison is made between the Ordnance Survey Map and the Admiralty Chart. Most of Mull rises well above the 500 ft.-contour, whereas a depth of 500 ft. is only met with in the surrounding seas in restricted areas, which themselves appear to be hollows over-deepened by ice-erosion.

To some extent the abrupt rise of Mull from the Continental Platform is due to the fact that its coastal cliffs locally correspond with the escarpment of the Tertiary lavas, particularly along the south-east coast. Such, however, cannot be urged as a general explanation : in the neighbourhood of Ulva, for instance, the Tertiary lavas reach far under sea to Staffa and the Treshnish Isles (Sheet 43) ; while, in the coastal cliffs south-west of Gribun, gneiss rises as high as 700 ft.

There is evidence of much oscillation in the late Tertiary sea-level. On the mainland, a considerable part of the South-West Highlands is included within the 800 ft. to 1000 ft. Plateau of Lorne. The summit-level of the Lorne Plateau is comparatively independent of geological structure, and is generally supposed to be in close connexion with the position of the sea during some stage of the Pliocene. It is difficult to resist the view that the eastern corner of Mull, extending a mile inland from the road connecting Loch Don and Loch Spelve, is a rather low continuation of the Lorne Plateau—unless much of it be referred to the lower platform of the Ross discussed in the following paragraph. Its interior margin cuts right across the course of many important Early Basic Cone-Sheets. Perhaps, indeed, it is not too much to claim that the greater part of Mull, with the exception of the central mountains and the Gribun peninsula, bears testimony to the same base-level of erosion as is responsible for Lorne. It has already been pointed out that central Mull rises above peripheral Mull on account of its greater hardness ; but some additional factor is required to explain why peripheral Mull so often approximates to the summit-level of the Lorne Plateau.

At a level distinctly lower than that of most of peripheral Mull, the western extremity of the Ross (Sheet 43) constitutes a rocky flat-topped ridge rarely rising more than 300 ft. above the sea. The greater part of this ridge is constituted of old gneiss and granite ; and it is tempting, at first sight, to see in it a relic of a resistant platform, from which Mesozoic sediments and Tertiary lavas have been stripped. This, however, cannot be the case, for the margin of the low ground is cut well within the lava-country, and includes the Ardtun Peninsula along with a strip reaching thence south-east towards Shiaba.
E.B.B.

### Notches and Caves, 100-160 ft. above High-Water-Mark (Fig. 65, p. 395).

The time-relations of the comparatively high platforms of the previous sections to the more or less submerged platform, also mentioned as responsible for our present coast-line, provide questions too difficult to be discussed at this juncture. One thing, however, is clear : at a late stage in Pre-Glacial times, when previous erosion had determined much of the present-day coastal form of the island, a submergence took place which has left pronounced traces from 100 to 160 ft. above the modern high-water-mark.

That the 100-160 ft. Pre-Glacial beach-notches of Mull and district are of later date than a much more extensive marine-erosion, with an upper limit more closely approximating the present sea-level, may be inferred from two main observations :—

(1) In the Gribun Peninsula south of the entrance to Loch na Keal (Sheet 43), and, again, along the coast of Loch Tuath northward to Caliach Point (Sheets 43 and 51), a Pre-Glacial notch at about 115 ft. above high-water is traceable as a minor, though very striking, feature of the coastal cliff. It is incredible that this notch could have originated as an extensive platform, and then be reduced to its present narrow dimensions by subsequent marine erosion at a lower level, without a much more pronounced loss of continuity.  W.B.W., E.B.B.

(2) At the head of Loch Scridain (Sheet 44), a Pre-Glacial notch is found at about 160 ft. above high-water cut into obvious pre-existing valley-sides, where it can be followed in and out of tributary hollows which descend below its level. The valley-sides lead down to a pronounced, moulded and striated, rock-platform, a part of which is covered at high tide.  E.M.A.

The Pre-Glacial notch at about 115 ft., is more continuously preserved than anything of the same kind at higher elevations. It was originally traced by Mr. Wright with the aid of a grant from the Royal Society of London[1]. He had previously, in 1907, discovered its counterpart in Colonsay (Sheet 35) during his work there for the Geological Survey. The subsequent examination of the Mull district by various members of the Geological Survey, after Mr. Wright had left Scotland, has confirmed his observations, and added to them three important records, all of which, curiously enough, belong to rather higher levels : these three are the notch at the head of Loch Scridain (Sheet 44), another at Kilchoan,

[1] W. B. Wright, 'On a Pre-Glacial Shore-line in the Western Isles of Scotland,' *Geol. Mag.*, 1911, p. 97.

Ardnamurchan (Sheet 52), and the sea-cave of Ulva (Sheet 43), to which last our attention was directed in 1920 by Mrs. Clark of Ulva House. Before considering these last three additions, an account will be given of the more familiar occurrences which seem to group themselves about the 115-ft. level.  E.B.B.

At the outset, it is important to state why this bench, or platform, of marine erosion is attributed to Pre-Glacial times rather than to the Late-Glacial submergence which has left such evident traces in the 100-ft. gravels, etc., of many parts of Scotland including Mull :—

(1) In the following descriptions, it will be shown that striæ have been preserved on the cliff behind the notch at Rudha nan Goirteanan, that ice-moulding of the platform is a feature of the Treshnish Isles, and that boulder-clay, capped by 100-ft. Late-Glacial gravel, has been found on the rock-platform at Crackaig. The Rudha nan Goirteanan and Treshnish Isles evidence is in keeping with what is very fully seen in Colonsay outside our area; and the Crackaig evidence is similarly repeated much more clearly in Islay. It must not be considered surprising that so few striæ have been found on what is termed the Pre-Glacial notch of the Mull district, for it happens that practically no striæ have been preserved anywhere in its neighbourhood irrespective of level.

(2) The locally pronounced featuring of the notch is curious when taken in conjunction with its tendency to disappear altogether. The conclusion is forced upon us that the notch has in many places been obliterated by erosion. Now the Late-Glacial 100-ft. beach-deposits have so often been spared by subsequent erosion, that, if the cliff of the notch we are considering belonged to the same Late-Glacial time, we should expect it to have a comparable distribution—but such is not the case. This evidence is in keeping, of course, with the local discoveries of boulder-clay between the Late Glacial beach-deposits and the Pre-Glacial platform where the two are found together. It is also in keeping with the modification of the cliff behind the notch into a slope (without corresponding scree) in the Dutchman's Cap of the Treshnish Isles.

In describing the 115-ft. notch, we shall begin with its more southerly occurrences in Mull, and, after working our way northwards along the coast, strike out for the islands; these latter, as a matter of fact, are in a sense always with us, prominent objects of the sea-scape.

*Mull, Gribun Peninsula.* Outside the mouth of Loch Scridain, above Macculloch's Tree, Rudha na h-Uamha (Sheet 43), a well-marked notch is traceable, cut in the lavas of the sea-cliff, with its inner angle now choked with scree. Levelling by Abney showed it to lie about 110 ft. above high-water-mark of ordinary spring tides; but this was at a point where the narrow platform was passing beneath scree, and its inner angle is probably a little higher, say 115 ft.

Farther north, beyond the great modern landslips of the Wilderness, the notch is manifest again, It is particularly clear at Rudha nan Goirteanan, where it is seen wholly cut in Moine gneiss. The angle is often choked with basalt-scree from the higher part of the cliffs behind. The only measurement taken for the platform was 98 ft., but the inner angle is probably a few feet higher. When first described, it was only possible to state that the gneiss of both platform and cliffs is somewhat rounded and moulded. Mr. Bailey has since found striæ running S.W. on the surface of the cliff. A little farther along the coast, there is another good example of the

388    Chapter XXXVI.—Post-Volcanic and Pre-Glacial.

notch to be seen above Stac Glas Bun an Uisge. It can also be recognized for half a mile east-north-east of Sgeir na Faoilinn. In both these localities, the notch is cut in the coastal cliff of Moine gneiss. Farther north-east above Port Uamh Beathaig, Mr. Bailey has traced what he takes to be the inner edge of the terrace for quarter of a mile, until lost sight of beneath a landslip. The notch here is cut in Trias, and its inner angle lies at 135 ft. above high-water-mark as determined by sextant.

*Mull, north of Loch na Keal.* The same notch reappears as a coastal feature north of Loch Tuath, where it can be followed almost continuously from Port Burg to Treshnish Point, just beyond the north edge of Sheet 43. The cliff above is generally about 200 ft. high, and the cliff below about 100 ft., often bathed at its foot by the sea. The whole is cut in the lava-series, and the level of the notch, like that of the modern shore, is influenced by the relative strength of the gently inclined basalt-flows. Where it is cut in soft slag overlying a massive band, it may reach a comparatively low level at one point, and a comparatively high level at another, depending on the height at which the junction between slag and solid occurs. Such range in level varies from about 100 to 125 ft. above high-water-mark. At the same time, the general persistence of level within these limits, for a distance of about 5 miles, clearly proves the marine origin of the bench. Mr. Richey has found a patch of boulder-clay on the platform at one point, the boulder-clay itself overlain by gravel attributable to the Late-Glacial 100 ft. beach. The locality is marked by a note on the one-inch Map (Sheet 43) just south of Crackaig, and the exposure occurs in the seaward cliff terminating the bench. Opposite a hollow in the Pre-Glacial cliff, east of Rudh' a' Chaoil, the surface of the platform has a distinctly ice-moulded appearance. It is a pity no striæ have been found on cliff or platform, but striæ are very poorly preserved everywhere in this part of the Mull district.

The southern shores of Calgary Bay (Sheet 51) show only poor remnants of the Pre-Glacial beach serving locally as a platform for the road. On the northern shores, however, the notch resumes the magnificent development exhibited farther south, and this it maintains round the coast of Mornish as far as Caliach Point. It is, in fact, a conspicuous object from the tourist-steamer plying between Oban, Staffa, and Iona. In one or two places, its surface has an appearance of ice-moulding. An Abney-level measurement gave 115 ft. about a mile south of Caliach Point; but similar variations, 105-120 ft. are met with here as farther south. Another estimate at the head of Calgary Bay gave 110 ft. above high-water-mark, with variations between 95 ft. and 115 ft.

*Gometra and Ulva.* Our remaining records lie outside Mull. In Gometra (Sheet 43), the platform is much dissected by hollows and geos, but the remnants of it maintain, on the whole, a fairly uniform level, and the old cliff behind is very well-preserved. The inner angle is distinctly marked for a distance of about a mile and a half around the west and south sides of the island. At its northern extremity, it is finely developed at a height of 105-118 ft. above high-water-mark. On the south coast, on both sides of Gometra House, measurement gave 110 ft. for the inner angle. Traced east, it appears to decline, and successive readings by Abney-level gave 105, 100 and 95 ft. within half a mile of Gometra House. The southern angle of Ulva, reaching out towards Little Colonsay, seems to be referable to the same platform; but much the most important evidence of Pre-Glacial marine erosion in this part of the district is afforded by the Ulva Cave to be described presently.

*Treshnish Isles.* Of the Treshnish Isles (Sheet 43), only two, the Dutchman's Cap (Bac Mòr, the equally descriptive Gaelic name, means Great Shield) and Lunga, rise high enough to show the Pre-Glacial shore-line.

In the Dutchman's Cap, a broad platform, belonging to the Pre-Glacial beach, forms the rim, and a central eminence, rising 284 ft. above O.D., forms the crown. The platform includes more than two-thirds of the whole, and is almost entirely below the 100 ft. contour. The platform slopes gently, and is higher at the north end. It tends to follow the surface of successive resistant lava-flows dipping gently towards the south or south-west. Measurement from the upper limit of *Balanus* on the rocky modern shore gave 75 or 80 to 90, or possibly 95 ft. for the inner angle of the platform round the south edge of the central hill. The general level of outlying portions of the platform here was 70 to 75 ft., rising nearly to 80 ft. in Bac Beag. At the north end, the inner angle lies at from 90 to 95 ft.; and, at the extreme north end of the island, the platform is about 97 ft. above the *Balanus*-limit. Everywhere, the platform is remarkably ice-moulded. No striæ have been observed, but *roches montonées* are obvious; they indicate ice-flow from the north-east or east-north-east.

Lunga, the largest of the Treshnish Isles, has a still more extensive platform. At the north end of the island, this platform tends to follow the top of a very massive lava. Its inner angle, at the ruined houses, is just 100 ft. above high-water-mark. Southward along the east side of the island, it rises steadily to 110 ft. or more. It then abandons the massive lava mentioned above, and starts again, south of Cruachan, at a lower horizon in the volcanic series, and only reaches to 80 or 85 ft. above high water. Along the north, east, and south of Lunga, ice-moulding of the platform from the north-east or east-north-east is very pronounced; though no striæ have been found.

*Iona.* Somewhat indeterminate traces of the platform reappear in Iona.     W.B.W.

The remaining occurrences to be noted all seem to occur at an unusually high level. The significance of this characteristic is at present doubtful. There is clearly opportunity for a difference of judgment in taking measurements of rather ill-defined features, such as inner angles of marine platforms; and there is also the possibility that the features under consideration may not belong all to the same epoch, or that, if they do, their subsequent history of elevation may differ in detail.     E.B.B.

*Loch Scridain.* A beautifully ice-moulded and striated rock-notch can easily be followed for about a mile along the hill-side north of Loch Beg at the head of Loch Scridain, where its course is laid down on the one-inch Map, Sheet 44. What is evidently the same notch can be picked up again as a scenic feature on the south side of the loch, above the Hotel, but is less well-defined, and has not been inserted on the one-inch Map. Readings with a sextant gave 160 ft. above high-water-mark. This is higher than any corresponding observation in the Hebrides, including even the floor-level of the Ulva Cave. The notch evidently marks an old beach-line because of its horizontality. It cannot be attributed to a glacial lake, as there is no corresponding deserted outlet which could have determined a persistent lake-level at this height. Moreover, it is incised in particularly hard rocks, demanding a length of time for its formation which seems out of keeping with a lake-hypothesis—it will be remembered that the Parallel Roads of Glen Roy are merely fashioned in morainic cover, not rock. The striæ which cross it conform with the striæ at a higher level, and are evidently the work of land-ice. It can scarcely be claimed that the beach is of Late-Glacial age, affected by a readvance of the land-ice of the period. Late-Glacial raised beach-deposits at about the 100-ft. level, are well-known in Mull; indeed, they occur at Rossal Farm close by ; but Mr. Wright has pointed out that these high-level Late-Glacial beach-deposits are never found in any but accidental connexion with rock-notches of appreciable dimensions. The Rossal gravels just mentioned, are, for instance, obviously later than the glaciation of the notch occurring west of their position.

E.M.A.

*Ulva Cave.* The Ulva Cave lies 300 yds. south-west of the summit of A' Chrannag, at the foot of a conspicuous basalt-lava escarpment. Its presence cannot be attributed to any local weakness in the basalt. It is manifestly a sea-cave. Its entrance is at present 10 ft. high and 25 ft. wide, and is partially blocked by a fall from the cliff. Inside, the cave is roughly circular in plan and is 48 ft. deep, measured inwards from the mouth, and 42 ft. broad. At its eastern end, it is 25 ft. high. The floor was examined by digging a trench 6 ft. long, and the same deep, about its middle. Boulder-clay was encountered immediately below a thin layer of animal-manure, and continued without change to where it appeared the rock-floor was reached—there was no intervening gravel. The included stones range exceptionally to a foot in diameter, and are subangular, a few showing striation. They consist of basalt, gneiss, granite, and Lower Lias fossiliferous limestone, in about equal proportion : so that, altogether, foreigners from the Sound of Mull and Morven far outnumber basalt derivable from Ulva or Mull. The matrix of the deposit varies from clay to loam. The clay-matrix often shows slickensiding, especially near stones ; and rootlets have frequently penetrated along the resultant smoothed surfaces. The height of the partially blocked cave-entrance, measured by sextant, is 155 ft. above high-water-mark. The floor, where met in the trench, is at 147 ft.

*Little Colonsay Cave.* A much less striking cave occurs 100 yds. north-west of the summit-cairn of Little Colonsay, at the foot of a prominent lava-scarp. Accurate measurement of level has not been made, but a fair approximation gave almost the same figure as for the Ulva Cave.

The dimensions of the cave are 15 ft. deep, 8 ft. broad, $3\frac{1}{2}$ ft. high. No boulder-clay was noticed.  G.A.B.

*Kilchoan, Ardnamurchan.* A level rock-notch is a conspicuous feature of the hill-slopes above Kilchoan on the Sound of Mull (Sheet 52). At its north-east end, it was found to stand 145 ft. above O.D., measured with reference to a convenient bench-mark. This might mean rather less than 140 ft. above high-water-mark, but still a few feet higher than is common on western Mull. The Pre-Glacial date of the notch, which was first investigated in 1921, follows with probability from the following observations :—

(1) No gravels are preserved in connexion with the notch. The highest gravels are those of the Late-Glacial 100-ft. beach exposed, for instance, at the Free Church Manse.

(2) The platform of the notch has suffered a remodelling by erosion, which does not seem to have affected the Late-Glacial and Post-Glacial beach-gravels, occuring as well-preserved flats at lower levels.  J.E.R.

## Gribun Landslip.

Mull includes several considerable landslips. The majority are clearly of Post-Glacial, or Late-Glacial, date, and are accordingly treated in the following chapter. The most massive landslip, however, is almost certainly Pre-Glacial. The Gribun road traverses its slopes a little north of the stream at Balmeanach (Sheet 43). The mass is in large measure a coherent rock-slip consisting of lavas and mesozoic sediments. Its total length is little under half a mile. Its Pre-Glacial age seems certain from the fact that it has lost all trace of landslip-featuring, although its tilted and somewhat broken constituents afford a striking enough scenic contrast to the escarpment from which they have broken away.

E.B.B.

# CHAPTER XXXVII.

## GLACIAL AND RECENT.

### INTRODUCTION.

IN one important particular of its glacial history, Mull occupies a position analogous to that of Skye. It includes a sanctuary, within which no mainland boulders have been found; though, beyond its limits, they occur in profusion (Figs. 64 and 65). Dr. Harker's [1] hypothesis of a local centre of dispersal seems to account satisfactorily for the sanctuary. It is supposed that local snowfall more than supplied the fluxional loss of ice which would have occurred if the surface of the general Scottish ice-sheet had assumed a smooth continuous slope above the mountainous obstruction of Mull. In consequence, a swelling in the ice-sheet always marked the site of the buried massif; and invasion of the sanctuary was prevented by outward flow from this subsidiary centre. The limit of mainland erratics, laid down in Figs. 64 and 65, roughly corresponds to the most restricted boundary of bottom-currents of ice originating over Mull.

In about half of Mull, the glacial record is entirely of ice-sheet type; and, while there are often plenty of erratics and striæ, there is surprisingly little boulder-clay or morainic drift. The story, even in this part of the district, is not a simple one. For instance, granite and gneiss boulders derived from Morven are stranded throughout the country north-west of Loch na Keal, including Ulva, and yet granophyre-boulders from Glen Forsa, or some neighbouring source, have found their way to Dervaig in Sheet 51. This crossing of boulder-tracks corresponds with divergent observations of striæ, and there can be no doubt that the currents of the ice-sheet varied greatly from time to time.

In the mountainous central region, in large measure included in Fig. 64, erratics and striæ both indicate a phase of valley-glaciation at the close of the glacial history of Mull; and, as usual in the Highlands, the valley-glaciers have left morainic deposits with a hummocky topography.

It has long been known, from faunal and other considerations, that the earlier raised-beach deposits of Scotland are Late-Glacial rather than Post-Glacial, and this is particularly clear in Mull.

It has also been recognized that the lowest well-marked raised beach of Scotland is Post-Glacial, and corresponds to an exceptionally long pause in the relative movement of sea and land, during which was effected most of the marine erosion that has

---

[1] A. Harker, 'Ice Erosion in the Cuillin Hills, Skye,' *Trans. Roy. Soc. Edin.*, vol. xl., 1901, p. 221.

# Introduction. 393

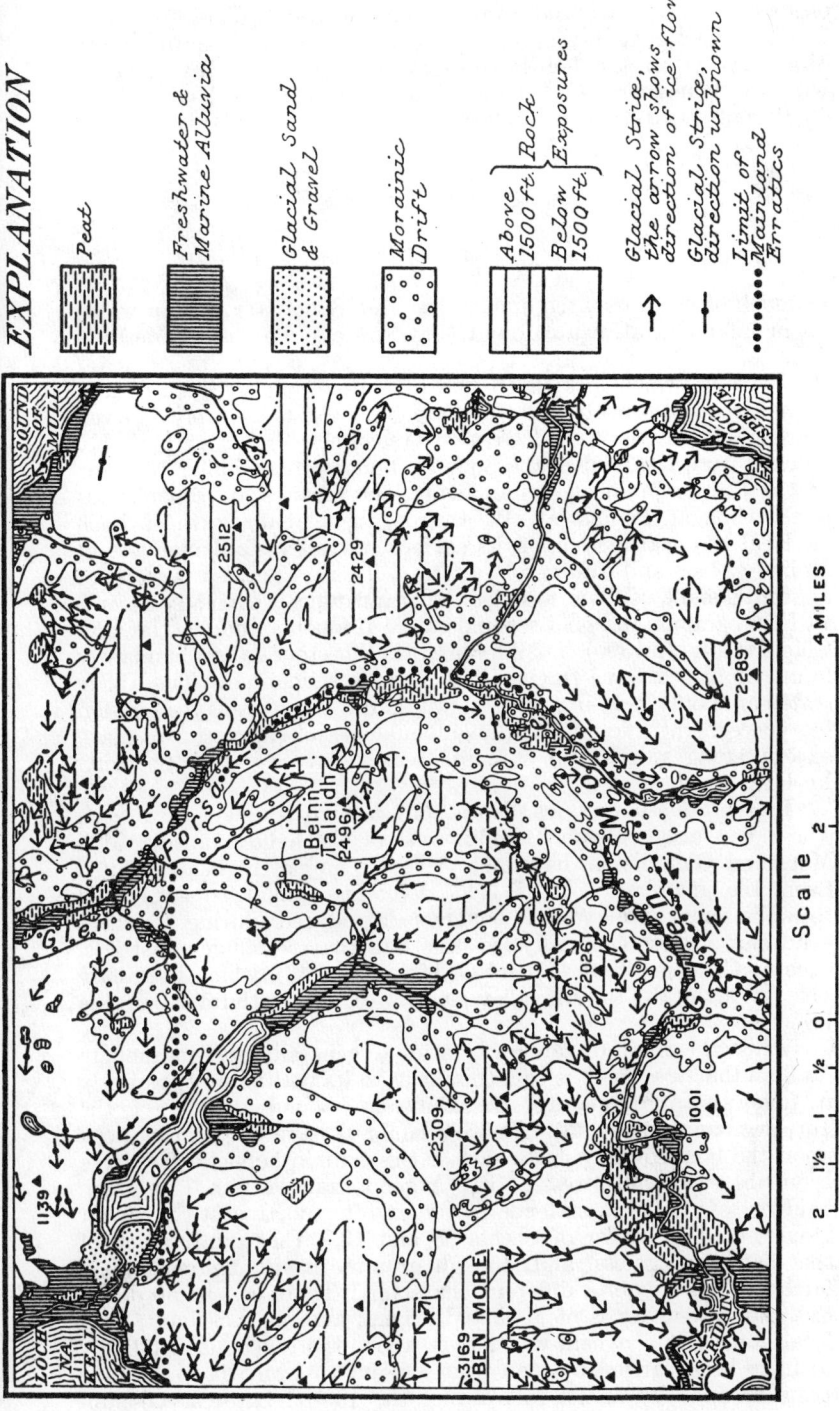

Fig. 64.

Map showing Superficial Deposits and Glaciation of Central Mull within limits indicated by note at left hand top corner of one-inch Map, Sheet 44.

occurred since the withdrawal of the ice-sheet. This low-raised beach is particularly well displayed at Oban and in south-eastern Mull. At Oban, archæological evidence brings it into relation with the commencement of the human record in Scotland.

Certain landslips are dealt with at the conclusion of the chapter.

## ICE-SHEET GLACIATION.

### ERRATICS.

IN much of its course, the limit of the Mull Sanctuary, within which no mainland erratics are found (Figs. 64 and 65), is wonderfully definite.

The hills which face the Sound of Mull, Dùn da Ghaoithe and its neighbours, are littered with foreign boulders, notably Morven granite, Moine gneiss, and Triassic sandstone—the latter probably plucked from the floor of the Sound. Southwards, similar boulders are well-represented on Beinn Fhada, and erratics are conspicuous in the Croggan Peninsula. Westwards, the plateau north of Loch na Keal has scattered upon its surface moderate-sized boulders of Moine gneiss and Morven Granite. There are also a few of quartzite, some of them rounded, but sandstone is virtually absent. In Ulva, granite and gneiss boulders are frequent, and, in the Pre-Glacial Cave (p. 390), fossiliferous Lias limestone was abundantly found along with one fragment of Triassic sandstone. The above statements only hold in a general way, and, where valley-glaciation has succeeded the ice-sheet, foreign boulders are seldom conspicuous, except along ridges and summits which have escaped cleansing by local ice-flow.

In the Sanctuary (Figs. 64 and 65), foreign boulders are absent. The ridges about Beinn Talaidh, west of the hollow uniting Glen More and Glen Forsa, have not yielded a single fragment derived from the mainland. It is held that this cannot be attributed to subsequent removal of boulders introduced during ice-sheet conditions, more especially as the Sanctuary includes districts, such as Carsaig (Sheet 44) and Gribun (Sheet 43), which were only able to maintain trivial valley-glaciers on the withdrawal of the ice-sheet.

While the limit of foreign erratics is generally well-marked in Mull, in the Ross there is some doubt as to its position. Apparently it follows, approximately, the Bunessan boundary-fault of the Tertiary lavas (Sheet 43). Foreign boulders have not been noticed upon the lava-country north-east of this fault, whereas a few have been observed south-west of it. A good example is a 1 ft. 6 in. boulder of grey hornblende-biotite-granite north-east of Scoor House, on the border of Sheets 35 and 43. The country-rock of the district is gneiss, and, though granite *in situ* figures largely farther west, it is of a different character. There can be no doubt the granite-boulder is of Mainland origin; and, as it is built, with local stones, into a field-wall 200 ft. above the sea, it is not likely to have been carried by man from a point to which it could have been brought by an ice-raft during the 100-ft. Late-Glacial submergence. Mainland boulders of granite are fairly common near Bunessan, but their low stations often render their mode of transport dubious.  E.B.B.

A few other interesting cases of ice-sheet distribution of boulders may be cited.

In general, the boulder-carry in the Morven district of Sheet 44 is from east to west. Erratics of Morven Granite are a feature of the basalt-plateaux, even west of Loch Aline; and blocks of quartzite and Lismore Limestone have been found on summits of the granite-area at An Sleaghach and Meall a' Chaoruinn. In apparent contradiction with this, a few erratics of Triassic sandstone and *Gryphea*-limestone also occur in the granite-area. These, however, cannot be attributed to a distribution from the west; for all the evidence negatives such a suggestion. It points, rather, to a submerged outlier of Mesozoic sediments somewhere along the complicated fault-belt that runs up Loch Linnhe.  G.W.L.

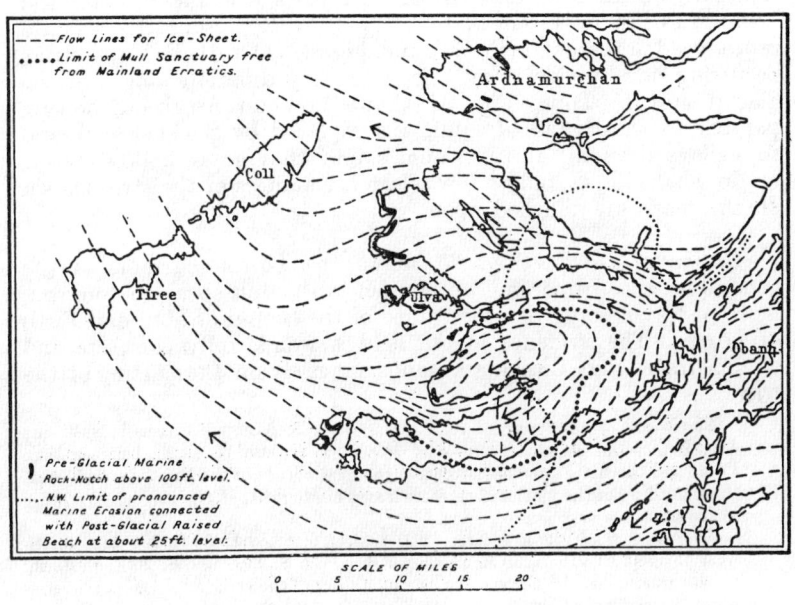

FIG. 65.

General Glaciation of District, and some Raised-Beach phenomena.

In the Oban district, granite-erratics, derived from the Etive complex, are well known. Such as are porphyritic have been derived from the central or Starav Granite of the complex; while the non-porphyritic belongs to the outer ring or Cruachan Granite.  E.B.B.

Attention has already been directed to the carry, in Mull, of granophyre-boulders westwards to Dervaig (Sheet 43). Granophyre-boulders are naturally also found well within the Sanctuary of Fig. 65. For instance, they are a fairly marked feature of the country south of Loch na Keal, as far as Balmeanach Farm in the Gribun district (Sheet 43). Probably Balmeanach marks the southern limit of ice which, farther east, traversed either the Glen Cannel or Beinn a' Ghràig Granophyres of Chap. XXXI. and XXXII.  E.B.B., J.E.R.

Another notable carry of boulders is illustrated in Iona, where great erratics of pink Ross of Mull Granite are everywhere conspicuous. Many of them lie above the limit reached by the sea in Late-Glacial times so that there can be no doubt of their transport by land-ice.  E.M.A.

Recognizable erratics are scarce in Coll and Tiree, except on the modern and Late-Glacial sea-beaches, where a veritable museum-selection is afforded. There is one very conspicuous black boulder of olivine-gabbro in Coll which must have been carried into position by land-ice. It measures $8 \times 7 \times 4$ ft., and stands on pale gneiss, 200 ft. above sea-level, on the rocky slopes of Ben Hogh, 300 yds. north-west of the head of Loch nan Cinneachan (Sheet 51). As will be shown presently, striæ prove that ice passing across Mull must have overridden both Coll and Tiree, so that it is strange how little Mull debris has been left on these islands. The gabbro-boulder may have come from Mull, or possibly from Ardnamurchan.

### STRIÆ.

Numberless striæ have been recorded on the six-inch Field-Maps, especially in that resistant part of Mull which is included within the Limit of Pneumatolysis of Plate III. (p. 91). Representative sets are reproduced in the one-inch Map and Fig. 64. On these Maps, striæ, occurring side by side, may belong to very different stages of the glacial story. Thus, Fig. 64 shows how an ice-sheet, flowing parallel to the Sound of Mull, was replaced by glaciers, restricted to valleys draining at right angles into that great hollow.

In what follows, only a few selected instances of the striæ shown on the maps are touched upon.

*Craignure Divide.* Fig. 65 shows the bottom-currents of the general ice-sheet diverging on contact with Mull so as to conform, more or less closely, with the course of the Sound of Mull and Firth of Lorne. The evidence in the field is wonderfully complete and demonstrates that Craignure stands precisely on the parting of the ways :—

*North-west of Craignure Bay.* Striæ have been noted directed N.W. at Java Point, and, in succession, W. 5 N., W. 2 N., and W. 10 S., between Java Point and the little stream that enters the west corner of the bay. In each case, the moulding of the glaciated surfaces leaves no doubt that the ice was streaming westwards.

*South-east of Craignure Bay.* Striæ occur directed E. 30 S. 200 yds. north-west of Sgeir Ruadh, S. 30 E. on Eilean Bàn, and S. 28 E. and S. 20 E. on Rudh' a' Ghuirmein. In all these cases, the moulding of the striated surfaces indicates a south-westwards flow.

The distance between the striæ pointing N.W. on Java Point and those directed E. 30 S. near Sgeir Ruadh is only 1 mile 300 yds. There are no striated surfaces known in the interval ; but on either side, whether up the Sound of Mull, or across Duart Bay, there are countless ice-moulded and striated surfaces referable to the general glaciation, and they also agree in putting the parting of the bottom-currents at Craignure.  E. B. B.

*North of the Sanctuary.* The general direction of the main ice-current passing more or less westwards across Mull, north of Craignure and the Sanctuary of Figs. 64 and 65, is vouched for by abundant striæ. At first, the flow is north of west, but, near Salen, a particularly strong current, directed south of west, was concentrated along the valley leading to Loch na Keal.  W.B.W.

A remarkable series of little striated hollows and winding grooves is a conspicuous feature of the south shore of Loch na Keal, between Knock and Scarisdale. The depressions have been cut in the rocks, irrespective of geological structure. Where, as is frequently the case, successive hollows are connected by a channel, they furnish miniature examples of rock-basins distributed along a valley. Appearances suggest that the main factor in their production has been running water: the hollows are almost certainly pot-holes; the winding channels stream-courses. Ice has been merely a modifying agent, as in the tortuous grooves on Kelley's Island, Lake Erie, figured in Chamberlin and Salisbury's *Geology*.[1] The clear manner, in which these pot-holes and grooves are exhibited on the shore of Loch na Keal, might suggest a marine origin; but less perfect examples have been noted at various altitudes in Mull up to 1000 ft. Probably, they originated through the operation of subglacial torrents, as suggested by Professor Holmquist in the case of analogous forms observed in Scandinavia, especially in the island of Utö, near Stockholm.[2]

The general westward flow of ice across northern Mull was not constant. It has already been pointed out that granophyre-boulders are fairly common near Dervaig (Sheet 51); and, in agreement with this, north-west striæ are occasionally met with throughout the district running parallel with the hollows of the Sound of Mull, Loch Frisa, and Loch Tuath. They are well-seen on the coastal platform near Dervaig itself, while, a little farther east, examples of eastward striæ are preserved. At Ulva (Sheet 43), close to the ferry, Mr. Burnett has found two sets of striæ (N.W. and S.W.) on the same rock-surface; but it is not possible to decide which is the earlier. J.E.R.

The south-westerly striæ of Ulva and neighbourhood afford evidence of a closing in of the ice-stream in the rear of the Mull Sanctuary. The same is shown by W. 20 S. striæ on moulded surfaces on Erisgeir (Sheet 43), and by S.W. or W.S.W. *roches montonées* on the Dutchman's Cap and Lunga of the Treshnish Isles. W.B.W., E.B.B.

*South of the Sanctuary.* Striæ, indicating a southward, and then south-eastward, flow of the main ice-sheet between Craignure and Loch Buie, are sufficiently numerous, as may be judged from Sheet 44 and Fig. 64. The reader must be warned, however, that the record of the ice-sheet, here, is more confused by extensive valley-glaciation, than in the more northerly district just described.

In the western part of the Ross of Mull, lying outside the Sanctuary of Fig. 65, very few striæ have been observed. Mr. Bosworth, has, however, noted two, one pointing N. 25 W., a little south of Bunessan, and another W.N.W. south of Loch Assapol (Sheet 43). Mr. Cunningham Craig has also recorded striæ W. 10-18 N. at Scoor House (Sheet 35). *Roches montonées* pointing N.N.W. are recognizable between the road and Loch na Làthaich, north of Bendoran Cottage.

In Iona, Professor Jehu[3] met with N.W. striæ south-west of Cnoc Druidean (Sheet 43).

[1] T. C. Chamberlin and R. D. Salisbury, 'Geology,' vol. iii., 1906, Fig. 485, p. 349.
[2] P. J. Holmquist, 'The Archean Geology of the Coast Regions of Stockholm.' *Guide 15 of the International Geol. Congress*, 1910, pp. 112-116.
[3] T. J. Jehu, 'The Archean and Torridonian Formations and the Later Intrusive Igneous Rocks of Iona,' *Trans. Roy. Soc. Edin.*, vol. liii., 1922, p. 166.

At the south-west end of Tiree, Mr. Simpson has found many striæ on Ben Hynish, which give readings between N. 20 W. and N. 35 W. They lie well above the limits of Late-Glacial submergence, and their regularity shows that they are a product of land-ice. These striæ are combined with little or no moulding, but taken in conjunction with the north-west glaciation of the eastern half of Coll and of Iona, they presumably indicate an ice-flow towards the north-north-west.

The general north-westerly, or north-north-westerly, glaciation of the Ross, and of Iona and Tiree, seems to be due to something more than a mere closing in of the ice-sheet behind the Mull Sanctuary, namely, the influence of ice proceeding from the north of Ireland. North-westerly striæ were long ago mapped by Mr. Wilkinson in the north-west corner of Islay. E.B.B.

*Within the Sanctuary.* As may be gathered from Fig. 64, the striæ of the Mull Sanctuary are in large measure attributable to the action of valley-glaciers in Late-Glacial Times. Thus, valley-glaciers at one time drained down Glen Cannel to unite on the bed of Loch Bà. At an earlier stage, however, there is clear evidence, from the moulding of all the passes leading down to Sleibte-coire, that ice from the upper portions of the Loch Bà drainage-system escaped south-westwards. (C.T.C.)

It would be easy to multiply the records of high-lying striæ referable to the ice-sheet, but only one more locality will be cited here. The summit-ridge of Beinn nan Gobhar, above Loch Buie (Sheet 44), carries, at one point, striæ pointing S.S.W. on well-marked *roches montonées*.

In the somewhat lower ground of the Ross extending west into Sheet 43, the striæ indicate two main directions of ice-flow. At one time, the peninsula appears to have been covered by ice flowing south from across Loch Scridain. At another, the ice-flow was more from east to west along a curved course leading south-west, then west, then north-west. Particularly complex crossings of striæ are a feature of the top of the sea-cliff near Dùnan na Marchachd (Sheet 44). It will be seen that the evidence of successive openings up and compressions of the lines of ice-flow, met with in this southern part of the Sanctuary, corresponds very closely with what is found north of Loch na Keal. E.B.B.

### LOCAL GLACIATION.

The evidence of valley-glaciation in Mull is threefold : (1) striæ and moulding in obvious relation to valley-systems ; (2) carry of boulders ; (3) moraines.

The contrast between valley-striæ and ice-sheet striæ is often very marked (Fig. 66, p. 404). This is true also of the carry of boulders. On these two grounds, it is clear that almost every important valley, bordered by ground higher than 1250 ft. above sea-level, was occupied by a glacier in Late-Glacial times ; whereas, the outlying plateaux lower than 1250 ft. show little or no trace of anything but ice-sheet glaciation.

Furthermore, hummocky morainic drift is only found where striæ and boulders indicate valley-glaciation. This rule is not peculiar to Mull, but is almost universal in the West Highlands, so that the two terms Valley-Glaciation and Moraine-Glaciation have long been used as synonyms. It is sufficiently obvious that the Valley-Glaciation Period must have been marked by some climatic factor that was wanting in the final stages of the Ice-Sheet Period. W.B.W., E.B.B.

The material of the hummocky moraines varies from a boulder-clay, without much striation of the boulders, to a rubble, in the production of which water has had a share. Such moraines are widespread in the district illustrated in Fig. 64, and are particularly impressive in the 'valley of the hundred hills' draining from Loch Fuaran north-westwards into Glen More.

A tendency for the hummocky moraines to assume a linear grouping is not infrequent; and, occasionally definite terminal and lateral moraines are recognizable :—

(1) The floor of Glen More (Sheet 44) above Craig Cottage is crossed by a series of fine terminal moraines concave towards the north-west.

(2) A beautiful boulder-moraine extends for 600 yds. between the 600 and 300 ft. contours on the east side of Abhuinn Coire na Feòla above Loch Spelve (Sheet 44). This moraine marks a retreat-position of the confluent glacier-systems of Loch Spelve and Loch Don. Other moraines belonging to the same system are discussed later in connection with the Late-Glacial Raised Beaches (p. 406). E.B.B.

(3) A well-defined terminal moraine crosses the junction of Sheets 43 and 44 at the mouth of Allt na h-Airidhe Brice, a tributary to Abhuinn Bail' a' Mhuilinn of Gleann Seilisdeir. Gleann Seilisdeir itself is partly occupied by drift, which has in places been moulded into oval hillocks parallel with the stream. These are probably drumlins; and, if so, must have been produced by a somewhat extensive glacier occupying Glean Seilisdeir prior to the formation of the Allt na h-Airidhe Brice moraine. The Gleann Seilisdeir glacier may have descended to sea-level, and was perhaps responsible for the absence of Late-Glacial Beaches at Kilfinichen. E.M.A.

The evidence of valley-glaciers given under (3) is interesting from its situation in regard to the high plateau of the Gribun Peninsula. Most of this plateau bears no trace of valley-glaciation, though its height is often 1500 ft. In the Croggan Peninsula (Sheet 44), an isolated glacier is indicated by moraines towards the head of Glen Libidil, where the neighbouring plateau scarcely reaches 1250 ft. There is little doubt that valley-glaciers throve increasingly in the eastern part of Mull. The confluent glaciers of Loch Spelve and Loch Don seem to be quite out of proportion to any which have left a record west of Ben More. E.B.B.

## Late-Glacial Raised Beaches.

Two of the marine beaches occurring in Mull are of demonstrably Late-Glacial date (p. 402): the older is about 100 ft. above high-water-mark; the younger about 75 ft. In addition, there is a much more prominent Post-Glacial raised beach standing about 25 ft. above high-water-mark. Fortunately, this Post-Glacial beach is often characterized by very pronounced rock-erosion, a feature not found in connexion with the other beaches of Mull. Occasionally, beaches intermediate in height between 25 and 75 ft. (mostly about 45 or 50 ft.) are met with in the island. Their date with reference to the glaciers, has not been determined on local evidence, but they will be considered for convenience in the present section.

Some of the beach-measurements cited below have been taken with reference to the upper limit reached by *Fucus caniculatus* on the modern beach, while others, reckoned from 'bench-marks' and 'spot-levels,' have reference to the Ordnance Survey datum. The former will be distinguished by the letters (h.w.m.), and the latter by (O.D.). The relation between the two sets of readings at the head of Loch na Keal was found to be 0 ft. (h.w.m.) = 6 ft. (O.D.).

Also, the high-water-mark of ordinary spring tides in the same locality lies approximately at $\frac{1}{2}$ or 1 ft. (h.w.m.) = $6\frac{1}{2}$ or 7 ft. (O.D.).

W. B. W.

*Loch na Keal* (Sheet 44).—Deltas belonging to high-raised beaches occur on the rivers of Derryguaig and Scarisdale. There is also a definite gravel-spit, perhaps at the 100-ft. level, above the road south of the last-mentioned river. J. E. R.

At the north-east angle of Loch na Keal, there is a well-marked gravel-terrace extending eastwards for about a mile towards Killichronan House. The following readings were taken for its coastal notch :—75, 80, 75, 70, 65 and 60 ft. (h.w.m.). A higher terrace at one point gave 95 and 100 ft. (h.w.m.) for its notch. Both terraces reappear north-east of Killichronan, the lower largely peat-covered. On either side of Killichronan, the 75-ft. terrace is locally cut in gravels of the 100-ft. terrace, and is therefore of later date. Along the main road, obvious beach-gravel shows through peat south-eastwards from Tòrr nan Clach. Its summit-level reaches 75-85 ft. (O.D.), and it appears to form a spit stretching across the valley from Torranlochain on the south-east. Reduced to high-water-mark, the level of the highest shingle of this beach is 78 ft. Behind the gravel spit, is bog at a lower level, 62 ft. O.D. (say 55 ft. above h.w.m.), which must have been a marsh at the time of the beach.

Nearer Salen, where the stream of Féith Bhàn crosses the road, a pit shows 5 ft. of well-stratified gravel with a top at 100 ft. (O.D.). There are thus two gravel-spits belonging to the group of Late-Glacial beaches in the pass between Salen and the head of Loch na Keal. These are to some extent obscured by peat, but the upper one was never of any great strength.

*Sound of Mull* (Sheet 44).—High raised beaches have not been recognized at the mouth of Glen Aros, and are absent at the entrance to Glen Forsa (p. 404).

As Fishnish Bay is approached, little road-metal pits afford exposures of beach-gravels among the rocks. The levels given on the map for the road where it passes over these gravels read, 90, 104, 92, and 76 ft. (O.D.). W. B. W.

High beaches are absent at the mouth of the Corrynachenchy stream. They reappear in the Fishnish Peninsula, and a gravel pit, near the bend of the road half-a-mile east of Balmeonach Cottage, is at about 100 ft. (O.D.). A little farther on, roadside gravel (on the line of the Bàn Eleanan Felsite) reaches 81 ft. (O.D.).

A beach-delta (? at 75 ft.) extends up Garmony Burn, but there is nothing corresponding in connection with Allt Achadh na Mòine. Between here and Altcrich, there are roadside patches of beach-gravel. At Scallastle River, high beaches are conspicuously absent.

In Java Peninsula, beach-gravels extend rather above 90 ft. (O.D.) ; and between Craignure and Duart Bay, roadside pits often show beach-gravels at about 100 ft. (O.D.).

Along the Morven coast, little is known of the Late-Glacial beaches. There is, however, a terrace with gravel about 40 ft. above the sea where the coastal road runs north-north-west near the edge of Sheet 44. There is also a conspicuous flat of sand and gravel at about the 100-ft. level at the head of Loch Aline (just outside Sheet 44).

G. W. L.

*Duart (Sheet* 44).—The finest development in Mull of the two main Late-Glacial beaches is met with in the Duart Peninsula, north-east of a line of sand and gravel mounds, which will be described in more detail presently as the Loch Don Sand-Moraine (p. 406). The higher Late-Glacial marine gravels of this neighbourhood are preserved in two outliers isolated by a strait floored with the lower gravels : one of these outliers occurs at Duart Point and measures 600 × 200 yds., and is shown on the six-inch map with a 'spot level' reading 99 ft. (O.D.), which means about 93 ft. (h.w.m.) ; the other is at Ardchoirk Farm and measures 1200 × 700 yds., and is of the same general level as the Duart Point outlier. The lower beach is more extensive and is much covered by peat. It can be traced for a couple of miles, and is bounded by a notch of erosion against the Loch Don Moraine (p. 406) and the outliers of the 100-ft. beach. There are sufficient 'levels' on the map to fix its margins approximately as at 80 ft. (O.D.). There can be no doubt that these two main beaches of Duart correspond with the 100-ft. and 75-ft. beaches of Loch na Keal. There is, in addition, a local delta-beach, near Camas Mòr, at a rather higher level than the well-defined Post-Glacial beach of the same locality.

*Firth of Lorne (Sheet* 44).—A feature of eastern Mull is the absence of Late-Glacial beaches in the hollow connecting Loch Don, Loch Spelve, Loch Uisg, and Loch Buie (p. 405).

The 75-ft. beach of Duart has been traced from the mouth of Loch Don to Port Donain, and its material, as seen in occasional exposures, consists of rather angular gravel.

Another patch is found bordering a stream at Gortenanrue, at the entrance of Loch Spelve.

In the Croggan Peninsula, the 75-ft. beach of Duart is well-developed at a stream south of Portfield ; and some high Late-Glacial beach is met with near the mouth of Glen Libidil. E.B.B.

At the north end of Loch Buie, high-lying beach-material is absent up Gleann a' Chaiginn Mhòir, though it forms the basis of the cultivated fields of Cameron Farm. Westwards, little is seen of high beaches, and their absence from Glen Byre is probably significant. G.V.W.

On the Mainland, Late-Glacial beaches have only been recognized near the head of Loch Feochan (p. 407). E.B.B.

*Loch Scridain (Sheet* 44).—High beach-gravels have been cultivated at Rossal Farm, at the head of Loch Scridain ; but extend no farther eastwards up Glen More (p. 407). G.V.W.

About half a mile south-west of Ardvergnish, there is a series of well-defined terraces occupying a small hollow and clearly belonging to the Late-Glacial beaches. They occur at levels of 45 ft., 70 ft., and 90 ft. (h.w.m.).

At the mouth of Allt na Coille Mòire, in the same neighbourhood, there are well-marked gravel beaches at 45 and 105 ft. (h.w.m.), the upper beach continuing for at least a couple of hundred yards west of the stream. Recent shingly gravel occurs at 4½ ft., and an older gravel spit of the modern shore reaches up to 7 ft. (h.w.m.). W.B.W.

High beaches are, however, absent about Kilfinichen Bay in connection with Abhuinn Bail a' Mhuilin draining Gleann Seilisdeir (p. 399). E.M.A.

On the south side of Loch Scridain, traces of high beaches are met with at the mouth of Glen Leidle. (B.L.).

*Ross of Mull (Sheets* 35 *and* 43).—A little rounded gravel with clayey layers occurs on a platform at 100 ft. (h.w.m.) forming Aoneadh Beag, on the south coast (Sheet 43). There is also a terrace-feature at about 60 ft. (O.D.) half a mile west of Tràigh nam Beach, on the north coast ; and, a little farther west, a gravelly deposit reaches to about the 100-ft. level at Ormsaig Farm. Raised beaches, however, are not well-marked in the Ross, except in the low tract which lies west of a curving line, starting at Eas Dubh, in the south, and passing by Loch Assapol to the northern shore at Bun an Leoib. In this tract, gravel-beaches are so striking a feature that details as to localities would be superfluous. They certainly include representatives of the 100- and 75-ft. Late-Glacial beaches of Duart and Loch na Keal, and also of another which lies at 40 or 50 ft., as well as the widespread 25-ft. Post-

Glacial beach. As will be emphasized later on, the 25-ft. Post-Glacial beach in this district is only occasionally marked by noteworthy rock-erosion.

There are numerous little pits in the high gravels. It is only necessary to mention a line of roadside pits near Salachran, north of Bunessan, and a couple within a mile of Bunessan along the main road towards Iona Ferry.

A series of measurements was taken on the south coast of Ardalanish Bay (Sheet 35). The 25-ft. Post-Glacial raised beach can here be identified with certainty, since it is backed, at one place, by a cliff in which there is a cave (Uamh Mhòr). Its inner angle lies at 22 ft. (h.w.m.). Above this, there are three fairly distinct beach-platforms. The lower two combine to give a sloping surface of gravel extending north to Ardachy House (Sheet 43) with a margin against rock, from 43 to 80 ft. (h.w.m.), according to locality. A spit, with a summit at 45 ft. (h.w.m.), bespeaks a temporary high-water-level about that height. This is corroborated by the occurrence of a coastal notch cut in gravel at 49 ft. (h.w.m.) on the west side of the bay. The gravel, in which this notch is cut, extends to 77 ft. (h.w.m.), but appears to correspond to a high-tide-level at approximately 63 ft. (h.w.m) Whatever its exact height, this gravel probably represents the 75-ft. Late-Glacial beach of Duart and Loch na Keal. Above it, in the Ardlanish-Ardachy district, there is a more or less peat-covered flat reaching 95 ft. (h.w.m.), which may be interpreted as belonging to the 100-ft. Late-Glacial beach of the localities just mentioned.

One of the first references to raised beaches in Mull is due to A. Stevenson,[1] who gives measurements from the inlet north-west of Camas Tuath, on the north coast. His readings are 25·3 ft. for the summit of an upstanding gravel-beach, and 40·5 ft. for a higher more doubtful terrace, both measured with reference to present high-water of spring tides. The 25·3-ft. measurement obviously belongs to the Post-Glacial beach ; the other to the approximately 49-ft. beach of the Ardlanish series.

*Iona.*—The Late-Glacial beaches are well-developed on the east side of Iona. At Dùn Cùl Bhuirg, on the west side of the island, the approximate 50-ft. beach is seen again with its inner angle at 60 ft. (h.w.m.), and its coarse storm-beach shingle thrown up to 65 ft., or even 70ft. The gravel on the modern shore reaches to 8 ft. (h.w.m.).

The hollow across the centre of the island has a number of terraces at 65 to 70 ft. (h.w.m.) ; and some less-defined higher ones at 95 ft.   W.B.W.

*Gribun Peninsular* (*Sheet* 43).—There is a delta-beach intermediate in level between 25 ft. and 100 ft. at the mouth of the stream at Tavool House, north of Loch Scridain. There is also a considerable spread of gravel, reaching 60 ft. or so above sea-level, at Balmeanoch Farm, four miles farther north. The farm-road north of Allt na Teangaidh runs along a spit of this beach.   E.B.B.

*Ulva Ferry and Loch Tuath* (*Sheet* 43).—There is a good development of beach-gravels up to 100 ft. (O.D.) on the mainland side of Ulva Ferry.   G.V.W.

At the mouth of the streams which enter the sea south of Beinn Duill, near Haunn, there is a fair amount of wave-worn shingle belonging to the Late-Glacial beaches. It reaches a height of 70 ft. (h.w.m.), whereas the shingle on the modern shore is thrown to 10 ft. This indicates an elevation of the beach amounting to 60 ft., or perhaps a little more.   W.B.W.

*Northern Mull* (*Sheets* 51 *and* 52).—The only other conspicuous terrace of high-level beach-gravels in northern Mull skirts the coast for a mile and a half southwards from Quinish Point.   G.V.W.

## Interplay between the Valley-Glaciers and Late-Glacial Sea.

The account of the Late-Glacial beaches of Mull, that has just been given, shows that they have a somewhat sporadic development. Except in the Ross of Mull, where, as in Iona, Coll, and Tiree, extensive rocky platforms have been exposed to the fury of the

[1] A. Stevenson, 'Notice of Elevated Beaches,' *Edin. New Phil. Journ.*, vol. xxix., 1840, p. 94.

Atlantic surf, these beaches seldom show any appreciable accumulation of gravel, unless in the vicinity of river-mouths. Moreover, they are, under no circumstances, backed by rocky cliffs peculiar to themselves. They are manifestly the records of comparatively brief periods of stability in the relations of land and sea ; and caution must be exercised in interpreting the vagaries of their distribution. It has long been realized that, in the highlands, the sea of the earlier raised beaches was often prevented by contemporary glaciers from extending into the more important glens of the country. Evidence of this sort is nowhere clearer than in Mull ; and a perusal of the descriptions given below will show that it is only to a limited extent of a negative character. Four principal Mull examples of valley-glaciers, continuing down to the sea-level of their day, are selected for discussion ; and their consideration is supplemented by a short account of glacial gravels in Glen Euchar, Lorne. All the instances dealt with are situated in Sheet 44. W.B.W., E.B.B.

*Loch Bà, Mull.*—Striæ on the shore and by the roadside, east of Rudha Àrd nan Eisirein, show that a glacier from Loch Bà once reached as low as present-day sea-level, travelling about W. 15 N., in marked contrast to the W. 25 S. flow of the main ice-sheet across the same rock-surfaces. For three-quarters of a mile south of the River Bà, a strip of ill-defined moundy moraines stretches from the Post-Glacial raised beach, near the burial ground, eastwards past Knock House to unite with the main morainic belt of the Loch Bà valley (Fig. 64, p. 393). These mounds carry felsite-blocks which have come from a mile farther up the valley. North of the river at Knock, there is an undulating flat of gravel, sometimes sinking into kettle-holes. Near the Post-Glacial raised beach, its height is about 35 ft. (O.D.), or about 6 ft. less above high-water mark ; but, at Loch Bà, it reaches to about 75 ft. (O.D.). A little east of Knock House, the River Bà crosses the gravel-spread, part of which extends south-eastwards as a peninsula with Benmore Lodge standing upon it. Its general height, near the Lodge, is about 70 ft. (O.D.), though it rises into mounds some four feet higher. A pit east of the Lodge shows 7 ft. of coarse sub-angular clean gravel with traces of steeply dipping stratification. Its included boulders are up to $1\frac{1}{2}$ ft. in diameter ; and blocks of felsite lying on neighbouring mounds measure as much as 4 ft. across. From Benmore Lodge northwards, the general level of the gravel sinks. Almost due north of the Lodge, close by the shore of the Loch, there is a remarkable bouldery ridge rising to 58 ft. (O.D.), some 10 ft. higher than the adjacent gravel-spread, and 17 ft. higher than the surface of Loch Bà.

It is obvious that the Loch Bà gravels are a fan discharged from a glacier which halted with its front standing where the Loch now finishes to the west. It is also clear, from its retention of its glacial form, that this fan must have been developed after the withdrawal of the sea responsible for the raised beaches already described at the mouth of the Scarisdale River, on the south-west, and at the head of Loch na Keal, on the north. Measurements of the latter show definite marine accumulations at 100 and 75 ft. above present high-water mark. In other words, the 100 and 75 ft.

## Chapter XXXVII.—Glacial and Recent.

raised beaches of Loch na Keal were already elevated when the Loch Bà Glacier stood facing the site of Benmore Lodge.

When one thinks of the exclusion of the 100-ft. sea from the Loch Bà valley, it is well to remember that the 100-ft. contour includes the valley-bottom for four miles south-east of Benmore Lodge.

*Glen Forsa, Mull.*—Similar evidence is furnished in Glen Forsa (Fig. 66): there are crossings of striæ belonging to the general

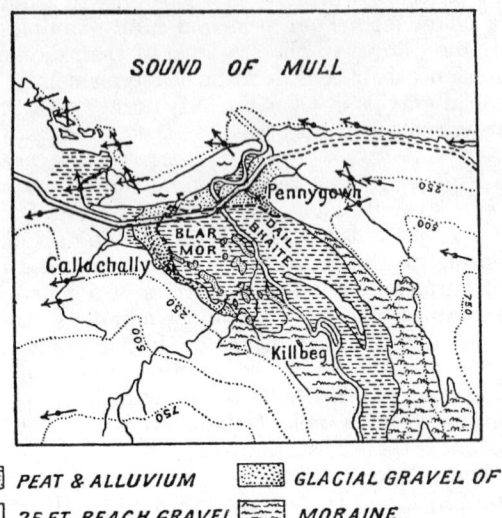

FIG. 66.

Sketch-map of the lower part of Glen Forsa, showing the eskers, gravel-fans and moraines of the local glaciation, as well as striæ of the general and local glaciations.

Quoted from *Summary of Progress for* 1909, p. 37.

and valley stages of glaciation; there are hummocky moraines restricted to the valley; and, near the mouth of the river, there is a gravel-fan reaching from a little over 45 ft. (h.w.m.) downwards with uniform slope till cut into a bank with its top at 34 ft. overlooking the Post-Glacial raised beach.

It is possible to lay down a line on Fig. 66, marking the position occupied by the glacier-front while this fan accumulated. There remains, in fact, a hollow cast of the glacier, now floored by the peat of Blàr Mòr and Dail Bhàite. Though this cast is not nearly so striking as that occupied by Loch Bà, a particularly beautiful feature is the occurrence within it of several little mounds and eskers rising to the same general level as the neighbouring fan. One of them, at Calachally Farm, terminates in the fan; and they all clearly mark the courses of subglacial streams connected with the formation of the frontal spread. The longest of these little eskers lies close to the river, and can be followed for quarter of a mile.

The hummocky moraines of Glen Forsa do not extend to the Post-Glacial raised beach, as at Loch Bà, but start at Killbeg Cottage, and continue thence for many miles through the valley-system of east-central Mull (Fig. 64); and it is important to realize that, for three miles east of the gravel-fan, the moraine-strewn bottom of Glen Forsa lies below the 100-ft. contour.

The gravel-fan, when considered in detail, is found to be composite, for its surface shows accumulation from two points of discharge. One of these coincides with the confluent-esker at Callachally Farm; the other is situated near Pennygown Farm. In both cases, a regular outward slope of the gravel bespeaks accumulation on dry land. The wonderfully clear definition of this composite fan, and also of its inner scarp, its tributary eskers, and the hummocky moraines behind, proves the further point that it can have suffered no submersion comparable in duration with those responsible for the 100-ft. and 75-ft. beach-gravels, found towards Loch na Keal, on the south-west, and Fishnish Bay, on the east. Accordingly, the Glen Forsa Glacier stood with its front near the sites of Callachally and Pennygown Farms, at some date later than the upheaval of the 100-ft. and 75-ft. beaches of the neighbourhood.

The halts of the Loch Bà and Glen Forsa glaciers, dealt with above, were probably synchronous. Presumably, both glaciers reached beyond the modern shore-line during the Late Glacial submersion-period. W.B.W.

*Loch Don and Loch Spelve, Mull.*—Valley-glaciers of eastern Mull were particularly strong in their development, and they combined in the low tract, which to-day is partially occupied by Loch Don and Loch Spelve, to yield a veritable piedmont glacier, the most considerable of its kind for many miles around.

The contrast between the direction of striæ referable to the piedmont and ice-sheet glaciations is pronounced in the Loch Don district: near the head of Loch a' Ghleannain, the piedmont glacier has left striæ running E. 17-15 N.; while, at Aird a' Chròtha on Loch Don, the ice-sheet striæ are directed rather south of S.E., and across the north-eastern corner of Loch Spelve they run almost due south.

Hummocky moraines, belonging to the piedmont glacier, are a very pronounced feature of the low ground skirting the road between Loch Don and Loch Spelve; and they continue recognizable as far east as Gualachaolish Farm, east of Loch Spelve.

Under these circumstances, the Loch Bà and Glen Forsa evidence would lead us to anticipate a widespread exclusion of the 100 ft and 75 ft. raised beaches from the low tract of Loch Don and Loch Spelve. As a matter of fact, this anticipation is realized under circumstances so well-defined that one can locally point to the actual position of the ice-front when the 75-ft. beach was raised.

For almost two miles, a row of sand and gravel mounds, the Loch Don Sand-Moraine, continues with little interruption from the roadside near Camas Mòr to Gorten Farm, at the mouth of Loch Don. The ridge is shown on the one-inch Map ornamented by a stipple. It rises occasionally to about 140 ft. (O.D.) ; and its irregularity of outline is in striking contrast to the platforms of the 100-ft. and 75-ft. beaches, spread out in front of it towards Duart Point. There are very few sections showing the inner construction of the ridge : at the north-east end, a road-side pit at 85 ft. (O.D.) exposes evenly-bedded gravel (probably beach-gravel), partly covered by boulder-clay ; along the road leading to Kilpatrick Farm from Lochdonhead, sand can be seen at one place passing steeply under boulder clay, and at another with bedding conforming in a general way to the shape of the ground and dipping under morainic rubble ; south of Kilpatrick Farm, big boulders occur on the surface of the sand and gravel, which, in a neighbouring pit, are seen to be flatly bedded ; and, at intervals along the Gorten sector of the ridge, the following notes have been taken :—

Irregular sands and clays and stony layers.
Steeply dipping sand passing under morainic rubble.
False-bedded gravel with morainic appearance.
Irregularly bedded silt and sand.

There can be no doubt that the Loch Don ridge is a moraine, heaped up during an advance of the piedmont glacier over the sand-strewn bottom of the Late-Glacial sea.

For a distance of a mile, in the neighbourhood of Kilpatrick Farm, the notch of the 75-ft. raised beach is cut in the outer face of the Loch Don Moraine. On the other hand, nowhere inside the moraine, is a vestige of this beach to be found. It is obvious then, that the piedmont glacier stood at the Loch Don Moraine when the elevation of the 75-ft. beach actually occurred.

Similar evidence is met with six miles farther south-west, at the other side of the piedmont glacier. Between Loch Spelve and Loch Uisg, there is a fan of glacial gravel, overlying disturbed shelly clay, save where the latter protrudes in well-formed morainic ridges. The disposition of these ridges indicates that they are the terminal moraines of a glacier which, for a time, occupied Loch Spelve ; and their material is obviously a marine clay derived from the bottom of the loch. The contained shells (mostly fragmental) belong to a glacial marine assemblage. A small collection was kindly determined by the late Mr. W. Evans :—

*Nuculana minuta* (Müll.).
*Nuculana tenuis* (Philippi).
*Astarte sulcata* (Da Costa).
   do. var. *elliptica* (Brown).
*Astarte compressa* (Mont.).
   do. ribbed=var *striata* ?
*Cyprina islandica* (Linné).
*Cardium sp.*
*Natica alderi?* (Forbes) or possibly *N. Pallida* (B. & S.).
*Balanus porcatus?* (Da Costa).

These shells and their matrix indicate that the Late-Glacial sea had access to Loch Spelve for a time. The readvance of the ice, responsible for the transport of the bottom marine clays and their rearrangement as moraines at Kinlochspelve, was very likely contemporaneous with the readvance that drove before it the sand-moraine of Loch Don.

The glacial gravel-fan, through which the Kinlochspelve moraines emerge, varies in height between about 100 ft. and 50 ft. (O.D.). A pit, by the main road north-west of the church, shows that, at their lower levels, the gravels were dropped with pronounced delta-bedding into standing water. This water evidently lay as a pool between the snouts of the Loch Spelve Glacier and a minor glacier occupying Loch Uisg. It cannot have been a branch of the sea connecting Loch Spelve through Loch Uisg with Loch Buie, for Loch Uisg has not been invaded by the sea since the ice melted; one can make this assertion, because no trace of a high-level delta faces any of the tributary gorges, which empty into the loch, although low-level deltas are prominently developed.   E.B.B.

*Loch Scridain and Glen More, Mull.*—For about two miles east of the high beach-gravels of Rossal Farm, the bottom of Glen More is a spacious flat lying below the 100-ft. contour, but nowhere is there any further trace of high beach-accumulations. The Coladoir River is so considerable a stream that it could not have failed to furnish a delta, had it issued into the Late Glacial sea responsible for the gravels at Rossal Farm. Under these circumstances, it is clear that the Coladoir Glacier must have extended to near Rossal Farm until after the withdrawal of the Late-Glacial sea.   (C.T.C.)

*Glen Euchar, Lorne.*—While the mountainous part of Mull, during the Late-Glacial period of submersion, nourished glaciers, that extended down all the more important glens right out into the sea, the remainder of the district, included in Sheet 44, presented a very different appearance. There was scarcely a glacier to be met with, and the sea bathed the coast-line uninterruptedly. The circumstances of the mainland are best illustrated in :—

(1) The development of 100-ft. beach deposits at the head of Loch Aline in Morven (p. 400).

(2) The high-level fluvio-glacial terraces met with along the River Euchar.

Where the River Euchar approaches the coast at Kilninver, there are two well-marked terraces of fluvio-glacial gravel, sand, and clay preserved above its low-lying alluvium and its deep rocky gorges. The highest of these terraces continues up Glen Euchar (markedly avoiding Glen Gallain) to Loch Scamadale (Sheet 45), altogether a distance of 5 miles. It ends about half way along the shores of Loch Scamadale, standing some 30 ft. above the loch, that is approximately 255 ft. above sea-level. Quite obviously, the terrace-gravels originated as an outwash-product of a tongue of ice, protruded, along Loch Scamadale, from a piedmont glacier occupying the low ground at the foot of Ben Cruachan. Near the coast at Kilninver, this terrace is still about 100 ft. above the sea ; and, accordingly, it seems necessary to suppose that one is dealing with a merging of fluvio-glacial gravel into 100-ft. beach deposits, a claim already advanced by Dr. Peach and Mr. Kynaston.[1] Half a mile north-east of Kilninver, a high terrace of unrounded gravel continues parallel with the coastal road for about a quarter of a mile, and undoubtedly represents one of the high glacial beaches of Mull.

Only one point of importance need be added to the brief statement published by Dr. Peach and Mr. Kynaston. At Kilninver, there are two high fluvio-glacial terraces. The lower one has not been developed for more than half a mile back from Kilninver. It probably corresponds with the 75-ft. Late-Glacial beach of Mull. The subsequent erosion of rocky gorges, and deposit of low-lying alluvium, were almost certainly initiated in Late-Glacial time, but the local evidence does not throw any particular light on this question.

E.B.B.

## Post-Glacial Raised Beach.

This raised beach is very much better marked through the greater part of Scotland than any of its companions. Its main distinguishing features, at first sight, are its comparatively low elevation, and the frequently pronounced wave-erosion which has accompanied its formation. Evidence, collected in other Scottish districts, shows that it was separated from its Late Glacial predecessors by a period of elevation allowing of extensive peat and forest growth on situations since depressed below sea-level ('submerged forest' of to-day); and that its shells possibly indicate a rather warmer climate than that which characterizes the modern sea. The chief points emphasized in the following account are :—

(1) The beach's striking development in the South-West Highlands, in which for this purpose we may include Oban and the south-eastern half of Mull ; and its comparative obscurity in the North-West Highlands, to which the other half of Mull, along with Coll, Tiree, Ardnamurchan, and Skye may be assigned.

(2) Its variations of elevation.

(3) Its archæological connections.

Before entering into detail, it is necessary to state that this particular beach has been differently elevated in different parts of Scotland. From this fact, there has arisen a certain amount of confusion in regard to nomenclature. Thus, in recent Geological Survey Memoirs on Sheets 19, 27, 32, 33, 35, and 53, the beach is spoken of as the 25-ft. raised beach, while in others dealing with Sheets 36 and 45 it is called the 50-ft. raised beach.

W.B.W., E.B.B.

### (1) CONTRASTED DEVELOPMENT OF THE BEACH IN THE SOUTH-WEST AND NORTH-WEST HIGHLANDS.

South-west of a line laid down in Fig. 65, the Post-Glacial beach, now under consideration, is generally represented by a fairly continuous rock-platform, often from 30-100 yds. in breadth. This

[1] B. N. Peach and H. Kynaston in 'The Geology of the Country near Oban and Dalmally,' *Mem. Geol. Surv.*, 1908, p. 165.

line, continued north-eastwards beyond the limit of Fig. 65, crosses the north-western shore of Loch Linnhe, of which the greater part, perhaps, belongs to the south-western province. South-eastwards, the boundary-line must pass outside Jura and Islay, and near to Colonsay; the exact position of Colonsay in this classification is a matter of opinion ; for, while it has spacious caves belonging to its Post-Glacial raised beach, it has scarcely any coastal cliff. This vagueness of definition is characteristic of the boundary zone all along its course. A gradual deterioration in the amount of rock-erosion is noticed on crossing the line from south-east to north-west. If this were merely local, it could be referred to local circumstances, for instance, to a variation in country-rock, or to a difference of exposure. It is only after experience has shown that the deterioration is definitely regional that one is driven to propound a regional explanation. It is suggested, on this basis, that the South-West Highlands enjoyed a more prolonged period of constant sea-level during the Post-Glacial submergence than the country farther north-west.

Very characteristic marine erosion-forms are to be encountered in the vicinity of Oban. Many of the roads, and the older part of the town, are situated on the coastal platform, cut in rock, with a high cliff behind it and an irregular covering of shingle on its surface. The platform is met with in Kerrera and all the other islands of the neighbourhood. Two of these are, indeed, named the Dutchman's and Shepherd's Hats, respectively, on account of their possessing central crowns surrounded by brims constituted of the Post-Glacial platform. (A quite different Dutchman's Cap has already been described, p. 389).

Let us contrast all this with what is found on the present-day shore. Glacial striæ are retained on some of the modern coastal rocks of Oban Bay. They are preserved from frost owing to their situation, but they have been exposed to wave-action ever since the upheaval of the Post-Glacial raised beach. This post-upheaval period of important wave-attack is long according to human standards. How much longer must have been the period of successful assault responsible for the Oban platform, which, as already stated, often ranges from 30 to 100 yds. in breadth !

Apart from the platform itself, and the conspicuous cliff behind, there are many other interesting features. For instance, a stack, reminiscent of the Old Man of Hoy, stands in front of the raised coastal cliff between Oban and Dunollie Castle ; and, in the other direction, beautiful undercutting is exhibited at the road-side, south-west of Dungallan House. The corresponding caves of the Oban district are chiefly remarkable for their archæological interest, which will be considered separately. At the present juncture, it is only necessary to refer to Uamh nan Columan, a cave above Rudha Tolmach, on Kerrera Sound, which is very commonly inhabited even to-day.

Caves are a common feature of the Post-Glacial raised cliff, wherever a fair amount of rock-erosion has been accomplished. The largest cave in Mull, Mackinnon's Cave, Gribun (Sheet 43), lies rather outside the limit assigned to the south-west district in Fig. 65 ; but it is presumably in large measure attributable to the

same sea as cut the Oban caves, for, though it is inaccessible at high tide, much of its bottom remains unsubmerged. Fingal's Cave, in Staffa, has its bottom submerged even at low water; and it would be pure speculation to assign any of its excavation to other than modern conditions.
E.B.B:

### (2) ELEVATION OF THE POST-GLACIAL BEACH.

The Post-Glacial raised beach is the only one of the series, for which even an approximate estimate of variation in altitude has been made. The deformation of this beach over its wide area of distribution in Scotland, Northern England, and Ireland is on the whole fairly well-known. It reaches its greatest height above the modern shore somewhere to the south-east of Loch Linnhe, and descends gradually on every side, reaching sea-level in a northward direction a little north of Caithness, probably in the Orkneys, and towards the south along the coast of the Wexford in Ireland. It decreases in altitude rather slowly towards the east and stands 10 or 12 ft. above high-water on the coast of Aberdeen, but on the north-west it drops more rapidly and is absent in Lewis and Harris. The average gradient is not more than an inch or two per mile, but in the last-mentioned direction it may be as much as four inches. It will be readily seen that, generally, within the limits of an Ordnance Sheet, very little change in altitude is observable. Moreover, determination of the change of level presents peculiar difficulties, and is liable to quite a number of errors. Ordnance altitudes, even in the rare instances where they are available, are of little use for the purpose, and the only reliable method is actual levelling. Even then, it is necessary to try and find localities where the conditions on the ancient and modern shores have been comparable, and, in such, to level from one shore to another, because individual gravel-deposits and rock-notches are found to have very varying relations to high-water-mark.

In the area dealt with in this Memoir, the beach exhibits quite an exceptional tilt, averaging as much as four inches to the mile from east to west. It has been elevated approximately 30 ft. at Oban, and only 20 ft. in the north and west of Mull and in Iona. Unfortunately, no measurements are available between Oban and Loch na Keal and Loch Scridain, in Mull.

*Oban.*—In Ganavan Bay, north of Oban (Sheet 44), a rock-notch and accompanying shingle-spit afford a very fair chance of estimating the elevation of the beach. The rock-notch is at 29 ft. (h.w.m.), and the shingle-spit, which is very large, reaches 40 ft. The recent shingle is only thrown to a height of 6 ft., but, being less abundant than that on the old shore, is hardly comparable. The raised rock-notch, on the other hand, may have been cut below high-water-mark, so that on the whole an elevation of from 30-32 ft. is indicated here.

South of Port a' Bhearnaig, at the north end of Kerrera (Sheet 44), there is a very well-marked rock-shelf backed by cliffs 20 to 40 ft. high and lying at from 25 to 27 ft. (h.w.m.), while a possibly comparable recent notch occurs at 1½ to 3 ft. below the same datum. An actual elevation of 27 to 30 ft. is thus indicated.

*Loch na Keal.*—At the head of Loch na Keal (Sheet 44), at the time of the 25-ft. beach, a massive shingle-spit was, as at the present day, built out from the southern shore confining behind it the River Bà, as far north as Drumlang Cottage. Later, the river broke through the spit, about a quarter of a mile north of the Burial

Ground ; but the modern shingle-spit has again deflected its outlet, so that it now enters the sea at the extreme northern corner of Loch na Keal, over a mile to the north of the temporary breach. Another break-through, however, is threatened at much the same point as before.

North-west of the Burial Ground, a shingle-beach, backed by a cliff of moraine, affords conditions comparable with those of the present shore, and thus gives an opportunity for an estimate of elevation. The top of the beach is at 27 to 28 ft. (h.w.m.), and the recent shingle is thrown to 5 to 6 ft., and occasionally goes a couple of feet higher. The amount of elevation thus indicated is 22 ft. ; it is certainly not more than 23 ft., or less than 20 ft.

*Loch Scridain.*—Near Dererach, at the head of Loch Scridain (Sheet **44**), the inner edge of the raised beach-gravel lies at 25 ft. (h.w.m.). The conditions appear to have been fairly comparable with those of the modern shore, the shingle on which is thrown up to 7 ft. Comparison therefore gives 18 ft. of elevation, but, as the raised beach-gravel is finer than that on the modern shore, this may be an under-estimate.

At Pennyghael, two miles farther south, several observations confirm the estimate of 18 ft. of elevation from this area, suggesting that the beach is a couple of feet lower here than on the western coast. These observations, however, hardly allow of deductions regarding a change of level of such small magnitude. Moreover, another measurement in the immediate neighbourhood gives a slightly higher reading, for, at Ardvergnish, the beach is preserved in a little stack and attains a level of 22 ft., while the gravel on the modern shore is thrown about 2½ ft. This would indicate an elevation of 19 or 20 ft.

*Ross and Iona.*—At Ardalanish, on the south coast of the Ross of Mull (Sheet **35**), a very fair estimate of the elevation of the beach was obtained, with sand and gravel at the inner angle occuring at 21 to 22 ft. (h.w.m.). The equivalent elevation might be 20 to 23 ft.

In Iona (Sheet **43**), the beach is fairly well developed as a terrace of deposition, and approximate estimates of the amount of elevation have been made. At Clachanach, the inner angle of the beach lies at about 20 ft. (h.w.m.), and south of the Free Church, and north of Dùn Cùl Bhuirg, similar readings were obtained. Near Calva, an opportunity of comparing shingle-accumulation on the raised beach and modern shore presented itself. The recent shingle reaches 5 ft. (h.w.m.), and that of the raised beach 26 or 27 ft. As the older shingle was more abundant than that of the modern shore, it was probably heaped to a greater height on the shore, so that a comparison would indicate an elevation of 18 to 22 ft. The shingle of the raised beach is ungrassed.

In Port an Fhir-bhréige, at the south end of the island, there is another rather remarkable deposit of ungrassed shingle which rises to 31 ft. (h.w.m.). The highest shingle-banks of the modern shore here reach 10-12 ft., and are comparable to those of the raised beach, thus indicating an elevation of 19 to 21 ft. Iona may therefore be regarded as lying approximately on the 20 ft. isobase of the Post-Glacial raised beach.

*Gribun.*—In the Gribun district (Sheet **43**), the beach is very irregularly developed, being largely controlled by the hard and soft beds of the Trias sandstones. The inner angle at the foot of the old cliff varies from 18-26 ft. (h.w.m.).

*Northern Mull.*—Several observations of level taken between Burg and Haunn, to the west of Kilninian (Sheet **43**), show that a rock-shelf, with local gravel, along this coast, lies at a fairly uniform level of 20 to 21 ft. (h.w.m.). Though no comparable notch is available on the modern shore, an elevation of from 19 to 22 ft. is indicated.

In Quinnish and Mornish (Sheet **51**) the beach is poorly developed, and determinations of elevation are difficult. At the head of Calgary Bay, the inner angle of the beach lies at a height of 20 ft. or so (h.w.m.). Dùn Bàn, on Mingary Àrd, is connected by a spit of gravel rising to 39 or 40 ft. (h.w.m.), and having an angle of erosion at its base at about 25 ft. It is not clear whether this spit is a product of the Post-Glacial shoreline. The amount of gravel on the modern shore is small. The beach may have an elevation of anything from 18 to 20 ft. in this district.

W.B.W.

### (3) ARCHÆOLOGY OF OBAN CAVES.

The platform of the Post-Glacial raised beach has furnished an excellent site for much of modern Oban, and excavations connected with the growth of the town have revealed five small but interesting caves, which had long lain sealed by fallen debris. Important accounts of these caves have been published between 1872 and 1898, notably by Sir William Turner and Dr. Anderson; and references to their papers will be found in the Bibliography, p. 423. Only the very briefest *resumé* of their results is here attempted.

The five caves are cited below in such fashion that they may be located by anyone visiting Oban; but their interesting features have, of course, disappeared, even where the caves themselves remain recognizable.

(1) Mackay Cave, at north end of Oban Bay, near Burn Bank House, where Nursery Road enters Strathaven Terrace.
(2) Gas-Works Cave.
(3) Distillery Cave.
(4) MacArthur Cave, near St. Columba's Church.
(5) Druimgarvie Cave, facing away from sea, near Railway Station.

All these caves were occupied by man unacquainted with the use of metals. The first four originated through direct sea-erosion, so that their occupation, in itself, denotes a withdrawal of the sea. In the case of the MacArthur Cave, there is suggestive, but debatable, evidence that this withdrawal had not proceeded very far before the cave was claimed by man. Accordingly, this particular cave is of special interest. The summary given below is based upon Dr. Anderson's account published in 1895.

The cave is 25 ft. long, and 16-20 ft. broad. Its bottom-deposits, from above downwards, were :—

Black vegetable mould, largely washed in from outside.
Shells with patches of ashes, wood-charcoal, and charred splinters of bones, the whole free from admixture with black mould or gravel, and extending over all the floor; thickness, 2 ft. 3 in. to 3 ft.
Gravel, composed entirely of small water-rolled stones; thickness, 6 in. to 1 ft. 6 in.
Partial layer of shells, thinning out towards sides and mouth of cave, and in several places presenting an irregular and patchy appearance in section, as if the shells had been deposited in heaps or pockets in the gravel; maximum thickness, 2 ft. 2 in.
Gravel, mixed with rock-fragments towards its bottom; thickness, 4 ft.

Human bones occurred in association with the black earth, along with quantities of small bones, presumably belonging to bats, rodents, and birds, and a few larger ones suggestive of cattle. There seems no reason for claiming a close connection between these remains and the shell-layers below.

The upper layer of shells was the refuse of cave-dwellers. The species represented were for the most part limpet, razor-shell, scallop, *tapes*, cockle, mussel, oyster, and periwinkle, and, occasionally, the larger and smaller whelk. Broken and splintered bones were interspersed among these shells; and it is obvious that bone-implements, many of which occur in the deposit, were actually made in the cave.

The gravel is described as a whole, in Dr. Anderson's account; but even its upper part is designated as clean-washed.

The lower layer of shells was in every way like the upper as regards its contained shells, bones, ashes, and implements; except that the shells and bones were more weathered.

The bones were found to belong to large red deer, roe-deer, ox, large pig, dog, badger, otter, and cat, and also fish. Crabs were represented by great claws alone, so that Dr. Peach suggests that the cave-men did not consider the rest of this creature fit for human food.

A few stone-implements were found, most of them fashioned from flint; bone and horn-implements were much more common; and included pins, awls, rubbers, and harpoons. Regarding the source of the flint, it is possible that flint, or silicified chalk, from Carsaig, Mull, may have been used; but it should be remembered that serviceable flint pebbles are characteristic of some of the beaches of the west, for instance those of Tiree and Iona.

In 1898, Dr. Anderson summed up his archæological researches thus :—

" It is evident that these three shell-mounds in Oronsay [Sheet 35] and the MacArthur and Druimgarvie Caves at Oban belong to the same archæological horizon,—a horizon which has not heretofore been observed in Scotland, but closely corresponding with the intermediate layers in the cavern of Mas d'Azil, on the left bank of the Arize in France, explored and described by M. Piette, and which he has reason to claim as filling up the hiatus that has been supposed to exist between the palæolithic and the neolithic." He continues by pointing out that the Azilian implements, in their home region, are " associated with existing fauna and with abundance of red-deer remains."

It is unnecessary, in this Memoir, to discuss whether a close resemblance of culture bespeaks actual contemporaneity in sites so distant as Mas d'Azil and Oban. We turn rather to what Dr. Anderson has styled the pertinent question whether the sea had wholly and finally left the cave when man first took possession. Dr. Anderson confessed himself very loth to admit an elevation of the Oban shore of 20 or 30 ft. since some stage of the Neolithic Period; and accordingly he put forward an alternative interpretation of the observed interstratification of human refuse and gravel. He suggested that the inhabitants, at first, occupied a saucer-shaped hollow in the cave-gravel, until this shingle eventually came to be trodden over the earlier part of the refuse. It is very difficult, however, to suppose that the refuse originally collected as a widespread, though incomplete, covering of the cave-floor, only at some later stage to be completely hidden by mere down-treading of the same gravel, as had long stood heaped about it in unstable equilibrium; and, accordingly, we do not accept Dr. Anderson's argument. In fact, he himself acknowledges that some of the associated investigators thought that the upper gravel must have been washed in over the deposit of food-refuse which was found beneath it; and, consequently, that the occupants of the cave must have been, for a time, driven out by the sea, soon afterwards to resume possession. Dr. Anderson points out that the buried refuse is not wave-sorted, and even retains its ashes ; so that those who disagree with him, and attribute the spreading of the upper gravel to sea-action, cannot lay claim to evidence of a period of resubmergence. It appears, rather, that the sea had withdrawn a little from the mouth of the cave, but was able, during an exceptional storm, to throw its shingle forward over the litter of refuse that had gathered on the floor.

Dr. Anderson's *à priori* objection to a Neolithic date for the final withdrawal of the Oban Post-Glacial sea to its present shore-line does not seem to carry much weight. There is, at any rate, abundant evidence that the shores of the extended Post-Glacial sea carried a Neolithic population in the Forth and Clyde districts of Scotland, and in the neighbourhood of Larne in the north of Ireland.
W.B.W., E.B.B.

## LANDSLIPS.

The steep coastal cliffs of Mull have been a source of many landslips, the more important of which are shown by ornament on the one-inch Map. A landslip of Pre-Glacial date near Gribun has already been discussed (p. 391) ; and a few words may here be added regarding a Late-Glacial landslip of the same district (Sheet 43). This particular landslip is about half a mile long parallel to the cliffs, and its irregular surface serves as the south-eastern boundary of the cultivated fields of Balmeanach Farm. Its Late-Glacial date is shown by the distance it lies in front of the great lava-escarpment ; for it is almost certain that the angle of this escarpment must have been choked by snow and ice to allow of so considerable a forward movement. This particular landslip is of the completely disintegrated type, and, accordingly, might be claimed with some propriety as a moraine.
E.B.B.

Other landslips of Mull and its neighbourhood, while perhaps of Post-Glacial date, are earlier than the Post-Glacial upheaval responsible for the raised beach of the previous section. A good example of this sort occurs at Inninbeg, east of Loch Aline (Sheet 44). It involves basalt-lavas and Liassic sediments, including the exposure of the low-tide reefs opposite the Keeper's Lodge. The notch of the Post-Glacial raised beach has been cut into it.
G.W.L.

Landslips, obviously of our own day, are very prominently exposed in the Wildnerness (Sheet 43).
E.B.B.

## CHAPTER XXXVIII.

### ECONOMICS.

#### Lignites of South-West Mull.

THE lignites of this district have already been dealt with in Chapter III., where further details were promised regarding the Beinn an Aoinidh, or Cadh' an Easa' seam (Sheet 44), which is the one that has received most attention in the past.

This seam outcrops on both sides of the valley drained by the small stream running past Airidh Mhic Cribhain. On the western side of the valley, it occurs at a distance of about 40 ft. below the base of a rather massive porphyritic lava, which fortunately can also be seen on the eastern, or Beinn an Aoinidh, side, thus witnessing to the identity of the coal-seam in its two separated outcrops. At Dearg Bhealach, or Cadh' an Easa', where the seam occurs at the top of the sea-cliff, one section reads as follows :—

|  | Inches. |
|---|---|
| Igneous rock . | — |
| COAL . | 2 to 6 |
| Carbonaceous shale . | 6 |
| BRIGHT COAL | 6 |
| Clay . | $\frac{1}{2}$ |
| BRIGHT COAL | 3 |
| HARD SPLINTY COAL (with brown streak) . | 6 |
| Carbonaceous shale . | 12 |
| Igneous rock . | — |

The coal is here very irregular, and the section alters from point to point. It was found about half a mile inland in the bed of the stream at Airidh Mhic Cribhain, where it was at least 18 in. thick.

On the Beinn an Aoinidh side of the valley, the coal is seen in the burn running to the north-west, which is shown in the one-inch Map, and named in the six-inch Map Allt a' Ghuaill (Burn of the Coal). It here immediately overlies an inclined sheet of dolerite, and is in a burnt condition, though at least 2 ft. in thickness. It is better developed in a parallel stream, nearer the cliff, which the writer has been told is the real Allt, or Eas, a' Ghuaill. Here, the seam is some distance above the intrusion, and measures at least 3 ft. of bright coal, with a clay-roof, and a hard layer of carbonaceous shale at base. Jameson mentions that an attempt to work the coal on Beinn an Aoinidh was made in the beginning of the eighteenth century,[1] and there have been various trials of more recent date. The coal has been used in the smithy at Pennyghæl, but the

---

[1] R. Jameson, 'Mineralogy of the Scottish Isles,' vol. i., 1800, p. 221; for account of Coal-Mining dating back to 1588, see J. M'Cormick, 'History of Mull,' *Celtic Monthly*, 1917, vol. xxv., p. 51, and published separately, 1923.

workings were all soon abandoned. Quite apart from the irregular nature of the seam, the difficulty of transport would be such as to make exploitation on any large scale, in all probability, impossible.

E.M.A.

These field-details may be supplemented by recourse to a valuable report prepared by Dr. W. Pollard in 1903, before the Geological Survey entered the district. It is drawn upon with the kind permission of the Duke of Argyll.

CHEMICAL REPORT ON SPECIMENS OF COAL FROM SOUTH-WEST MULL.

I. (No. 114). Ardtun, one-inch Map, Sheet 43 (Specimen from top of box).
II. (No. 115). Same as I. (Specimen from bottom of box).
III. (No. 116). Cadh' an Easa', South Shore, Sheet 44.
IV. (No. 117). Eas Dubh, Shiaba, Sheet 43.
V. (No. 118). Gowanbrae, Bunessan, Sheet 43.

PROXIMATE ANALYSIS.

|  | I. | II. | III. | IV. | V. |
|---|---|---|---|---|---|
| Moisture | 14·95 | 15·69 | 11·38 | 8·23 | 5·52 |
| Volatile matter | 25·09 | 24·12 | 33·35 | 30·40 | 42·70 |
| Fixed Carbonaceous Residue | 25·03 | 24·46 | 38·33 | 25·91 | 41·50 |
| Ash | 34·93 | 35·73 | 16·94 | 35·46 | 10·28 |
| Total | 100·00 | 100·00 | 100·00 | 100·00 | 100·00 |
| Total Sulphur | 2·92 | 1·55 | 1·05 | 1·28 | 1·66 |

| ULTIMATE ANALYSIS. | III. | V. |
|---|---|---|
| Carbon | 54·55 | 67·27 |
| Hydrogen | 3·73 | 4·98 |
| Oxygen | 12·03 | 10·32 |
| Nitrogen | 0·80 | 1·02 |
| Combustible Sulphur | 0·57 | 0·61 |
| Ash | 16·94 | 10·28 |
| Moisture | 11·38 | 5·52 |
| Total | 100·00 | 100·00 |
| Sulphur in Ash | 0·48 | 1·05 |
| Calculated Calorific Value | 5144 | 6702 |

COMPOSITION OF THE COMBUSTIBLE MATTER, i.e., OF COAL, EXCLUSIVE OF SULPHUR, ASH, AND MOISTURE.

| | | |
|---|---|---|
| Carbon | 76·71 | 80·46 |
| Hydrogen | 5·25 | 5·96 |
| Oxygen | 16·92 | 12·36 |
| Nitrogen | 1·12 | 1·22 |

The coals are of the Lignitous Class, in some respects similar to those found at Bovey, Devon. It should be stated that freshly mined specimens would probably contain more moisture than is shown above, as the samples received were packed in wooden boxes, which would allow of drying.

Dr. Pollard's conclusions regarding the analysed specimens are as follows :—

(1) The percentage of ash is high. In Nos. I., II., and IV., it amounts to 35 per cent., so that these are really Carbonaceous Shales or highly impure coals. No. V., is much better in this respect, as it contains only 10 per cent., a quantity not much greater than is found in some well-known Scottish coals.

(2) The coals are also sulphurous, though not excessively so.

(3) Their high percentage of volatile matter and low percentage of carbon give them a low heating power, and probably make them smoky in burning. They contain from 5 to 15 times as much ash as the lignite of Bovey. It is not likely there will be much demand for them as fuel, unless the best qualities can be carefully separated by picking ; even then, the demand could only be local.

## PETROLEUM.

Dr. Heddle quotes a record of petroleum impregnating a zeolite from Beinn an Aoinidh (Sheet 44). Locally, the Beinn an Aoinidh Coal has been invaded by intrusions ; and no doubt this has led to natural distillation of petroleum. There is, however, no probability whatever of finding petroleum in Mull on a commercial scale.
E.B.B.

## GRAPHITE.

The occurrence of graphite in Mull has occasionally received notice, and an attempt was made at one time to work some graphitic material to the west of Dererach, on the north shore of Loch Scridain (Sheet 44). All that was found must have been contained in large xenoliths, or blocks isolated in igneous rock, and the few other known occurrences are probably of the same nature (see Chapter XXIV.). It is very unlikely that there is anywhere enough of this material to pay exploitation.
E.M.A.

## IRON-ORE.

Iron-ore in commercial quantities has not been found in Mull, though certain minor occurrences of magnetite have from time to time attracted attention. These belong to the Corra-bheinn and Ben Buie Gabbros, respectively, of Chapter XXII.

In the Corra-bheinn Gabbro, on the hillside nearly half a mile west of Cruachan Dearg, there occur lenticular patches, which appear to contain an unusually large proportion of magnetite. None of the patches observed exceeds 12 ft. in length, and 3 ft. in thickness, but they have a habit of occurring near one another along the same strike.
(C.T.C.)

In the banded complex, which forms the margin of the Ben Buie Gabbro west of Loch Fuaran (p. 245), there are seams consisting chiefly of magnetite, but their individual thickness is only about 4 inches.
G.V.W.

## Chapter XXXVIII.—Economics.

### SAPPHIRE.

The occurrence of sapphires, easily recognizable in hand-specimens, is a feature of many of the xenoliths described in Chapter XXIV. Exceptionally, they measure half-an-inch across, but are thin and plate-like and full of inclusions, and it is very doubtful whether they have any commercial value except as curiosities. The most easily found localities are those on the north shore of Loch Scridain (Localities 9-13, p. 272) and at Rudh' a' Chromain (Locality 50).

E.M.A.

### ROAD-METAL.

There is no difficulty in finding potential road-metal for local purposes. The only difficulty arises in making use of it economically where traffic does not justify expensive quarrying. The most important road-metal quarry of the district is in a basalt, or andesite, lava of Old Red Sandstone age on the south shore of Loch Feochan. Most of the Mull roads are metalled from little pits in moraine, raised beach-gravel, well-jointed basalt-dykes, or rotted basalt-lava; and the result is not very satisfactory.

If south-west Mull were more conveniently situated, there can be little doubt that its columnar basalts would supply an export industry of road-metal. The material is almost certainly of good quality, and its jointing is such as to facilitate quarrying. The only two localities in south-west Mull, which might possibly afford opportunities for shipping material derived from columnar basalts, are the Ardtun Peninsula, north of Bunessan, and Tavool, on the north shore of Loch Scridain (Sheet 43). Better shipping accommodation is, of course, to be had at Loch Aline (Sheet 44) and Tobermory (Sheet 52); and the basalt of Calve Island, at the mouth of Tobermory Bay, is somewhat columnar.

### DIATOMITE.

In 1851, the Duke of Argyll drew attention to an occurrence of diatomite, or kieselguhr, near Loch Bà (Sheet 44); and, in 1853 and 1854, Dr. W. Gregory published descriptions of its contained diatoms (*see* Bibliography). Since then, the deposit has naturally attracted attention from the point of view of exploitation. Unfortunately, however, it is of very limited extent, and thin; so that it cannot be described as promising. Details obtained by Mr. Tait are as follows :—

The deposit is the filling of a small pool, now grown over with rushes and moss. It lies in a channel connecting Benmore Lodge and Knock; and, in time of flood, a stream issues from Loch Bà and passes through the pool. It lies just west of a fence bounding a plantation, 500 yds. east of Knock. Its maximum measurements are 100 yds. by 70 yds. In two pits dug in 1915, rather near the edge of the basin, diatomite was found from 10 inches to 1 ft. 3 in. in thickness. It rested on grey gritty clay, and was covered by about 2 ft. of peat. The diatomite or 'white clay,' is known to the people of the neighbourhood, who used to use it as white-wash.

## BRICK-CLAY.

Marine clays, similar in character to those used for brickmaking in the Forth and Clyde districts, are exposed as ridges from under gravel in the fields of Kinlochspelve Farm (Sheet 44). There is plenty of material in a position which would allow of easy working. Where seen, however, the clay contains shell-fragments, which interfere to some extent with its quality. E.B.B.

## SHELL-SAND.

Many of the Western Isles, including Iona, have extensive deposits of shell-sand, either on their beaches, or forming coastal dunes. The natural distribution of this material by wind has helped greatly in furnishing a fertile soil. In Mull, there are such accumulations in the western part of the Ross, for instance, east of Port Uisken (Sheet 35); but much the most important is to be found at Calgary Bay (Sheet 51). Here, an endless supply of sand, containing rather more than 70 per cent. $CaCO_3$ is available, and could be easily shipped for local purposes. G.V.W.

## OTHER ECONOMIC MINERALS.

The district has some other economic resources, such as :— Granite in the Ross; slate in Kerrera and Seil; Limestone in Lismore and Inchkenneth; Sandstone at Loch Feochan, Carsaig, and Bloody Bay; and Glass-Sand at Loch Aline; but these belong to the Pre-Tertiary rocks and are dealt with in a memoir entitled the Pre-Tertiary Geology of Mull, Loch Aline, and Oban. E.B.B.

# APPENDIX I.

## LIST OF GEOLOGICAL SURVEY PHOTOGRAPHS.

### By R. LUNN and W. MANSON.

Acid Vein with Xenoliths.—2716.
Basalt-Lavas.—2671, 2673, 2686-7, 2706, 2715, 2725, 2727-9, 2732-3, 2736-9, 2744-5, 2747-8, 2751-6, ? 2759; (Pillow-Lavas, by Mr. G. V. Wilson and Mrs. A. M. Bailey, Quarter-Plates, 1, 2, 3).
Beinn Bheag Gabbro.—2704.
Beinn Mheadhon Felsite.—2677-8.
Ben Buie Gabbro.—2707-8.
Carsaig Arches.—2734-5.
Channels due to Marine Erosion.—2729.
Chilled Margins.—2632, 2633, 2637, 2646, 2719-2721.
Columnar Basalt Lavas.—2732-3, 2736-8, 2751, 2753-6, ? 2759.
Columnar Basalt Intrusion.—2730.
Cone-Sheets.—2692-2700, 2702-5, 2707-8, 2717-2721, 2726.
Dykes.—2629, 2631-2633, 2637, 2651.
Faults.—2674, 2725.
Flint-Conglomerate.—2684-5.
Glen More Ring-Dyke.—2704.
Loch Uisg Gabbro.—2708-2711.
Loch Uisg Granophyre.—2709-2712, 2714.
Macculloch's Tree.—2736-7.
Pipe-Amygdales.—2745, 2752.
Pre-Glacial Marine Platform.—2746, 2757-8.
Pre-Glacial Marine Cave.—2749, 2750.
Pre-Tertiary Rocks of Lorne.—2613-2616, 2618-2628, 2634-2660, 2663-2665.
         ” of Mull and Morven.—2671-6, 2679-2683, 2725, 2739-2744.
Raised Beaches.—2610-2612, 2617, 2629-2631, 2661, 2662, 2666-2669, 2687, 2690, 2712-2714, 2723, 2724.
Sill.—2646.
Surface Agglomerates and Tuffs.—2731, 2734-5.
Tertiary Folding.—2679, 2687, 2689, 2691, 2706.
Trap-Features.—2748, 2751.
Vent-Agglomerates.—2698-2701, 2707-8.
Xenolithic Layer in Lava.—2722.

Copies of these half-plate photographs are deposited for reference in the London and Edinburgh Offices of the Geological Survey, and prints and lantern-slides are supplied at a fixed tariff.

# APPENDIX II.

## BIBLIOGPAPHY.[1]

1774. PENNANT, T.  A Tour in Scotland and Voyage to the Hebrides, MDCCLXXII. 4to, Chester. (And later editions). [Mull, Part I., pp. 406-409.]
1774. BANKS, J., *in* A Tour in Scotland and Voyage to the Hebrides, by T. PENNANT.  Part I., pp. 300-312.  [Account of Staffa, 5 plates.]
1780. TROIL, U. VON.  Letters on Iceland.  8vo, Lond.  [Staffa, pp. 266-293.]
1790. MILLS, A.  Some Account of the Strata and Volcanic Appearances in the North of Ireland and Western Islands of Scotland.  *Phil. Trans. Roy. Soc.*, vol. lxxx., p. 73.
1791. McNICOL, D.  United Parishes of Lismore and Appin.  *Statistical Account of Scotland*, vol. i., p. 494.  [Marl, Deer Horns.]
1795. CAMPBELL, D.  Parish of Kilfinichen and Kilviceuen.  *Statistical Account of Scotland*, vol. xiv., p. 184.  [Rocks and Stones—Mines.]
1797. SAINT-FOND, FAUJAS B.  Voyage en Angleterre, en Écosse et aux Îles Hébrides.  Paris.  (English Translation, 2 vols., 8vo, Lond., 1799; Edition with geological notes by Sir A. Geikie, 1907).
1800. JAMESON, R.  Mineralogy of the Scottish Isles; with Mineralogical Observations made in a Tour through different parts of the Mainland of Scotland, and Dissertations upon Peat and Kelp.  2 vols., 4to, with maps and plates.  Lond. and Edin.  [Mull, vol. i., pp. 202-235.]
1814. MACCULLOCH, J.  On Staffa.  *Trans. Geol. Soc.*, vol ii., p. 501.
1818. DANIELL, W.  Illustrations of the Island of Staffa, in a series of views, accompanied by topographical and geological descriptions.  Obl. fol., Lond.
1819. MACCULLOCH, J.  A Description of the Western Islands of Scotland including the Isle of Man.  3 vols., London.  [Vol. i.: pp. 552-580, Mull; vol. ii.: pp. 1-22, Staffa; pp. 59-79, General Comparison of the Trap Islands; vol. iii.: Pls. 6, 7, 20, 21; Maps, pp. 63, 73.]
1819. PLAYFAIR, J.  Geographical and Statistical Description of Scotland. vols. Edinburgh.  [Vol. ii., pp. 289-294, Mull, Staffa, Ulva, Inch Kenneth.]
1820. BOUÉ, A.  Essai Géologique sur l' Écosse.  8vo, Paris, no date.
1821. COMPTON, EARL (later MARQUIS OF NORTHAMPTON).  Description of the Rocks which occur along a portion of the South Coast of the Isle of Mull. *Trans. Geol. Soc.*, vol. v., pp. 369-374, Pls. 19-21.
1829. MURCHISON, [Sir] R. I.  Supplementary Remarks on the Strata of the Oolitic Series, and the Rocks associated with them, in the Counties of Sutherland and Ross, and in the Hebrides.  *Trans. Geol. Soc.*, 2nd ser., vol. ii., pp. 353-368; [Fig. 3, Pl. xxxv. is a horizontal section of the South Coast of Mull.]
1831. PANCKOUCHE, C. L. F.  L'Ile de Staffa, et sa grotte basaltique; dessinées et decrites par C. F. L. PANCKOUCHE.  Fol., Paris.
1840. MACCULLOCH, J. (posthumous).  A Geological Map of Scotland.  Published by order of the Lords of the Treasury by S. ARROWSMITH.  Scale, four miles to one inch.
1840. STEVENSON, A.  Notice of Elevated Sea-Beaches.  *Edin. New Phil. Journ.*, vol. xxix., p, 94.

[1] Compiled by D. Tait.

1844. NICOL, J.  Guide to the Geology of Scotland.  8vo, Edin.  [Mull, p. 229.]
1845. CAMPBELL, D.  Parish of Kilfinichen and Kilviceuen.  *New Statistical Account of Scotland*, vol. vii., pp. 303-304.  [Geology and Mineralogy; Mines.]
1845. M'LEOD, J.  Parish of Morvern.  *New Statistical Account of Scotland*, vol. vii., pp. 169-171.  [Geology, etc., Mines, Quarries.]
1850. ARGYLL, DUKE OF.  A Tertiary Fossiliferous Deposit underlying Trap in the Island of Mull.  [Notice of.]  *Edin. New Phil. Journ.*, vol. xlix., p. 350.
1851. ARGYLL, DUKE OF.  On a Fossiliferous Deposit underlying Basalt in the Island of Mull.  *Rep. Brit. Assoc.*, Edinburgh, 1850, (Sections) p. 70.
1851. ARGYLL, DUKE OF.  Notice of a Fossiliferous Deposit underlying Basalt in the Island of Mull.  [Abstract.]  *Proc. Roy. Soc. Edin.*, vol. iii., p. 21.
1851. ARGYLL, DUKE OF.  On Tertiary Leaf-Beds in the Isle of Mull [3 Figs.].  With a Note on the Vegetable Remains from Ardtun by E. FORBES [3 Pls.].  *Quart Journ. Geol. Soc.*, vol. vii., pp. 89-103.
1851. ARGYLL, DUKE OF.  (Letter to Dr. GREGORY about Diatomite from an old channel between Loch Bà and the sea).  *Proc. Roy. Soc. Edin.*, vol. iii., p. 58.
1851. ROSE, A.  Notice of the Recent Discovery of Plumbago or Graphite in the Island of Mull, Hebrides.  *Rep. Brit. Assoc.*, Edinburgh, 1850, (Sections) p. 102.
1852. SCOTT, A. J.  Analysis of Indian Ores of Manganese, and of some Scottish Zeolites.  *Edin. New Phil. Journ.*, vol. liii., p. 282.  [Scolecite from Mull.]
1853. GREGORY, W.  On the Species of Fossil Diatomaceæ found in the Infusiorial Earth of Mull.  *Proc. Roy. Soc. Edin.*, vol. iii., pp. 176-177, and *ibid.* pp. 204-306.  [Additional observations.]
1853. STRENG, A.  Beitrag zur Theorie der vulkanischen und plutonischen Gesteinbildung.  *Poggendorff's Annalen der Physik und Chemie*, band xc, p. 103.  [Staffa analysis p. 114.]
1854. GREGORY, W.  Additional Observations on the Diatomaceous Deposit of Mull.  *Quart. Journ. Micro. Sci.*, vol. ii., p. 24.
1856. BEDFORD, E. J.  Notice of some Raised Beaches in Argyllshire.  *Quart. Journ. Geol. Soc.*, vol. xii., p. 167.
1856. HARKNESS, R. and BLYTH, J.  On the Lignites of the Giant's Causeway and the Isle of Mull.  *Edin. New. Phil. Journ.*, vol. iv., New Ser., pp. 304-312; and *Rep. Brit. Assoc.*, Cheltenham, 1856, (Sections), p. 66.  [Plate with microstructure—Analyses of Lignites.]
1858. NICOL, J.  Geological Map of Scotland and Explanatory Note.  Scale:—One inch to ten miles.
1859. NICOL, J.  On the Slate Rocks and Trap Veins of Easdale and Oban.  *Quart. Journ. Geol. Soc.*, vol. xv., p. 110.
1861. GEIKIE, [Sir] A.  On the Chronology of the Trap-Rocks of Scotland.  *Trans. Roy. Soc. Edin.*, vol. xxii., pp. 633-653, Pl. xxxviii (Map).
1862. GEIKIE, [Sir] A.  On the Chronology of the Trap-Rocks of Scotland.  *Proc. Roy. Soc. Edin*, vol. iv., p. 309.  Additional Observations on the Chronology of the Trap-Rocks of Scotland.  *Proc. Roy. Soc. Edin.*, vol. iv., p. 453.
1862. MURCHISON, Sir R. I. and GEIKIE, [Sir] A.  First Sketch of a New Geological Map of Scotland, with Explanatory Notes.  Edin.  Scale:—$\frac{3}{4}$ inch to 20 miles.
1864. BOUÉ, A.  Ueber die saülenförmigen Gesteine, einige Porphyr-districte Schottlands, so wie über die vier Basaltgruppen des nördlichen Islands und der Hebriden.  *Sitz. K. Akad. Wiss. Wien.* (N.F.), band. xlix, pp. 439-452.
1865. GEIKIE, [Sir] A.  The Scenery of Scotland viewed in connection with its Physical Geology.  1st Ed. 1865; 2nd Ed. 1887; 3rd Ed. 1901.
1867. GEIKIE, [Sir] A.  On the Tertiary Volcanic Rocks of the British Islands.  *Proc. Roy. Soc. Edin.*, vol. vi., p. 71.  [Abstract in *Geol. Mag.*, Dec. i, vol. iv., p. 316.]
1868. GEIKIE, [Sir] A.  Address by the President (of Section C, Geology).  *Rep. Brit. Assoc.*, Dundee, 1867, (Sections) pp 49-54.

1868. ARGYLL, DUKE OF. On the Granite and other Rocks of Ben More, from a letter addressed to Prof. Phillips. *Rep. Brit. Assoc.*, Dundee, 1867, (Sections) p. 55. [Abstract in *Geol. Mag.*, vol. iv., p. 553.]

1868. SMYTH, T. On the Geology of the Coasts of Antrim and Londonderry, and on the age of the Giant's Causeway; being Observations made in the North of Ireland in the Autumns of 1865 and 1866. *Trans. Geol. Soc. Edin.*, vol. i., pp. 68-81.

1869. ARGYLL, DUKE OF. Iona. *Good Words* for 1869, pp. 535-543, 614-620, 708-714. Reprinted separately, Lond., 1870. Eighth Ed., Edin. 1894.

1870. TATE, R. and HOLDEN SINCLAIR, J. On the Iron-Ores associated with the Basalts of the North-east of Ireland. *Quart. Journ. Geol. Soc.*, vol. xxvi., p. 151. [Refers to Mull.]

1871. ZIRKEL, F. Geologische Skizzen von der Westküste Schottlands. *Zeitschr. d. deutschen geologischen Gesellschaft*, band xxxiii., pp. 1-124, Pl. i-iv.

1871. GEIKIE, [Sir] A. On the Tertiary Volcanic Rocks of the British Islands. First Paper. *Quart. Journ. Geol. Soc.*, vol. xxvii., p. 279. [Mull referred to incidentally.]

1872. SCROPE, G. P. Volcanoes. 8vo, Lond. 2nd Ed. [Staffa, p. 99.]

1872. TURNER, WILLIAM. On Human and Animal Bones and Flints from a Cave at Oban, Argyllshire. *Rep. Brit. Assoc.*, Edinburgh, 1871, (Sections) p. 160.

1873. JUDD, J. W. Note on the Discovery of Crataceous Rocks in the Islands of Mull and Inchkenneth. *Rep. Brit. Assoc.*, Brighton, 1872, (Sections) p. 115.

1874. JUDD, J. W. The Secondary Rocks of Scotland. Second Paper. On the Ancient Volcanoes of the Highlands and the Relations of their Products to the Mesozoic Strata. [With Map and Sections of Mull.] *Quart. Journ. Geol. Soc.*, vol. xxx., p. 220, Plates xxii-xxiii. [Abstract *Geol. Mag.*, Dec. ii, vol i., p. 68 and 139.]

1874. GEIKIE, J. The Great Ice Age and its Relation to the Antiquity of Man. 8vo, Lond.; 2nd Ed. 1877; 3rd Ed. 1894.

1875. GRIEVE, J. and MAHONY, J. A. Notice of Fossils from the Leaf-beds of Mull and Analysis of the Matrix. *Proc. Nat. Hist. Soc. Glasgow*, vol. ii., p. 22.

1876. GEIKIE, [Sir] A. Geological Map of Scotland, Edin. and Lond. Scale:— One inch to ten miles; Edition in 1910.

1878. JUDD, J. W. The Secondary Rocks of Scotland. Third Paper. The Strata of the Western Coast and Islands. *Quart. Journ. Geol. Soc.*, vol. xxxiv., pp. 660-743.

1879- GARDNER, J. S. and ETTINGSHAUSEN, C. VON. A Monograph of the British
1886.     Eocene Flora. 2 vols. 4to, Lond., Palæontographical Society. [Vol. i., Plate xiii., Figs. 5, 5a, 6, 6a.; Vol. ii., The Basaltic Formation of Ireland and Scotland, pp. 77-82; The Scottish Eocene Basalts, pp. 103-106. The Break between the Cretaceous and Eocene Rocks in the British Area, pp. 130-136. Plates x., xxv., xxvi., xxvii.]

1880. GEIKIE, [Sir] A. The Lava-Fields of North-Western Europe. *Nature*, vol. xxiii., pp. 3-5. [Reprinted in *Geological Sketches at Home and Abroad*, pp. 274-285; 1882.]

1880. MILNE HOME, D. Fifth Report of Boulder Committee, Staffa. *Proc. Roy. Soc. Edin.*, vol. x., p. 120.

1881. KOCH, W. E. Notes on Mull and its Leaf Beds. *Trans. Geol. Soc. Glasgow*, vol. vii., p. 52.

1881. KENDALL, J. W. The Iron Ores of Antrim. *Trans. N. of Eng. Inst. Min. Eng.*, vol. xxx., pp. 107-13. [Indirect reference.]

1882. MILNE HOME, D. Traces in Scotland of Ancient Water-lines. 8vo. Edin. [Coast of Lorne and Mull, pp. 19-21.]

1883. JUDD, J. W. and COLE, G. A. J. On the Basalt-glass (Tachylyte) of the Western Isles of Scotland. *Quart. Journ. Geol. Soc.*, vol. xxxix., pp. 444-465.

1884. DUNS, J. Notes on Boulders in Island of Mull. (To Boulder Committee.) *Proc. Roy. Soc. Edin.*, vol. xii., p 217.

1884. MILNE HOME, D. Tenth and Final Report of Boulder Committee, Mull. *Proc. Roy. Soc. Edin.*, vol. xii., p. 807.
1885. GARDNER, J. S. On the Evidence of Fossil Plants regarding the Age of the Tertiary Basalts of the North-East Atlantic. *Proc. Roy. Soc.*, vol. xxxviii., pp. 14-23.
1886. BELL, D. Notes on the Geology of Oban. *Trans. Geol. Soc. Glasgow*, vol. viii., p. 116.
1886. GARDNER, J. S. Second Report on the Evidence of Fossil Plants regarding the Age of the Tertiary Basalts of the North-East Atlantic. *Proc. Roy. Soc.*, vol. xxxix., pp. 412-415.
1886. JUDD, J. W. On the Gabbros, Dolerites, and Basalts, of Tertiary Age, in Scotland and Ireland. *Quart. Journ. Geol. Soc.*, vol. xlii., pp. 49-89, Plates iv-vii.
1887. WHITEHOUSE, C. The Caves of Staffa. *Scottish Geographical Mag.*, vol. iii., pp. 497-521
1887. GARDNER, J. S. On the Leaf Beds and Gravels of Ardtun, Carsaig, etc., in Mull. *Quart. Journ. Geol. Soc.*, vol. xliii., pp. 270-300. Figs. and Plates xiii-xvi. (Leaves). Notes by COLE, G. A. J.
1888. TEALL, [Sir] J. J. H. British Petrography, with special reference to the Igneous Rocks. Lond. [Mull, pp. 171, 186, 275, 319 ; Pls. x., xvi.].
1888. COLE, G. A. J.. On Some Additional Occurrences of Tachylyte in Mull *Quart. Journ. Geol. Soc.*, vol. xliv., p. 300.
1888. KENDAL, P. F. Preliminary Notes on some Occurrences of Tachylyte in Mull. *Geol. Mag.*, Dec. iii., vol. v., p. 555.
1888. GEIKIE, [Sir] A. The History of Volcanic Action during the Tertiary Period in the British Isles. *Trans. Roy. Soc. Edin.*, vol. xxxv., pp. 21-184 ; Pl. i. ; Figs. 1, 16, 18, 19, 28, 41, 43-50, 60.
1889. JUDD, J. W. On the Growth of Crystals in Igneous Rocks after their Consolidation. *Quart. Journ. Geol. Soc.*, vol. xlv., p. 178 ; Pl. viii.
1889. JUDD, J. W. The Tertiary Volcanoes of the Western Isles of Scotland. *Quart. Journ. Geol. Soc.*, vol. xlv., p. 187-218.
1889. GEIKIE, [Sir] A. Recent Researches into the Origin and Age of the Highlands of Scotland and the West of Ireland. Lecture delivered at the Royal Institution, June 7th, 1889. *Nature*, vol. xl., pp. 299 and 320.
1890. JUDD, J. W. On the Propylites of the Western Isles of Scotland, and their relation to the Andesites and Diorites of the District. *Quart. Journ. Geol. Soc.*, vol. xlvi., pp. 341-385.
1891. MCDOUGALL, J. The Leaf-Caves of Mull. *Trans. Geol. Soc. Glasgow*, vol. ix., p. 286.
1892. GEIKIE, Sir A History of Volcanic Action in the British Isles. From the end of the Silurian Period down to Older Tertiary time. Presidential Address, Geol. Soc. *Quart. Journ. Geol. Soc.*, vol. xlviii., Proc. pp. 60-179. [Tertiary, pp. 162-168.]
1894. ARGYLL, DUKE OF. Iona, 8vo, Edin. [This first appeared in *Good Words* for 1869. Reprinted separately, Lond. 1870, and later editions. Geology, pp. 126-142.]
1895. HEDDLE, M. F. On the Occurrence of Tachylyte at Loch Scredan, Mull. *Trans. Geol. Soc. Glasgow*, vol. x., p. 80.
1895. ANDERSON, J. Notice of a Cave recently discovered at Oban, containing Human Remains, and a refuse heap of Shells and Bones of Animals, and Stone and Bone Implements. *Proc. Soc. Antiq. Scot.*, vol. xxix., p. 211.
1895. TURNER, W. On Human and Animal Remains found in Caves at Oban, Argyllshire. *Proc. Soc. Antiq. Scot.*, vol. xxix., p. 410.
1896. METCALFE, A. T. The Tertiary Lava-Fields of the West Coast of Scotland. *44th Ann. Rep. Nottingham Nat. Soc.*, pp. 1-12.
1896. GEIKIE, Sir A. The Latest Volcanoes of the British Isles. *Trans. Geol. Soc. Glasg.*, vol. x., pp. 179-197.
1897. GEIKIE, Sir A. The Ancient Volcanoes of Great Britain. Vol. ii. Roy. 8vo, London.
1898. ANDERSON, J. Notes on the contents of a small Cave or Rock Shelter at Druimvargie, Oban. *Proc. Soc. Antiq. Scot.*, vol. xxxii., p. 298.

1899. CURRIE, J. On an Iona Erratic containing Withamite. *Trans. Geol. Soc. Edin.*, vol. vii., p. 115,
1899. CURRIE, J. The Minerals of the Tertiary Eruptive Rocks of Ben More, Mull. *Trans. Geol. Soc. Edin.*, vol. vii., pp. 223-229.
1901. HEDDLE, M. F. The Minerology of Scotland, posthumous, edited by J. G. GOODCHILD, Roy. 8vo, 2 vols., Edinburgh.
  [Vol. i.—Graphite, p. 2; ?Silver, p. 10; Quartz, p. 47; Chalcedony, p. 56; Calcite, p. 130;
  Vol. ii.—Pectolite, p. 31; Olivine, p. 52; Epidote, pp. 65, 207; Prehnite, p. 70; Gyrolite, pp. 78, 79; Apophyllite, p. 82; Xonotlite, p. 82; Tobermorite, p. 83; Heulandite, p. 84; Epistilbite, p. 86; Stilbite, p. 90; Laumontite, p. 92; Chabazite, p. 95; Analcite, p. 98; Natrolite p. 104; Scolecite, p. 107; Mesolite, p. 109; Faröelite, p. 112; ?Kirwanite, p. 145; Petroleum, p. 186; Saponite and Celadonite, p. 211.]
1903. Summary of Progress of the Geological Survey of Great Britain for 1902. [HARKER, A., p. 113.]
1904. HARKER, A., with notes by CLOUGH, C. T. The Tertiary Igneous Rocks of Skye. *Mem. Geol. Surv.* [Refers to Mull, pp. 4, 10, 26, 55, 106, 112, 153, 170, 327, 334, 335, 338, 362.]
1908. KILROE, J. R. On the Occurence and Origin of Laterite and Bauxite in the Vogelsberg. *Geol. Mag.*, Dec. v., vol. v., p. 534. [Indirect reference.]
1908. Summary of Progress of the Geological Survey of Great Britain for 1907, pp. 64-70. [Section across Loch Don Anticline, p. 65.]
1909. Summary of Progress of the Geological Survey of Great Britain for 1908, pp. 55-58.
1909. HARKER, A., *in* the Geology of the Seaboard of Mid Argyll. *Memoirs of the Geological Survey*. (Explanation of Sheet 36). [Mull, p. 79.]
1910. Summary of Progress of the Geological Survey of Great Britain for 1909, pp. 26-38. [Map of South-East Mull, p. 28; Map of Loch Bà, p. 32; Fossils, p. 57.]
1910. MURRAY, Sir JOHN, and PULLAR, LAURENCE. Bathymetrical Survey of the Fresh Water Lochs of Scotland. Vol. ii, p. 171-176; vol. v., Pls. 65-67.
1910 BOSWORTH, T. O. Wind Erosion on the Coast of Mull. *Geol. Mag.*, N.S., Dec. v., vol. vii., pp. 353-355, Pls. xxviii. and xxix.
1911. Summary of Progress of the Geological Survey of Great Britain for 1910, pp. 27-40. [Map of Loch Bà, p. 33; Sections parallel to Loch Bà, p. 34; Section, River Gaodhail, p. 36.]
1911. CUNNINGHAM-CRAIG, E. H.; CLOUGH, C. T.; and FLETT, J. S., *in* The Geology of Colonsay and Oronsay with part of the Ross of Mull. *Memoirs of Geological Survey*. (Explanation of Sheet 35, with part of Sheet 27.) [Mull Tertiaries, Chap. xv.]
1911. WRIGHT, W. B. Preglacial Shoreline in the Western Isles of Scotland. *Geol. Mag.*, p. 104.
1912. Summary of Progress of the Geological Survey of Great Britain for 1911, pp. 32-37.
1912. BAILEY, E. B. A Mull Problem; The Great Tertiary Breccia. *Geol. Mag.*, p. 517.
1913. Summary of Progress of the Geological Survey of Great Britain for 1912, pp. 36-49. [Sketch Maps:— (1) Area between Rossal and Loch Buie, p. 39; (2) Beinn a' Ghràig and Beinn Fhada, p. 42; (3) South-East Mull, p. 46; (4) Sapphire-localities, p. 48. Rock-Analyses, pp. 68-70.]
1913. BAILEY, E. B. A Mull Problem: The Great Tertiary Breccia, *Rep. Brit. Assoc.*, Dundee, 1912. (Sections), p. 459. [*See* 1912.]
1914. Summary of Progress of the Geological Survey of Great Britain for 1913, pp. 43-54. [Sketch Maps :— (1) Loch Spelve and Loch Buie, p. 44; (2) Loch Bà Ring-Dyke, p. 49.]
1914 HALLIMOND, A. F. Optically uniaxial Augite from Mull. *Min. Mag.*, vol. xvii., p. 97.
1914. WRIGHT, W. B. The Quaternary Ice Age, 8vo., Lond. [Mull, pp. 60, 369.]
1914. MACNAIR, P. Argyllshire and Buteshire. *Cambridge County Geographies.*

1915. M'LINTOCK, W. F. P.   On the Zeolites and Associated Minerals from the Tertiary Lavas around Ben More, Mull.   *Trans. Roy. Soc. Edin.*, vol. li. pp. 1-33 and 3 Plates.
1915. Summary of Progress of the Geological Survey of Great Britain for 1914, pp. 34-42.   [Sketch maps:— (1) Loch Bà Felsite Ring-Dyke, p. 36 ; (2) Distribution of Pillow- Lavas, p. 40.   Leaf Beds, p. 61.]
1916. Summary of Progress of the Geological Survey of Great Britain for 1915. [Rock-Analyses, pp. 23-27.]
1916- M'CORMICK, J.   History of Mull including the Legends and Traditions of the
1917.   Islands.   *The Celtic Monthly*, vol. xxiv., Chaps. i-x. ; vol. xxv., Chaps. xi- xix. ; vol. xxv., p. 51.   [Leaf Beds and Account of Coal Mining dating back to 1588.], pp. 92-93 [Granite and Granite-Quarries.]
1917. ANDERSON, E. M., and RADLEY, E. G.   The Pitchstones of Mull and their Genesis ; with Notes on the Rock-Species Leidleite and Inninmorite, by H. H. THOMAS and E. B. BAILEY.   *Quart. Geol. Soc.*, vol. lxxi., pp. 205-217.
1921. Summary of Progress of the Geological Survey of Great Britain for 1920, pp. 36-43.   [Section :— Bloody Bay, p. 37.]
1921. COCKERELL, T. D. A.   Some British Fossil Insects.   *The Canadian Entomologist*, vol. liii., No. 1, pp. 22-23.   [*Carabites scoticus*, n.sp. from Leaf-Bed.]
1922. THOMAS, H. H.   On Certain Xenolithic Tertiary Minor Intrusions in the Island of Mull (Argyllshire) ; with Chemical Analyses by E. G. RADLEY. *Quart. Journ. Geol. Soc.*, vol. lxxviii., pp. 219-260, and 5 Plates [microphotos].
1922. JEHU, T. J.   The Archæan and Torridonian Formations and the Later Intrusive Rocks of Iona.   *Trans. Roy. Soc. Edin.*, vol. liii., pp. 165-187.
1922. MACKINDER, H. J.   Britain and the British Seas, with maps and diagrams. 2nd Ed. Oxford.   1st Ed. 1906.   [Mull, pp. 22, 50, 73, 74, 76, 86, 292 ; Sound of Mull, pp. 73, 132.]
1923. M'CORMICK, J.   The History of Mull, including the Legends and Traditions of the Island.   Glasgow.   [This appeared in *The Celtic Monthly*, in parts (1916-17).]
1923. MACLEAN, J. P.   History of Mull, embracing Description, Climate, Geology, Etc.   Greenville, Ohio.
1924. BAILEY, E. B.   The Desert Shores of the Chalk Seas.   *Geol. Mag.*, p. 102.
1924. BOWEN, N. L., GREIG, J. W., and ZIES, E. G.   Mullite, a silicate of alumina. *Journ. Washington Acad. Sci.*, vol. xiv., pp. 183-191.

# INDEX.

ABERDEEN (Mainland), 410.
A' Bhog-àiridh, 56° 29' N., 5° 55' W., 158, 208.
Abhuinn an t-Sratha Bhàin, 56° 27' N., 5° 48' W., 159, 179.
—— Bail' a' Mhuilinn, 56° 24' N., 6° 5' W., 104, 272, 399, 401.
—— Barr Chailleach, 56° 25' N., 5° 42' W., 121.
—— Coire na Feòla, 56° 23' N., 5° 47' W., 169, 207, 399.
—— Lìrein, 56° 26' N., 5° 43' W., 121.
—— Loch Fhuaran, 56° 23' N., 5° 55' W., 384.
—— na h-Uamha, 56° 27' N., 6° 1' W., 362.
—— nan Tòrr, 56° 21' N., 6° 3' W., 269, 273, 279.
Abietineæ, 70, 74, 75, 78.
Achadh Forsa, 56° 33' N., 5° 46' W., 60.
A' Chioch, 56° 26' N., 6° 0' W., 127, 228, 289, 290, 362 ; A' Chioch Sheets, 289, 290, 362.
A' Chrannag, 56° 28' N., 6° 10' W., 390.
Acid *and* Intermediate Cone-Sheets 7, Chap. XIX. ; Analyses, 19, 20 ; *see* Early and Late Acid Cone-Sheets.
—— Lavas, 45, 196, 198.
—— Magma, Explosions, 5, 30, 135, 187, 188, 196-8, 209, 345.
—— Magma-Type, 20, 21 ; Analyses, 20.
—— Residuum, Chemical Action, 18, 19, 47, 218, 256, 324, 326, 328-330, 355.
Actinolite, 352, 354 ; *see* Hornblende.
Aegerine, 26, 190.
Aegerine-Augite, 21, 26, 139, 190, 192, 193, 318, 334, 348, 350.
Age-Relations, *see* Sequence.
Agglomerate *and* Explosions, 5, 6, 8, 10, 30 ; at Surface, 37, 45, 121, 168, 169, Chap. XV., 232 ; in Vents, 8, 36, 37, 38, 45, 51, 129, 158, 159, 167, 168, 179-181, 186-189, Chap. XVI., 211, 221, 236, 243, 247, 292, 294, 341, 345, 363-6.
Àird a' Chròtha, 56° 25' N., 5° 40' W., 405.
—— Ghlas, 56° 19' N., 5° 59' W., 57-59.
Aird Kilfinichen, 56° 23' N., 6° 3' W., 103.
—— na h-Iolaire, 56° 23' N., 6° 12' W., 56, 109.

Àiridh Fraoch, 56° 19' N., 6° 3' W., 273.
—— Mhic Cribhain, 56° 18' N., 6° 5' W., 64, 100, 273, 415.
Aix (France), 75.
Alaska, 82.
Albite *and* Albitization, 3, 50, 95, 104, 128, 141, 143, 147, 148, 150, 153, 161-164, 170, 208, 210, 214, 219, 225, 227, 228, 232, 233, 254, 256, 286, 296, 300, 301, 304, 316, 318, 325, 326, 328, 329, 346, 347, 349, 352, 353, 367, 374, 375.
Alkaline Intrusions, 5, Chaps. XIV., XXXV.
—— Magma-Series, 26-28; Analyses, 27.
*Alnus*, 80.
*Alnites ? Macquarrii*, 70, 78.
Allival (Rum) : Allivalite Analysis, 23.
Allivalite, 242, 248, 249 ; Allivalite-Eucrite Magma-Series (with Analyses), 21-23.
Allt Achadh na Mòine, 56° 29' N., 5° 47' W., 212, 245, 400.
—— a' Choire Bhàin, 56° 26' N., 5° 52' W., 340, 346.
—— a' Ghoirtein Uaine, 56° 22' N., 5° 49' W., 207.
—— Àiridh nan Caisteil, 56° 22' N., 6° 12' W., 111, 112.
—— a' Mhàim, 56° 22' N., 5° 58' W., 188.
—— a' Mhucaidh, 56° 24' N., 6° 4' W., 104, 272.
—— an Dubh-choire, 56° 28' N., 5° 45' W., 196, 198, 227 ; Craignurite Cone-Sheet Analyses, 19.
—— an Eas Dhuibh, 56° 30' N., 5° 52' W., 217, 339.
—— an Fhìr-eòin, 56° 18' N., 6° 10' W., 107.
—— Ardnacross, 56° 34' N., 6° 0' W., 119.
—— Chaomhain, 56° 20' N., 6° 7' W., 272.
—— Cnoc nam Piob, 56° 19' N., 6° 4' W., 101.
—— Coire Fraoich, 56° 28' N., 5° 50' W., 130.
—— Coire nan Gabhar, 56° 27' N., 6° 0' W., 367.
—— Dubh Dhoire Theàrnait, 56° 32' N., 5° 39' W., 293, 379.
—— Ghillecaluim, 56° 24' N., 5° 56' W., 235.
—— Leacach, 56° 32' N., 5° 44' W., 60, 293.

Allt Molach, 56° 25′ N., 5° 51′ W., Chap. XXIX., 376.
—— Mòr, 56° 34′ N., 6° 16′ W., 117.
—— Mòr, 56° 29′ N., 5° 52′ W., 339.
—— Mòr, 56° 29′ N., 6° 3′ W., 358.
—— Mòr Coire nan Eunachair, 56° 30′ N., 5° 50′ W., 129, 206.
—— Mòr na h-Uamha (Rum) : Eucrite Analysis, 23.
—— na Coille Mòire, 56° 24′ N., 6° 2′ W., 272, 289, 376, 401.
—— na Crìche, 56° 22′ N., 6° 10′ W., 279.
—— na h-Airidhe Brice (not named on Sheet 43), 56° 24′ N., 6° 6′ W., 399.
—— na h-Eiligeir, 56° 26′ N., 5° 55′ W., 217, 340.
—— nam Fiadh, 56° 24′ N., 6° 1′ W., 272.
—— nan Clàr, 56° 28′ N., 5° 52′ W., 130, 158, 163, 180, 216, 256, 298.
—— nan Leòthdean, 56° 30′ N., 5° 54′ W., 343.
—— na Samhnachain, 56° 32′ N., 5° 43′ W., 60, 293.
—— na Socaich, 56° 33′ N., 5° 43′ W., 60.
—— na Teangaidh, 56° 25′ N., 6° 8′ W., 56, 59, 108, 402.
Altcrich, 56° 29′ N., 5° 45′ W., 223, 228, 229, 400.
Am Binnein, 56° 24′ N., 6° 0′ W., 291.
—— Mìodar, 56° 32′ N., 5° 46′ W., 60.
Amygdales, 35, 36, 48, 50, 95, 97, 104, 120, 123, 124, 127, 128, 131, 132, 139-143, 152, 153, 154 ; Analyses, 34 ; see Vesicles.
Analcite, 14, 16, 30, 117, 138, 139, 141, 142, 192, 193, 368, 380, 381, 382, 425.
Analyses, 15, 17-20, 23, 24, 26, 34, 263, 416.
An Càrnais, 56° 30′ N., 5° 53′ W., 343.
An Coileim, 56° 20′ N., 5° 55′ W., 59, 178.
—— Coire, 56° 28′ N., 5° 59′ W., 345.
—— Coireadail, 56° 24′ N., 5° 54′ W., 314.
*Anchonemus*, 71.
An Cruachan, 56° 27′ N., 5° 58′ W., 217, 321, 341 ; *see* Cruachan Augite-Diorite.
ANDERSON, E. M., Part-Author, E. M. A. ; *see also* 6, 7, 12, 19, 52, 264, 269, 271, 287, 426.
—— J., 412-414, 424.
—— TEMPEST, 98.
Andesite, *see* Augite-Andesite.
An Dùnan, 56° 19′ N., 5° 57′ W., 64, 103 ; An Dùnan (Carsaig) Leaf-Bed, 2, 64, Chap. IV.
An Eiligeir, 56° 25′ N., 5° 43′ W., 160.
—— Garradh, 56° 19′ N., 5° 52′ W., 59.
—— Gearna, 56° 26′ N., 6° 2′ W., 128, 152-3 ; Epidote *and* Garnet Analyses, 34.

Angiospermæ, 68, 70, 78-86.
An Leth-'onn, 56° 23′ N., 5° 59′ W., 124.
Anorthite, 249, 251, 255, Chap. XXIV., 283.
Anorthoclase, 193, 375 ; *see* cryptoperthite.
An Sithean, 56° 21′ N., 5° 54′ W., 124.
—— Sleaghach, 56° 32′ N., 5° 38′ W., 395.
Antrim, 43, 44, 55 ; *see* Ireland.
Aoineadh Beag, 56° 18′ N., 6° 6′ W., 108, 401.
Aoineadh Thapuill, 56° 23′ N., 6° 12′ W., 181.
Apatite discussed, 28, 328, 329.
Apophyllite, 425.
Appin (Mainland), 42.
Applecross (Mainland), 359.
*Araucaria*, 89 ; *A. excelsa*, 77.
Araucarineæ (?), 70, 76, 77, 78.
*Araucarites Sternbergii*, 77.
Archæology of Oban Caves, 11, 394, 412-414.
Arcuate (Concentric) Folds, 4-6, 37, 52, 168, 173-179, 180, 181, 184, 222, 232.
Ardachoil, 56° 25′ N., 5° 43′ W., 37, 240, 241.
Ardachy, 56° 18′ N., 6° 15′ W., 402.
Ardalanish, 56° 17′ N., 6° 15′ W., 402, 411.
Ardchoirk, 56° 26′ N., 5° 39′ W., 121, 145, 401.
Ardmore Point, 56° 39′ N., 6° 8′ W., 292, 294.
Ardmucknish (Mainland), 271, 380.
Ardnacross, 56° 34′ N., 6° 0′ W., 26, 37, 39, 186, 187, 192, 193 ; Trachyte Plug Analysis, 27.
Ardnamurchan (Mainland), 45, 53, 60, 172, 292, 301, 391, 408.
Ardtornish House, 56° 31′ N., 5° 45′ W., 65.
Ardtun Peninsula, 56° 20′ N., 6° 14′ W., 2, 40, 41, 44, 47, 61-63, 65, Chap. IV., 101, 108, 109, 145, 183, 265, 386, 416, 418 ; Ardtun Leaf-Beds, 2, 38, 43, 44, 53-55, 61-63, Chap. IV., 97.
Ardura, 56° 24′ N., 5° 45′ W., 169, 240, 241.
Ardvergnish, 56° 24′ N., 5° 59′ W., 125, 126, 401, 411.
Arfvedsonite, 26, 190.
ARGYLL, *see* DUKE OF ARGYLL.
Arid Climate, 2, 97 ; *see* Desert.
Arinasliseig, 56° 25′ N., 5° 48′ W., 212, 214.
Arize (France), 413.
Arla, 56° 34′ N., 5° 59′ W., 119 ; Brunton Type Dyke Analysis, 17.
Aros House, 56° 37′ N., 6° 3′ W., 58.
Arran (Hebrides), 47.
Assimilation, 30, 38, 154, 157, 162, 167, 169, 170, 210, 212, 219, 255, 266, Chap. XXIV., 336, 347, 349, 350, Chap. XXXIII., 365.

*Astarte compressa*, 406 ; *A. sulcata*, 406 ; *A. sulcata* var. *elliptica*, 406.
Atanikerdluk (Greenland), 80, 82, 84, 85.
Auchnacraig, 56° 24′ N., 5° 40′ W., 58, 121.
Augite-Andesite (Leidleite and Inninmorite), Chaps. XXIII.-XXV. ; Analyses, 19, 263 ; Definitions, 280-283.
Augite-Andesite Pebbles at Ardtun, 63.
Augite-Diorite, 6, 18, 36, Chap. XVIII., 244, 316, 318, 321, 327-329, 349, 350, 355.
Augite-Phenocrysts, 16, 18, 25, 138, 148, 335, 346, 347, 349, 359, 369 ; *see* Enstatite-Augite.
Auto-Intrusion of Lavas, 39, 96, 104, 112, 113.
Auto-Pneumatolysis, 18, 50, 128, 141, 142, 300, 301, 304.
Australia, 87, 109.
Anvergne (France), 45, 101.
Axis of Symmetry, 6, 51, 216, 242, 298, 332, 339, 344.
Azilian, 11, 413.

Bac Beag, 56° 27′ N., 6° 29′ W., 389.
Bac Mòr (Dutchman's Cap), 56° 27′ N., 6° 28′ W., 140, 387, 389, 397, 409.
Bailemeonach, 56° 30′ N., 5° 48′ W., 206, 400.
Bailey, A. M., 420.
—— E. B., Part-Author, E. B. B. ; *see also* 3, 5, 7, 9, 10, 51, 52, 68, 95, 259, 264, 328, 341, 378, 381, 387, 388, 425, 426.
—— W. H., 68, 75, 89.
Baking, *see* Contact-Alteration.
*Balanus porcatus*, 406.
Balmeanach, 56° 25′ N., 6° 8′ W., 56, 57, 59, 108, 114, 391, 395, 402, 414.
Balnahard, 56° 26′ N., 6° 8′ W., 57.
Balure, 56° 23′ N., 5° 46′ W., 184.
Banded Gabbro, 243, 252.
—— Granulite, 38, 252.
Bàn Eileanan, 56° 30′ N., 5° 46′ W., 36, 58, 167, 173, 176, 400.
Banks, J., 40, 421.
Barachandroman, 56° 22′ N., 5° 47′ W., 37, 197, 198.
Barkevikite, 26, 191.
Barnacarry Bay, 56° 21′ N., 5° 33′ W., 379.
Barrow, G., 32.
Basalt ; Lavas, 2, 6, 30, 36-39, Chaps. II., V.-X. ; Intrusions, 4, 8, 10, 30, 36-39, Chaps. II., XI., XXIII.-XXVIII., XXXIV. ; Analyses, 15, 17, 24.
Beach River, 56° 20′ N., 6° 5′ W., 273, 384.
Bearraich, 56° 22′ N., 6° 11′ W., 107.
Bedford, E. J., 422.
Beerbachite, 252.

Beetles, 53, 61.
Beheaded Drainage, 384.
Beinn a' Bhainne, 56° 20′ N., 5° 50′ W., 123.
—— a' Ghràig, 56° 28′ N., 5° 59′ W., 128, 341, 342, 345 ; Beinn a' Ghràig Granophyre Ring-Dyke, 9, 10, 21, 36, 43, 46, 48, 50, 185, 223, 235, 290, 300, Chap. XXXII., 351, 366, 395 ; Analysis, 20 ; Caps to Ring-Dyke, 10, 36, 48, 342, 343, 345 ; *see* Contact-Alteration.
—— a' Mheadhoin, 56° 25′ N., 5° 55′ W., 158, 219.
—— an Aoinidh, 56° 18′ N., 6° 3′ W., 41, 64, 100, 415, 417.
——an Lochain, 56° 21′ N., 6° 1′ W., Inninmorite Sheet Analyses, 19, 263.
—— an Lochain, 56° 26′ N., 6° 7′ W., 288.
—— Bheag, 56° 27′ N., 5° 51′ W., 66, 159, 242, 296 ; Beinn Bheag Gabbro, 8, 38, 203, 223, Chap. XXII., 311, 313.
—— Bheag, 56° 22′ N., 5° 54′ W., 246.
—— Bheag, 56° 28′ N., 5° 59′ W., 349 ; Granophyre Ring-Dyke Analysis, 20.
—— Bheag, 56° 26′ N., 5° 46′ W., 160, 179.
—— Bhearnach, 56° 27′ N., 5° 48′ W., 132, 134, 148, 158, 179, 235, 236.
—— Bhearnach, 56° 26′ N., 5° 47′ W., 223, 236.
—— Bhùgan, 56° 20′ N., 6° 7′ W., 272.
—— Bhuidhe, 56° 34′ N., 6° 16′ W., 117.
—— Bhuidhe, 56° 29′ N., 5° 54′ W., 341.
—— Chàisgidle, 56° 26′ N., 5° 53′ W., 7, 9, 181, 208, 242, 299, 300, Chap. XXIX., 332, 333, 337, 339, 340, 341, 346, 368.
—— Chàrsaig, 56° 20′ N., 5° 58′ W., 239.
—— Chreagach, 56° 19′ N., 6° 7′ W., 273.
—— Chreagach, 56° 19′ N., 6° 1′ W., 140, 273.
—— Chreagach Bheag, 56° 29′ N., 5° 51′ W., 129, 200, 203, 206, 207, 221, 236, 238, 239.
—— Chreagach Mhòr, 56° 29′ N., 5° 51′ W., 96, 129, 130, 209, 216, 222, 234, 238, 239, 363.
—— Duill, 56° 32′ N., 6° 20′ W., 402.
—— Fhada, 56° 27′ N., 6° 0′ W., 9, 126-128, 152, 153, 290, 313, 321, 341, 342, 344, 345.
—— Fhada, 56° 24′ N., 5° 50′ W., 37, 132, 134, 179, 180, 235, 294, 394.
—— Iadain (Morven), 60.
—— Mheadhon, 56° 28′ N., 5° 49′ W., 202, 206, 211, 214 ; Felsite, 5, 36, 37, 195, 203, 206, 207, Chap. XVII., 222.
—— na Croise, 56° 21′ N., 5° 57′ W., 124 ; Scolecite Analysis, 35.

Beinn na Duatharach, 56° 27' N., 5° 53' W., 131, 158, 181, 208, 209, 335, 340.
Beinn na Duatharach Gabbro, 158, 161; Analysis, 24.
—— na h-Uamha. 56° 29' N., 5° 55' W. 318.
—— nam Feannag, 56° 22' N., 5° 57' W., 123.
—— nam Meann, 56° 29' N., 5° 48' W., 206, 214, 363.
—— nan Gabhar, 56° 27' N., 5° 59' W., 339-342, 345.
—— nan Gobhar, 56° 25' N., 5° 59' W., 65, 188.
—— nan Gobhar, 56° 21' N., 5° 55' W., 124, 125, 398.
—— nan Lus, 56° 30' N., 5° 55' W., 158, 161, 318, 340, 343.
—— na Sròine, 56° 20' N., 5° 46' W., 60, 122.
—— Talaidh, 56° 27' N., 5° 51' W., 66, 132, 158, 179, 217, 228, 238, 256, 296, 298, 299, 300, 311, 313, 315, 321, 340, 372, 373, 376, 394; see Talaidh Type.
BELL, D., 424.
—— W., 43.
Ben Buie, 56° 22' N., 5° 53' W., 180, 199, 200, 202, 203, 206, 207, 221, 222, 224, 241, 243, 246, 248, 252, 292, 294, 296, 369; Ben Buie Gabbro (Eucrite), 5, 8, 10, 21, 23, 30, 38, 51, 167, 168, 180, 181, 199, 200, 203, 222, 223, 232, 235, 236, 241, Chap. XXII., 317, 383, 417; Analysis, 23.
—— Cruchan (Mainland), 407.
Benderloch (Mainland), 42.
Bendoran, 56° 19' N., 6° 16' W., 397.
Ben Hogh, 56° 38' N., 6° 36' W., 396.
—— Hynish, 56° 27' N., 6° 56' W., 398.
—— Lee (Skye); Dolerite Sill Analysis, 15.
Ben More, 56° 25' N., 6° 1' W., 25, 27, 36, 38, 45, 49, 65, 92-94, 96, 97, 104, 125-128, 141-144, 152, 157, 161, 181, 187, 193, 195, 199, 202, 228, 259, 268, Chap. XXVI., 341, 344, 361-363, 383, 399; see Mugearite-Lavas.
Benmore Lodge, 56° 28' N., 5° 58' W., 345, 403, 404, 418; Granophyre Ring-Dyke Analysis, 20.
Ben Nevis (Mainland), 7, 51.
BERRY, E. W., 74, 87, 89.
Betula, 80, 81; B. alba, 81.
Betulaceæ, 70, 78-81.
Bibliography (general), 421; (palæobotanical), 89.
Bifurcation, see Upward Branching of Ring-Dykes.
Big-Felspar Basalt, 37, 93, 103, 124, 125, 126, 130, 146, 147, 160, 178, 196; Big-Felspar Dolerite and Gabbro, 37, 159, 164, 199, 206.
Binnein Ghorrie, 56° 18' N., 6° 4' W., 273.
Biotite; Primary, 21, 63, 187, 193, 251, 255, 368, 369; in Contact and Mixed

Rocks, 49, 129, 152-155, 160, 162, 170, 209, 214, 253, 317, 325, 336, 348, 352-354, 363.
Bìth Bheinn, 56° 27' N., 5° 55' W., 131, 158, 162, 331, 335.
BLAKE, G. S., 328.
Blàr Mòr, 56° 31' N., 5° 55' W., 404.
Bloody Bay, 5° 39' N., 6° 6' W., 94, 108, 116, 118, 419.
BLYTH, J., 422.
Boehmeria antiqua, 70, 89.
Bole, 43, 53, 97, 100, 103-105, 107, 115, 127, 129, 186.
Bones and Bone-Implements, 412, 413.
BONNEY, T. G., 264.
Bostonite, 26, 38, 65, 185, 187-9, 193, 194, 208, 279, 375, 380.
BOSWORTH, T. O., 61, 397, 425.
BOUÉ, A., 43, 421, 422.
Bovey (Devon), 416, 417.
Boulder-Clay, 387, 388, 390, 392, 406.
BOWEN, N. L., 28, 31, 32, 268, 426.
Bow River (Canada), 85.
Bràigh a' Choire Mhòir, 56° 31' N., 5° 58' W., 26, 37, 186-7, 191-2; Trachyte Plug Analysis, 27.
Breapadail, 56° 26' N., 5° 55' W., 340.
Breccia, see Agglomerate.
BREMNER, A., 384.
Brick-Clay, 419.
Brimishgan, 56° 27' N., 6° 5' W., 259.
BRÖGGER, W. C., 190,191.
Brunton Dyke (North of England), 372; Brunton Type, 285, 287, 359, 370, 371, 372; Analyses, 17; Definition, 285.
Buchite, 266, Chap. XXIV.; Analysis, 34.
BUCKMAN, S. S., 184.
Bun an Leoib, 56° 20' N., 6° 11' W., 401.
Bunessan, 56° 19' N., 6° 14' W., 36, 38, 44, 65, 67, 111, 183, 394, 395, 397, 402, 416, 418.
Burg, 56° 32' N., 6° 16' W., 411.
Burgh, 56° 22' N., 6° 10' W., 97, 111; see Macculloch's Tree.
BURNETT, G.A., Part-Author, G. A. B.; see also 397.

CADELL, H. M., 47, 384.
Cadh' an Easa', see Tràigh Cadh' an Easa'.
Caisteal Sloc nam Ban, 56° 24' N., 6° 10' W., 56, 108, 115.
Caithness (Mainland), 377, 378, 410.
Caldera, North-West, 131, 158, 173, 235, 238, 333, 339, 340; South-East, 3, 4, 12, 52, 66, 97, 98, 132-135, 149, 150, 165, 169, 173, 175, 179, 180, 234, 236; see Cauldron-Subsidence.
Calgary, 56° 35' N., 6° 16' W., 117, 140, 388, 411, 419.
Caliach Point, 56° 36' N., 6° 19' W., 292, 294, 386, 388.

Callachally, 56° 31′ N., 5° 55′ W., 343, 404, 405.
Calva, 56° 21′ N., 6° 24′ W., 411.
Calve Island, 56° 37′ N., 6° 2′ W., 118, 418.
Camas an Lagain, 56° 30′ N., 6° 9′ W., 115.
Camas an t-Seilisdeir, 56° 23′ N., 5° 45′ W., 167.
Camas Lèim an Taghain, 56° 33′ N., 5° 33′ W., 379.
Camas Mòr, 56° 27′ N., 5° 41′ W., 177, 184, 401, 406.
Camas Tuath, 56° 20′ N., 6° 17′ W., 402.
Cameron, 56° 21′ N., 5° 53′ W., 232, 401.
CAMPBELL, D., 412, 422.
Camptonite, 10, 16, 37, 38, 140, 271, 317, 361, 368, Chap. XXXV.
Canna (Hebrides), 42, 172; Doleritic Mugearite Sill Analysis, 27.
Cantal (France), 43.
Caolas na h' Àirde, 56° 32′ N., 5° 46′ W., 60.
Capull Corrach, 56° 20′ N., 6° 10′ W., 272.
*Carabites scoticus*, 71.
*Cardium*, sp., 406.
Càrn Bàn, 56° 24′ N., 5° 41′ W., 177, 178, 236, 295.
—— Mòr, 56° 34′ N., 6° 14′ W., 117.
Carolina (America), 74, 89.
*Carpinus*, 80; *C. Betulus*, 80, 88.
Carraig Mhic Thòmais, 56° 22′ N., 6° 12′ W., 108.
—— Mhòr, 56° 19′ N., 5° 57′ W., 102.
Carsaig, 56° 19′ N., 5° 59′ W., 2, 20, 27, 35, 36, 38, 44, 53, 57, 59, 63, 94, 95, 100, 101, 103, 108, 109, 140, 173, 175, 181, 185, 187, 189, 193, 267, 273, 287, 369, 394, 413, 419; Carsaig Leaf-Bed, see An Dùnan.
—— Arches, 56° 18′ N., 6° 3′ W., 54, 63, 102, 103, 108, 140.
*Castanea*, 86; *C. Ungeri*, 82.
Cauldron-Subsidence, 7, 9, 173, 180, 333, 340, 341; see Caldera.
*Caulinites fecundus*, 74.
Caves, Pre-Glacial, 387, 390, 391; Post-Glacial, 11, 409, 410, 412-414.
Ceann an Tùir, 56° 30′ N., 5° 53′ W., 343.
Celadonite, 49, 425.
Central Lava Group, 30, 38, 91, Chaps. V., VIII., IX., X., 179, 181.
—— Lava Types, 30, Chaps. X., XI., 241, 285, 290, 359, 369; Analyses, 17, 24; Definitions, 92, 136, 145-149.
—— Magma-Types, 16-18, 23-25; Analyses, 17, 24.
—— Subsidence, 46, 98, 168, 179, 339, 341, 342.
—— Volcano, 4, 45, 46.
Centrally Inclined Sheets, see Cone-Sheets.

Centres, 7, 9, 234, 296, 298, 299, 306, 307, 333, 337, 338.
*Cephalotaxus*, 62.
Cervicorn Augite, 18, 219, 290, 303, 305, 327, 328, 371, 373; Definition, 303.
Chabazite, 425.
Chalcedony, 425.
CHAMBERLIN, T. C., 397.
Chalk, 2, 39, 43, 44, 54-59, 62.
China, 86.
Chilled Margins, of Intrusions, 8, 24, 32, 47, 96, 186, 192, 193, 214, 221, 223, 224, 234, 239, 246, 247, 260, 266, 267, 269, 288; of Lavas, 96, 98, 101, 104, 112, 125, 132, 134, 150.
Chlorite, produced, 49, 50, 127, 128, 139-141, 143, 144, 160, 162, 163, 171, 214, 215, 220, 228, 233, 239, 254, 304, 305, 316, 318, 319, 325, 349, 352, 367, 372; destroyed, 152, 154, 155, 162, 214, 336, 353, 354.
CHRIST, H., 73, 89.
Chromite, 250.
*Citrus*, 88.
Clachaig, 56° 27′ N., 5° 57′ W., 158, 162, 208, 335, 368, 375; see Glen Clachaig.
Clachanach, 56° 20′ N., 6° 23′ W., 411.
Clachan Bridge, 56° 19′ N., 5° 35′ W., 37, 365.
Clachandhu, 56° 26′ N., 6° 8′ W., 55.
CLARK, F. W. (Mrs), 387.
Climate, 2, 53, 54, 89.
Clino-Enstatite, see Enstatite-Augite.
CLOUGH, C. T., Part-Author, posthumous, (C.T.C.); see also, 2, 7, 9, 10, 47, 51, 52, 150, 188, 241, 255, 266, 270, 271, 325, 328, 341, 377, 425.
Clyde (Mainland), 359, 379, 380, 414, 419.
Cnoc a' Bhràgad, 56° 20′ N., 6° 0′ W., 287.
—— a' Chrònain, 56° 20′ N., 5° 51′ W., 231.
Cnocan Buidhe, 56° 18′ N., 6° 2′ W., 273.
Cnoc Damh, 56° 30′ N., 5° 49′ W., 176.
—— Druidean, 56° 19′ N., 6° 25′ W., 397.
—— na Dì-chuimhne, 56° 29′ N., 6° 5′ W., 369.
—— na Faolinn, 56° 23′ N., 5° 46′ W., 165-167, 170.
—— Reamhar, 56° 20′ N., 6° 5′ W., 273.
Coal, see Lignite.
COCKERELL, T. D. A., 67, 71, 89, 426.
Coill' a' Bhealaich Mhòir, 56° 20′ N., 5° 48′ W., 230.
—— a' Chaiginn, 56° 21′ N., 5° 54′ W., 125.
—— an Aodainn, 56° 26′ N., 5° 54′ W., 203, 332, 346.
Coillenangabhar, 56° 20′ N., 6° 9′ W., 272.

Coille na Sròine, 56° 28' N., 5° 57' W., 340-342, Chap. XXXIII.
Coir' a' Charrain, 56° 25' N., 6° 5' W., 272.
—— a' Mhàim, 56° 25' N., 5° 55' W., 299, 313, 314.
—— a' Mhadaidh (Skye); Olivine-Gabbro Analysis, 24.
—— an t-Sailean, 56° 25' N., 5° 57' W.: Gabbro-Granophyre Ring-Dyke, detached Part of Glen More Ring-Dyke, 51, 243, 299, 313, 314, 321, 327, 328 ; Analyses, 29.
Coirc Bheinn, 56° 25' N., 6° 4' W., 103, 104, 128, 287.
Coire an Ùruisge, 56° 30' N., 5° 54' W., 343.
—— Bàn, 56° 33' N., 6° 8' W., 358.
—— Bearnach, 56° 26' N., 5° 46' W., 239.
—— Buidhe, 56° 20' N., 5° 59' W., 287; Felsite Analysis, 20; Small-Felspar Dolerite Analysis, 24.
—— Ghaibhre, 56° 27' N., 5° 50' W., 223, 243, 299, 311, 313, 316, 321.
—— Gorm, 56° 23' N., 5° 49' W., 37, 134.
——Mòr, 56° 27' N., 5° 46' W., 37, 169, 173, 178, 195, 196, 198 ; Coire Mòr Syncline, 37, 94, 98, 121, 159, 160, 168, 173, 178, 195-198, 203, 206.
—— Mòr, 56° 25' N., 5° 54' W., 158, 217.
—— na Caise, 56° 21' N., 5° 43' W., 122.
—— na Feòla, 56° 23' N., 5° 48' W., 223, 241, 247.
—— na Lice Duibhe, 56° 25' N., 5° 55' W., 158.
—— nam Fiadh, 56° 35' N., 6° 2' W., 358.
—— nam Muc, 56° 30' N., 5° 52' W., 343.
—— nan Dearc, 56° 27' N., 5° 44' W., 198.
—— nan Each, 56° 26' N., 5° 48' W., 212.
—— Slabhaig, 56° 32' N., 5° 42' W., 293.
Coir' Odhar, 56° 25' N., 6° 0' W., 127, 134, 295.
Coladoir River, 56° 24' N., 5° 56' W., 124, 407.
COLE, G. A. J., 38, 46, 47, 62, 63, 149, 265, 423, 424.
Coll (Hebrides), 380, 396, 398, 402, 408.
COLOMB, —, 84.
Colonsay (Hebrides), 386, 387, 409.
Colorado (America), 74, 84.
Columnar Lavas, 3, 16, 38, 44, 56, 61, 62, 65, 93, 94, 101, 102, 108-113, 115, 118, 145, 146, 418.
Compact Central Type, 149 ; Analyses, 17.
Composite Intrusions, 6, 8, 17, 32, 37, 212, 221, 223, 224, 266-268, 279, 286, 287, 291, 356, 375, 376 ; Definition, 8.
COMPTON (EARL), 52, 140, 270, 421.
Concentric Folds, see Arcuate Folds.
Cone-Sheets, 6, 7, 11, 30, 50, 51, Chaps. XIX., XXI., XXVIII.; Analyses, 17, 19, 20.
Coniferales, 43, 69, 70, 74-78, 89, 97, 111, 113.
Contact-Alteration, by Beinn a' Ghràig Granophyre, 35, 36, 50, 152-153, 161, 162, 353 ; Ben Buie Gabbro, 170, 234, 246 ; Corra-bheinn Gabbro, 155, 209, 234, 363 ; Glen Cannel Granophyre, 162, 163, 209, 331, 335, 336 ; Knock Granophyre, 36, 129, 152 ; Loch Uisg Granophyre, 8, 37, 123, 153-155, 198, 231 ; 'S Airde Beinn Dolerite, 157, 160-161.
—— of Basalt-Lavas, 8, 35, 36, 49, 50, 97, 128, 151-155, 336 ; dykes, 366.
Cordierite, 52, 198, 229, 266, Chap. XXIV.
Corra-bheinn, 56° 25' N., 5° 56' W., 173, 175, 246, 296, 298, 299, 313, 321, 363 ; Corra-bheinn Gabbro, 8, 38, 51, 187, 203, 207, 209, 210, 235, 241, Chap. XXII., 298, 299, 363, 417.
Còrrachadadh, 56° 29' N., 5° 53' W., 239, 339.
Corrom (Delta-Watershed), 384.
Corrynachenchy, 56° 30' N., 5° 50' W., 227, 229, 400.
Corundum, see Sapphire.
Corylites, 89 ; C. hebridica, 70, 78-80 ; C. Macquarrii, 79.
Corylus, 79-80 ; C. americana fossilis, 81 ; C. Avellana, 80 ; C. Macquarrii, 79, 80, 81 ; C. rostrata, 80 ; C. rostrata fossilis, 81.
Cowal (Mainland), 47.
Crackaig, 56° 32' N., 6° 18' W., 387, 388.
Craig, 56° 24' N., 5° 55' W., 38, 134, 243, 245, 246, 321-324, 399 ; Craig Porphyrite, 245, 248, 253.
CRAIG, E. H. C., see CUNNINGHAM CRAIG, E. H.
Craignure, 56° 28' N., 5° 42' W., 36, 37, 58-60, 106, 120, 173, 176, 222, 223, 228, 396, 397, 400 ; Craignure Anticline, 36, 37, 167, 173, 176, 177 ; Granophyre Cone-Sheet Analysis, 20.
Craignurite, Chap. XIX., 281, 282, 284, 291, 300, 304, 318, 328, 329, 348, 373, 374 ; Analyses, 19, 29 ; Definition, 227.
Crater-Lake, 3, 48, 51, 66, 98, 99, 180.
Creachan Mòr, 56° 18' N., 6° 3' W., 101.
Creach Beinn, 56° 23' N., 5° 49' W., 173, 223, 234, 235, 236, 238, 243, 246, 247.
—— Bheinn Bheag, 56° 22' N., 5° 50' W., 247.
—— Bheinn Lodge, 56° 22' N., 5° 48' W., 207.

Creag an Fheidh, 56° 21′ N., 6° 4′ W., 100.
Creagan Mòr, 56° 24′ N., 5° 42′ W., 236.
Creag Dhubh, 56° 27′ N., 5° 57′ W., 331.
—— na Còmhla, 56° 22′ N., 5° 55′ W., 243, 245.
—— na h'Iolaire, 56° 24′ N., 5° 52′ W., 132, 199, 292, 294; Creag na h' Iolaire Felsite, 5, 211, 212, 215, 294 : Creag na h'Iolaire Sheets, 211, 292, 294; Quartz-Gabbro Analysis, 29.
—— nam Fitheach, 56° 24′ N., 6° 9′ W., 107.
Cretaceous, 54-56, 58, 60, 74, 85, 87.
Crinanite, 359, 368; Analysis, 15; Definition, 16.
Croggan, 56° 23′ N., 5° 43′ W., 36, 37, 59, 60, 94, 122, 173, 177, 394, 399, 401.
Cruachan, 56° 29′ N., 6° 25′ W., 389.
—— Augite-Diorite (An Cruachan, 56° 27′ N., 5° 58′ W.), Chap. XVIII.
—— Beag, 56° 26′ N., 5° 55′ W., 181.
—— Dearg, 56° 26′ N., 5° 56′ W., 203, 246, 255, 296, 298, 300, 301, 304, 340, 417; Cruachan Dearg Type, 241, 285, Chap. XXVIII., 374; Definiton, 304; Silica-Percentages, 18; Talaidh Type Cone-Sheet Analysis, 17.
—— Granite (Mainland), 395.
—— Mìn, 56° 19′ M., 6° 8′ W., 273.
Cruach Choireadail, 56° 24′ N., 5° 54′ W., 131, 132, 134, 148, 150, 164, 199, 299, 303, 314, 315, Chap. XXX.; Porphyritic Central Type Lava and Differentation-Column Analyses, 24, 29.
—— Doire nan Guilean, 56° 23′ N., 5° 56′ W., 148; Porphyritic Central Type Lava Analysis, 24.
—— Inagairt, 56° 20′ N., 5° 56′ W., 124, 144.
—— nan Con, 56° 22′ N., 5° 56′ W., 123.
*Cryptomeria japonica*, 76; *C. Sternbergii*, 70, 76.
Cryptoperthite, 190, 193, 195, *see* Anorthoclase.
Cuillins (Skye), 51, 162, 392; Dolerite Cone-Sheet *and* Olivine Gabbro Analyses, 15, 24.
Culliemore, 56° 22′ N., 6° 8′ W., 272, 279.
CUNNINGHAM CRAIG, E. H., 52, 226, 268, 270, 397, 425.
Cupressineæ, 70, 75, 77.
*Cupressinoxylon*, 70, 75, 77, 78, 81; *C. MacHenryi*, 75.
*Cupressites MacHenryi*, 70, 75, 78.
*Cupressus Pritchardi*, 75.
CURRIE, J., 49, 50, 140, 142, 425.
—— J. (Mrs.), 140.
*Cyprina islandica*, 406.
Cyrena, 61.

DAIL Bhàite, 56° 31′ N., 5° 54′ W., 404.
Dakota (America), 74, 84, 85.
DALY, R. A., 9, 30.
DANIELL, W., 421.
Dark Lavas of Ben More, 94, 126.
*Davidia*, 86.
DAWSON, W., 85, 89.
Dearg Bhealach, 56° 18′ N., 6° 5′ W., 64, 415.
—— Phort, 56° 20′ N., 6° 22′ W., 183.
—— Sgeir, 56° 22′ N., 6° 12′ W., 140, 265; Pectolite Analyses, 34.
Dererach, 56° 24′ N., 6° 1′ W., 126, 411, 417.
Derryguaig, 56° 27′ N., 6° 4′ W., 127, 400.
Derrynaculen, 56° 23′ N., 5° 56′ W., 94, 125, 148, 173, 178, 234, 239, 246; Derrynaculen Granophyre, 4, 5, 38, Chap. XII., 173, 180, 243, 245, 246, 263; Porphyritic Central Type Lava Analysis, 24.
Dervaig, 56° 35′ N., 6° 11′ W., 39, 115, 117, 358, 395, 397.
Desert Phenomena, 38, 39, 52, 54-58.
DESMAREST, M., 40.
Devitrification, *see* Pitchstones.
Devon (England), 416.
DEWEY, H., 149, 150.
Diatomite, 418, 422.
Dicotyledones, 61, 62, 67, 70, 78-86, 112.
Differentiation, 9, 28, 30, 33, 38, 51, 216, 217, Chap. XXX.; Differentiation-Column Analyses, 29.
Dishig, 56° 27′ N., 6° 3′ W., 289, 361.
Disko Island (Greenland), 84.
Doir' a Mhàim, 56° 26′ N., 5° 50′ W., 132, 135, 313, 321.
Doire Daraich, 56° 30′ N., 5° 40′ W., 293.
—— Dorch, 56° 31′ N., 5° 51′ W., 129.
Dolerite, Chaps. XI., XXI., XXIII.-XXIX., XXXIV.; Analyses, 15, 17, 24,
*Doliostrobus*, 77.
Double-Tier Jointing, 38, 44, 101, 109, 111, 112, 118, 145, 146.
Drainage-System, 384, 385.
Druimgarvie Cave (Oban), 412, 413.
Druim Mòr, 56° 24′ N., 5° 29′ W., 294.
—— na Crìche (Skye) : Mugearite Sill Analysis, 27.
Drumlang, 56° 29′ N., 5° 59′ W., 410.
Drynoch (Skye) : Basalt Lava Analysis, 15.
Duart, 56° 27′ N., 5° 39′ W., 37, 58, 120, 175, 177, 360, 396, 400, 401, 406; Duart Bay Syncline, 37, 173, 175.
DUKE OF ARGYLL, 1, 43, 44, 53, 61, 62, 65, 68, 69, 89, 265, 418, 422-424; Present DUKE, 416.
Dùnan na Marchachd, 56° 18′ N., 6° 1′ W., 273, 398.
Dùn Bàn, 56° 38′ N., 6° 13′ W., 411.

2 E

Dùn Bhuirg, 56° 21′ N., 6° 10′ W., 114, 139.
—— Breac, 56° 24′ N., 5° 59′ W., 187, 188.
—— Cùl Bhuirg, 56° 20′ N., 6° 25′ W., 402, 411.
—— da Ghaoithe, 56° 27′ N., 5° 47′ W., 37, 179, 211, 236, 362, 396.
Dungallan House, 56° 25′ N., 5° 29′ W., 409.
Dunite, 249.
Dùn Leathan, 56° 37′ N., 6° 13′ W., 117.
—— Mòr, 56° 29′ N., 6° 7′ W., 157, 288.
Dunollie Castle, 56° 26′ N., 5° 29′ W., 409.
Dùn Ormidale, 56° 23′ N., 5° 31′ W., 375.
DUNS, J., 423.
Dùn Scobuill, 56° 22′ N., 6° 6′ W., 103.
Dutchman's Cap (Bac Mòr), 56° 27′ N., 6° 28′ W., 140, 387, 389, 397, 409.
—— Hat (Bach Island), 56° 23′ N., 5° 36′ W., 409.
Dykes (Veins of Older Authors), 4, 7, 10, 30, Chaps. II., XXXIV., XXXV., 383; Analyses, 15, 17, 35; Explosions, 10, 37, 364-366; *see* Side-Step *and* Swarm.
Dynamics of Intrusion, 11.

EARLY Acid Cone-Sheets, 7, 8, 18, 30, 36, 37, 168, 173, 181, 187, 195, 203, 206, 207, 211, Chap. XIX., 232, 236, 246, 362.
—— Basic Cone-Sheets, 7, 8, 14, 30, 36, 37, 51, 96, 165, 167, 173, 179, 181, 187, 195, 203, 206, 207, 211, 212, 216, 222, 232, Chap. XXI., 245-247, 292, 295, 296, 299, 341, 362, 363.
—— Rhyolites and Felsites, 5, 30, 36, Chaps. XV.-XVII.
—— Gabbros, 4, 8, 30, Chaps. XI., XXII.
—— Granophyres, 4, 30, 37, 48, Chap. XII.
Eas Dubh, 56° 18′ N., 6° 8′ W., 64, 108, 401, 416.
Economics, Chap. XXXVIII.
EDWARDS, W. N., 68, 73.
Eigg (Hebrides), 47, 95, 172.
Eilean a' Bhaird (Canna): Mugearite Sill Analysis, 27.
—— an Fheòir, 56° 22′ N., 6° 4′ W., 272.
—— Bàn, 56° 19′ N., 6° 15′ W., 183, 266, 271, 272.
—— Bàn, 56° 28′ N., 5° 41′ W., 396.
—— Buidhe, 56° 19′ N., 5° 36′ W., 365.
—— Feòir, 56° 29′ N., 6° 0′ W., 356, 370, 374.
—— Mòr, 56° 21′ N., 5° 52′ W., 232.

Eilean Musdile, 56° 27′ N., 5° 36′ W., 293.
—— na Beitheiche, 56° 33′ N., 5° 52′ W., 292.
—— Rudha an Ridire, 56° 30′ N., 5° 42′ W., 175.
—— Trianach, 56° 27′ N., 5° 40′ W., 173, 176, 177.
*Elatocladus Campbelli*, 70, 77.
Elgol (Skye), 193.
Enclaves, *see* Xenoliths.
ENNOS, F. R., 13, 15, 17, 20, 23.
Enstatite, *see* Rhombic Pyroxene.
Enstatite-Augite (Uniaxial Augite, Clino-Enstatite), 18, 21, 26, 52, 190, 208, 227, 228, 241, 284, 286, 293, 302, 346-348, 374; Analysis, 34.
Eocene, 2, 44, 53, 55, 69, 71, 74-77, 80, 84, 85, 89, 383.
Eorsa, 56° 28′ N., 6° 5′ W., 289, 290.
Epidote, 3, 48, 49, 95, 96, 104, 120, 121, 123, 127-131, 141, 143, 147, 148, 153, 155, 162, 164, 170, 206, 207, 214, 215, 232, 236, 254, 255, 367, 425; Analysis, 34.
Epistilbite, 425.
*Equisetum*, 62, 89; *E. Campbelli*, 70, 71; *E. maximum*, 71.
Equisetales, 70-72.
Erratics, 41, 392-396.
Erie (Colarado), 74.
Erisgeir, 56° 25′ N., 6° 15′ W., 172, 397.
Eskdale Muir (Mainland), 263.
Eskers, 36, 404, 405.
Etive (Mainland), 395; Etive Dyke-Swarm, 10, 52.
ETTINGSHAUSEN, C. von, 69, 75, 87, 89, 423.
Eucrite, Analyses, 23; *see* Ben Buie Gabbro.
Eutectic Structure, 276, 277, 300, 303, 365.
EVANS, W., 406.
Explosions, *see* Acid Magma *and* Agglomerate.
EYLES, V. A., 280, 380.

FAGACEÆ, 70, 82.
*Fagus*, 81; *F. castaneæfolia*, 82; *F. dentata*, 82; *F. pristina*, 86.
FALCONER, J. D., 302, 303, 325, 328.
Faröelite, 140, 425.
Faults, Chap. XIII.
Fayalite, 152, 154, 155.
Féith Bhàn, 56° 30′ N., 5° 58′ W., 400.
Felsites, 5, 7, 9, 30, Chaps. XVII., XIX., XXIII., 286, 291, Chap. XXIX., 321, 327, Chaps. XXXI., XXXII., XXXIV; Analysis, 20.
Felstones (of Judd), 45, 46, 49.
Feorlin Cottage, 56° 20′ N., 5° 59′ W., 57, 59, 273.
Ferns, 62, 69.
Filicales, 70, 73, 74.
*Filicites hebridicus*, 69, 70.

Fingal's Cave, 56° 26′ N., 6° 20′ W., 108, 111, 113, 145, 410.
Fionna Mhàm, 56° 24′ N., 6° 9′ W., 115, 145.
Fionn Aoineadh, 56° 22′ N., 6° 12′ W., 288.
Fionn-Chrò (Rum): Mugearite Analysis, 27.
Firth of Lorne, 56° 23′ N., 5° 37′ W., 122, 175, 401.
Fishnish, 56° 31′ N., 5° 49′ W., 106, 129, 176, 206, 227, 279, 359, 400, 405.
Fissure-Eruption, 4, 44, 46-48, 99.
Fladda, 56° 30′ N., 6° 23′ W., 140.
FLETT, J. S., 15, 16, 52, 144, 149, 150, 266, 268, 271, 368, 377-381, 425.
Flint : Conglomerate, 44, 54, 58, 62-64 ; Implements, 413.
Florida (America), 73.
Folds, 4, 5, 38, 168, Chap. XIII.
FORBES, E., 1, 44, 53, 68-71, 73, 77-79, 83, 86, 89.
Forth (Mainland), 414, 419.
Fort Union (America), 74, 84.
Fossils, see Leaves, Molluscs, Shells, Shelly Clay, Trees.
Fountainhead, 56° 33′ N., 5° 44′ W., 60.
Fourche Mountains (America), 190, 191,
Fourchite, 379.
France, 413.
Frank Lockwood's Island, 56° 18′ N., 5° 50′ W., 360.
FRENCH, J. W., 12.
Fredriksvarn (Norway), 190.
FUCHS, J. N., 35.

GABBRO, 4, 8, 9, 30, 37, 38, 44, Chaps. XI., XX., XXI., XXII., XXIX., XXX.; Analyses, 24, 29.
Gamhnach Mhòr, 56° 19′ N., 5° 58′ W., 185, 186, 189-191, 193, 269 ; Syenite Analysis, 27.
Ganavan, 56° 26′ N., 5° 28′ W., 410.
Gaodhail, 56° 29′ N., 5° 53′ W., 36, 181, 216, 217, 219, 224, 239, 296, 298, 301, 304, 305, 332, 341, 384 ; Gaodhail Augite-Diorite, 6, Chap. XVIII.
Garabal Hill (Mainland), 32.
Garbh Shlios, 56° 23′ N., 5° 55′ W., 253.
GARDNER, J. S., 1, 2, 44, 53, 55, 61-63, 67, 69, 70-76, 79, 80, 82-86, 89, 423, 424.
Garmony, 56° 30′ N., 5° 47′ W., 36, 37, 400.
Garnet, 50, 124, 128, 153, 161, 367 ; Analysis, 34.
Gasteropods, 61.
GEHLEN, A. F., 35.
GEIKIE, A., 1, 3, 4, 8, 10, 44, 46-49, 64, 92, 94, 95, 152, 156, 172, 252, 260, 263, 344, 383, 422-424.
—— J., 423.
Georgia (America), 89.
Gheal Gillean Type (Skye), 370.

Giant's Causeway, 40, 146 ; Basalt Analysis, 17.
GILBERT, G. K., 48.
*Ginkgo*, 62 ; *G. adiantoides*, 70, 74.
Ginkgoales, 70, 74.
Glac a'Chlaonain, 56° 30′ N., 5°57′ W., 181, 343, 345.
Glacial, 11, Chap. XXXVII.; Fans, 36, 403-406 ; Landslip, 414 ; Raised Beaches, 11, 392, 399-408 ; Shells, 37, 406.
Glaciated Pot-Holes, 37, 396, 397.
Glac Mhòr, 56° 38′ N., 6° 6′ W., 118.
Glais Bheinn, 56° 32′ N., 5° 42′ W., 105, 106, 183, 293.
Glas Bheinn, 56° 23′ N., 5° 58′ W., 200, 204, 207, 223 ; Glas Bheinn Granophyre, 4, 5, 38, Chap. XII., 173, 179, 180, 199, 200, 203, 206, 207, 232, 243, 341.
Glas Eileanan, 56° 30′ N., 5° 43′ W., 175.
Glasgow (Mainland), 328.
Glass-Sand, 419.
Gleann a' Chaiginn Mhòir, 56° 22′ N., 5° 51′ W., 38, 168, 248, 400.
—— Doire Dhubhaig, 56° 26′ N., 6° 5′ W., 103.
—— Dubh, 56° 24′ N., 6° 0′ W., 126, 187, 363.
—— Lìrein, 56° 26′ N., 5° 45′ W., 37, 121, 196, 198, 214.
—— Mhic Caraidh, 56° 32′ N., 6° 6′ W., 358.
—— Seilisdeir, 56° 24′ N., 6° 5′ W., 384, 399, 401.
—— Sheileach, 56° 24′ N., 5° 29′ W., 294.
—— Sleibhte-coire, 56° 24′ N., 5° 48′ W., 37.
Glen Aros, 56° 32′ N., 6° 1′ W., 104, 385, 400.
Glenbyre, 56° 20′ N., 5° 54′ W., 59, 94, 124, 125, 239, 401.
Glen Cannel, 56° 27′ N., 5° 54′ W., 36, 223, 224, Chap. XXXI., 340, 385, 398 ; Glen Cannel Granophyre, 9, 21, 36, 181, 185, Chap. XXXI., 341, 344, 345, 361, 363, 366, 375, 385, 395 ; see Contact-Alteration.
—— Clachaig, 56° 27′ N., 5° 56′ W., 217, 255, 321, 332, 339, 340, 341, 346 ; see Clachaig.
—— Coe (Mainland), 7, 9, 12, 42, 51, 341.
Glenelg (Mainland), 377, 378.
Glen Euchar, 56° 19′ N., 5° 29′ W., 403, 407.
—— Forsa, 56° 29′ N., 5° 53′ W., 36, 37, 134, 148, 158, 159, 181, 203, 217, 222, 224, 234, 235, 239, 248, 256, 296, 298, 321, 339, 341-343, 346, 374, 384, 385, 394, 400, 404, 405.
Glen Gallain, 56° 19′ N., 5° 29′ W., 407.
Glengorm Castle, 56° 38′ N., 6° 10′ W., 157.

Glen Leidle, 56° 21′ N., 6° 0′ W., 100, 123, 125, 384, 401 ; *see* Leidleite.
—— Libidil, 56° 20′ N., 5° 47′ W., 177, 230, 231, 232, 366, 399, 401.
Glen More, 56° 23′ N., 5° 59′ W., to 56° 24′ N., 5° 45′ W., 9, 29, 36, 37, 51, 123, 125, 158, 165, 188, 195, 197, 199, 234, 235, 239, 243, 245, 246, 255, 298-300, Chaps. XXIX., XXX., 373, 384, 394, 399, 401, 407 ; Glen More Gabbro-Granophyre Ring-Dyke, 28, 38, 299, 300, 309, 311, 312-315, Chap. XXX., 337 ; Analyses, 29.
Glen Strath Farrar (Mainland), 377.
*Glyptostrobus europæus*, 70.
Gneiss-Fragments in Agglomerate, 5, 45, 180, 181, 187, 188, 196, 200, 201, 207, 243, 341.
Goat Cave, 56° 26′ N., 6° 20′ W., 111.
GOEPPERT, H. R., 75, 76, 84, 90.
Goirtein Driseach, 56° 21′ N., 6° 5′ W., 273.
Gometra, 58° 29′ N., 6° 17′ W., 38, 39, 107, 114, 388, 389.
GOODCHILD, J. G., 50, 140, 425.
Gorten, 56° 26′ N., 5° 39′ W., 406.
Gortenbuie, 56° 26′ N., 5° 54′ W., 334.
Gortendoil, 56° 23′ N., 5° 42′ W., 122.
Gowanbrae, 56° 19′ N., 6° 14′ W., 65, 416.
Granite, 419.
Granophyre, 4, 5, 7-10, 30, 36-38, 43, 48, 49, 50, Chaps. XII., XIX., XX., XXIX.-XXXIII.; Analyses, 20.
Granular Structure, 47, 137, 146, 148, 149, 154, 189, 192.
Granulitic Structure, 38, 154, 155, 160-162, 198, 209, 210, 214, 219, 248, 256, 316, 336, 348, 352, 354.
Graphite, Chap. XXIV., 417, 425.
Gravitational Differentiation, 9, 28, 33, 38, 51, 216, 217, 310, 311, 313, 314, 318, Chap. XXX.; Analyses, 29.
GRAY, ASA, 73.
Great Glen Fault (along Loch Linnhe), 177, 184, 395.
Greenland, 69, 74, 76, 77, 80, 89.
Gregarious Dykes, 47.
GREGORY, J. W., 149.
—— W., 418, 422.
*Grewia crenulata*, 70.
Gribun, 56° 26′ N., 6° 8′ W., 39, 41, 45, 49, 55-57, 59, 62, 101, 102, 107, 108, 114, 115, 140, 183, 257, 259, 265, Chap. XXVI., 380, 383, 385-387, 391, 394, 402, 409, 411.
GRIEVE, J., 68.
Grinnel Land (Canada), 80.
Grob a' Chuthaich, 56° 23′ N., 5° 41′ W., 120.
Ground-Mass defined, 280.
Gruline, 56° 29′ N., 5° 58′ W., 36, 345.
Gualachaolish, 56° 23′ N., 5° 42′ W., 405.
Guibean Uluvailt, 56° 25′ N., 5° 58′ W., 187, 207, 209, 210.

Gymnospermæ, 69, 70, 74-78.
Gyrolite, 34, 425.

HALLE, T. G., 77, 90.
HALLIMOND, A. F., 35, 51, 52, 138, 284, 425.
Hare Island (Greenland), 84, 85.
Häring (Tyrol), 75.
HARKER, A., 8, 13, 15, 21, 23, 24, 27, 32, 33, 50, 51, 63, 95, 96, 101, 144, 146, 152, 155, 193, 216, 248, 249, 252, 324, 351, 359, 370, 372, 378, 392, 425.
HARKNESS, R., 422.
Harris (Hebrides), 410.
HATCH, F., 47.
Haunn, 56° 33′ N., 6° 20′ W., 402, 411.
Hawaii, 49 ; *see* Kilauea.
HEDDLE, M. F., 35, 38, 49, 50, 140, 141, 265, 270, 417, 424, 425.
*Hedysarum*, 88.
HEER, O., 69, 74, 76, 77, 79, 80, 82, 83, 85, 89, 90.
Hercynite-Pleonaste, 276 ; Analysis, 35.
HERTZ, H. R., 12.
Heulandite, 50, 142, 425.
High Water Mark compared with Ordnance Datum, 400.
History of Research, Chap. II., 67-70 ; *see* Bibliography.
HOLDEN SINCLAIR, J., 423.
HOLLAND, T. H., 35.
HOLLICK, A., 74, 90.
HOLMQUIST, P. J., 397.
HOLTTUM, R. E., Part-Author of Chap. IV., R. E. H. ; *see also* 2, 44, 53, 55.
Homberg (Germany), 76.
HOME, D. MILNE, 423, 424.
HOPKINS, W., 47.
Hornblende : Primary, 21, 42, 169, 170, 189, 191, 193, 347, 361, 368, 380, 382 ; Secondary, 49, 141, 161, 164, 219, 220, 225, 232, 241, 251, 254-6, 304, 327-9, 334, 346, 348, 353, 366, 367 ; in Contact-Zones, 50, 128, 152, 153, 155, 162, 163, 319, 335, 336, 348, 354, 363 ; in Hybrids, 21, 210, 219, 317, 318, 347, 352, 353, 355.
HORNE, J., 384.
Horn-Implements, 413.
HULL, E., 16.
Hummocky Moraines, 392, 399, 403, 404.
HUTTON, J., 40, 43.
Hybrids, 10, 36, 210, 212, 216, 219, 316, 324, 325, 330, 342, 349, Chap. XXXIII.
Hypersthene, *see* Rhombic Pyroxene.

ICE-SHEET, 11, 392, 398.
Iceland, 3, 4, 7, 10, 40, 46, 47, 84.
IDDINGS, J. P., 109.

Inch Kenneth, 56° 27′ N., 6° 9′ W., 108, 172, 181, 380, 419.
Inclined Sheets, 6, 50, see Cone-sheets.
Inninbeg, 56° 31′ N., 5° 45′ W., 414.
Inninmore Bay, 56° 30′ N., 5° 42′ W., 61, 105, 181, 183, 281, 293 ; Inninmore Fault, 37, 181-183, 379, 383 ; Inninmore Sheets and Dykes, 37, 374..
Inninmorite (after Inninmore Bay), 52, 165, 170, Chap. XXIII., 272, 273, 279, Chap. XXV., 293, 346, 374 ; Analyses, 19, 35, 263 ; Definition, 282.
Insecta, 71.
Intermediate Cone-Sheets, see Acid Cone-Sheets.
—— to Sub-Acid Magma-Type, 18-20 ; Analyses, 19.
Intersertal Structure defined, 280.
Iona, 56° 20′ N., 6° 24′ W., 36, 38, 108, 111, 146, 172, 378, 388, 389, 397, 398, 402, 410, 411, 413, 419.
Iona Ferry, 56° 20′ N., 6° 23′ W., 36, 402.
Ireland, 68, 75, 76, 172, 398, 410, 414; see Antrim.
Iron-Ore, 417.
Ishriff, 56° 25′ N., 5° 50′ W., 38, 132, 298, 384 ; Ishriff Granophyre Ring-Dyke, 299, 300, 309, 313, 315, 316 ; Compact Central Type Basalt Lava Analysis, 17.
Islay (Hebrides), 387, 408.
Itinerary, 36.

JAMES, A. V. G., 109.
JAMESON, R., 41, 415, 425.
Japan, 5, 73, 76.
Java Point, 56° 29′ N., 5° 42′ W., 175, 176, 396, 400.
Jedburgh Type (Mainland), 136.
JEHU, T. J., 378, 397, 426.
JOHNSON, SAMUEL, 40.
JONES, T. R., 55, 58.
JUDD, J. W., 1, 4, 32, 35, 45, 46, 48, 51, 55, 57, 58, 60, 101, 109, 156, 383, 423, 424.
Jura (Hebrides), 409 ; Crinanite Dyke Analysis, 15.

KALAHARI Desert (South Africa), 54.
Kellan Mill, 56° 29′ N., 6° 3′ W., 358.
Kelley's Island (America), 397.
KENDALL, J. W., 423.
—— P. F., 47, 424.
Kerrera, 56° 24′ N., 5° 33′ W., 42, 409, 410, 419.
—— Ferry, 56° 24′ N., 5° 31′ W., 375.
Kettle-Holes, 403.
Kieselguhr, 418.
Kilauea (Hawaii), 3, 4, 49, 99.
Kilchoan, 56° 42′ N., 6° 6′ W., 386, 391.

Kilfinichen, 56° 23′ N., 6° 3′ W., 103, 140, 399, 401 ; Xonotlite Analysis, 35.
Killbeg, 56° 30′ N., 5° 54′ W., 405 ; Killbeg Ring-Dyke, 36, 343.
Killean, 56° 24′ N., 5° 42′ W., 166.
Killichronan, 56° 30′ N., 6° 0′ W., 400.
Killiemore House, 56° 23′ N., 6° 3′ W., 272.
Kilninian, 56° 32′ N., 6° 14′ W., 114, 411.
Kilninver, 56° 20′ N., 5° 31′ W., 37, 184, 407, 408.
Kilpatrick, 56° 27′ N., 5° 40′ W., 406.
KILROE, J. R., 425.
Kinloch House, 56° 22′ N., 6° 1′ W., 272.
Kinloch Hotel or Inn, see Kinlochscridain.
Kinlochscridian (Head of Loch Scridain), 56° 23′ N., 6° 0′ W., 38, 39, 122, 144, 366, 390 ; Mugearite Lava Analysis, 27.
Kinlochspelve, 56° 22′ N., 5° 48′ W., 37, 123, 178, 207, 407, 419.
Kintallen (not the Kintallenite locality), 56° 33′ N., 5° 58′ W., 105 ; Analyses of Brunton Type Tholeiite Dykes, 17.
KITCHIN, F. L., 68.
Kirwanite, 425.
Knock, 56° 29′ N., 5° 59′ W., 129, 342, 361, 396, 403, 418. Knock Granophyre Ring-Dyke, 9, 21, 36, 48, 300, Chap. XXXII., 361, 367 ; Analysis, 20 ; see Contact-Alteration.
KNOWLTON, F. H., 74, 84, 90.
KOCH, W. E., 69, 423.
KRASSER, F., 85, 90.
KRAUS, G., 75.
Kunstadt (Moravia), 85.
KYNASTON, H., 364, 365, 408.

LABRADORITE Analysis, 34.
Laccolith, 6, 48, 257.
LACROIX, A., 154, 268, 269.
Lag a' Bhàsdair, 56° 26′ N., 6° 0′ W., 126.
Laggan Bay, 56° 29′ N., 6° 8′ W., 115.
—— Lodge, 56° 21′ N., 5° 50′ W., 233.
Lainne Sgeir, 56° 34′ N., 6° 18′ W., 117.
Lake Erie (America), 397.
Landslips, 391, 414.
Lantern-Slides, 420.
Laramie (America), 84, 85, 89.
Larne (Ireland), 414.
Late Acid Cone-Sheets, 222, 223, 228.
—— Basic Cone-Sheets, 6-8, 17, 30, 36, 38, 51, 216, 235, 238, 245-247, 290, 294, Chap. XXVIII., 306, 307, 309, 310, 311, 313, 315, 333, 344, 345, 355, 362 ; Analyses, 17, 18.
—— Glacial Raised Beaches, 11, 392, 399-408.

Laterite, 102.
Laumontite, 142, 155, 425.
LAURENT, L., 80, 81, 90.
Lavas, 2-4, 30, Chaps. II., V.-X., 188, Chap. XV.; Analyses, 15, 17, 24, 27.
Leac an Staoin, 56° 21′ N., 5° 57′ W., 125.
Leaves (Plants), 2, 43, 53, 61-63, 64, 65, 67, Chap. IV., 97, *see* An Dùnan *and* Ardtun Leaf-Beds.
LEE, G. W., Part-Author, G. W. L.; *see also* 54, 61, 68, 94.
—— W. T., 90.
Leidleite (after Leidle River), 52, Chap. XXIII., 269, 279, Chap. XXV., 288, 293, 300, 354, 373, 374; Analyses, 19, 263; Definition, 281.
Leidle River, 56° 21′ N., 6° 0′ W., 125, 273; *see* Glen Leidle.
Lerags, 56° 22′ N., 5° 30′ W., 294, 375.
LESQUEREUX, L., 74, 81, 84, 85, 90.
Lewis (Hebrides), 410.
Liath Dhoire, 56° 28′ N., 5° 53′ W., 339.
LIGHTFOOT, B., Part-Author, absent, (B. L.); *see also* 52, 123, 124.
Lignites (Coals), 38, 40, 41, 44, 53, 60, 62-65, 97, 111, 270, 274, 415-417; Analyses, 416.
Limestone, 419.
LINE, J., 82.
Linlithgow (West Lothian), 302, 303, 325, 328, 330.
Lismore, 56° 30′ N., 5° 32′ W., 184, 292, 293, 379, 380, 381, 419.
Little Colonsay, 56° 27′ N., 6° 15′ W., 113, 140, 389, 391.
Loch a' Charraigein, 56° 19′ N., 6° 9′ W., 272.
—— a' Ghleannain, 56° 25′ N., 5° 41′ W., 175, 177, 405.
—— Airdeglais, 56° 23′ N., 5° 51′ W., 132, 200, 203, 236, 295, 314, 384.
—— Aline, 56° 33′ N., 5° 46′ W., 37, 53, 59, 60, 65, 105, 172, 281, 292, 293, 359, 362, 379, 395, 400, 407, 414, 418, 419.
Lochaline, 56° 32′ N., 5° 47′ W., 36, 37; Basalt Lava Analyses, 15.
Loch a' Mhàim, 56° 29′ N., 5° 50′ W., 211.
Lochan an Daimh, 56° 19′ N., 5° 51′ W., 123.
—— an Doire Dharaich, 56° 26′ N., 5° 42′ W., 184, 222.
Loch an Eilein, 56° 24′ N., 5° 51′ W., 314.
—— an Ellen, 56° 24′ N., 5° 51′ W., 313, 314, 322.
Lochan na Cille, 56° 33′ N., 5° 49′ W., 187.
—— nam Ban Uaine, 56° 27′ N., 5° 53′ W., 335, 363.
—— Tana, 56° 22′ N., 5° 57′ W., 125.
Loch Assapol, 56° 18′ N., 6° 12′ W., 107, 273, 379, 397, 401.

Loch Bà, 56° 28′ N., 5° 57′ W., 6, 7, 36, 48, 51, 131, 201, 209, 216, 217, 298, 300, 306, 307, 332, 333, 337, 339-346, 361, 385, 398, 403-405, 418. Loch Bà Felsite Ring-Dyke, 7, 9, 21, 30, 36, 48, 51, 132, 181, 203, 208, 239, 300, 306, 307, 318, 321, 332, 333, 335, Chap. XXXII., 351, 361, 366, 375; Analysis, 20.
—— Baile a' Ghobhainn, 56° 32′ N., 5° 29′ W., 293.
—— Beg, 56° 23′ N., 6° 0′ W., 141, 363, 390.
—— Buie, 56° 21′ N., 5° 52′ W., 37, 38, 94, 122-124, 173, 175, 178, 230-233, 239, 397, 398, 401, 407, 417.
—— Caol, 56° 19′ N., 6° 17′ W., 183.
—— Don, 56° 25′ N., 5° 39′ W., 37, 120, 121, 173, 177, 222, 224, 228, 360, 362, 383, 385, 399, 401, 405, 406; Loch Don Anticline, 37, 58, 173, 177, 178, 183, 184, 222, 383; Loch Don Sand-Moraine, 37, 401, 406, 407; Mudstone Analysis, 34.
—— Don Bridge, 56° 26′ N., 5° 41′ W., 177.
Lochdonhead, 56° 26′ N., 5° 41′ W., 37, 38, 145, 406.
Loch Feochan, 56° 21′ N., 5° 30′ W., 294, 364, 401, 418, 419.
Loch Frisa, 56° 34′ N., 6° 6′ W., 118, 157, 358, 397.
Loch Fuaran, 56° 22′ N., 5° 55′ W., 168, 188, 243, 248, 317, 399; Loch Fuaran Banded Granulites, 38, 245, 252, 253.
—— Linnhe, 56° 33′ N., 5° 31′ W., 395, 409, 410; Loch Linnhe (Great Glen) Fault, 177, 184, 395.
—— Meadhoin, 56° 36′ N., 6° 7′ W., 156.
—— nan Cinneachan, 56° 37′ N., 6° 35′ W., 396.
—— na Creitheach (Skye), 193, 194.
—— na Géige, 56° 20′ N., 6° 0′ W., 273.
—— na Keal, 56° 28′ N., 6° 3′ W., 9, 36, 37, 43, 103-105, 107, 125, 127, 128, 257, 288, 356, 358, 360, 361, 367-370, 374, 385, 386, 388, 392, 395-398, 400-405, 410, 411.
Loch na Làthaich, 56° 20′ N., 6° 15′ W., 61, 183, 266, 397.
—— Scamadale (Mainland), 407.
—— Scavaig (Skye), 172.
—— Scridain, 56° 22′ N., 6° 4′ W., 8, 35, 36, 38, 41, 53, 61, 65, 100, 103, 108, 113, 114, 124-126, 139, 140, 175, 181, 187, 188, 228, 260, 265, 266, 269, 272, 273, 278, 284, 363, 366, 376, 384, 386, 387, 390, 398, 401, 402, 407, 410, 411, 417, 418; Loch Scridain Sills, 8, 38, 52, 183, 186, Chaps. XXIII.-XXV., 288; *see* Xenoliths.
—— Sguabain, 56° 24′ N., 5° 50′ W., 134, 197, 235, 299, 314.
—— Slappin (Skye), 172.

Loch Spelve, 56° 23′ N., 5° 44′ W., 36, 38, 98, 120, 122, 123, 159, 165-167, 169, 170, 173, 175, 177, 178, 180, 183, 184, 197, 198, 201, 231, 234, 235, 236, 239, 240, 295, 366, 370, 384, 385, 399, 401, 405-407 ; Loch Spelve Anticline, 37, 168, 173, 175, 178, 179, 195, 236.
—— Teàrnait, 56° 33′ N., 5° 40′ W., 293.
—— Tuath, 56° 31′ N., 6° 14′ W., 39, 107, 113, 115, 386, 388, 397, 402.
—— Uisg, 56° 21′ N., 5° 49′ W., 37, 38, 122, 173, 178, 230-232, 294, 401, 407 ; Loch Uisg Gabbro or Dolerite, 8, 37, Chap. XX.; Loch Uisg Granophyre, 8, 30, 37, 122, 123, 173, 195, 198, Chap. XX. ; *see* Contact-Alteration.
Lòn Bàn, 56° 28′ N., 5° 53′ W., 217.
—— Reudle, 56° 31′ N., 6° 18′ W., 113.
Long Island (America), 74.
Lorne, District east of Firth of Lorne, including Kerrera, 42, 44, 184, 292, 293, 356, 385, 396, 403, 407.
Lough Neagh (Ireland), 75.
Lunga, 56° 29′ N., 6° 25′ W., 140, 389, 397.
LUNN, R., 420.
Lussa River, 56° 25′ N., 5° 48′ W., 165, 169, 184, 212, 236, 384.

MACARTHUR Cave, Oban, 412-414.
M'CORMICK, J., 415, 426.
MACCULLOCH, J., 1, 41-43, 48, 52, 77, 90, 111, 383, 421 ; Macculoch's Tree, 56° 22′ N., 6° 12′ W., 3, 38, 39, 43, 65, 77, 97, 111, 113, 145, 387.
MCDOUGALL, J., 424.
Mackenzie River (Canada), 80, 84,
MACKINDER, H. J., 384, 426.
Mackinnon's Cave, 56° 25′ N., 6° 9′ W., 409.
MACLEAN, J. P., 426.
M'LEOD, J., 422.
M'LINTOCK, W. F. P., 35, 49, 50, 128, 141-143, 152, 153, 368, 426.
MACNAIR, P., 425.
McNICOL, D., 421.
M'QUARRIE, MURDOCH, 43.
Magmatic-Sequence, 6, 7, 30-33, 45, 46, 48, 49, 51.
Magma-Series *and* Magma-Types (with Analyses), 13-28 ; Definition, 13.
Magnetite, 47, 417.
MAHONY, J. A., 68.
Malcolm's Point, 56° 18′ N., 6° 3′ W., 63.
Màm a' Chullaich, 56° 31′ N., 5° 40′ W., 183.
—— Lìrein, 56° 27′ N., 5° 46′ W., 212.
—— an Tiompain, 56° 25′ N., 5° 53′ W., 315.
MANSON, W., 181, 420.
Maol Buidhe, 56° 30′ N., 5° 52′ W., 129, 341.

Maol Buidhe, 56° 26′ N., 5° 56′ W., 340.
—— Bhuidhe, 56° 30′ N., 5° 55′ W., 343.
—— na Coille Mòire, 56° 24′ N., 6° 3′ W., 272.
—— nam Fiadh, 56° 26′ N., 5° 50′ W., 132, 158, 299, 311, 313, 362, 366,
—— nan Damh, 56° 25′ N., 6° 1′ W., 49, 50, 127, 128, 152, 153, 289.
—— nan Uan, 56° 27′ N., 5° 44′ W., 121.
—— na Samhna, 56° 21′ N., 5° 58′ W., 228.
—— Odhar, 56° 22′ N., 5° 47′ W., 165.
—— Tobar Leac an t-Sagairt, 56° 24′ N., 5° 52′ W., 315.
—— Uachdarach, 56° 25′ N., 5° 54′ W., 158, 315.
Marblehead (America), 193.
Marginal Tilt, 173, 175.
Martha's Vineyard (America), 74.
Maryland (America), 74.
Mas d'Azil (France), 413.
Massachusetts (America), 193.
Matavanu (Savaii Islands, Pacific), 98
MAUFE, H.B., Part-Author, absent, (H. B. M.) ; *see also* 7, 9, 10, 52, 341, 379.
Meall a' Chaoruinn, 56° 32′ N., 5° 38′ W., 395.
—— an Fhìar Mhàim, 56° 31′ N., 6° 5′ W., 374.
Meallan Fulann, 56° 26′ N., 6° 20′ W., 111.
Meall na Caorach, 56° 33′ N., 6° 0′ W., 104.
—— nan Capull, 56° 23′ N., 5° 49′ W., 243, 247.
MENZEL, P., 80, 90.
Mesolite, 140, 425.
Mesostasis, defined, 280.
METCALF, A. T., 424.
Mezen (Mont Mézenc, France), 43.
MILLS, A., 40, 421.
MILNE HOME, D., 423, 424.
Mingary Àrd, 56° 38′ N., 6° 13′ W., 411.
Miocene, 44, 68, 69, 74, 76, 80, 82, 84, 85, 89.
Mishnish, 56° 38′ N., 6° 8′ W., 116.
Molluscs, 61 ; *see* Shells.
Monadh Beag, 56° 25′ N., 5° 51′ W., 132, 311, 318, 321 ; Compact Central Type Basalt Lava Analysis, 17.
Monchiquite, 16, 377-379.
Monocotyledons, 62.
Montana (America), 74.
Moraines, 37, 38, 159, 392, 399, 400, 406, 407, 414.
Mornish, 56° 36′ N., 6° 16′ W., 116, 388, 411.
Morven, District N. of Sound of Mull, 36, 37, 45, 55, 56, 60, 61, 105, 106, 172, 278, 356, 360, 362, 373, 374, 379, 380, 382, 390, 392, 394, 400, 407 ; Plateau Type Basalt Lava Analyses, 15.

Mount Genèvre (France), 149.
Muck (Hebrides), 172.
Mudstone, 53-61, 203, 271, 274; Analysis, 34.
Mugearite Lavas, 2, 36, 44, 48, 65, 92-94, 104, 115, 121, 123-127, 129, 130, 143-145, 185, 341, 344, 361, 362, 380; Plug, 115, 288; Analyses, 27.
Mull, Island, south of Sound of Mull and west of Firth of Lorne; Mull Swarm of Dykes, 10, 37, 48, 52, 359-362, 366, 379, 380; see Gregarious and Radial Dykes.
Mullach Glac an t-Sneachda, 56° 20′ N., 6° 3′ W., 269, 273; Leidleite Sill Analyses, 19, 263.
Mullite, 35, 268; see Sillimanite.
Multiple Intrusions, 298, 356, 364; Definition, 32.
MURCHISON, R. I., 421, 422.
MURRAY, J., 425.
Myrtaceæ, 86.

NA Bachdanan, 56° 29′ N., 5° 56′ W., 131, 208, 341.
—— Binneinean, 56° 28′ N., 5° 55′ W., 331, 335, 336.
—— Liathanaich, 56° 21′ N., 6° 16′ W., 183.
*Natica alderi* ?, 406.
Na Torranan, 56° 30′ N., 6° 8′ W., 115, 288.
Natrolite, 14, 16, 30, 139, 140, 425.
Nebraska (America), 73.
Nepheline, 16, 26, 140, 191, 192, 378.
NEWBERRY, J. S., 74, 81, 82, 84, 87, 90.
Newfoundland, 73.
New Mountain (Japan), 5.
NICOL, J., 422.
Non-Porphyritic Central Lava Types, 30, 145-149, 290; Analyses, 17.
—— Central Magma Type, 16-18, 30, 31; Analyses, 17.
North of England, 319, 372, 410.
North-West Highlands, 408, 409.
Norway, 190.
*Nuculana minuta*, 406; *N. tenuis*, 406.
Nuns' Pass, 56° 19′ N., 6° 0′ W., 57, 63, 267, 273; Xenolith Analyses, 34.
Nuts, 80.

OAK, 61.
Oban, 56° 25′ N., 5° 28′ W., 36-38, 41, 42, 105, 120, 293, 364, 379, 388, 292, 394, 395, 408-410, 412-414, 419.
Obsequent Drainage, 384.
Ocellar Structure, 140, 317, 361, 369, 377, 380, 382.
Old Red Sandstone distinguished from Tertiary Lavas, 42-44.
—— Man of Hoy (Orkney), 409.
Oligocene, 80.

Olivine: Destruction by Pneumatolysis, 3, 95, 120, 131, 141, 147, 148, 161, 164, 239, 366; Difference in Pale and Dark Groups of Ben More, 126; Nodules, 369; see Fayalite.
Olivine-Poor *and* Olivine-Free Basalts, Tholeiites, Dolerites and Gabbros, 16-18, 23-25, 30, 51, Chaps. V., VIII.-XI., 241, Chaps. XXIII.-XXX., XXXIV.; Analyses, 17, 24, 29.
Olivine-Rich Basalts, Dolerites, Gabbros, Eucrites, and Allivalites, 14-16, 21-23, 30, 51, Chaps. V.-VIII., X., XXI., XXII., 286, 289, 295, Chap. XXXIV; Analyses, 15, 23.
OMORI, F., 5.
*Onoclea*, 61, 69, 73, 74, 89; *O. hebridica*, 70, 73, 74; *O. inquirenda*, 74; *O. sensibilis*, 73.
Ophimottling, 126, 138, 144, 146, 232, 240, 295; Definition, 138.
Ordnance Datum compared with High Water Mark, 400.
Orkney Islands, 377, 378, 410.
Ormsaig, 56° 20′ N., 6° 8′ W., 401.
Oronsay (Hebrides), 413.
Orval (Rum); Basalt Lava Analysis, 15.
Ouachitite, 16, 378.
Overlying Formation (of Macculloch), 41.

*Pagiophyllum Sternbergi*, 70, 76.
Palæontology, see Beetles, Leaves, Molluscs, Shells.
Pale Lava Group of Ben More, 44, 48, 92, 94, 122, 124, 125, 126, 128, 341; Petrology of included Basalts, 126.
PANKOUCHE, C. L. F., 421.
Parallel Roads of Glen Roy (Mainland), 390.
PASSAGE, S., 54.
PEACH, B. N., 364, 365, 378, 384, 408, 413.
Pectolite, 425; Analyses, 34.
PENNANT, T., 40, 421.
Pennycross, 56° 19′ N., 5° 59′ W., 57; Basalt Lava Analysis, 15.
Pennyghael, 56° 22′ N., 6° 1′ W., 411, 415; Inninmorite Sheet Analysis, 19.
Pennygown, 56° 31′ N., 5° 54′ W., 343, 405.
Perlitic Structure, 335, 336.
Petrographical Province, 16, 21, 47.
Petroleum, 417, 425.
Petrological Introduction, 13-33.
Phenocryst Analyses, 34.
Phonolitic Trachyte, 192.
Photographs, 420.
*Phyllites*, 70, 86; *P. ardtunensis*, 70, 85, 86; *P. platania*, 70, 85.
*Phyllocladus asplenioides*, 87; *P. subintegrifolius*, 87.

Picotite, 250.
PIETTE, E., 413.
Pillow-Lavas, 3, 4, 24, 37, 38, 48, 51, 66, 96-98, 121, 132-134, 148, 149, 150, 180, 243, 290, 313.
Pimply (Pustular) Weathering, 95, 100, 105, 107, 117, 120; explained, 138.
*Pinites*, 70, 74, 75; *P. Pritchardi*, 75.
*Pinus palæostrobus*, 75; *P. Strobus*, 75.
Pipe-Amygdales, 39, 98, 114, 115, 122.
Pitchstone, 8, 18, 38, 41, 52, Chaps. XXIII., XXV., 288, 373; Analyses, 19, 263.
Plantae incertae sedis, 70, 86-88.
Plants, *see* Leaves.
Platanaceæ, 70, 83-85.
*Platanites hebridicus*, 62, 70.
*Platanus*, 89; *P. aceroides*, 83-85; *P. aceroides latifolia*, 84; *P. Guillelmæ*, 84, 85; *P. hebridica*, 70, 83-85; *P. nobilis*, 84, 85; *P. occidentalis*, 84, 85; *P. primæva*, 85; *P. Raynoldsii* var. *itegrifolia*, 84.
Plateau Lava Group, 30, 38, 53, 54, Chaps. V.-VIII., X.; Definition, 91-94.
—— Lava Types, 3, 14, 16, 30, 47, 51, Chaps. V., X., 286, 289, Chap. XXXIV.; Definition, 92, 136.
—— Magma-Type, 14-16, 30; Analyses, 16.
Plateaux, 385, 386.
PLAYFAIR, J., 421.
Pleonaste, *see* Hercynite-Pleonaste.
Pliocene, 80, 385.
Pneumatolysis, 3, 10, 18, 35-38, 45, 47, 49, 50, 54, 59, 65, 94-97, 106, Chaps. VIII., IX., 141-143, 147, 148, 150, 169, 198, 239, 249, 295, 300, 361, 366, 367, 396.
Podocarpineæ, 70, 76, 77.
*Podocarpus*, 69, 89; *P. borealis*, 70; *P. Campbelli*, 70, 76; *P. eocaenica*, 70.
Point Bonita (California), 149.
POLLARD, W., 11, 13, 15, 23, 24, 27, 271, 416, 417.
Porphyrite, 227, 242, 248, 284, 361.
Porphyritic Central Lava Types, 30, 146, 148-150, 163, 164, 241, 285, 293, 369; Analyses, 24.
—— Central Magma Type, 23-25, 30, 31; Analyses, 24.
Porphyry, 361.
Port a' Bhearnaig, 56° 25' N., 5° 30' W., 410.
—— a' Ghlinne, 56° 20' N., 5° 46' W., 37, 122, 365, 366.
—— an Fhasgaidh, 56° 26' N., 6° 20' W., 111.
—— an Fhir-bhréige, 56° 18' N., 6° 26' W., 411.
—— an Tobire, 56° 32' N., 5° 58' W., 105.
—— Burg, 56° 31' N., 6° 15' W., 115, 388.

Port Donain, 56° 24' N., 5° 40' W., 37, 58-60, 120, 121, 175, 177, 183, 384, 401.
Portfield, 56° 22' N., 5° 42' W., 401.
Port Mòr, 56° 20' N., 6° 9' W., 266, 271, 272.
—— na Croise, 56° 21' N., 6° 10' W., 266, 284.
—— nam Faochag, 56° 27' N., 6° 15' W., 113.
—— nam Marbh, 56° 23' N., 5° 40' W., 60.
—— na Muice Duibhe, 56° 21' N., 5° 43' W., 60.
—— na Tairbeirt, 56° 24' N., 5° 39' W., 60, 369.
—— Uamh Beathaig, 56° 25' N., 6° 9' W., 388.
—— Uisken, 56° 17' N., 6° 13' W., 419.
Post-Glacial Landslips, 414; Marine Notches and Caves, 11, 408-414.
Post-Volcanic, 10, 11, Chaps. XXXVI., XXXVII.
Pre-Glacial *and* Post-Volcanic, 10, 11, Chap. XXXVI.; Landslip 391; Marine Notches and Caves, 11, 38, 39, 386-391.
Prehnite, 140, 143, 153, 425.
Preshal More *and* Beg (Skye), 146.
Primary Formation (of Macculloch), 41.
Priosan Dubh, 56° 22' N., 5° 56' W., 125.
Propylite, 49, 300, *see* Pneumatolysis.
*Protophyllocladus*, 87; *P. subintegrifolius*, 88.
Pteridophyta, 70-74.
Puy, 45, 46.
Pulaskite, 190, 191.
PULLAR, LAURENCE, 425.
Pyrites, *see* Sulphides.

QUARTZ-Dolerites *and* Gabbros, 7, 9, 18, 241, 281, Chaps. XXVIII.-XXX., 372, 373, 376; Analyses, 17, 29; *see* Gravitational Differentiation, Linlithgow, Lothians, Scottish Lowlands, Sgulan Type, Talaidh Type.
Quartz-Reactions, 154, 229, 274, 365.
*Quercites greenlandicus*, 70.
*Quercus*, 89; *Q. greenlandica*, 70, 82, 89; *Q. platania*, 85, 89; *Q. prinus*, 82.
Quinish, 56° 37' N., 6° 12' W., 36, 116, 117, 411.
Quinish Point, 56° 38' N., 6° 14' W., 402.

RADIAL Dykes, 359, 360.
RADLEY, E. G., 13, 15, 17, 19, 20, 24, 27, 29, 35, 50, 52, 426.
Raised Beaches, 11, 37-39, Chaps. XXXVI., XXXVII.
Rannoch River, 56° 33' N., 5° 42' W., 60.

## Index.

RANSOME, F. L., 149.
Raton (America), 84.
Recent, 11, Chap. XXXVII.
Red Weathering, see Rusty Weathering.
REID, CLEMENT, 61, 67.
—— CLEMENT (Mrs.), 67, 68, 88.
Réidh Eilean, 56° 21′ N., 6° 28′ W., 108, 146, 172.
REIMAN, H., 80, 90.
Remelt, 32, 33.
RENDLE, A. B., 68.
*Rhamnites lanceolatus*, 70; *R. major*, 70, 86; *R. multinervatus*, 70, 86.
Rhombic Pyroxene (Enstatite or Hypersthene): Primary, 18, 21, 164, 190, 226, 227, 241, 249, 250, 251, 268, 269, 270, 281-283, 287, 288, 293, 302; Secondary in Contact-Products, 154, 155, 198, 209, 252; in Hybrids *and* Mixed Rocks, 229, 232, 233, 245, 254, 274, 317, 325, 352-355.
Rhyolite, 196, 198, 201, 208, 209, Chap. XVII., 335, 339, 340, 346, 366; Analysis, 20.
RICHEY, J. E., Part-Author, J. E. R.; *see also* 9, 50, 51, 216, 218, 339, 367, 388.
RICHTHOFEN, F. VON, 46.
Riebeckite, 26, 191, 192, 194.
Ring-Dykes, 7-9, 11, 12, 38, 51, Chaps. XXIX., XXX., XXXII.; Analyses 20, 29; Definition, 306.
Ring-Fractures, 6, 9, 11, 12, 51, 340, 341; *see* Caldera.
River Bà, 56° 29′ N., 5° 59′ W., 410.
—— Clachaig, *see* Glen Clachaig.
—— Euchar, *see* Glen Euchar.
—— Forsa, 56° 29′ N., 5° 53′ W., 217, 339, 343; *see* Glen Forsa.
Road-Metal, 418.
ROBERTSON, C., 35.
ROGERS, A. W., 54.
ROSE, A., 52, 422.
ROSENBUSCH, H., 191, 280, 381.
Ross (Mainland), 377; *see* Ross of Mull.
Rossal, 56° 23′ N., 5° 59′ W., 124, 187, 193, 194, 390, 401, 407; Rossal Type, 193, 194, 374.
Ross of Mull, Peninsula south of Loch Scridain, 41, 172, 183, 184, 257, 272, 273, 379, 385, 386, 394, 395, 398, 401, 402, 411, 419.
Rudha an t-Sasunnaich, 56° 31′ N., 5° 44′ W., 293.
—— Àrd Ealasaid, 56° 32′ N., 5° 58′ W., 105.
—— Àrd nan Eisirein, 56° 29′ N., 6° 0′ W., 403.
Rudh' a' Chaoil, 56° 32′ N., 6° 20′ W., 388.
—— a' Chromain, 56° 18′ N., 6° 0′ W., 38, 52, 186, 187, 193, 266-269, 271, 273, 279, 286, 287, 418; Talaidh Type Tholeiite Sill Analysis, 17.
Rudha Dearg, 56° 32′ N., 5° 48′ W.,
105; Plateau Type Basalt Lava Analysis, 15.
—— Dubh, 56° 19′ N., 5° 56′ W., 124.
Rudh' a' Ghuirmein, 56° 28′ N., 5° 41′ W., 396.
—— a' Glaisìch, 56° 34′ N., 5° 59′ W., 186.
Rudha Gorm, 56° 33′ N., 5° 58′ W., 119.
Rudh' Àird a' Chaoil, 56° 24′ N., 5° 43′ W., 239.
Rudha Mhàirtein, 56° 22′ N., 5° 42′ W., 122.
—— Mòr, 56° 29′ N., 6° 5′ W., 105.
—— na Caillich (Skye), 377.
—— na Faing, 56° 24′ N., 5° 43′ W., 178.
—— na h-Uamha, 56° 22′ N., 6° 12′ N., 81, 111, 112, 387; Staffa Type Basalt Lava Analysis, 17.
—— na Leip, 56° 38′ N., 6° 3′ W., 118.
—— na Mòine, 56° 28′ N., 6° 2′ W., 127.
—— nan Gall, 56° 38′ N., 6° 4′ W., 116, 118.
—— nan Goirteanan, 56° 23′ N., 6° 12′ W., 387.
—— na Sròine, 56° 28′ N., 5° 42′ W., 177.
Rudh' an Fheurain, 56° 23′ N., 5° 32′ W., 375.
—— an t-Sean-Chaisteil, 56° 35′ N., 5° 59′ W., 186, 192; Trachyte Plug Analysis, 27.
Rudha Tolmach, 56° 24′ N., 5° 30′ W., 409.
—— Tràigh Gheal, 56° 20′ N., 5° 44′ W., 360.
Rum (Hebrides), 21, 45, 172, 248, 359, 383; Basalt Lava, Eucrite, *and* Mugearite Sill, Analyses, 15, 23, 27.
Rusty Weathering, 95, 100, 102-108, 115, 117, 118, 120, 123, 127, 129, 131.

SAILEAN Mòr, 56° 30′ N., 6° 16′ W., 114.
SAINT FOND, F. B., 40, 421.
'S Aìrde Beinn, 56° 37′ N., 6° 7′ W., 39, 46, 156, 157, 160; *see* Contact-Alteration.
Salachran, 56° 20′ N., 6° 13′ W., 402.
Salen, 56° 31′ N., 5° 57′ W., 36, 38, 39, 104, 106, 117, 129, 130, 131, 181, 186, 288, 289, 296, 343, 385, 396, 400; Trachyte Plug Analysis, 27; Salen Type, 146, 228, 287, 293, 294, 359, 367, 370-372, 376, 379; Analysis, 17; Definition, 285.
SALISBURY, R. D., 397.
Salite-Structure, 47, 169, 218, 226, 256, 302, 327, 329, 350, 352, 353, 373.
Sanctuary (Glacial), 11, 392-8.
Sand-Infillings of Lava-Cavities, 97, 115.
Sandstone, 419.

## Index.    443

Sanidine, 190.
Sanidinophyre, 63.
Saponite, 425.
SAPORTA, G. de, 75, 86, 90.
Sapphire (blue Corundum), 8, 35, 52, 59, 113, Chap. XXIV., 418.
Saskatchewan (Canada), 73.
Savary Glen, 56° 33' N., 5° 50' W., 187, 293.
Scallastle, 56° 29' N., 5° 44' W., 121, 279.
—— Bay, 56° 29' N., 5° 45' W., 120, 129, 167, 170, 221, 223, 227, 228, 359, 362, 363.
—— River, 56° 28' N., 5° 44' W., 121, 400; Craignurite Cone-Sheet Analyses 19.
Scandinavia, 397.
Scarba (Hebrides), 378.
Scarisdale River, 56° 28' N., 6° 0' W., 289, 361, 367, 396, 400, 403.
Scenery, 95, 115, 173, 242, 339, 342, 358, 383.
Schiller-Structure, 47, 251, 352.
SCHIMPER, W. P., 85, 90.
SCHINDEHÜTTE, G., 76, 90.
Schossnitz (Silesia), 84.
Scobull, 56° 22' N., 6° 6' W., 265, 272.
Scolecite, 49, 97, 104, 128, 140, 142, 143, 153, 155, 425; Analyses, 34.
Scoor, 56° 18' N., 6° 10' W., 273, 394, 397.
SCOTT, A. J., 35, 422.
Scottish Carboniferous Lavas, 16, 136, 138, 144.
—— Lowland Quartz-Dolerites, 226; see Linlithgow.
—— (South of Scotland) Arenig Pillow Lavas, 150.
—— (South of Scotland) Dykes, 263.
Screen, 9, 51, Chap. XXIX., 337, 342, 343, 345, 348, 351; Definition, 306.
SCROPE, G. P., 44, 101, 109, 423.
Seabank Villa, 56° 23' N., 6° 4' W., 103, 269, 272.
Seanvaile, 56° 23' N., 5° 45' W., 167, 169, 170, 370.
Secondary Formation (of Macculloch), 41.
Sediments, 2, Chap. III.
Segregation-Veins in Lavas, 16, 31, 39, 92, 105, 114, 117, 118, 121, 138-142, 145, 377; in Intrusions, 31, 189, 253, 323, 377.
Seil, 56° 19' N., 5° 37' W. (Island, the name is printed at margin of Sheet 44), 379, 419.
Sequence, 6, 7, 23, 30, 31, 33, 43, 46, 48, 49, 51, 93, 143, 167, 168, 173, 186-188, 195, 211, 221, 231, 235, 245, 247, 259, 279, 289, 299, 309, 310, 333, 344, 360-362.
Sequoia, 78; *S. Langsdorfii*, 70, 76, 77; *S. sempervirens*, 76.
Sequoiineæ, 70, 75, 76.
*Sequoiites* (?) *Langsdorfi*, 70, 75, 76, 78.

SEWARD, A. C., Part-Author of Chap. IV., A.C.S.; see also 44, 53, 55, 71, 75, 77, 87, 90.
Sgeir a' Chaisteil, 56° 30' N., 6° 25' W., 140.
—— na Faoilinn, 56° 24' N., 6° 11' W., 388.
—— nan Gobhar, 56° 29' N., 5° 43' W., 175.
—— Ruadh, 56° 28' N., 5° 41' W., 396.
Sgiath Ruadh, 56° 32' N., 5° 58' W., 105.
Sgulan Mòr, 56° 25' N., 5° 52' W., 318, 368; Sgulan Type, 316, 318, 319.
Sgùrr Dearg, 56° 26' N., 5° 47' W., 37, 132, 134, 158-160, 163, 164, 173, 175, 178, 180, 184, 196, 199, 235, 236; Sgùrr Dearg Vents, 37, 179, 196, Chap. XVI., 211, 212.
—— Mhòr, 56° 19' N., 5° 57' W., 64.
—— of Eigg (Eigg), 95.
Shatter-Belt, 384.
Sheath-and-Core Structure, 52, 260, 261-264, 266, 267, 269, 288.
Sheets, see A' Chioch Sheets, Cone-Sheets, Creag na h' Iolaire Sheets, *and* Sills.
Shells, 406, 412, 413; see Molluscs.
Shell-Sand, 419.
Shelly Clay, 37, 406.
Shepherd's Hat (Eilean nan Gamhna), 56° 25' N., 5° 32' W., 409.
Shiaba, 56° 18' N., 6° 9' W., 64, 386, 416.
Side-Step of Dykes, 358.
Silicified Chalk, 2, 39, 54-59, 62, 413.
Sillimanite, 266, Chap. XXIV.; 'Sillimanite' of Loch Scridian Xenoliths is really Mullite, see 268; Analysis of Mullite-Bearing Xenolith, 34.
Sills, 8, Chaps. XXIII.-XXVII., 361, 362; Analyses, 15, 19, 20, 24, 27, 263; Sills discussed in relation to Lavas, 8, 95, 96, 101.
SIMPSON, J. B., 280, 380, 398.
SINCLAIR, G. M., 109.
—— J. H., 423.
Skye (Hebrides), 4, 32, 45, 47, 49, 50, 51, 95, 144, 146, 152, 155, 162, 172, 193, 248, 252, 324, 359; Basalt and Dolerite, Olivine-Gabbro, *and* Mugearite, Analyses, 15, 24, 27.
Slac nan Sgarbh (Jura): Crinanite Dyke Analysis, 15.
Slate, 419.
Sleat (Skye), 377.
Sleibhte-coire, 56° 25' N., 5° 57' W., 187, 188, 199, 200, 203, 206-208, 210, 248, 398.
Sligachan River (Skye): Olivine-Gabbro Analysis, 24.
Sloc an Neteogh, 56° 31' N., 6° 17' W., 114, 115.
Slochd, 56° 22' N., 6° 7' W., 261, 265, 272.

Sloc nan Uan, 56° 18′ N., 5° 39′ W., 379.
Small-Felspar Basalts *and* Dolerites, see Porphyritic Central Lava Type, *also* 37, 156, 158, 159, 235, 246, 359, 365, 370 ; Analyses, 24.
Small-Felspar Rhyolites *and* Felsites, 196, 197, 206, 208, 209, 223, 227, 228, 335, 346, 374 ; Analysis (Loch Bà), 20.
SMYTH, T., 423.
Soay (off Skye), 172.
Socach a' Mhàim, 56° 26′ N., 5° 52′ W., 315.
SOLANDER, ——, 40.
Solfataras, 49, 50 ; *see* Pneumatolysis.
Sound of Kerrera, 56° 24′ N., 5° 31′ W., 375.
Sound of Mull, 56° 30′ N., 5° 45′ W., 36, 37, 58, 65, 94, 104, 105, 106, 129, 167, 170, 175, 176, 186, 212, 221-224, 227, 279, 296, 363, 390, 391, 396, 397, 400.
South Africa, 54.
South-West Highlands, 408, 409.
Speinne Beag, 56° 34′ N., 6° 2′ W., 375.
Sphene, 153, 225, 334, 350.
Spheroidal Weathering, 48, 95, 100, 105, 107, 120.
Spherulitic Structure, 285, 334, 346, 348, 349, 374, 375 ; *see* Variolitic Structure.
Spinel, 229, 249, 252, 266, Chap. XXIV., 369 ; Analysis, 34.
Sròn Daraich, 56° 24′ N., 6° 0′ W., 363.
—— Dubh, 56° 23′ N., 5° 53′ W., 150, 245, 253.
—— Gharbh, 56° 22′ N., 5° 49′ W., 123.
—— nam Boc., 56° 29′ N., 5° 56′ W., 335, 340, 342, Chap. XXXIII.
Stac Glas Bun an Uisge, 56° 24′ N., 6° 11′ W., 39, 388.
—— Mhic Mhurchaidh, 56° 21′ N., 6° 28′ W., 108.
Staffa, 56° 26′ N., 6° 20′ W., 3, 30, 36-40, 43, 44, 94, 102, 107, 108-111, 140, 145, 172, 385, 388, 410 ; Scolecite Analysis, 34 ; Staffa Type, 16, 32, 93, 101, 145, 146 ; Analyses, 17 ; Definition, 146.
Starav Granite (Mainland), 395.
STEVENSON, A., 402, 421.
Stilbite, 140, 425.
Stockholm (Sweden), 397.
Stone-Implements, 413.
Stoping, 341, 344 ; *see* DALY.
STRENG, A., 17, 421.
Striae, 387, 392, 396-398.
Subaerial Eruptions, 43-45, 97.
Subsequent Drainage, 384.
Succession, *see* Sequence.
Sulphides, 49.
Summary of Progress of the Geological Survey, 50, 425, 426.
Sutherland (Mainland), 377.
Swarm of Dykes, 10, 47, 52 ; *see* Etive *and* Mull Swarms.

Syenite, 26, 38, Chap. XIV., 369, 380 ; Analysis, 27.
—— (old sense), 44.
SYMES, R. G., 364.

TABLE of Lorne, 56° 32′ N., 5° 41′ W., 293.
Tachylite (Basalt-Glass), 38, 46, 47, 49, 260, 265, 285, 288, 300, 356 ; Silica-Determinations, 18, 47, 265.
TAIT, D., Author of Bibliography, 421 ; *see also* 52, 53, 61, 67, 121, 124, 126, 129, 132, 267, 379, 418.
Talaidh Type (after Beinn Talaidh), 162, 217, 218, 219, 226, 241, 284-287, Chap. XXVIII., 327, 329, 359, 370, 372, 373 ; Analysis, 17 ; Definition, 284, Chap. XXVIII.
TATE, R., 423.
Tavool, 56° 22′ N., 6° 9′ W., 43, 78, 103, 108, 109, 111, 113, 145, 181, 272, 279, 418 ; Tavool Tree, 78, 113.
*Taxites* (?) *Campbelli,* 70, 77.
*Taxodium* (*Glyptostrobus*) *heterophyllum,* 62.
*Taxus,* 76-78.
Teanga Bhàn, 56° 22′ N., 5° 48′ W., 207.
TEALL, J. J. H., 16, 46, 150, 252, 319, 372, 381, 424.
Tectonics, 4, Chap. XIII.
Teschenite, 377.
*Thinnfeldia,* 87.
Tholeiite, 101, 146, 214, 224, 228, Chaps. XXIII.-XXVIII., 319 ; Analyses, 17 ; Definitions, 280, 284, 285.
THOMAS, H. H., Part-Author, H. H. T. ; *see also* 35, 51, 52, 188, 269, 278, 426.
THOMSON, S., 35.
Thomsonite, 140, 153.
THORODDSEN, T., 3, 4, 7.
Through-Valley, 384.
Time-Relations, *see* Sequence.
Tiroran, 56° 23′ N., 6° 5′ W., 103, 261, 265, 272.
Tiree, 56° 32′ N., 6° 44′ W., 172, 396, 398, 402, 408, 413.
Tobermorite (after Tobermory), 425 ; Analyses, 34.
Tobermory, 56° 37′ N., 6° 4′ W., 36, 39, 46, 58, 60, 115, 118, 156, 157, 358, 418 ; Tholeiite, Trachyte, Labradorite, *and* Tobermorite, Analyses, 17, 27, 34.
Toll Doire, 56° 30′ N., 5° 57′ W., 343, 345, 349, 350.
Tòm a' Choilich, 56° 21′ N., 6° 2′ W., 264 ; Inninmorite Sill Analyses, 19, 263.
Tom a' Chrochaire, 56° 31′ N., 6° 0′ W., 104.
—— na Gualainne, 56° 24′ N., 5° 51′ W., 132, 197.
Tomsléibhe, 56° 28′ N., 5° 52′ W., 216, 217, 224, 384.

Tòn Dubh-sgairt, 56° 23′ N., 6° 11′ W., 140, 181, 183.
Torness, 56° 26′ N., 5° 49′ W., 134, 212, 223, 224, 235, 311, 314; Torness Felsite, 5, 199, Chap. XVII., 236.
Torosay (detached), S.W. corner of Gribun Peninsula, 49.
—— Castle, 56° 27′ N., 5° 41′ W., 58, 60.
Tòrr a' Ghoai, 56° 23′ N., 5° 55′ W., 188, 246.
Torranlochain, 56° 30′ N., 5° 58′ W., 400.
Torrans, 56° 21′ N., 6° 4′ W., 273.
Tòrr Mòr, 56° 20′ N., 6° 14′ W., 65.
—— Mòr, 56° 28′ N., 6° 12′ W., 108.
—— Mòr, 56° 20′ N., 6° 11′ W., 272.
—— na h-Uamha, 56° 25′ N., 5° 57′ W., 187, 207, 208, 210, 299, 300, 313, 314.
—— nan Clach, 56° 30′ N., 5° 59′ W., 400.
Trachyte, 26, 37, 39, 61, 143-145, Chap. XIV., 228, 374, 375, 380; Analyses, 27.
Traigh Bhan Sgoir, see Tràigh Bhàn na Sgurra.
Tràigh Bhàn na Sgurra, 56° 17′ N., 6° 10′ W., 31, 52, 266, 268, 270, 273, 379.
—— Cadh' an Easa', 56° 18′ N., 6° 5′ W., 64, 100, 102, 273, 415-417.
—— nam Beach, 56° 21′ N., 6° 6′ W., 181, 401.
Trap-Features, 3, 43, 95, 96, 100, 103, 105, 107, 115, 120, 123, 127, 129, 131.
Trees, 3, 38, 39, 43, 65, 77, 78, 97, 111-113.
Treshnish Isles, 56° 29′ N., 6° 25′ W., 107, 140, 172, 385, 387, 389, 397.
—— Point, 56° 33′ N., 6° 20′ W., 117, 388.
Tridymite, 198, 267, 274, 365.
TROIL, U. von, 40, 421.
TURNER, W., 412, 423, 424.
Tynemouth Dyke (North of England), 319.
Tyrol, 75.

UILLT Gharbha, 56° 23′ N., 5° 54′ W., 243.
Uisgeacha Geala, 56° 24′ N., 5° 55′ W., 246.
Uisge Fealasgaig, 56° 21′ N., 5° 59′ W., 273.
*Ulmus effusa*, 86.
Ultrabasic Rocks, 32.
Uluvalt, 56° 24′ N., 5° 58′ W., 125.
Ulva, 56° 29′ N., 6° 13′ W., 36, 39, 107, 108, 115, 139, 140, 172, 385, 387-390, 392; Ulva Cave, 390, 391, 394.
Ulva Ferry, 56° 29′ N., 6° 9′ W., 108, 115, 157, 397, 402.
Uniaxial Augite, see Enstatite-Augite.

Upper Achnacroish, 56° 27′ N., 5° 42′ W., 58.
—— Druimfin, 56° 36′ N., 6° 3′ W., 358.
Upward Branching (Bifurcation) of Ring-Dykes, 9, 311, 313, 314.
Usu San (Japan), 5.
Utö (Sweden), 397.

VALLEY-Glaciers, 392, 398, 399; in Relation to the Sea, 11, 402-408.
Variolitic Structure, 18, 19, 24, 25, 98, 121, 138, 147-150, 163, 164, 180, 281, 282, 284, 285, 290, 300, 304, 305, 352, 365, 374.
Variolites, Silica-Percentages, 18.
Vents, see Agglomerate.
Vesicles invaded by mesostasis, 150, 282, 319; invaded by froth, 98, 149, 150.
Vesuvius (Italy), 3.
Vivarais (France), 41.
Vogesite, 381, 382.

WAHL, W., 190, 284.
WARD, L.F., 84, 90.
Water of Andesites, stony and glassy (pitchstones), 263, 264; Analyses, 263.
Weevil, 71.
WERNER, A. G., 41.
West Lothian, see Linlithgow.
Wexford (Ireland), 410.
WHITEHOUSE, C., 424.
WHYMPER, E., 80, 84, 90.
Wilderness, 56° 23′ N., 6° 12′ W., 56, 59, 108, 109, 140, 387, 414.
WILKINSON, S.B., 398.
WILLIAMS, J. F., 190, 191.
WILSON, G. V., Part-Author, G. V. W.; see also 5, 51, 52, 420.
Wrench-Fault, 37, 183, 311, 385.
WRIGHT, W. B., Part-Author, W. B. W.; see also 6, 9, 51, 208, 351, 386, 390, 425.
Wyoming (America), 84.

XENOLITHS of Loch Scridain, 8, 31, 38, 43, 52, 59, 113, 259, 265-267, Chap. XXIV., 287, 417, 418; Analyses, 34.
—— of other districts, 212, 214, 219, 223, 229, 245, 246, 268, 288-290, 317, 325, 329, 335, 336, Chap. XXXIII., 365, 369, 370.
Xonotlite, 425; Analysis, 34.

ZIRKEL, F., 44, 45, 423.
Zeolites, 45, 138, 139, 152, 153, 155, 161, 193, 270; Analyses, 34.

# MAPS AND MEMOIRS RELATING TO THE GEOLOGY OF THE COUNTRY NORTH OF THE HIGHLAND BORDER FAULT.

Published by the Geological Survey of Great Britain.
(Scottish Branch, 33 George Square, Edinburgh).

## MAPS.

### GEOLOGICAL MAP OF THE BRITISH ISLANDS.

Scale 25 miles to the inch, second edition, colour printed, 1912, 2s.; uncoloured, 1s.

### QUARTER-INCH GEOLOGICAL MAP OF SCOTLAND.

Colour printed (Scale of four miles to one-inch).

Sheet 6. Caithness, Central and Eastern Parts. (1920). 3s.
,, 12. Perthshire, Forfarshire, Kincardineshire, &c. (1910). 3s.
,, 13. Argyllshire, W. Part of Mainland, and Islands of Colonsay, Islay, Jura, &c. (1921). 3s.
,, 16. Campbeltown District, Arran, &c. (1907). 3s.

### ONE-INCH GEOLOGICAL MAP OF SCOTLAND. (*Colour printed, with solid and Drift Geology*).

Assynt District, Parts of Sheets 101, 102, 107, 108. (1923). 3s.

Arran, Isle of, Parts of Sheets 13 and 21. (1910). 3s.

Sheet 28. Argyllshire. Knapdale, Jura, North Kintyre, Loch Tarbert, Loch Sween, Loch Caolisport, and West Loch Tarbert. (1911). 3s.
,, 35. Argyllshire. Colonsay, with part of the Ross of Mull. (1911). 3s.
,, 36. Argyllshire. The Seaboard of Mid-Argyll, Loch Melfort, Loch Crinan. (1917). 3s.
,, 37. Argyllshire. (Loch Avich, Loch Awe (S. half), Loch Goil, Loch Long (Central), Inverary, Lochgoilhead. (1922). 3s.
,, 43. Argyllshire. Staffa, Iona, and W. Mull. (Engraving).
,, 44. Argyllshire. Island of Mull, Morven (S. part), Lismore Island, Kerrera, Mainland from Oban to south of Kilninver. (1923). 3s.
,, 53. Inverness-shire (S. part), Argyllshire (N.E. part). Fort William, Ben Nevis, Ballachulish, Appin, Glen Etive. (1921). 3s.
,, 54. Inverness-shire. (S.E. part), Argyllshire (N.E. part), Perthshire (N.W. part). Loch Treig, Loch Ericht, Loch Rannoch. (1923). 3s.
,, 60. Inverness-shire. Rhum, Canna, Eigg, Muck, Oigh-sgeir. (1917). 3s.
,, 64. Inverness-shire, Perthshire, Aberdeenshire, Banffshire (parts of). Kingussie, Glen Feshie Forest, Forest of Atholl (N. part), Glen Tilt. (1913). 3s.
,, 65. Aberdeenshire (S.W.), Forfarshire (N.W. part), Perthshire (N.E. part). Balmoral Castle, Ballater, &c. (1911). 3s.
,, 70. Inverness-shire. West-Central Skye, with Sligachan, Glenbrittle House, Cuillin Hills, Island of Soay. (1913). 3s.
,, 71. Inverness-shire. Glenelg, Lochalsh and S.E. part of Skye. (1910). 3s.
,, 74. Inverness-shire (E. part), with parts of Nairnshire, Elginshire, Banffshire. Tomatin, Grantown-on-Spey, Aviemore, &c. (1917). 3s.
,, 81. Ross-shire (part of S.W.), Inverness-shire and Islands of Raasay. Rona, Portree Bay, and part of N.E. Coast of Skye, Applecross, &c. (1921). 3s.
,, 82. Ross-shire (Central), Inverness-shire (N.W. part). Lochcarron, Glencarron, Torridon, &c. (1913). 3s.

Sheet 83. Ross-shire (S. part), Inverness-shire (N. part). Dingwall, Inverness (W. part), Beauly, &c. (1918). 3s.
,, 84. Nairnshire, Inverness-shire (N.E. part), Elginshire (W. part). Nairn, Forres, Dava Station, &c. (1923). 3s.
,, 86. Aberdeenshire (N.W. part), Banffshire (N.E. part). Huntly, Turriff, &c. (1923). 3s.
,, 92. Ross-shire (N.W. part). Loch Maree, Kinlochewe, Loch Broom (S. part), &c. (1913). 3s.
,, 93. Ross-shire, Cromartyshire (E.). Glen Beag, Kincardine, Alness, Evanton, &c. (1912). 3s.
,, 96. Aberdeenshire (N.W. part), Banffshire (N. part). Cullen, Portsoy, Banff. (1923). 3s.
,, 102. Sutherlandshire (S. part), Ross-shire (N. part). Lairg, Strath Oykell, Lower Loch Shin. (Engraving).
,, 103. Sutherlandshire (E. part). Helmsdale, Strath Brora, Golspie, Dornoch, Rogart. (Engraving).
,, 110. Caithness (S. part). Sarclet, Lybster, Latheron, Dunbeath. (1914). 3s.
,, 116. Caithness (N. part). Thurso, Dunnet, Watten, Wick. (1914). 3s.

ONE-INCH GEOLOGICAL MAP OF SCOTLAND. (*Hand Coloured*).

Sheet 12. Argyllshire. S. half of Kintyre, with Campbeltown, Mull of Kintyre, &c. (1894). 11s.
,, 19. Argyllshire. S. half of Islay, with Port Charlotte, Bowmore, Port Ellen. (1898). 10s. 3d.
,, 20. Argyllshire. N. half of Kintyre, Gigha Island, and part of Islay. Ardmore Point, Glenbarr, Carradale. (1896). 9s. 6d.
,, 21. Buteshire. Ayrshire (portion of N.W.), Argyllshire, Islands of Bute, Arran (Central and N. part) and Cumbraes, Skipness district in Kintyre. (Solid, 1901, 14s.; Drift, 1906, 14s.).
,, 27. Argyllshire. N. half of Islay, Oronsay, S.W. Jura, Portaskaig, Rhuvaal Lighthouse, &c. (1900). 8s.
,, 29. Argyllshire. S. half of L. Fyne, Kyles of Bute, L. Striven, lower portion of Firth of Clyde. (1892). 14s.
,, 30. Renfrewshire (N. part), Dumbartonshire (S. part), &c. Greenock, Helensburgh, Dumbarton. (1878). 14s.
,, 38. Argyllshire (part of E.), Dumbartonshire (N. part), Perthshire (S.W. part), &c. Aberfoyle, Callander, Luss, Arrochar. (1901). 14s.
,, 39. Perthshire and Clackmannanshire (S. parts), Stirlingshire (N.E. part). Stirling, Callander, Dunblane, Braco, Auchterarder. (1882). 14s.
,, 45. Argyllshire (Central). Oban, Dalmally, Loch Awe, Loch Etive. (1909). (Solid, 14s., and Drift, 14s.).
,, 46. Perthshire (W. part), Argyllshire (N.E. part). Tyndrum, Killin, Balquhidder. (1900). 14s.
,, 47. Perthshire (Central). St. Fillans, Crieff, Inver. (1888). 14s.
,, 48. Perthshire (E. part). Forfarshire (S.W. part), and Fifeshire (N.W. part). Dunkeld, Perth, Cupar Angus, Dundee, Cupar. (1883). 11s.
,, 55. Perthshire (N. part). Kinloch Rannoch, Aberfeldy, Pitlochry, Blair Athole. (Solid, 1902, 14s.; Drift, 1906, 14s.).
,, 56. Perthshire (N.E. part), Forfarshire (W. part). Blairgowrie, Alyth, Kirriemuir. (1895). 14s.
,, 57. Forfarshire (E. part), Kincardineshire (S. part). Forfar, Edzell, Brechin, Montrose. (1897). 10s. 3d.
,, 66. Kincardineshire (W. part), Forfarshire (N.E. part), Aberdeenshire (S. part). Banchory, Aboyne, Laurencekirk. (1897). 14s.
,, 67. Kincardineshire (E. part). Stonehaven, Inverbervie, Portlethen. (1898). 5s. 9d.
,, 75. Inverness-shire (part of E.), Elginshire (S.E. corner), Banffshire (S. part), Aberdeenshire (W. part). Tomintoul, Glenlivet, &c. (1895). 14s.
,, 76. Aberdeenshire (Central), Kincardineshire (part of N.W.). Inverurie, Muir of Rhynie, Tarland. (1886). 13s. 3d.

Sheet 77. Aberdeenshire (S.E. part), Kincardineshire (N.E. part). Aberdeen, Newburgh, Fintray, Maryculter. (1885). 8s.
„ 85. Elginshire (E. part), Banffshire (Central), Aberdeenshire (part of W.). Keith, Rothes, Dufftown, Aberlour, Dallas. (1898). 14s.
„ 87. Aberdeenshire (N.E. part), and Banffshire (detached portions). Peterhead, Ellon, Methlick. (1885). 13s. 3d.
„ 91. Ross-shire (part of W.). Gairloch, Poolewe, Aultbea. (1893). 14s.
„ 94. Ross-shire (N.E. part), Cromartyshire, Sutherlandshire (parts of), &c. Skibo Castle, Tain, Tarbat Ness, Cromarty. (1889). 8s. 9d.
„ 95. Elginshire (N. part), Banffshire (N.W. part). Elgin, Lossiemouth, Fochabers, Buckie. (1886). 5s. 3d.
„ 97. Aberdeenshire (N. part), Banffshire (N.E. corner). Fraserburgh, New Aberdour, &c. (1881). 4s. 6d.
„ 100. Ross-shire (part of N W.). Udrigle, Achagarve. (1888). 2s. 6d.
„ 101. Ross-shire (part of N.W.), Cromartyshire (detached portion), Sutherlandshire (part of S.W.). Ullapool, Inverkirkaig, &c. (1892). 14s.
„ 103. Sutherlandshire (E. part). Helmsdale, Golspie, Dornoch, Rogart. (1896). 8s. 9d.
„ 107. Sutherlandshire (part of W.). Lochinver, Inchnadamph, Scourie. (1892). 14s.
„ 113. Sutherlandshire (N.W.). Cape Wrath, Rhiconich. (1886). 5s. 3d.
„ 114. Sutherlandshire (N. part). Durness, Tongue, &c. (1889). 13s. 3d.
„ 115. Sutherlandshire (N. E. part), Caithness-shire (N.W. part). Farr, Melvich, Reay, Strath Halladale. (1898). 14s.

SIX-INCH SHEETS.

With geological lines, uncoloured, price 5s. each. Prices for hand-coloured copies can be obtained on application to the Director General, Ordnance Survey Office, Southampton.

Sutherlandshire, Illustrating the Structure of the N.W. Highlands.
 5. Durness. (1892).
 71. Inchnadamph, Loch Assynt. (1891).
Inverness-shire (Skye).
 38. Treen, portion of Glen Drynock, Alldearg House, Grula, &c. (1905).
 39. Sligachan Inn, Glamaig, &c. (1905).
 44. Glen Brittle, Loch Brittle, &c. (1905).
 45. Loch Coruisk, Strathaird House, &c. (1905).

MEMOIRS.

Caithness, The Geology of. Sheets 110 and 116, with parts of 109, 115, and 117. (1914). 4s.
Cowal. The Geology of Cowal, including the part of Argyllshire between the Clyde and Loch Fyne. Sheets 29, and parts of 37 and 38. (1897). 6s.
Islay, The Geology of. Sheets 19 and 27, with the western part of Sheet 20. (1907). 2s. 6d.
Mull, Loch Aline, and Oban, The Pre-Tertiary Geology of. Sheet 44, and parts of Sheets 35, 43, 45, and 52. (1924).
Mull, Loch Aline, and Oban, The Tertiary and Post-Tertiary Geology of. Sheet 44, and parts of Sheets 43, 51 and 52. (1924).
North-West Highlands of Scotland, The Geological Structure of. Parts of Sheets 71, 81, 82, 91, 92, 101, 102, 107, 108, 113 and 114. (1907). 10s. 6d.
Skye, The Tertiary Igneous Rocks of. Sheets 70 and 71, and parts of 80 and 81. (1904). 9s.
The Mesozoic Rocks of Applecross, Raasay, and North-East Skye. Part of Sheet 81. (1920). 6s.

SPECIAL REPORTS ON THE MINERAL RESOURCES OF GREAT BRITAIN.

Vol. XI. Iron Ores (continued). The Iron Ores of Scotland. (1920). 10s.
Vol. XVII. The Lead, Zinc, Copper and Nickel ores of Scotland. (1921). 7s. 6d.
Vol. XXIV. Cannel Coals, Lignite and Mineral Oil in Scotland. (1922). 2s.

MEMOIRS DESCRIPTIVE OF THE ONE-INCH GEOLOGICAL SHEETS.
(For counties *see* under one-inch maps).

Sheet 21. Central and Northern Districts of Arran; S. part Island of Bute; the Cumbraes, Ayrshire (part of N.W. coast), District of Skipness. (1903). 4s.
,, 28. Knapdale, Jura and North Kintyre. (1911). 3s.
,, 35. Colonsay and Oronsay, with part of the Ross of Mull. (1911). 2s. 3d.
,, 36. Seaboard of Mid-Argyll. (1909). 2s. 3d.
,, 37. Mid Argyll. (1905). 3s. *See also Memoirs on Cowal.*
,, 43. Staffa, Iona, and Western Mull (in preparation), *See also Memoirs on Mull.*
,, 44. *See Memoirs on Mull.*
,, 45. The Country near Oban and Dalmally (1908). 2s. 6d. *See also Memoirs on Mull.*
,, 51. *See Memoirs on Mull.*
,, 52. *See Memoirs on Mull.*
,, 53. Ben Nevis and Glen Coe. (1916). 7s. 6d.
,, 54. Corrour and the Moor of Rannoch. (1923). 4s.
,, 55. Perthshire. The Country round Blair Athole, Pitlochry, Aberfeldy. (1905). 3s.
,, 60. The Small Isles of Inverness-shire. (1908). 4s. 6d.
,, 64. Upper Strathspey, &c. (1913). 2s.
,, 65. Braemar, Ballater and Glen Clova. (1912). 2s. 6d.
,, 70. West Central Skye, with Soay. (1904). 1s. *See also Memoir on Skye.*
,, 71. Glenelg, Lochalsh and the South-East part of Skye. (1910). 3s. 6d. *See also Memoirs on Skye and N.W. Highlands.*
,, 74. Mid-Strathspey and Strathdearn. (1915). 2s. 6d.
,, 75. West Aberdeenshire, Banffshire, parts of Elgin and Inverness-shires. (1896). 1s. 6d.
,, 76. Central Aberdeenshire. (1890). 1s.
,, 81. *See Memoirs on N.W. Highlands, and Mesozoic Rocks of Applecross.*
,, 82. Central Ross-shire. (1913). 2s. 3d. *See also Memoir on N.W. Highlands.*
,, 83. Beauly and Inverness. (1914). 2s.
,, 84 and part of 94. Lower Findhorn and Lower Strath Nairn. (1923). 5s.
,, 85. Lower Strathspey. (1902). 1s. 6d.
,, 86 and 96. Huntly, Turriff, and Banff. (1924). 6s. 6d.
,, 87. Aberdeenshire and Banffshire (parts of). (1886). 9d.
,, 91. *See Memoir on N.W Highlands.*
,, 92. The Fannich Mountains, &c. (1913). 2s. 6d. *See also Memoir on N.W. Highlands.*
,, 93. Ben Wyvis, Carn Chuinneag, Inchbae, &c. (1912). 4s.
,, 97. Northern Aberdeenshire, Eastern Banffshire. (1882). 4d.
,, 101. *See Memoir on N.W. Highlands.*
,, 102. Strath Oykell, Lower Loch Shin (in preparation). *See also Memoir on N.W. Highlands.*
,, 103. Brora, &c. (in preparation).
,, 107. *See Memoir on N.W. Highlands*
,, 108. ,, ,, ,,
,, 110. ,, ,, *Caithness.*
,, 113. ,, ,, *N.W. Highlands.*
,, 114. ,, ,, ,,
,, 115. ,, ,, *Caithness.*
,, 116. ,, ,, ,,
,, 117. ,, ,, ,,

PHOTOGRAPHS OF GEOLOGICAL SUBJECTS, CATALOGUE OF.

Series B. Whole Plate, and C. Half-Plate. (1910). 6d.
(The photographs illustrate the geology and scenery of various parts of Scotland, and may be seen at the Survey Office. Prints, lantern slides and enlargements can be supplied. Tariff of prices on applications.)